WATER CENTRIC
SUSTAINABLE
COMMUNITIES

WATER CENTRIC SUSTAINABLE COMMUNITIES

Planning, Retrofitting, and Building the Next Urban Environment

Vladimir Novotny,
Jack Ahern, and
Paul Brown

JOHN WILEY & SONS, INC.

Published by John Wiley & Sons, Inc., Hoboken, New Jersey
Published simultaneously in Canada

For general information about our other products and services, please contact our Customer Care
Department within the United States at (800) 762-2974, outside the United States at (317) 572-3993 or
fax (317) 572-4002.

Wiley also publishes its books in a variety of electronic formats. Some content that appears in print may
not be available in electronic books. For more information about Wiley products, visit our web site at
www.wiley.com.

Library of Congress Cataloging-in-Publication Data:

Novotny, Vladimir, 1938-
    Water centric sustainable communities: planning, retrofitting,and building the next urban
environment / Vladimir Novotny, John Ahern, Paul Brown.
        p. cm.
    Includes index.
    ISBN 978-0-470-47608-6 (cloth); ISBN 978-0-470-64282-5 (ebk); ISBN 978-0-470-64283-2 (ebk);
    ISBN 978-0-470-64284-9 (ebk); ISBN 978-0-470-94996-2 (ebk); ISBN 978-0-470-95169-9 (ebk);
    ISBN 978-0-470-95193-4 (ebk)
    1. Municipal water supply.   2. Water resources development.   3. Sustainable development.
4. Urban runoff – Management.   5. Watershed management.   I. Ahern, John, 1949–
II. Brown, Paul, 1944–   III. Title.
    TD346.N68 2010
    628.109173′2 – dc22   2010005121

CONTENTS

PREFACE   xii

**I HISTORIC PARADIGMS OF URBAN WATER/STORMWATER/
   WASTEWATER MANAGEMENT AND DRIVERS FOR CHANGE   1**

   I.1   Introduction / 1
   I.2   Historic Paradigms: From Ancient Cities to the 20th Century / 5
         I.2.1   First Paradigm / 8
         I.2.2   Second Paradigm / 9
         I.2.3   Third Paradigm / 15
         I.2.4   Fourth Paradigm / 25
         I.2.5   The Impact of Automobile Use / 32
         I.2.6   Urban Sprawl / 38
         I.2.7   The Rise of New Great Powers Competing for Resources / 40
   I.3   Drivers for Change towards Sustainability / 42
         I.3.1   Population Increases and Pressures / 44
         I.3.2   Water Scarcity Problems and Flooding Challenges
                 of Large Cities / 49
         I.3.3   Greenhouse Emissions and Global Warming Effects / 51
         I.3.4   Aging Infrastructure and the Need to Rebuild and Retrofit / 59
         I.3.5   The Impossibility of Maintaining the Status Quo and
                 Business as Usual / 60
   I.4   The 21st Century and Beyond / 65
   References / 68

II URBAN SUSTAINABILITY CONCEPTS   72

   II.1   The Vision of Sustainability / 72
   II.2   The Sustainability Concept and Definitions / 73
          II.2.1   A New (Fifth) Paradigm Is Needed / 73
          II.2.2   Definition of Pollution / 76
          II.2.3   Sustainability Definitions / 80
          II.2.4   Economic versus Resources Preservation Sustainability / 82

v

II.2.5   Sustainability Components / 85

II.2.6   The Environment and Ecology / 87

II.2.7   Living within the Limits in the Urban Landscape / 90

II.2.8   The Economy / 94

II.3   Towards the Fifth Paradigm of Sustainability / 97

II.3.1   Emerging Sustainable Urban Water/Stormwater/Used Water Systems / 99

II.3.2   Triple Bottom Line—Life Cycle Assessment (TBL—LCA) / 104

II.3.3   Water Reclamation and Reuse / 106

II.3.4   Restoring Urban Streams / 108

II.3.5   Stormwater Pollution and Flood Abatement / 110

II.3.6   Urban Landscape / 113

II.4   Cities of the Future—Water Centric Ecocities / 114

II.4.1   Drainage and Water Management / 114

II.4.2   Microscale Measures and Macroscale Watershed Goals / 116

II.4.3   Integrated Resource Management Clusters—Ecoblocks of the Cities of the Future / 120

II.4.4   Interconnectivity of Clusters—Spatial Integration / 123

II.5   Ecocity/Ecovillage Concepts / 124

References / 129

III   PLANNING AND DESIGN FOR SUSTAINABLE AND RESILIENT CITIES: THEORIES, STRATEGIES, AND BEST PRACTICES FOR GREEN INFRASTRUCTURE   **135**

III.1   Introduction / 135

III.1.1   Achieving Sustainability / 135

III.1.2   Sustainability through Urban Planning and Design / 137

III.2   Ecosystem Services / 138

III.2.1   Concepts / 138

III.2.2   The Non-Equilibrium Paradigm / 141

III.3   Planning for Resilient and Sustainable Cities / 143

III.3.1   Ecosystem Service Goals and Assessments / 143

III.3.2   Resilience Strategies / 144

III.3.3   Scenario Planning / 155

III.3.4   Transdisciplinary Process / 157

III.3.5   Adaptive Planning / 157

III.4   Best Practices for Green Infrastructure / 158

III.4.1   SEA Street Seattle / 159

III.4.2   Westergasfabriek Park, Amsterdam / 162

III.4.3   Staten Island Blue Belt, New York / 162

III.4.4   Ecostaden (Ecocities): Augustenborg Neighborhood
and Western Harbor, Malmö, Sweden / 164

III.5   Discussion / 170

References / 171

IV   STORMWATER POLLUTION ABATEMENT AND FLOOD CONTROL—STORMWATER AS A RESOURCE   177

IV.1   Urban Stormwater—A Problem or an Asset? / 177

IV.1.1   Problems with Urban Stormwater / 177
IV.1.2   Current Urban Drainage / 182
IV.1.3   Urban Stormwater Is an Asset and a Resource / 184
IV.1.4   Low Impact Development (LID) / 186

IV.2   Best Management Practices to Control Urban Runoff for Reuse / 189

IV.2.1   Soft Surface Approaches / 190
IV.2.2   Ponds and Wetlands / 201
IV.2.3   Winter Limitations on Stormwater Management
and Use / 212
IV.2.4   Hard Infrastructure / 216
IV.2.5   ID Urban Drainage—A Step to the Cities of the Future / 218

References / 222

V   WATER DEMAND AND CONSERVATION   228

V.1   Water Use / 228

V.1.1   Water on Earth / 228
V.1.2   Water Use Fundamentals / 232
V.1.3   Municipal Water Use in the U.S. and Worldwide / 235
V.1.4   Components of Municipal Water Use / 239
V.1.5   Virtual Water / 240

V.2   Water Conservation / 241

V.2.1   Definition of Water Conservation / 241
V.2.2   Residential Water Use / 241
V.2.3   Commercial and Public Water Use and Conservation / 249
V.2.4   Leaks and Other Losses / 251

V.3   Substitute and Supplemental Water Sources / 252

V.3.1   Rainwater Harvesting (RWH) / 252
V.3.2   Gray Water Reclamation and Reuse as a Source
of New Water / 256
V.3.3   Desalination of Seawater and Brackish Water / 260
V.3.4   Urban Stormwater and Other Freshwater Flows as
Sources of Water / 266

References / 268

VI   WATER RECLAMATION AND REUSE   **272**

VI.1   Introduction / 272
VI.2   Water Reclamation and Reuse / 274
  VI.2.1   The Concept / 274
  VI.2.2   Reclaiming Rainwater and Stormwater / 279
  VI.2.3   Water-Sewage-Water Cycle—Unintended
    Reuse / 280
  VI.2.4   Centralized versus Decentralized Reclamation / 281
  VI.2.5   Cluster Water Reclamation Units / 282
VI.3   Water Quality Goals and Limits for Selecting
  Technologies / 286
  VI.3.1   Concepts / 286
  VI.3.2   Landscape and Agricultural Irrigation / 289
  VI.3.3   Urban Uses Other Than Irrigation and Potable
    Water Supply / 293
  VI.3.4   Potable Reuse / 297
  VI.3.5   Groundwater Recharge / 300
  VI.3.6   Integrated Reclamation and Reuse—Singapore / 304
References / 308

**VII   TREATMENT AND RESOURCE RECOVERY
UNIT PROCESSES**   **311**

VII.1   Brief Description of Traditional Water and Resource
  Reclamation Technologies / 311
  VII.1.1   Basic Requirements / 311
  VII.1.2   Considering Source Separation / 312
  VII.1.3   Low-Energy Secondary Treatment / 315
  VII.1.4   New Developments in Biological Treatment / 324
VII.2   Sludge Handling and Resource Recovery / 329
  VII.2.1   Types of Solids Produced in the Water Reclamation
    Process / 331
  VII.2.2   A New Look at Residual Solids (Sludge) as a
    Resource / 334
VII.3   Nutrient Recovery / 336
VII.4   Membrane Filtration and Reverse Osmosis / 339
VII.5   Disinfection / 340
VII.6   Energy and GHG Emission Issues in Water
  Reclamation Plants / 346
VII.7   Evaluation and Selection of Decentralized Water Reclamation
  Technologies / 348
  VII.7.1   Closed Cycle Water Reclamation / 348
References / 354

VIII ENERGY AND URBAN WATER SYSTEMS—TOWARDS NET ZERO CARBON FOOTPRINT   358

   VIII.1 Interconnection of Water and Energy / 358

      VIII.1.1 Use of Water and Disposal of Used Water Require Energy and Emit GHGs / 358
      VIII.1.2 Greenhouse Gas Emissions from Urban Areas / 360
      VIII.1.3 The Water-Energy Nexus on the Regional and Cluster Scale / 362
      VIII.1.4 Net Zero Carbon Footprint Goal for High-Performance Buildings and Developments / 365

   VIII.2 Energy Conservation in Buildings and Ecoblocks / 371

      VIII.2.1 Energy Considerations Related to Water / 371
      VIII.2.2 Heat Recovery from Used Water / 379

   VIII.3 Energy from Renewable Sources / 380

      VIII.3.1 Solar Energy / 380
      VIII.3.2 Wind Power / 387

   VIII.4 Energy from Used Water and Waste Organic Solids / 392

      VIII.4.1 Fundamentals / 392
      VIII.4.2 Biogas Production, Composition, and Energy Content / 394
      VIII.4.3 Small and Medium Biogas Production Operations / 397
      VIII.4.4 Anaerobic Upflow Reactor / 398

   VIII.5 Direct Electric Energy Production from Biogas and Used Water / 399

      VIII.5.1 Hydrogen Fuel Cells / 400
      VIII.5.2 Microbial Fuel Cells (MFC) / 403
      VIII.5.3 Harnessing the Hydraulic Energy of Water/Used Water Systems / 406

   VIII.6 Summary and a Look into the Future / 408

      VIII.6.1 A New Look at the Used Water Reclamation Processes / 408
      VIII.6.2 Integrated Resource Recovery Facilities / 411

   VIII.7 Overall Energy Outlook—Anticipating the Future / 416

      VIII.7.1 A Look into the Future 20 or More Years Ahead / 416
      VIII.7.2 Is Storage a Problem? / 421

   References / 422

IX RESTORING URBAN STREAMS   **427**

   IX.1 Introduction / 427

      IX.1.1 Rediscovering Urban Streams / 427
      IX.1.2 Definitions / 437

IX.2 Adverse Impacts of Urbanization to Be Remedied / 438

    IX.2.1 Types of Pollution / 438

    IX.2.2 Determining Main Impact Stressors to Be Fixed by Restoration / 443

    IX.2.3 Effluent Dominated and Effluent Dependent Urban Water Bodies / 447

IX.3 Water Body Restoration in the Context of Future Water Centric (Eco)Cities / 453

    IX.3.1 Goals / 453

    IX.3.2 Regionalized versus Cluster-Based Distributed Systems / 455

    IX.3.3 New Developments and Retrofitting Older Cities / 457

IX.4 Summary and Conclusions / 476

References / 479

X PLANNING AND MANAGEMENT OF SUSTAINABLE FUTURE COMMUNITIES   **482**

X.1 Integrated Planning and Management / 482

    X.1.1 Introduction / 482

    X.1.2 Footprints / 484

X.2 Urban Planning / 487

    X.2.1 Ecocity Parameters and Demographics—Population Density Matters / 488

X.3 Integrated Resources Management (IRM) / 493

    X.3.1 Sustainability / 493

X.4 Clusters and Ecoblocks—Distributed Systems / 497

    X.4.1 The Need to Decentralize Urban Water/Stormwater/Used Water Management / 497

    X.4.2 Distribution of Resource Recovery, Reclamation and Management Tasks / 499

    X.4.3 Cluster Creation and Size / 503

    X.4.4 Types of Water/Energy Reclamations and Creation of a Sustainable Urban Area / 505

X.5 System Analysis and Modeling of Sustainable Cities / 514

    X.5.1 Complexity of the System and Modeling / 514

    X.5.2 Triple Bottom Line (TBL) Assessment / 518

X.6 Institutions / 525

    X.6.1 Institutions for Integrated Resource Management / 526

    X.6.2 Enhanced Private Sector / 532

    X.6.3 Achieving Multibenefit System Objectives / 533

References / 535

XI ECOCITIES: EVALUATION AND SYNTHESIS   **539**

XI.1   Introduction / 539
XI.2   Case Studies / 542

XI.2.1   Hammarby Sjöstad, Sweden / 542
XI.2.2   Dongtan, China / 549
XI.2.3   Qingdao (China) Ecoblock and Ecocity / 556
XI.2.4   Tianjin (China) / 560
XI.2.5   Masdar (UAE) / 566
XI.2.6   Treasure Island (California, U.S.) / 573
XI.2.7   Sonoma Mountain Village (California, U.S.) / 579
XI.2.8   Dockside Green / 585

XI.3   Brief Summary / 588
References / 590

APPENDIX   **595**
INDEX   **597**

PREFACE

There is a growing belief among engineers, planners, and scientists that the function and purpose of our urban water infrastructure needs a radical redefinition. The conceptual models employed in most cities have not changed much since Roman times. They comprise rapid-conveyance piped systems that keep land relatively dry most of the time, provide a supply of potable water, and use water to carry away human and industrial wastes for disposal. Managing water quality, both for potable and disposal purposes, is generally accomplished by removing contaminants at the beginning or end of the pipe. Periodic flooding is controlled with additional structural barriers that rapidly drain urbanized areas towards downstream locations. Importantly, these systems have always been integrated into the built environment of buildings and streets—and largely taken for granted in terms of their functional role.

The calls for "Water centric sustainable communities," "Cities of the Future," "Sustainable Future Communities" may sound today more like futuristic dreams than a potential reality. But the future is always an extension of history and change is forced, and guided, by new incremental discoveries and stresses. Now is the time when serious stresses such as population increase and migration into cities and global climatic changes have emerged as serious issues of global concern. Infrastructure in old cities has deteriorated, and the U.S. is beginning to understand, and confront, the consequences of suburban sprawl in terms of infrastructure requirements, energy costs and pollution. The time has come to look for and implement new concepts for urban planning and design.

Historically, cities started as walled villages or settlements on or near a water source. Without water, life in the cities could not be sustained. Water also provides for cleaning and hygiene, transportation, irrigation of crops and gardens, defense, and transportation. In the "Cities of the Future" context, one has to look to, and learn from, the past about the importance of water in cities. Past successes can provide inspiration and continuity for the future and also can reveal what happens when water management is abused and/or water is lost. The first integrated, modern water/used water and stormwater management system can be dated to the second millennium B.C. in the Minoan civilization on the Mediterranean island of Crete. The Minoan cities had stone paved roads and offered water and sewage disposal to the upper-class. Water was brought to the cities from wells and clean mountain waters by aqueducts, rainwater was collected and stored in underground cisterns, and water was distributed to the fountains and upper class villas by advanced water systems using clay pipes. There is archeological evidence of Minoan use of flushed bathrooms. Wastewater and storm water were collected in sanitary and storm sewers. Many of these technologies were then adopted by Greek and Roman civilizations (see Chapter I) who improved the systems. One could call this a semi-sustainable linear system with rain water reclamation and reuse. Water was stored in underground cisterns and flowed by gravity to the users. The achievements in water collection, transport, distribution, public health aspects (flushing toilets, public baths) of these ancient civilizations were so advanced that they can only be compared to the water management systems in the developed countries at the end of the nineteenth century. Hence, the future builds on the past and, as will be shown throughout this book, many "new" and proposed technologies of the future such as rainwater and stormwater recycling are technologies of the distant past that were forgotten.

The fast-conveyance drainage infrastructure conceived by Minoans and Romans and reintroduced into the growing cities of Europe during the industrial revolution and into the U.S. cities in the second half of the nineteen century has produced great gains in protecting public health and safety—by "eliminating" unwanted, highly-polluted runoff and sewage. However, the acceptance of fast drainage conveyance systems caused disruption of urban hydrology and polluted most surface waters. Many urban surface waters such as lakes and wetlands were drained or filled for urban uses. In the middle of the twentieth century, surface water quality in cities became unbearable, magnified by introduction of industrial chemicals and fertilizers into the environment. The pollution and disappearance of urban water bodies eliminated all possibilities of on-site reuse but uncontrolled reuse of untreated and later partially treated wastewater continued in cities located downstream. Deteriorating water quality, diminishing flows and population increase led most cities to rely on long transfers of water and regional linear wastewater disposal without reuse. Even after passing groundbreaking water pollution control regulations in the developed countries such as the Clean Water Act in the U.S. and the Water Framework Directive in the European Community countries at the end of the last millennium and billions spent on costly "hard" solutions like sewers and treatment plants, water supplies and water quality remain a major concern in most urbanized areas worldwide. A large portion of the pollution is caused by the predominant elements of the built urban landscape: a preference for impervious over porous surfaces; fast "hard" conveyance infrastructure rather than "softer" approaches like ponds and vegetation; and rigid stream channelization instead of natural stream courses, buffers and floodplains. Because the hard conveyance and treatment infrastructure were designed to provide protection for a five to ten year storm, these systems are often unable to safely deal with the extreme events and sometimes fail with serious or catastrophic consequences.

In the second half of the last century, the world witnessed a rapid emergence and growth of megacities. According to the United Nations, in 1950 there were eight cities in the world that had population of more than 5 million and the terms "megapolis" or "megalopolis" were added to the dictionary. In 2000 the number of megalopoli increased to thirty three and soon (in 2015) we may count close to fifty megalopoli, most of them in the developing world. And this trend will continue. Many of the large urban areas in developing countries lack adequate infrastructure for providing clean potable water and safe disposal of waste. Water and food safety and shortages are major problems which, if the current trends continue, will magnify in the future. China alone will build urban habitats for 300 million people in the next twenty five to thirty years! The Twenty-first century will see the most extensive building and rebuilding of urban settlements driven by population increase and in-migration from rural areas to the cities. At the same time, in developed countries, the water infrastructure, mostly underground, is crumbling, leaking, and being overloaded by undesirable inflows of polluted and "clean" water. Cities are running out of landfill space for solid waste and disposal sites for sludge from treatment plants. At the beginning of this century, in Naples (Italy) garbage and solid waste stayed on the streets for months because no landfills were suitable for the disposal! People in large cities in the developing world have to rely on bottled or boiled water because of insufficient and contaminated water supply.

Awareness of the effects of the excessive atmospheric emissions of green house gases (GHG) (carbon dioxide, methane, nitric oxides, and other gases) on global climatic changes has became a major concern. The consequences of the expected global climatic changes have been scientifically proven. The global community must both reduce GHG emissions and adapt to the changes that cannot be avoided. The expected global climatic changes include increased global warming that will cause more extreme weather, terrestrial glacier and polar ice melting, and melting of permafrost in arctic (tundra) forests. Melting of terrestrial glaciers will increase sea levels and impact coastal communities. Providing water, electric energy, and fuel for transportation and heating has a major impact on global climatic change. The water-energy nexus and its impact on water availability and global climatic change is considered as a major footprint of urbanization along with water and food security, hydrological and ecological effects, nutrient (phosphate) management and future availability for growing crops safely and without severe impacts on water quality. Excessive nutrient losses into receiving waters cause eutrophication and hypertrophy exemplified by massive algal bloom rendering surface water supplies unusable. The microorganisms forming the massive algal blooms prefer warmer temperatures; hence, the emergence of algal blooms may increase with global warming.

Today, the demands of rapid urbanization and depleted or degraded resources drive us to look for totally new systems where used water is recycled, rainwater is harvested, peak stormwater flows slowed down, and discharges of pollutants to remote receiving waters from both pipes and land use are significantly reduced or eliminated entirely. These objectives alter the fundamental functions of the system. This system is intertwined with built environment, transporation, urban landscape, ecology, and living.

Broadly stated, the natural and built environment within urban watersheds is being reconfigured to restore hydrological and ecological functions, provide for the water needs of the community, and maintain the health of people and habitat—with less reliance on energy-intensive, ecologically damaging imported supplies or exported surpluses and waste products that use water for carriage. These system-level changes, which entail significantly greater levels of integration, are emerging in various forms throughout the world.

This book documents the wide spectrum of technological advances that will contribute to that new paradigm. It calls for the integration of technological advances with a radically reformed vision of what constitutes a healthy urban environment. It is premised on the notion that a city's relationship with the natural world is more complex than simply importing needed resources and disposing of its waste, while accomplishing the multiple transformations, both economic and social, that constitute a city's primary functions.

The concepts of the new paradigm of sustainable water centric ecocities have been emerging for the last fifteen years in environmental research and landscape design laboratories in several countries of Europe (Sweden, Germany, United Kingdom, Netherlands), Asia (Singapore, Abu Dhabi, Saudi Arabia, China, Japan and Korea), Australia, USA (Chicago, Portland, Seattle, Philadelphia, San Francisco) and Canada (British Columbia, Great Lakes). This paradigm is based on the premise that urban waters are both the lifeline of cities and the focus of the sustainable cities movement. The evolution of the new paradigm of urbanization ranges from the microscale "green" buildings, subdivisions or "ecoblock" to macroscale ecocities and ecologically reengineered urban watersheds, incorporating transportation, food production and consumption and neighborhood urban living. Defining an urban ecoregion concept, focusing on the entire sustainable water cycle, starting with the water supply sources and ending with the wastewater and solid waste recycle and reuse, is becoming a necessity since a city and its water and waste management cannot be separated from its potable water sources and cannot have an unsustainable adverse impact on downstream users, and cities.

Under the new paradigm, used water and discarded solids become a resource that can provide energy in a form of electricity, biogas, hydrogen, fertilizer, raw materials for reuse, and heat. Under this new paradigm, the terms "wastewater" or "waste" become misnomers and are replaced with "used water," "reclaimed water," and "resource recovery." This change can be accomplished by a hybrid (partially decentralized) or even fully decentralized water/storm water/used water system leading to on-site water reclamation and reuse, energy and nutrient recovery and other benefits. The future of sustainable urbanization, arguably, is to switch from heavy energy use and large GHG emissions to "carbon neutrality" or "net zero" carbon effects. The integrated systems do not just include water and used water, integration also includes urban and suburban transportation, heating and cooling, ecology, protection and rediscovery of urban surface and ground water systems, leisure, culture and recreation of citizens. The core concepts of integration of urban water, resources, and energy management are: (a) there are no wastes—only resources, and (b) optimization of resource value requires an integration of water and energy in addition to ecological and social resilience. This change will lead to revitalization of older cities and retrofitting them with sustainable energy, frugal water infrastructure and developing ecologically-healthy water systems. Integrated designs should be based upon maximizing economic, ecological and social equity value and also restoring and protecting ecosystem function damaged by past economic development. Public health benefits would also be considerable. Changing the city environment by laying sidewalks connecting residential areas to schools and shopping, building recreational paths along clean streams and impoundments, making surface water suitable for canoeing, kayaking and swimming, and providing plots for community gardens could change urban life styles to more healthy living. When sidewalks are built and roads are made narrower, people start walking. In one community in Minnesota which implemented such changes people increased their life expectancy by several years (Newsweek, February 15, 2010).

At one level this book attempts to pull together and document all of the component parts and approaches that can contribute to the new paradigm. Ultimately, to be successful however, the authors believe readers must be moved to see the new potential and possibilities that fundamentally different design objectives offer. It is intended for professionals and practitioners in all of the institutions and communities that influence the slow transformation of urban form and function.

Because engineers, scientists, and planners are always constrained by the expectations and requirements imposed by legal, economic, and social institutions, it is important that any new model of urban water management be widely understood and embraced by the citizens, elected officials, local authorities, regulators, developers, businesses, and many others who influence the urban process in every community. When those of us who have a hand on the controls of urban development agree upon and work towards the achievement of new fundamental objectives, it will allow the next generation of practitioners to apply the creative energy, innovation, and integrated solutions needed for a sustainable future. We hope this book contributes to a better understanding of tomorrow's design objectives, as well as providing insights into the emerging tools available to accomplish them.

These topics are both complex and difficult. They take us out of our professional comfort zone and sit us down with new faces from new communities. But as Thomas Kuhn wrote in his groundbreaking book, *The Structure of Scientific Revolutions* (The University of Chicago Press, 1996), the change will not be easy. He defined the concept of "paradigm shifts" as

> Because it demands large-scale paradigm destruction . . . the emergence of new theories is generally preceded by a period of pronounced professional insecurity . . . Failure of existing rules is the prelude to a search for new ones. (Kuhn 1996, pp. 67–68)

Our mission is to promote and foster that search for new approaches and new rules. If this book contributes to the understanding of what exists, what is being explored, and what can be achieved in the future, the authors will have achieved their objectives.

This book is interdisciplinary, covering the water:energy nexus with its effects on urban development, water supply, drainage, ecology, GHG emissions and system integration. It presents an analysis of the comprehensive problem of unsustainable past urbanization practices and offers hopeful solutions for the sustainable cities of the future. It covers the history of water/stormwater/used water paradigms and driving forces for change (Chapter I) followed by the development of the new (fifth) sustainable ecocity paradigm in Chapter II. Chapters III and IV deal with the old/new urban drainage and best management practices in the context of urban planning and design. The "old/new" term refers to the fact that some drainage concepts call for change from the underground piping infrastructure and impervious urban surfaces, to systems resembling the historic surface drainage and previous hydrology. They are based on switching from fast conveyance drainage employing underground sewers and surface concrete lined channels to storage and infiltration-oriented naturally-looking drainage systems. Chapters V and VI present water conservation, reclamation and reuse systems. In the U.S., the key is to significantly reduce wasting water which in large portions of the country leads to water shortages, and implementing costly and environmentally unsustainable water transfers or high cost desalination. The switch from traditional energy demanding used water and management of residual solids to energy producing and less water demanding new concepts are described in Chapters VII and VIII. Chapter VIII covers energy saving technologies and producing alternate/renewable energy that complement integrated water/resources management. It also presents a proposal for integrated resource recovery facilities converting used water and solids into reclaimed water, biogas, hydrogen, heat and electric energy and recovers nutrients and residual organic solids that can be reused. Stream restoration and daylighting are highlighted in Chapter IX. Clean urban streams and lakes, even small ones, are a dominant part of the urban landscape to which people seem to gravitate, to live on or near, and to use for recreation and enjoyment. They are the backbone of the integrated urban water systems, recipients of clean and highly treated water flows and sources of water for reuse. In the future, clean urban streams could even become a source of energy. Chapter X integrates all the pieces and defines the ecocity, i.e., a water centric sustainable carbon neutral community that balances social, and economical aspects, and at the same time restores or recreates healthy terrestrial and aquatic urban ecology. The ecocity is a place where it is good to live, work and walk around, and to enjoy culture and recreation. Chapter XI presents the goals, unifying concepts and parameters of several built or planned ecocity developments in Sweden, China, United Arab Emirates, and the U.S.

Acknowledgments

This book was conceived as a follow-up of the proceedings of the Wingspread Workshop on the Cities of the Future held in 2006 at "Wingspread" the Frank L. Wright-designed conference center operated by The Johnson Foundation (Cities of the Future, Novotny and Brown, IWA publishing, 2007). This was one of the first instances where the notion of Cities of the Future was discussed extensively by a group of international experts. The discussion continued and COF has now become the major international initiative sponsored by UNESCO and several international professional organizations and associations. At the beginning of this century, several ecocity projects were conceived and some are now being realized. The idea of producing a textbook that could be used in classrooms and as a manual for landscape architects, urban planners, ecologists, civil and environmental engineers and graduate students in these specializations became a logical outcome of this fast spreading worldwide initiative.

A grant provided by CDM (Cambridge, Massachusetts) to the primary author which enabled the bulk of writing and collecting the materials for the book is acknowledged and appreciated. CDM also provided materials and photos of their project sites in the U.S. and Singapore. The authors are also in debt to many contributors to the book who provided materials for the book and permission to use and quote their figures, photos and writing. Dr. Glen Daigger, Senior Vice-president and Chief technology officer of CH2M-Hill Corporation (Denver, Colorado) provided comments, text and graphic materials for the book. Extensive writing and graphic materials were also provided by Patrick Lucey and Cori Baraclough of the Aqua-Tex Scientific Consulting in Victoria (British Columbia) and Herbert Dreiseitl, Principal of Atelier Dreiseitl in Uberlingen (Germany). Numerous other companies and agencies such as Siemens, Arup, Public Utility Board (Singapore), Masdar Development Co. (UAE), Sonoma Mountain Village, City of San Francisco, Sunwize Technologies, GlashusEtt and City of Stockholm (Sweden), Milwaukee Metropolitan Sewerage District and UNESCO's SWITCH project provided graphic art and information on their ideas, projects and products.

The authors also greatly appreciate the international cooperation with many colleagues interested and involved in COF research and implementation who are now working together in the International COF Steering Committee established and appointed by the leadership of the International Water Association (Paul Reiter, Executive Director and Dr. Glen Daigger, IWA President). The committee leaders are Paul Brown (CDM) and Professor Kala Vairavamoorthy (University of Birmingham, UK) who also directs the UNESCO sponsored SWITCH project on the Cities of the Future.

I

HISTORIC PARADIGMS OF URBAN WATER/STORMWATER/ WASTEWATER MANAGEMENT AND DRIVERS FOR CHANGE

I.1 INTRODUCTION

Since the onset of urbanization millennia ago, cities were connected to water resources, which were their lifeline. Without this connection to water, there would be no cities and, ultimately, no life. When water became scarce, cities were abandoned, and sometimes entire civilizations vanished, as exemplified by the history of the indigenous Hohokam and Anasazi peoples living in the southwestern U.S. in the 15th century, in communities of more than a thousand people—communities that lasted for about a thousand years, but were abandoned, most likely because of extensive drought and the failure of their irrigation systems. Obviously, there were several reasons other than water scarcity causing ancient cities to become ghost towns, then ruins, and finally archeological excavations, centuries or millennia later. Some were related to loss of soil fertility caused by a lack of water for irrigation or poor irrigation practices, which resulted in famine; epidemics of water-borne diseases; exhaustion of the natural resource that was being extracted; or contaminated water, for example, by lead in ancient Rome. Water scarcity is sometimes a result of poor city management and institutions that were inadequate to deal with the multiplicity of conflicting uses and demands for water. Urban waters provided navigation, fish and other seafood, power to mills, laundry, recreation for kings and other nobility, defense during siege by invading armies, and religious significance in some countries and cultures (e.g., India) where certain bodies of water are worshiped.

Water also cleans cities; in historic cities, rainfall washed away the deposits on the streets containing garbage, manure from animals, and human fecal matter. Rainfall and ensuing runoff were—and still are, in many urban areas in some countries—the main and often the only means of disposal of accumulated malodorous solids. During antiquity and the Middle Ages, rivers in sparsely settled rural areas were clean and abundant with fish. In contrast, the environment of ancient, medieval, and post–Industrial Revolution cities was generally filthy and polluted. Terrible epidemics plagued medieval cities, exacerbated by wars and famine. In one medieval epidemic, during a prolonged continental war in the 17th century, 25% of the entire European population vanished.

The situation of urban water resources during the 19th century and in the first half of the 20th century worsened. As cities became industrialized, pollution from industries and loads from reinvented flushing toilets in households (communal flushing toilets were known and used by ancient civilizations of Greece and Rome millennia ago) discharged without treatment into streams resulted in bodies of water devoid of oxygen and smelly due to hydrogen sulfide emanating from decomposing anoxic sediments and water. The response of city engineers and planners was to put the streams out of sight—that is, cover them and/or turn them into combined sewers. In general, until the 20th century, the water environment was not a major interest of architects, builders, or rulers/governments of cities. The people living in the ancient and medieval cities were obviously afraid of epidemics, but the connection between polluted water and diseases was not made until the second half of the 19th century.

The impairments in many urban rivers are caused by the typical characteristics of the urban landscape: a preference for impervious over porous surfaces; fast "hard" conveyance drainage infrastructure, rather than "softer" approaches such as ponds and vegetation; and rigid stream channelization instead of natural stream courses with buffers and floodplains. Under the current paradigm of urbanization, the hard conveyance and treatment infrastructure was designed to provide protection from storms occurring on average once in five to ten years; hence, these systems are usually unable to safely deal with extreme events and prevent flooding, and they sometimes fail with serious consequences. In addition, in many urban river systems, excessive volumes of water are being withdrawn and often transferred long distances, creating bodies of water with insufficient or no flow in some locations, and bodies of water overloaded with effluent and/or irrigation return flows in other areas.

In the mid-2000s, tsunamis and hurricanes struck coastal urban areas, creating catastrophes of enormous proportions. Although these events have occurred throughout history, the human and economic costs of these events were unprecedented. It became painfully evident that the current typical urban landscape and its drainage infrastructure could not cope with these hydrologic events, and the consequences were thousands of lives lost, the suffering and dislocation of survivors during and after these events, and hundreds of billions of dollars in damages. Given that coastal cities are among the fastest growing areas in the world, it is essential to address these problems. Under the circumstance of extreme flows, the current underground urban drainage is almost inconsequential (Figure 1.1), and the hydrologic connection with the landscape is fragmented or nonexistent, providing little buffering protection. Scientific predictions indicate that the frequency and force of extreme hydrologic events will increase with global warming (SPM, 2007; Emanuel, 2005; IPCC, 2007).

On the other side of the hydrological spectrum, many cities, not only in arid zones, are running out of water for satisfying the needs of people. The balanced biota has

Figure 1.1   Impact of Hurricane Katrina in New Orleans (Louisiana) in 2005. Urban infrastructure and human response failed.

disappeared from urban bodies of water because of insufficient flow and has either been replaced by massive growths of pollution tolerant undesirable species (sludge worms, massive blooms of cyanobacteria, and other algal species) or disappeared completely. In the 20th century some cities withdrew so much water that rivers downstream from the withdrawal dried up. The traditional response by urban planners and water engineers was to tap water resources from increasingly larger distances. Bringing water from large distances is not a new concept; Romans built aqueducts up to 50 kilometers long, and the Byzantine Empire brought water to its capital city of one million people from up to 400 km (250 mi) away.

Much progress had been accomplished by the end of the 20th century in the U.S. and other developed countries, but despite the progress made in the U.S., many of the nation's urban water bodies still do not meet the chemical, physical, and biological goals established by the U.S. Congress in the early 1970s. Current research indicates that progress is not only unsatisfactory, but that it may, in fact, have stalled. The fast-conveyance drainage infrastructure conceived of in Roman times to eliminate unwanted, highly polluted runoff and sewage has produced great gains in protecting public health; however, in spite of billions spent on costly "hard" solutions such as sewers, treatment plants, pumping, and long-distance transfers, the safety of water supplies and water quality for aquatic life and human recreation still remain major concerns in most urbanized areas.

At the end of the 20th century, calls for achieving sustainability or green development grew strong and became a mantra for individuals, nongovernmental organizations, and some politicians. Terms for and opinions on "green development and technology," "smart growth," "low- or no-impact development," "LEED- or ISO-certified development," "sustainable development," and "sustainability" appeared in large numbers in the scientific literature, media articles, and feature shows, sometimes linked to "global warming" and "greenhouse emissions." Urban planners have also been promoting green- and brownfield developments. There are at least one hundred definitions of sustainability in the literature (Dilworth, 2008). Most of them have certain common intra- and intergenerational denominators—that is, human beings have the responsibility not to damage and/or overuse resources, so future generations will have the same or better level of resources, and one group's or nation's use of the resources cannot deprive others from the same rights of use (see Chapter II). Hence, sustainability means balancing economic, social, and environmental needs in an intragenerational context. Because the resources are not unlimited and some are nonrenewable, at the present pace of overuse some could be exhausted in less than one hundred years. It appears impossible, with the available and limited resources, that the rate of consumption and (over)use of resources by some, but not all, people living in developed countries could be extended to the entire and growing population of the world. Therefore, changes are coming, and the goal is to achieve a new, more equitable balance. The rate of water consumption and the magnitude of pollution are directly linked to the use of resources. On the other hand, new and better water and environmental management, reuse of resources and byproducts of urban life, and maintenance or restoration of natural resources will have many beneficial impacts on the health, living environment, economy, and social well-being of people that will extend far beyond the boundary of the cities. It has also been realized that not only natural resources and water are involved; a major component of sustainability is energy consumption and related greenhouse gas emissions causing global warming.

Cities have a significant relevance for sustainable development. McGranahan and Satterthwaite (2003) listed three major reasons why cities are playing a major role: (1) Today more than half of the world population is living in cities, and the proportion of the urban population will be increasing in the future. Cities also concentrate the largest amount of poor people. (2) Urban centers concentrate most of the world's economic activities such as commerce and industrial production, and, as a result concentrate most of the demand for natural resources and generate most waste and pollution. (3) Cities have the largest concentration of the middle class and wealthy people who work, but do not necessarily live, there; hence, a lot of energy is required for commercial activities and for people living in or commuting to the cities. Cities also impose a large demand on the power generated in fossil fuel power plants, the water brought from large distances, and the food produced in distant, often foreign, farms. Cities also require energy for moving wastes to treatment and disposal sites. All of these activities not only require energy but also emit large quantities of greenhouse gases (GHG).

At the end of the 20th century, and even more so in this century, it has become evident that the urban water infrastructure cannot cope with increasing stresses—and that, in the new millennium, this infrastructure could crumble because of its age and the inherent deficiencies of traditional designs. Now there is widespread movement towards a new interdisciplinary understanding of how the water infrastructure and natural systems must work in harmony to provide fundamental needs, and this move-ment is ready for success. Urban sustainability concepts and efforts at the beginning of the new millennium were still fragmented, and the role of water resources and wa-ter management was perceived differently by landscape architects, urban planners, developers, urban ecologists, and civil and water resources engineering communities. For landscape architects and developers, urban water resources provided attractions for development. For urban ecologists, development was a cause of environmental degradation. For urban planners, surface water resources often represented an ob-stacle to development and transportation; covering urban streams and bringing them underground used to provide additional space for development, urban roadways, and parking. The civil and environmental engineering community was caught in between. Consequently, the term "watercentric urbanism" had different meanings for these communities. For some urban planners and architects, urban waters often are associ-ated with visual attraction or, in extreme cases, spaces that can be covered and used for more development. This concept could be called "water-attracted development" that can range from clearly unsustainable and vulnerable beachfront developments to city developments, providing visual enjoyment of water and an access to secondary recreation. In this book, "water centric" urbanism means that urban waters are the lifeline of cities, that they must be managed, kept, and/or restored with ecological and hydrological sustainability as the main goal to be achieved. Obviously, such wa-ters would be attractive for a sustainable green development, including protection of riparian zones.

I.2 HISTORIC PARADIGMS: FROM ANCIENT CITIES TO THE 20TH CENTURY

The word "paradigm" is derived from the Greek word *paradeigma* ($\pi\acute{\alpha}\rho\acute{\alpha}\delta\epsilon\iota\gamma\mu\acute{\alpha}$), which means an example or comparison. A paradigm is a model that governs how ideas are linked together to form a conceptual framework, in this case a framework by which people build and manage cities and water resources. A paradigm is first based on logic, common sense, and generational experience, and later on scientific knowledge. It is derived by a discourse in the political domain; science alone may not be the primary determinant of a paradigm. A wrong or outdated paradigm may persist because of tradition, lack of information about the pros and cons of the outdated paradigm, or lack of resources to change it. At the same time, our conceptual models of these systems and our understanding of how they should function and relate to one another have been improving. There are at least four recognizable historical models or paradigms that reflect the evolution and development of urban water resources management; these are outlined in Table 1.1.

Table 1.1 Historic paradigms of urban water/stormwater/wastewater management

Paradigm	Time Period	Characterization	Quality of Receiving Waters
I. Basic water supply	B.C. to Middle Ages; still can be found in some developing countries	Wells and surface waters for water supply and washing; qanads constructed in some parts of the world; streets and street drainage for stormwater and wastewater; animal and often human fecal matter disposed onto streets and into surface drainage; privies and outhouses for black waste; most street surface pervious or semipermeable; roofs often thatched or covered with sod.	Excellent in large rivers; in small and middle-sized streams, poor during large rains, good in between the rains. Pollutants of concern: most likely pathogens because of animal fecal matter on the streets.
II. Engineered water supply and runoff conveyance	Ancient Crete, Greece, and Rome; cities in Europe in the Middle Ages until the Industrial Revolution in the 19th century	Wells and long-distance aqueducts for public fountains, baths (Rome) and some castles and villas; some treatment of potable water; wide use of capturing rain in underground cisterns; medium imperviousness (cobblestones and pavers); many roofs covered with tiles; sewers and surface drainage for stormwater; some flushing toilets in public places and homes of aristocracy discharging into sewers, otherwise privies and outhouses for black waste; animal and sometimes human fecal matter disposed onto streets and into surface drainage; no wastewater treatment.	Excellent to good in large rivers, poor to very poor in small and medium urban streams receiving polluted urban runoff contaminated with sewage; widespread epidemics from waterborne and other diseases. Pollutants of concern: pathogens, lead (in Roman cities because of widespread use of lead, including pipes), BOD of runoff.

III Fast conveyance with no minimum treatment	From the second half of 19th century in Europe and U.S., later in Asian cities, until the second half of the 20th century in advanced countries, still persisting in many countries	Wells and long-distance aqueducts for water supply; potable water mostly from surface sources treated by sedimentation and filtration; wide implementation of combined sewers in Europe and North America; beginning of widespread use of flushing toilets; conversion of many urban streams into underground conduits; initially no or only primary treatment for wastewater, secondary treatment installed in some larger U.S. and German cities after 1920s; after 1960 some smaller communities built lower-efficiency secondary treatment; paving of the urban surface with impermeable (asphalt and concrete) surfaces; swimming in rivers unsafe or impossible.	Poor to very poor in all rivers receiving large quantities of untreated or partially treated wastewater discharges from sewers, runoff discharged into sewers, and combined sewer overflows; rivers sometimes devoid of oxygen, with devastating effects on biota; Cuyahoga River on fire in Cleveland; waterborne disease epidemics diminishing due to treatment of potable water. Pollutants of concern: BOD, DO, sludge deposits, pathogens.
IV Fast conveyance with end of pipe treatment	From the passage of the Clean Water Act in the U.S. in 1972 to present	Gradual implementation of environmental constraints resulting in mandatory secondary treatment of biodegradable organics; regionalization of sewerage systems; additional mandatory nitrogen removals required in European Community; recognition of nonpoint (diffuse) pollution as the major remaining problem; increasing concerns with pollution by urban and highway runoff as a source of sediment, toxics, and pathogens; increasing focus on implementation of best management practices for control of pollution by runoff; emphasis on nutrient removal from point and nonpoint sources; beginning of stream daylighting and restoration efforts in some communities.	Improved water quality in places where point source pollution controls were installed; due to regionalization, many urban streams lost their natural flow and became effluent dominated; major water quality problems shifted to the effects of sediment, nutrients, toxics, salt from de-icing compounds, and pathogens; biota of many streams recovered, but new problems with eutrophication and cyanobacteria (blue-green algae) blooms emerged.

I.2.1 First Paradigm

This paradigm of water management of ancient cities was characterized by the utilization of local wells for water supply and exploitation of easily accessible surface water bodies for transportation, washing, and irrigation; streets were used for conveyance of people, waste products, and precipitation. The ancient Mediterranean civilizations of Greece and their cities were built on sound engineering principles that incorporated sophisticated water supply systems and drainage. Athens in 500 B.C. had public and private wells and surface drainage (Figure 1.2). Several hundred years later, Romans conquered Greece and adopted and improved their water/stormwater systems. However, urban runoff of ancient and medieval cities was not clean: it carried feces from animals and from people.

The archeological excavations in Pompeii and Herculaneum in Italy (two Roman cities covered by ash during the Vesuvius eruption in 79 A.D.) and elsewhere provide a vivid testimony of the water engineering and management that was typical for the late period of the first paradigm. Figure 1.3 shows a major street in Pompeii which indicates that streets were used for collection and conveyance of urban runoff polluted by animal feces, overflows from fountains, and wastewater from the houses. Human fecal waste was not disposed into the street drainage.

Figure 1.2 Drainage systems in ancient Athens (ca. 500 B.C.). This 1 m x 1 m surface drainage channel is located in the agora (gathering place) of the ancient Greek metropolis (Photo V. Novotny).

Figure 1.3   The Via Abbondanza in the Roman city of Pompeii near Naples in Italy. Stepping stones document that the street was used for drainage. The street also had water fountains conveniently located along the street so citizens and merchants did not have to go far for water. Overflow from the fountains washed the streets (Photo V. Novotny).

As cities grew and local wells could not provide enough water, more sophisticated water designs allowed water to be brought from larger distances by underground delivery systems called *qanads*, constructed in southeast Asia, North Africa, and the Middle East. Typically with qanads, a large well was dug by manual labor at the foothills of nearby mountains providing abundant water, and the well was connected by a gravity flow tunnel with the city, where it provided water to the population and irrigation of crops. Some qanads brought water from distances as far as 40 kilometers, and the wells and tunnel were dug more than 100 meters deep (Cech, 2005). Cech also noted that qanads are still used today in the Middle East and parts of China.

I.2.2 Second Paradigm

As water demand increased and easily accessed local groundwater, rain, and surface supplies became insufficient to support life and commerce, the second paradigm emerged in growing ancient and medieval cities: the engineered capture, conveyance, and storage of water. This period is characterized by more advanced engineered water systems that brought water from large distances to the cities. As the economies of the states and cities—driven by slave labor—were increasing, water resources became more important for commercial and military navigation, and canals were built around the cities to enhance defense. The beginning of the Middle Ages is usually associated with the conquest of the western Roman Empire by barbarians and the subsequent

Figure 1.4 One of the largest Roman aqueducts, Pont du Gard in southern France (former Roman province of Gallia), which is today a UNESCO heritage site (Photo V. Novotny).

abolishment of slavery in most European countries. The eastern part of the Roman Empire became the Byzantine Empire and continued for another 900 years.

Over the centuries the Romans developed extensive systems for water distribution which relied both on wells and on elaborate systems of providing clean water brought from nearby mountains. The first Roman aqueduct was constructed in 312 B.C. (Cech, 2005). The aqueducts of ancient Rome brought water from mountains as far away as 50 kilometers (Figure 1.4). Water was stored in tanks and underground cisterns and distributed by lead or baked clay pipes to fountains, public baths, public buildings, and the villas of the aristocracy. Fountains were located evenly all over the towns so that each homeowner who did not have a private water supply could reach the fountains without any difficulty. Water supply pipes were laid along the streets, providing water continuously to fountains, each with an overflow directed onto the street surface. As shown on Figure 1.3, the street also provided drainage of stormwater (Nappo, 1998). Figure 1.5 shows an example of a house in Pompeii built with a courtyard (atrium) in the middle, where the rain-collecting cistern was located and all roof runoff was directed. The practice of rain harvesting and storing rainwater in cisterns was also typical in many ancient and medieval cities and is still common in many communities in dry Mediterranean regions and elsewhere.

Romans were not much concerned with the disposal of wastewater, as long as it did not pose a great nuisance. Paved streets in most cases were continuously washed

Figure 1.5 Atrium of a large house in Pompeii. Roof rainwater was directed into the basin in the center, from which it was directed into an underground cistern. Overflow was conveyed to the street (Photo V. Novotny).

by the overflow from the fountains and by rainwater. In Roman cities common people washed themselves in public baths, which were also a place for socializing. In Pompeii and other cities, laundry was done in commercial laundries and cleaning shops. Some cities also had communal flushing toilets. To handle pollution of urban runoff and the flow of wastewater from baths and public buildings, sewers were invented. This invention allowed polluted street flows and wastewater to be conveyed underground to the nearest rivers. The Roman sewer, the Cloaca Maxima, has been functioning for more than two thousand years (Figure 1.6); however, sewers were installed a thousand years later in other European cities.

In contrast, in medieval cities of Europe (with the exception of Muslim regions of Spain and the Balkans), common people and even the nobility had poor personal hygiene, rarely took baths, and had no showers. As a result, domestic per capita water use in medieval European cities was much smaller than in Roman cities or modern cities, most likely at the level that today would be considered a minimum daily use. Most excreta and fecal matter were disposed on site in outhouses and latrines. Like

Figure 1.6 Outlet of the Roman sewer, the Cloaca Maxima (Largest Sewer) into the Tiber River. The sewer is functioning today, but a barrier was installed to prevent entry because of security concerns.

those of ancient cities, street surfaces were polluted by fecal matter and trash. Solid waste deposits on streets of medieval Paris were sometimes 1 meter high, and night chamber pots were generally emptied into street drainage.

Ancient Rome and medieval Constantinople had populations of about one million at their height, while medieval London, Paris, Amsterdam, and Prague had populations in tens of thousands, at most, and Berlin was a village. Constantinople (present-day Istanbul in Turkey) on the shores of the Bosporus was the capital of the Byzantine Empire, which lasted until the 15th century A.D., and for more than a thousand years it was the center of East European and Mediterranean civilization. After the conquest of Rome by barbarians in the 5th century, it was the cultural and commercial center of the world. This city inherited—and improved upon—Roman culture and engineering know-how when the Roman Empire split into its eastern and western parts. Its water system was similar to that of Rome, using aqueducts to provide fresh water, but also relying heavily on private and public rainwater harvesting and cisterns. The longest aqueduct (400 km) was built in the 4th and 5th centuries to provide water to this megalopolis. Water was stored in more than one hundred cisterns throughout the city that provided 800,000 to 900,000 m^3 (211 to 238 mg) of storage. In the 7th century the city built its largest underground cistern (Figure 1.7).

Figure 1.7 This underground Basilica cistern, capable of storing 80,000 m^3 (21.1 mg) of water, was built at the beginning of the seventh century in Constantinople, the capital of the Byzantine Empire (present-day Istanbul in Turkey) (photo V. Novotny).

Another large medieval city with more than 200,000 inhabitants was Venice (in present-day Italy), which was a center of the powerful Venetian Republic (697–1795 A.D.), competing with the Byzantine Empire over the dominance of the Mediterranean region. The city is located on 118 small islands inside the 500-km^2 Lagoon of Venice and is known for its famous canals. Historically, Venice relied on private and public wells and fountains, and all sewage was discharged directly into the canals. Essentially, the Republic of Venice, including its other cities (Padua, Verona), operated its water and wastewater disposal using the concepts of the first paradigm, although it periodically dredged the canals within the city to remove accumulated sludge. The city also built a network of canals on the mainland surrounding the lagoon and relocated two major rivers outside of the lagoon to prevent its siltation. The historic city of Venice, which today has about 80,000 permanent residents and many thousands of tourists, still discharged all wastewater into its canals with minimum treatment at the end of the last millennium. Since the beginning of the 21st century, low-level distributed treatment has been implemented in the historic city.

A pipeline system delivering water to London from the Thames River and nearby springs was built at the beginning of the 13th century, and by the end of 18th century, major European cities had a water distribution system that relied on public fountains

Figure 1.8   Caesar Fountain in Olomouc in the Czech Republic, sculpted and built in 1725 (Photo V. Novotny).

and deliveries of water by pipelines to individual houses. Many public fountains in medieval cities were pieces of art (Figure 1.8). For most of the medieval era, water supply pipelines were made of baked clay or wood (Figure 1.9), and were replaced by cast iron later in the 19th century. Large sewers were of masonry. In some cities water to individual houses was provided by private water vendors (Cech, 2005). Most houses, however, had only one faucet with a sink. Sewers were not common, and many smaller and even middle-sized cities in Europe did not have sewers until the 20th century. The use of standpipes and/or private vendors for water distribution can still be found in many undeveloped countries. The end of the second paradigm could be dated to the middle of the 19th century, when the servitude of rural people to their feudal masters in Europe and slavery in the U.S. were broken, which resulted in a massive population migration into cities. This was the beginning of the Industrial Revolution, which shifted the economic power to the cities, away from the landholding nobility who had held the rural population in servitude (or slavery).

Figure 1.9 Making wood pipes for the medieval water supply systems (courtesy: Museum of Water Supply in Prague, Czech Republic).

I.2.3 Third Paradigm

Beginning in the first half of the 19th century, the freed rural population migrated to cities and joined the labor force in rapidly expanding industries, then run more by steam and fossil fuel (dirty) energy than the clean water or air energy (water wheels or wind mills) typical for small industries and mills during the second paradigm. This change, along with the ensuing rapid expansion of cities, increased urban pollution dramatically. In the second half of the 19th century, sewers were accepting domestic and sometimes industrial black sewage loads. However, most industries clustered near the rivers discharged effluents directly into streams without treatment. Because urban water bodies served both for water supply and wastewater disposal, sewage cross-connection and contamination of wells and potable water sources caused widespread epidemics of waterborne diseases.

The third paradigm of urban water and wastewater management added a massive investment in building sewers, in trying to cope with the pollution of urban surface waters. Urban water bodies were becoming unbearably polluted and a serious threat to public health. Other monumental projects included flood controls by stream straightening, lining, and ultimately covering; building thousands of reservoirs for water supply and hydropower; navigation river projectsand canals. Even today, $30–$40 billion (in 2000 dollar value) are spent annually on new dams worldwide (Gleick, 2003), and monumental cross-country canals and water transfers are being built or planned, such as a canal bringing water from the water-rich Yangtze River to Beijing and other cities located in the water-poor Northeast of China. A transcountry canal is planned in the Republic of Korea.

Since the end of the nineteenth century communities were building combined sewers and treatment plants for potable water, as engineering methods to solve the problem of pollution of surface waters. Flushing toilets changed the way domestic fecal matter was disposed. Until then collection tanks and pits in outhouses and latrines were emptied periodically by private haulers. The introduction of flushing toilets and bathrooms conveyed fecal and other wastewater into the newly built or existing stormwater sewers. The goal of pollution control was *fast conveyance* of wastewater and urban runoff out of sight from the premises to the nearest body of water.

Wastewater treatment, at the end of the 19th century and beginning of the 20th century, was limited to sedimentation and self-purification in the receiving water bodies. This was not even remotely sufficient to resolve the nuisance problem with sewage discharges. One solution was to pump sewage and apply it onto fields for crop irrigation, which was practiced in the late 1800s around London, Berlin, Paris, and Sydney (Cech, 2005), Mexico City (Scott, Zarazua, and Levine, 2000), in China, and in many other locales. In the late 1800s, septic tanks and leaching fields were first used in the United States, and these are still used today in places without sewerage. In Europe in the early 1900s, sedimentation of solids in sewage and anaerobic digestion of the deposited solids were done in septic tanks (Figure 1.10) known as Imhoff tanks (commemorating German pioneer of sanitary engineering Karl Imhoff). Activated sludge plants, trickling filters, and sewage lagoons were invented in the early 1900s.

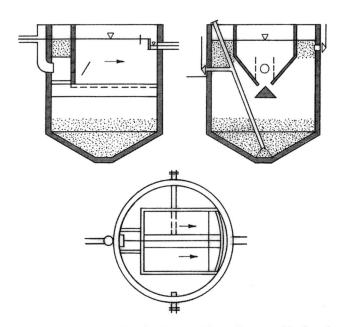

Figure 1.10   The Imhoff tank combined primary settling with anaerobic digestion of settled sludge. It was invented by Karl Imhoff in Germany at the beginning of the 20th century. The tank has an aerobic settling compartment in the middle, anaerobic sludge digestion in the lower part, and scum-collecting volume on the top (Replotted from Novotny et al., 1989).

Figure 1.11 Converting Mill Creek into a sewer in Philadelphia (PA) in 1883 (photo provided by the Philadelphia Water Department Historical Collection).

At the same time, covering streets and other areas of cities with impervious pavements was preventing rainfall infiltration and, concurrently with the increased withdrawals of water from streams, depriving urban streams of the base flow needed for dilution of pollutant loads between the rains. During dry weather, some streams carried mostly sewage and became *effluent dominated* (see Chapter IX). The solution was to put small and medium-sized urban streams out of sight and convert them to combined sewers (Figure 1.11). The aim of these *fast conveyance urban drainage systems* (sewers, lined and buried streams) was to remove large volumes of polluted water as quickly as possible, protecting both public safety and property, and discharging these flows without treatment into the nearest receiving body of water. Almost all sewers, even the old ones originally designed to carry heavily polluted urban runoff from streets, were combined—that is, they carried a mixture of sewage, infiltrated groundwater, and stormwater flows. Over a relatively short period of fifty to one hundred years, most of the urban streams disappeared from the surface, as shown on Figure 1.12.

In the absence of effective treatment technologies that would remove putrescible pollution from sewer outfalls and heavily polluted urban runoff (most of the street traffic was still by horse-drawn wagons and coaches), city engineers resorted to grandiose projects to alleviate the pollution problems. In Boston, Massachusetts, several square kilometers of the tidal marsh of the Charles River estuary called the Back Bay, plagued by standing sewage pools, were filled between 1857 and 1890 and converted to upscale urban development that more than doubled the size of the city

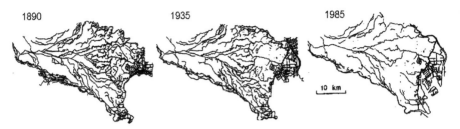

Figure 1.12   Disappearance of streams in the Tokyo (Japan) Metropolitan area (Courtesy Prof. Horoaki Furumai, 2007).

at that time. Approximately at the same time, a large tributary of the Charles River named Stony Brook was causing a nuisance and threatening public health. Because of public health regulation for sewer discharge points, lowlands in the neighborhoods into which the brook was discharging became terminal sewage pools. Periodic epidemics swept through the city regularly. Raw sewage from Stony Brook flowed directly into the tidal Back Bay, with environmentally destructive results. Historian Cynthia Zaitzevsky (1982) describes the effect of sewage on the Back Bay: "...the residue lay on the mud flats, baking odiferously in the sun. Eventually it became incorporated into the mud. Under these conditions, the last vestiges of the salt marsh could not remain healthy for long. When the park commissioned a survey of the area in 1877, animal life was no longer able to survive in the waters of the Back Bay." As a result, a 12-kilometer stretch of the brook through the city was buried and converted into large box culverts. Only names such as Stony Brook Park or Stony Brook subway and train station remain, and most of the Boston population does not even know that a medium-sized historic river existed in the city 150 years ago. Figure 1.13 shows the old gate house where Stony Brook went underground. After sewer separation in 2002, the relatively clean water originating in a headwater nature conservancy area upstream is now flowing in a double culvert storm sewer, while a large portion of a once very lively and important part of the city that used to surround the brook has deteriorated. The gate house shown in the figure is gone today, but the river is still underground. The practice of burying small and medium streams and converting them into subsurface sewers was common to almost every city in the world, ranging from small to large.

Because of the poor sanitation and discharges of untreated wastewater into groundwater and surface water bodies, terrible epidemics of waterborne diseases plagued the urban population throughout the Middle Ages until the end of the 19th century. The cholera epidemics in Chicago (Illinois) in the late 1800s, caused by contamination of the city's water intake from Lake Michigan, led the city government to commission the building of an engineering marvel, the Chicago Sanitary and Ship Canal (CSSC), finished in 1910. The canal reversed the flow of the Chicago River, which had originally flowed into Lake Michigan, diverting it into the Des Plaines River (Figure 1.14) that flows, after becoming the Illinois River, into the Mississippi River (Macaitis et al., 1977; Novotny et al., 2007). In this canal and the Des Plaines River, all sewage and most of the overflows from the combined sewers

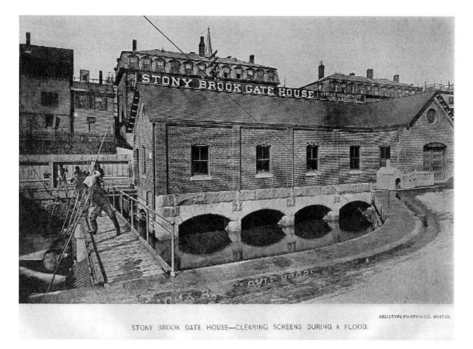

STONY BROOK GATE HOUSE—CLEARING SCREENS DURING A FLOOD.

Figure 1.13   The gate house with bar racks through which Stony Brook in Boston (Massachusetts) entered underground into 12-km- long culverts more than one hundred years ago. *Source:* Charles Swift, BostonHistory.TypePad.com)

(CSOs) are diverted into the Illinois River, a tributary of the Mississippi River, and do not contaminate the water intakes in Lake Michigan. The CSSC is now one of the largest inland shipping waterways, larger than the Suez Canal, and the Lower Des Plaines River is also the largest effluent dominated body of water in the world (see Chapter IX).

The third paradigm period had numerous other pollution catastrophes due to unregulated or poorly regulated point source discharges and absolutely no controls of diffuse (nonpoint) pollution. Severe cases of painful and deadly mercury and cadmium poisoning of fishermen in Japan were reported in the 1960s. Minamata mercury poisoning disease was first discovered in Japan in 1956, and another outbreak occurred in 1965. As a result of fish contamination, thousands died and tens of thousands were infected. As a result of point pollution, many streams were dead, smelly water bodies with sludge deposits that could only harbor dense populations of sludge worms (Krenkel and Novotny, 1980).

By the end of the 19th century, people began to understand that unsanitary living conditions and water contamination contributed to disease epidemics. This new awareness prompted major cities to take measures to control waste and garbage. In the United States, industrial chemicals and wastes, including sulfuric acid, soda ash, muriatic acid, limes, dyes, wood pulp, and animal byproducts from industrial mills, contaminated waters. In the industrial U.S. Northeast and Midwest, and also in

Figure 1.14 The effluent dominated Des Plaines River in Joliet, Illinois, after the confluence with the Chicago Sanitary and Ship Canal. It has become one of the largest inland waterways in the U.S. Photo V. Novotny.

industrial Europe, almost all major and middle-sized rivers were severely affected by pollution. New pollutants such as household detergents formed foam on the weirs 5 meters or more thick. The Cuyahoga River in Cleveland, Ohio, which flows into Lake Erie, became so polluted that the river caught on fire (Figure 1.15) several times between 1936 and 1969. The fire was due to floating debris and a thick layer of oils floating on the surface of the river.

In the mid-1850s, Chicago built the first major primary treatment plant to treat its sewage in the United States. From 1880 until well into the second half of the 20th century, water pollution control efforts in the U.S. and industrialized countries of Europe focused on removal of objectionable solids, disease-causing pathogens, and oxygen-demanding organic substances (BOD) that were turning receiving water bodies into unsightly, oxygen-deprived black-colored smelly streams or pools. During the third paradigm period, primary and later secondary wastewater treatment technologies were introduced in several cities but did not address the overall, *uncontrolled water-sewage-water cycle* (Imhoff, 1931; Lanyon, 2007; Novotny, 2007) in which water in an upstream community is converted to sewage, discharged into a receiving water body, and reused downstream as potable water by another community (see Chapter IX). Dissolved oxygen concentrations preventing fish kills provided guidance for estimating the waste-assimilative capacity of streams. The

Figure 1.15   The fire of the Cuyahoga River in Cleveland, Ohio, in 1952. *Source:* Cleveland Press Collection, Cleveland State University Library.

primary reason for installation of treatment plants by some communities was protection of public health and avoidance of nuisance from unsightly and odorous anoxic urban waters.

Increasing imperviousness. Paving the cities and roads dates back to ancient Greece and Rome (see Figures 1.2 and 1.3). However, in medieval cities only important streets and plazas were paved with cobblestone pavement that, hydrologically, had relatively large depression storage (about 1 cm) for storing rainwater, and was partially pervious. Many side streets and squares had unpaved dirt surfaces. Roofs were obviously impervious, many covered with tiles or wood shingles, although in the early times, some were thatched or even covered with sod (in Scandinavia).

The practice of using relatively smooth concrete and asphalt pavements on a large scale dates to the first half of the 20th century. Unlike the stone pavements of ancient and medieval cities, modern pavements are highly impervious and have relatively small depression storage, of about 1 to 2 mm, to capture and evaporate rain. Some portions of historic cities became almost completely impervious.

Dropping groundwater table and subsidence. As a result of imperviousness, infiltration into underground sewers, and sump pumps draining deep construction sites, basements, and underground garages and tunnels, hydrology of urban watersheds has changed by reducing groundwater recharge by infiltration, thus increasing surface urban runoff. Consequently, peak flood flows in surface and sub-surface drainage increased by 4 to 10 times (Novotny, 2003), while the groundwater recharge by infiltration diminished. Because of the dropping groundwater table, many cities built on filled wetlands—such as a large portion of Boston and Philadelphia in the U.S., Venice in Italy, Mexico City, parts of Tokyo and Osaka (Japan)—have serious problems with subsidence of their historic buildings built with wood piles foundations. Palaces were built on wood piles in Venice and elsewhere for almost a thousand years, and as long as the wood piles are submerged in groundwater they do not rot. Under these conditions, untreated wood piles can be considered to be permanently durable as long as the water table remains above the tops of the piles, and the wood and surrounding soil remain saturated. However, if the groundwater level drops below the top of the piles, plant growth and insects can attack the wood, and a greatly increased supply of oxygen, combined with moisture and moderate temperatures, facilitates the growth of fungi. Grubs or wood borers, termites, and other insects may also attack the "exposed" wood (Aldrich and Lambrechts, 1986). In Tokyo, large-scale problems with subsidence due to groundwater mining were observed first in 1914 and continued with increased intensity thereafter. Ground subsidence caused destruction of many buildings. Countermeasures against ground subsidence started in the 1960s, and the rate has slowed (Furumai, 2008).

Urban flooding. Building storm and combined sewers could not alleviate urban flooding problems, and increased imperviousness made it worse. Storm sewers are traditionally designed to carry flows resulting from storms that have a recurrence interval of once in five to ten years, and the capacity of combined sewers is generally six times the dry weather flow. This means that every rain with an intensity of approximately 3 mm/hour will result in an overflow (Metcalf & Eddy, Inc., 2003; Novotny, 2003). In part of Tokyo (Japan), which is highly impervious, floods occur with a frequency of once in two years. Because the land in cities became highly valuable for development, cities encroached into floodplains and, to minimize flooding, streams were straightened, diked, and lined to increase their velocity and capacity to carry more flow. Figure 1.16 shows the Los Angeles River, which today is a concrete fast-flow flood conveyance channel. Increasing velocity during high flows created adverse safety problems, and the streams became sometimes deadly to children playing or falling in them. The answer to this problem was fencing off the streams. Streams lined with concrete or similar materials (masonry) cannot support aquatic life, and the result is ecologically almost the same as putting them underground. Rivers converted into flood conveyance channels also received overflows from combined sewers and stormwater runoff (Figure 1.17). Lining streams and building sewers did not resolve the flooding problems. At best the problem was moved and accentuated downstream. In almost every large city, some rivers and streams were covered to make a space for parking lots and other developments (see Chapter IX).

In the 1960s, the public was rising in protest against the excessive pollution of the environment. Rivers on fire, black streams devoid of oxygen (a black color is given to water by sulphuric bacteria that thrive in anoxic waters), the stench of anoxic waters and sludge all reached a point that people could not bear. In London, summer sessions of Parliament had to be canceled because of the bad smell emanating

Figure 1.16   Los Angeles River. It was once a natural river, but it was converted into a lifeless flood conveyance channel with no connection to population living nearby. The river in some sections is a perennial effluent dominated channel; in some other sections, it has no dry-weather flow. *Source:* US Army Corps of Engineers.

from the Thames River. In 1962, Rachael Carson's famous book *Silent Spring* was published, describing the consequences of the contamination of flora and fauna by chemical pesticides. The "silence" was due to the disappearance of birds, dying because their body tissue had been contamination by DDT and other pesticides. In the U.S., the third paradigm period culminated in the passage of the Federal Water Pollution Control Act Amendments of 1972 (Clean Water Act) making end-of-pipe treatment mandatory.

Environmental awakening, which marks the end of the third paradigm, occurred years later in some advanced countries, and many developing countries (e.g., China, India, Brazil) are now, at the beginning of the third millennium, recognizing that unrestricted urban development leads to environmental catastrophes. The factors that affected the direction of urban water/stormwater/wastewater development and management during the third paradigm were mainly in the category of economic development, hampered by the lack of technologies for water and wastewater purification and the public's lack of awareness of alternatives to poor environmental quality. In the 1950s and before, the smokestacks of factories were a sign of progress, and land

Figure 1.17 Lincoln Creek in Milwaukee, Wisconsin, was a concrete-lined channel receiving combined sewer overflows, before restoration in the 1990s. Photo V. Novotny.

for development was abundant. Effective environmental restrictions were few in the more advanced countries of the West and North, and there were none in the undeveloped countries of the East and South. Institutional infrastructures, such as nationwide pollution control authorities with regulatory and enforcement powers, were nonexistent or included under public health departments or ministries. Protecting public health and avoiding deadly epidemics were the main goal of government agencies and the mission of sewerage utilities formed in large cities in the U.S. in the first half of the 20th century.

The *global warming problem* was either unknown or unrecognized during much of the 20th century. Surprisingly, the coming of a new ice age was widely published by some media in the 1970s. Nonpoint (diffuse) pollution by urban runoff was not recognized as a problem until the late 1960s, when the American Public Works Association (APWA, 1969) published a study identifying the pollution problem of urban runoff. The existence of rural nonpoint pollution was denied by the farming community. The problem of eutrophication was also recognized only at the end of the third paradigm period in the 1960s (Rohlich, 1969), in spite of the fact that Lake Erie and other water bodies were dying because of the excessive loading by phosphorus, caused mainly by agricultural runoff and use of phosphate-containing detergents.

The third paradigm era can be characterized as one of continued rapid economic development with goals of maximizing profits on the microscale and growth of the grossnational product on the macroscale, in countries under the capitalist economic system. In these countries, before 1970, the tools for remedies were restricted to protests and litigation, finally leading to a paradigm change embedded in the Clean Water Act, the Safe Drinking Water Act, and several other important laws passed by the U.S. Congress. In Central and Eastern European countries under the socialist systems, the tools of protest and litigation—as well as the concept of profit—were not available; hence, the only goal of the planned economies in those countries in the second half of the 20th century was increased industrial and agricultural production based on often unrealistic governmental quotas.

I.2.4 Fourth Paradigm

The passage of the Clean Water Act (CWA) by the U.S. Congress in 1972, over the president's veto, was the necessary impetus to change the paradigm for water and wastewater management in the United States. However, the Act passed by the U.S. Congress has also had worldwide effects because many countries adopted some of its provisions and/or used it for the development of their own water and pollution management, and legislative control acts. At the end of the 20th century, the European Parliament enacted the Water Framework Directive (WFD). Although water quality standards in most of Europe, based on a long tradition, were formulated in a form different from those of the U.S., the goals of pollution abatement shifted from protecting the public from diseases and death to broader goals of protecting the well-being of people and aquatic biota and promoting safety for those using bodies of water for recreation (see Novotny (2003)).

Meeting these goals required massive investments in building treatment plants, for achieving safe drinking water quality, and wastewater treatment facilities that would bring the receiving water into compliance with the more stringent water quality standards formulated and enacted according to the goals of the CWA: the attainment and protection of the *physical, chemical, and biological integrity of the nation's waters* and *providing conditions for safe primary and secondary recreation in and on the waters*. Integrity of water bodies was defined as *"a balanced, adaptive community of organisms having a species composition and diversity comparable to that of natural biota of the region"* (Karr et al., 1986). Physical integrity is usually interpreted as habitat conditions suitable for maintaining a balanced aquatic biota.

Drinking water protection was included in the Safe Drinking Water Act, and the provisions of both this act and the CWA were combined and reflected in the surface water quality standards. Implementing best available treatment technologies became mandatory for point sources. Nonpoint pollution controls in the U.S. have been voluntary, but mandatory nonpoint pollution abatement has been enacted in the European Community, Japan, and Korea.

Hence, the period between the enactment of the CWA in the U.S. and the present time has comprised the *fourth paradigm* of urban water management and protection, in which both point and increasingly diffuse sources of pollution were considered and addressed in many separate and discreet initiatives. This paradigm could also be called the *end-of-pipe control* paradigm, because the predominant point of control of both point and diffuse pollution is where the polluted discharge from the fast conveyance system (sewer or lined channel) enters the receiving water body. Pollution by urban runoff and other diffuse sources was recognized as a problem only about 30 to 40 years ago and was included in the CWA. In the U.S. mandatory but somewhat inefficient urban and highway storm waste discharge permitting was enacted at the end of 1990 and is currently slowly being implemented.

In the U.S., after the passage of the Clean Water Act in 1972 (fourth paradigm), the new massive program of building treatment plants was based on the "economy of scale" characterized by large regional treatment facilities with long-distance transfers of wastewater over smaller local plants. Local treatment plants built before 1970 were mostly rudimentary primary only plants, or low-efficiency trickling filter facilities, or aerobic/anaerobic lagoons (sewage stabilization ponds). In most cases these plants were unable to meet the goals of the Clean Water Act. The new large-scale activated sludge treatment facilities offered better efficiency capable of meeting the more stringent effluent standards and were managed by highly skilled professionals.

Under the fourth paradigm, the urban water/wastewater management systems are once-through flow type; that is, potable water from wells and surface water bodies is treated, brought—often from large distances—to the city where it is used, and converted to wastewater which is then collected in sewers and conveyed by interceptors to a regional public wastewater treatment facility whereby the treated effluent is then discharged into a receiving body of water. Reuse is currently still rare and was almost nonexistent before the year 2000. A notable exception in the U.S. is the system of wastewater reuse in Tucson, Arizona (see box). After 2005, Southern California found itself in a sewer drought, and several cities began implementing water conservation and wastewater reclamation and reuse, including Los Angeles, Orange County (California), and others.

WASTEWATER RECLAMATION IN TUCSON

The city of Tucson, located in the arid region of southern Arizona, is a fast-growing metropolis that in 2000 had 843,746 inhabitants, based on the U.S. Census Bureau. When the city was a desert outpost at the end of the 19th century, there were perennial rivers flowing from surrounding mountains, providing a water supply. In the second half of the 20th century, the city outgrew its surface water supply, the rivers became ephemeral, and the city began mining water from the underlying aquifer. The aquifer became the source of water, and the water table was dropping rapidly. At great cost, a canal connecting the city with the Colorado River was built. Tucson is today reclaiming 90% of its effluent and reusing it for irrigation of golf courses, parks, and other uses.

The long-distance transfers of water and wastewater dramatically changed the hydrology of the impacted watershed. Surface waters became flow deficient after withdrawals, and the water bodies receiving the effluent discharges downstream then

Figure 1.18 Deep tunnel storage of CSOs and SSOs in Milwaukee, Wisconsin. The stored mixture of rainwater and sewage is pumped for treatment. A similar but much larger tunnel system was excavated in Chicago, Illinois. Courtesy Milwaukee Metropolitan Sewerage District (MMSD).

became effluent dominated. However, even today, the problems with combined and sanitary sewer overflows (CSOs and SSOs) have not been and most likely will not be fully mitigated in the near future. These overflows have to be captured and stored in expensive mostly underground storage facilities (Figure 1.18) and subsequently treated. For example, many kilometers of 12-meter-diameter interceptors known as "deep tunnel" were built in Milwaukee, Wisconsin, and Chicago, Illinois, storing millions of cubic meters of a mixture of stormwater and wastewater from captured CSOs and SSOs (Table 1.2). The stored sewage/water mixture is pumped into the regional treatment plants. Each underground pumping station in Milwaukee and Chicago uses several pumps that are among the largest ever built, and the cost of energy for pumping is high. Similar tunnels with a large pumping station were completed at the end of the 20th century in Great Britain and in 2008 in Singapore. The long-distance water/wastewater transfers from source areas over large distances also require electric energy for pumping, treating (e.g., aeration), and transporting

Table 1.2   Parameters of the Milwaukee and Chicago deep tunnel storages for CSO and SSO flows (various sources from the Milwaukee Metropolitan Sewerage District and the Metropolitan Water Reclamation District of Greater Chicago)

System	Milwaukee (WI)	Chicago (IL)
Capacity (million m^3)	2.0 (in 2009)	9.1 (in 2008)
Length (km)	46	175.3
Diameter (meters)	5.2–9.7	5.2–9.7
Depth underground (meters)	100	73–106
Cost (US$, 1990 level)	1 billion +	3 billion

treatment residuals to their point of disposal. This use of energy contributes to greenhouse gas (GHG) emissions. The volume of "clean" groundwater water infiltration and illicit inflows (I-I) into sanitary sewers has to be pumped and treated with the sewage and uses more energy. The I-I inputs could, during wet weather, more than triple the volume of dry-weather wastewater flows in sewer systems and overwhelm treatment plants (Metcalf & Eddy, Inc., 2003; Novotny et al., 1989).

Results of point source pollution controls. Control of point pollution discharged from municipal and industrial sources is mandatory under the provisions of the Clean Water Act in the U.S., the Water Framework Directive in all EC countries, in Japan, Australia, and several other countries that had environmental catastrophes during the third paradigm period. As a result, the water quality of many streams in these countries has improved. The Cuyahoga River in Cleveland is still polluted, but it will not catch on fire. Fish returned to the Thames River in London decades ago, and the Vltava (Moldau) River downstream of Prague (Czech Republic) is not black anymore because of a lack of oxygen. The Charles River in Boston, which only a couple of decades ago was ranked as one of the most polluted rivers (it was acutely dangerous to fall in), had its first "swimming" days in 2007. On the other side of the world, however, the Yangtze River in China is heavily polluted, and in India pilgrims take annual baths in the Ganges River, which has a very high content of pathogens and is also heavily polluted.

Aquatic life has returned even to some effluent dominated rivers. A study of the largest effluent dominated river in the world, the Des Plaines River southwest of Chicago (Figure 1.14), found that fish have returned in spite of the fact that 90% of the medium flow and almost all of the low flow is the treated effluent from the Metropolitan Water Reclamation District of Greater Chicago (Novotny et al., 2007). Unfortunately, in the first decade of this millennium, invasive river carp (white carp, a native of Siberian rivers released into the Mississippi River) have overpopulated the Illinois and Des Plaines Rivers and decimated the ecological balance. An expensive electrical barrier is now keeping the pesky large fish from entering the Great Lakes via the Chicago Sanitary and Ship Canal.

The focus of point source pollution controls has now shifted from removing biodegradable organic pollution, suspended solids, and pathogens—the three original fundamental compounds in the National Pollution Discharge Elimination

System (NPDES) permits—to adding toxic compounds, nutrients, and other pollutants to the controlled (permitted) pollutants. In Europe, installation of Bardenpho treatment facilities has been required for cities discharging wastewater into the European coastal waters, especially those of the North, Baltic, and Black Seas, which is most of Europe. The Bardenpho system, developed by James Barnard (2007) in the 1980s, uses both aerobic and anoxic units to achieve high-degree removals of biochemical oxygen demand (BOD) and nutrients (nitrogen and phosphorus), in contrast to the traditional activated sludge plants (still prevalent in the U.S.), based on the technology of the first half of the last century, that remove BOD and suspended solids and only a small percentage of nutrients (see Chapter VII). Typical removal efficiencies of the current Bardenpho systems are 95% for BOD and suspended solids, 75–85% for nitrogen, and 85–95% for phosphorus, yielding effluent concentrations of typical municipal wastewater of 10 mg/L BOD, 5 mg/L nitrogen, and 1 mg/L of phosphorus (Metcalf & Eddy, Inc., 2003; Sedlak, 1991). These low effluent concentrations are achieved with less costly chemical additions; also, the process requires less aeration oxygen than the traditional activated sludge process. Hence, conversion of conventional activated sludge plants to the Bardenpho systems would have a positive impact on energy consumption and GHG emissions. Aeration is highly energy-demanding and has a significant carbon emission footprint. Taking a new look at the anaerobic digestion processes as treatment units either in digesters or upflow anaerobic sludge blanket (UASB) units, producing energy instead of using energy, may be the future of sustainable used water reclamation, whereby used water (not wastewater anymore) is a resource rather than waste (see Chapter VIII).

Introducing membrane filters after biological treatment results in effluent quality comparable to or better than the quality of many receiving water bodies (Barnard, 2007). Chapters VII and VIII will describe the most advanced yet affordable treatment methodologies that can be used for water reclamation, not just treatment.

Diffuse (nonpoint) pollution abatement. While the point controls have been implemented in the U.S. on a wide scale, progress with abatement of pollution caused by urban runoff has been slow in most cities and notably also on the nation's highways. Urban runoff has been found responsible for more than half of the remaining water quality problems in the U.S. Urban and highway runoff is responsible for the major part of pollution by toxic metals, polyaromatic hydrocarbons (PAHs), and salinity (from de-icing chemicals used to keep streets and roads free of snow and ice during winter driving). Urban runoff also contains pathogens and coliform bacteria that may cause violations of water quality standards; however, these may not be of human origin. Nevertheless, water quality regulations do not provide relief from the standards just because the bacterial contamination may be from animals or birds.

Best management practices (BMPs) for control of urban and highway runoff have been developed and used in many communities (see Chapters III and IV). They have been described in books by Field, Heaney, and Pitt (2000), Novotny (2003), and in many state and U.S. EPA manuals. Manuals for sustainable urban drainage systems (SUDS) have been published in the United Kingdom. SUDS and urban/highway BMPs have the same goals and cover similar practices, design, and implementation. It should be noted that the BMPs category is more broad; BMPs deal with diffuse pollution caused by precipitation and other causes and are not focused only on urban drainage. Environmental engineering classification divides BMPs into the following categories (Novotny, 2003; Oregon State University, 2006):

1. Prevention (soil conservation; ban on pollution substances—such as lead in gasoline—or on persistent pesticides; public education; change of drainage inlets, such as curbs, gutters, and drains)
2. Source controls (street sweeping, erosion control and soil conservation practices, litter removals)
3. Hydrologic modification (porous pavement, enhancing surface infiltration and retention, and evapotranspiration)
4. Reducing of delivery (increasing attenuation) of pollutants carried by surface runoff or shallow groundwater flow from the source area to the receiving waters (infiltration road shoulders, biofiltration and infiltration in swales)

Many ingenious best management practice systems for controlling urban runoff pollution and flow were developed in the last 20 years of the last century. Chapter IV will describe BMPs in more detail, focusing on new developments. Figure 1.19 shows the concepts of infrastructure of an experimental sewer system (ESS) that was a comprehensive drainage system developed and tested by the Tokyo Metropolitan Sewerage Agency (Fujita, 1984; Furumai, 2007). The goal of the system is to minimize flows and pollution from the combined sewer overflows. Because the Tokyo metropolitan area is built on a thick layer of permeable volcanic deposits, the motto of the program is that only sewage should be directed into sewers, and all surface flow should infiltrate. The system, installed early in the 1980s but never fully implemented in the entire city, consists of pervious pavements, permeable (perforated) street gutters, infiltration trenches and wells, and special manholes that provide storage and sedimentation. Each manhole is connected to an infiltration pipe. The ideas were very sound and implementable, and research continues.

Chapter IV describes the current concepts of the use of BMPs and change of philosophy over the last 30 years. The original philosophy behind the BMPs designs and implementation was to remove pollutants from the runoff flow after the fact, without addressing the factors that cause pollution generation.

During the fourth paradigm period (from the 1970s until today), regulations for controls of point sources were enacted and worked quite well. U.S. industries and municipalities obeyed the effluent standards and implemented technologies that would comply with the standards but, in most cases, would not go beyond. Only rarely did dischargers take initiatives to go beyond compliance with the effluent limitation expressed by permits. In the U.S., the Clean Water Act initially authorized subsidies to municipalities for building certain types of infrastructure, but this program was sometimes counterproductive because municipalities were bound to technologies that brought subsidies and avoided innovations for which subsidies were unavailable; hence, reclamation and reuse systems were rare in the 20th century. Notable exceptions included the production of Milorganite, which is a commercially

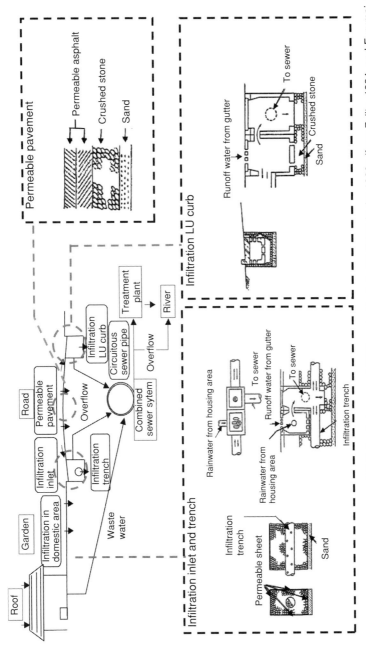

Figure 1.19 Experimental sewer system (ESS) for drainage and CSO control installed in Tokyo in the 1980s (from Fujita, 1984, and Furumai, 2007, public domain sources).

31

distributed fertilizer produced from sludge by the Milwaukee Metropolitan Sewerage District, and effluent reuse in several communities located in the arid parts of the U.S., in Beijing (China), in Israel, and elsewhere where water scarcity is a problem. Implementing diffuse (nonpoint) pollution control programs has relied on persuasion and incentives.

Later during the fourth paradigm period, pollution prevention was added as a factor, which then led to promotion of the concepts of reuse and recycling (Mihelcic et al., 2003). Reuse and recycling also became a popular method for reducing pollution loads from industries that used high volumes of water, such as paper mills and oil and sugar refineries. The new millennium has also brought a new look at the role of BMPs or SUDS in making a change to a new paradigm, the paradigm of sustainability. The key was the realization that BMPs—not hard infrastructures whose only purpose is to remove pollutants—are a part of the landscape, and that the landscape itself can provide buffering and attenuation; that is, it can become a part of the BMP train (Novotny and Hill, 2007; Novotny, 2007). This approach and concept will require the interdisciplinary efforts of urban planners, landscape architects, and experts in urban ecology and biology, along with environmental engineers and planners.

In general, the current fourth paradigm in the U.S., the European Community, Japan, Australia, Singapore, and a handful of other countries is continuing economic development with environmental restrictions, controls, and regulations—which, however, are still ineffective to guarantee that the legislative goals of attaining and maintaining the integrity and sustainability of water and air sources will be met. Trends that show emissions causing global warming are still increasing; new pollutants and problems such as nutrient enrichment leading to obnoxious algal blooms, toxic discharges, and new emerging pollutants—such as pharmaceutical residuals and endocrine disruptors—are growing, and the serious problems with legacy pollution in sediments have not been abated.

I.2.5   The Impact of Automobile Use

The introduction of automobiles at the beginning of the 20th century was slowly changing the way people lived, but it was not until 1960 that automobiles began to have a major impact on urbanism, hydrology, pollution, and greenhouse emissions. Before 1950, the main means of urban transportation and commuting were trains, electric light rail, and buses that were plentiful and convenient. Until the 1950s, long-distance travel by cars was actually difficult because of poor, often unpaved and narrow roads. In spite of their automobile ownership, many people lived in the cities and used automobiles far less frequently than today. A large majority of families in less developed countries owned only one automobile or no automobile. The change came with the building of arterial and ring freeways after World War II, opening distant rural areas to the development of urban subdivisions. Suddenly, people were able to buy a piece of land, build their dream house, and commute by car. The use of automobiles for commuting to work from distant subdivisions and the increasing living standards of suburban commuters spurred the need for more than one automobile per family, larger building lots, and houses with two- and later three-car garages.

Figure 1.20   Typical freeway congestion/traffic jam (Courtesy and copyright Comstock, Inc., 2000, from Texas Transportation Institute, Texas A&M). Highway runoff and especially snowmelt (in colder regions) are highly polluted and a major source of oil and grease, toxic metals, and organics. Runoff from urban streets also contains coliforms and pathogens.

Another result was traffic jams during rush hours in many cities worldwide, with an opposite effect—that is, traffic is often slowed to a standstill, resulting in freeway congestion (Figure 1.20).

Figure 1.21 shows the data on U.S. car ownership and miles driven. The Nielsen Company data reveal that nearly 9 in 10 Americans owned a car in 2000, making it the world's largest ownership in terms of car penetration and absolute numbers. Furthermore, the U.S. has enjoyed an increase of 8 percentage points in car ownership over the past five years—the highest recorded growth globally. Saudi Arabia follows the U.S. with the second-highest car ownership (86%). Because of the availability of cheap gasoline until the mid-2000s, U.S. car owners tend to drive larger cars with a lower number of miles/gallon (kms/liter) than their European counterparts, who pay higher prices for fuel. After 2008, the rate of car purchases dramatically decreased and high mileage automobiles (hybrids or plug-ins) became more popular.

Figure 1.21 also shows the increased rate of purchasing automobiles and the annual distance driven by the drivers between 1960 and 1980, which was also the period of building freeways, from and around the cities. This and inexpensive gasoline (the gas during that period was selling for less than $0.15/liter ($0.6/gallon)) were causes of the movement of middle- and upper-class people from the cities into suburbs,

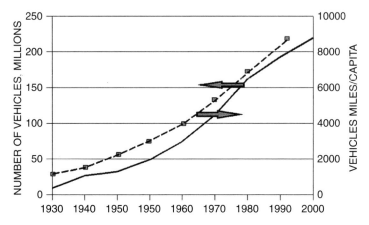

Figure 1.21 Trends in vehicular ownership and mileage driven per year (data from The Nielsen Co. statistics); 1 mile = 1.608 km.

resulting in urban sprawl. As the proximate zones surrounding the cities became saturated with subdivisions, new subdivisions were built farther and farther away, and from the end of the 20th century until 2008 this trend continued. People were driving larger distances and spending more and more time traveling to their places of work in the cities. As a consequence of this massive relocation of the U.S. population from the cities to the suburbs, the city core areas deteriorated, shopping moved from downtown stores to suburban malls, and only deprived mostly minority populations remained in the dilapidated central cities surrounded by low-density urban sprawl developments. During and after the recession of 2008–2009, car purchases dropped, and many urban sprawl subdivisions in the U.S. Southwest became ghost towns.

The sole reliance on automobile traffic had many water, water quality, and, obviously, carbon (greenhouse) emission consequences. For one, in addition to impervious roads, shopping malls, schools, apartment buildings, industries, and office buildings all use exceedingly large areas for parking. The parking lots, mostly impervious, occupy much larger areas than the establishments they serve (Heaney, 2000). The net result is that the combination of low-density residential zones, with connecting roads, and parking in residential, commercial, and industrial areas generates more than three times more urban runoff per family than the population living in the cities in pre-automobile times (before 1960). These parking and road areas are typically mandated by the authorities.

In a modern city, water infrastructure is closely tied to transportation and moving people from the places they reside to places of work and leisure. Streets and highways are the places where most of the drainage and water supply infrastructure is located, starting with drainage ditches, channels, and swales and ending with sewers. On the other hand, traffic emission, street litter and animal fecal deposits, road work, dripping oil in parking lots, and pavement wear are very significant sources of pollution. Impervious surfaces of roads and walkways also have significant hydrologic effects on flow. In some cases, roads were built over covered streams or abutted closely to the streams.

The emissions of pollutants from vehicular traffic are typically given in pollutant mass vehicle mile or kilometer. This implies that if the number of cars and the kilometer driven per year increase linearly, the pollutant loads from roads due to traffic emissions increase exponentially. Vehicular emissions include (Shaheen, 1975; Novotny et al., 1997; Novotny, 2003; Sansalone and Glenn, 2002):

Oil and grease (chemical oxygen demand (COD), BOD)
Acidity (mainly from acidified nitrous oxide from tailpipe exhaust during rain)
Toxic priority pollutants
    Toxic metals
    Polyaromatic hydrocarbons (PAHs)
    Petroleum hydrocarbons
    Asbestos
    Toxic metals (cadmium, copper, lead, nickel, and zinc)
Other pollutants related to automobile traffic are:
De-icing salts containing sodium, chloride, metals, cyanides, and PAHs

Roadways, especially those with high traffic densities exceeding 50,000 vehicles per day, are a major source of toxic pollution in highway runoff, which, in a typical large city, exceeds that from all sewage.

Carbon (greenhouse) gas emissions by vehicular traffic. Vehicular traffic is also the major source of CO_2 emissions. The U.S. EPA (2008) estimated that 1 liter of gasoline produces 2.33 kilograms of CO_2 (19.4 lbs per 1 gallon). This value is consistent with the Intergovernmental Panel on Climate Change (IPCC, 2007) report. Based on the data in Figure 1.21, each driver driving 16,000 kilometers (10,000 miles) per year in 2000 with a car that had an average mileage (fuel consumption) of 8.6 km/liter (20.3 miles/gallon) would emit

$$CO_2 \text{ emission} = \frac{16000 \text{ [km/year]}}{8.6 \text{ [km/liter]}} \times 2.33 \text{ [kg } CO_2/\text{liter]}$$
$$= 4335 \text{ [kg } CO_2/\text{year/car]}$$

or 4.3 metric tons (4.8 U.S. tons) per driver per year. In addition to carbon dioxide, automobiles produce the GHGs methane (CH_4) and nitrous oxide (N_2O) from the tailpipe, as well as fluorocarbons (HFC) from leaking air conditioners. These emissions are related to kilometers (miles) driven rather than to fuel consumption and account for about 5% of the GHG emissions, while carbon dioxide accounts for 95%. Hence the GHG emissions per an average car in 2000 would have been roughly 4.5 metric tons.

Are We Running Out of Oil and Energy? In 2007–2009, oil prices increased dramatically as a result of the limits on the production and refinery capacity. Other effects on the price of oil included speculation, the fall of the value of the U.S. dollar, and increasing demand in other countries, notably China, India, and Eastern European countries, that are rapidly catching up to the developed countries—and new developing countries like Mexico, the Republic of Korea, and Indonesia will join them. It must be pointed out, however, that the U.S. is by far the largest consumer of oil, to satisfy the needs for automobile fuel, heating oil, and raw materials for chemical industries. The price of a barrel of oil on the world market, in 2003 under $30, increased in 2008, resulting in gasoline prices of more than $4/gallon ($1.3/liter), but it dropped back to less than $3.00/gallon ($0.75/liter), and the financial worldwide crisis of 2008–2009 brought the price of the oil back below $100/barrel.

Experts are estimating that the availability of easily extractable oil and gas is reaching its peak now (Figure 1.22) and that by the end of the century, if the pre-2008 trend had continued, most of the energy driving the economy would be derived from coal. Coal can be converted to synthetic fuel, as occurred during World War II in Germany and Central Europe, but this would be at the price of higher energy consumption, more GHG emissions, and more pollution (Chomat, 2008).

Other experts point out that the world, for foreseeable future (one hundred years ahead), will not run out of oil (Schipper et al., 2001; Schipper, 2008). However, the rate of producing fuel from oil is stagnant and will be decreasing, and oil is becoming more and more expensive. Apparently, there is still oil underground in tar

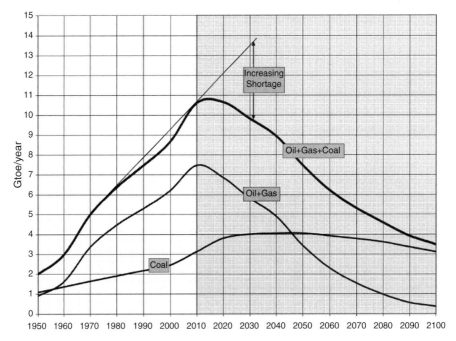

Figure 1.22 Total fossil fuel availability worldwide (from Chomat, 2008, reproduced with permission).

sand deposits, but the extraction and availability of "cheap" and easily extractable oil are diminishing. There is also a problem with oil security and with competition for oil from the new emerging powers, whose billions of people can now or in the near future afford an automobile.

As happened in Europe two decades ago, when a large tax was imposed by the governments on gasoline sales to reduce demand, $4/gallon ($1.05/liter) in the U.S. in 2008 was a threshold that long-distance drivers from far away subdivisions would tolerate. It should be noted that European drivers pay approximately 75% more for fuel. Many U.S. drivers commuting large distances did not have a commuting alternative, and few were able to rapidly switch to more efficient cars and abandon gas-guzzling SUVs, but, slowly, change was coming, and for the first time, gasoline use dropped in 2008. A *Time* magazine article (Ripley, 2008) reported a study that showed that housing values in cities and neighborhoods that required long commutes and provided few transportation alternatives to private automobiles were falling more precipitously than in more central, compact, and accessible places, thus slowing the rate of urban sprawl.

However, prices of $1/liter ($4/gallon) or more may also have some environmentally undesirable consequences, such as promoting alternate fuel from corn, which not only is environmentally unsustainable but leads to great increases of food prices, which has a devastating impact in poor countries. It also renews calls for drilling for oil offshore and in other natural areas, and for extracting oil from tar sands, which requires a lot of water and energy. Off-shore drilling may have catastrophic environmental consequences as exemplified in 2010 in the Gulf of Mexico oil rig explosion and its impact on ecology and economy of Louisiana and other Gulf states.

More promising and realistic is switching from gasoline and diesel fuel to vehicles run on electricity or hydrogen, and using public transportation provided by electric trains, light rail, or electric (trolley) buses. This method of transport has been available and widespread in Europe and, in the US, for example, in San Francisco for more than one hundred years and would bring great environmental benefits.

In 2008–2010, worldwide use of non-fossil energy was much smaller and represented only about 17% of the total energy use. This, however, may vary from country to country. For example, Brazil derives a significant portion of its energy needs from the Itaipu hydroelectric dam on the Paraná River, which also provides all electric needs to Paraguay. Similarly, in China, a significant amount of energy is derived from the Three Gorges Dam project on the Yangtze River. France derives most of its energy needs from nuclear and hydro power. Austria also has very significant and dominant energy sources in hydropower, and lately wind. In general, the worldwide use of non-fossil (renewable) energy in 2005–2010 was (Chomat, 2008):

Nuclear	5.2%
Renewable	11.3%
Hydropower	5.3%
Biomass – wood	5.2%
Geothermal	0.6%
Wind powered	0.25%
Solar powered	0.15%

It is clear now that to balance energy needs with availability, demand has to decrease (by conservation), and non-fossil energy production has to increase. Renewable energies are "clean" energy sources. Nuclear energy sources do not emit greenhouse gases; however, due to their lower efficiency of conversion of fuel energy into electric energy, heat emitted from the plants by their cooling systems is significantly greater than that from the traditional fossil-fueled plants, which leads to more thermal pollution of waters receiving cooling water discharges (Krenkel and Novotny, 1980), and the problem of safe disposal of radioactive spent fuel has not been fully resolved as of 2010. Chapter VIII discusses the water-energy nexus and has more discussion on the future energy outlook, including the increase of renewable electric energy production.

I.2.6   Urban Sprawl

After the beginning of the Industrial Revolution, in the second half of the 19th century, cities were expanding rapidly to accommodate population migration from rural areas to provide labor to expanding industries. This expansion progressed in various forms, by annexing villages surrounding the cities, building new settlements for blue-collar workers and white-collar industrial managers. This was also a period of building mansions for suddenly super-rich industrial and railroad magnates. This period of urban expansion lasted in the U.S. and European cities until 1950. Most progressive cities in the U.S. and Europe, and also some cities in Asia, provided public intra- and intercity transportation. For example, before 1950, Boston (Massachusetts) had a dense network of electric streetcars, and a subway system was built there more than one hundred years ago. Electric trolleys and trains were also widely available in the suburbs (Figure 1.23). Chicago (Illinois), Northern Indiana, and Milwaukee (Wisconsin) were interconnected by interurban electric trains, and the same was true for New York, St. Louis (Missouri), Philadelphia and Pittsburgh (Pennsylvania), and countless other larger cities in North America, Europe, and elsewhere. In the U.S., most of the intra- and intercity public transportation based on electric trains and buses was abandoned by the 1960s, while in Europe and elsewhere it was retained and modernized. Currently, China has embarked on a massive program of building an electric rapid interurban transit system and intraurban subway, light rail, and electric bus lines.

At the beginning of the second half of the 20th century, the migration pattern and expansion of the cities had changed. While the population in cities outside of the U.S. remained in the cities and nearby suburbs, and the cities developed effective and affordable means of public intra- and intercity public transportation, affluent middle- and high-income families in the U.S. moved from the central cities into the suburbs. The primary mode of transportation switched to automobiles, and the good public transportation was reduced, switched to buses, or disappeared completely. Today, many suburbs are far from the city, and people commute by automobiles, sometimes spending several hours each day commuting. In a typical U.S. suburb, lots are large, many contain large mansion type housing with very high energy use, and three-car garages are not uncommon. A modest renewal of light rail transportation

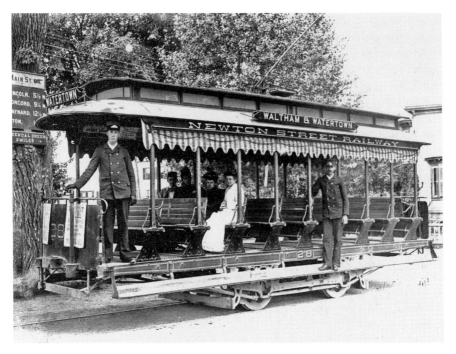

Figure 1.23   Public transportation in Boston, Massachusetts, suburbs in 1893 (Courtesy of the Newton Historical Museum, Newton, Massachusetts). These suburban and interurban electric light rails around Boston mostly disappeared after the 1950s.

in some major U.S. cities occurred at the end of the 20th century. The movement of people to distant low-density suburbs without adequate water, sewage disposal, and transportation infrastructures is called *urban sprawl*.

The movement of the U.S. population in the 20th century was generally attributed to the building of freeways, multi-lane highways, and automobile use. However, such migration to a suburbia far away from the cities where people worked did not happen on such a large scale in other countries, in spite of building freeways, sometimes earlier than in the U.S. (e.g., in Germany). It is a known fact that living in the center of Paris, London, Rome, Prague, Vienna, or any major European city is highly desirable, and the value of real estate is at a premium. Outside of the U.S., it is in the medium- to high-density suburbs where lower-income people reside and use mostly public transportation to commute to their workplaces. Consequently, people living in urban areas outside the U.S. drive far less, commute far more by public transportation, and have smaller cars. The result of the massive urban sprawl in the second half of the 20th century in some major U.S. urban areas was a partial collapse of the central cities, the demise of public transportation, and the conversion of many commercial downtowns after 5:00 P.M. into ghost towns inhabited by cleaning crews, conventioneers, and tourists. Cities were losing their population dramatically, and as more affluent people moved to suburbs, the cities lost their tax base, and infrastructure deteriorated. By early 2000, Detroit (Michigan) lost $1/2$ of its population and large areas of the city were converted into abandoned wasteland.

At the end of the 20th century, Dittmar (1995) made a prediction and assumptions about the future of automobile use and its effect on the pollution of urban runoff, which was corroborated by Heaney (2000), obviously reflecting the uptrends of automobile ownership and miles driven at the end of the 20th century. Dittmar stated that it was not the American driver who was choosing exurbia and long-distance driving, and that the government was not simply responding by building more roadways and parking lots, and making it possible to buy land in rural areas for subdivisions. He pointed out that people were responding to a set of signals the society was giving them by building freeways, ring roads, and beltways, subsidizing free parking and suburban development through water, stormwater, and wastewater utility infrastructure, and providing tax incentives that favored suburban living and suburban home ownership. For one thing, the main original reason for building cross-country freeways in the U.S. in the fifties was not to bring people to the suburbs, but to move the military and raw materials quickly for national defense reasons, during war. However, cheap gas, deteriorating central cities, and disappearing public transport were giving people signals favoring urban sprawl.

Urban sprawl puts a large demand on energy because of cooling and heating large homes, and on water resources used for lawn irrigation. Low-density suburbs often rely on private on-site sewage disposal by septic tanks, which have a high rate of failure. It is the middle- and higher-income households with two or more automobiles, living in low-density suburbs, that have the highest consumption of resources, far more than those with similar incomes living in cities (McGranahan and Satterthwaite, 2003; Newman, 1996). Hardoy, Mitlin, and Satterthwaite (2001) made an assessment that "one particularly wealthy, high-consumption individual or household with several large automobiles, a large inefficiently heated or cooled home, and with frequent use of air travel (for pleasure and/or work) can have a more damaging global ecological impact than thousands of urban dwellers in informal settlements (shanty towns) in low income nations."

I.2.7 The Rise of New Great Powers Competing for Resources

For most of the 19th, and especially the 20th, centuries, the U.S., Western Europe, and Japan had most of the industries and were the wealthiest. The U.S. has been the greatest industrial power. However, in the last 20 years, tremendous demographic, political, educational, and economic changes have occurred throughout the world, especially in the countries that have been identified as developing countries. The end of the colonial era in Asia and Africa in the middle of the last century, the end of the Cold War, international cooperation through the United Nations and other international organizations, and the increased impact of nongovernmental national and international organizations have all had a significant impact on the standards of living and health care in many countries that previously suffered epidemics, famine, and unjust subjugation to colonial and military/occupational rules. Improved health care and nutrition have also resulted in rapid, almost exponential, population growth in these countries, massive migration from rural areas to cities, and, sometimes, a reversal of progress.

The increased living standards are not uniform, they are more favorable in South and East Asia, where Japan, the Republic of Korea, Malaysia, China (Hong Kong and surrounding provinces in China, Shanghai and surroundings, Tianjin province, and the capital province of Beijing), parts of Thailand, Singapore, parts of India, and some countries in the Middle East have made great advances. In all rapidly advancing Asian countries, as happened a few decades before in Western Europe and the U.S., rapid economic progress has been accompanied by deterioration of the environment on the local and regional scale, and adverse impacts on public health and living conditions due to pollution.

At the end of World War II, all Asian, African, and Latin American countries were undeveloped, relying primarily on manual labor in cottage industries, agriculture, and commerce. Most of their people were living in rural areas. Japan's economy was destroyed by the war. At the end of the 1940s and in the 1950s, China was ending its revolution and civil war, and its economy was decimated and cut off from the industrialized world. During the "Great Leap Forward," which was a forceful policy of attempting to bring about a rapid economic advancement in communal systems, between 1959 and the first part of the 1960s, food production actually decreased and China had a Great Famine, during which millions of people died of starvation. The Great Leap Forward was followed by the "Cultural Revolution" in the later 1960s and the 1970s. These were periods of massive relocation into newly established rural communes, which decimated the economy. The period of fast economic growth in China started after the economic (and also political) reforms of 1984. India ended British colonial rule and gained independence in 1947. Today, Japan, the Republic of Korea, China, and India are the major Asian countries competing for resources, and their total greenhouse gas emissions are among the largest in the world. These countries will also have the largest increases of urban population in the upcoming years (see section I.3.1).

The annual gross domestic product (GDP) growth of China after the Cultural Revolution, in the years after the economic reform from 1984 to 2005, was 9.6%, and it reached 11.9% in 2006 (Barboza, 2007), but, as a result of the financial crisis of 2008–2009, the growth has decreased to about 9%. In the same period, the economies of the U.S. and some European Union countries were at a standstill, or a recession. Since 1984 China's economy doubled itself more than three times. By 2002 the Chinese economy was about 8.5 times what it had been at the beginning of the economic reforms in the 1980s. China already has the world's third-largest economy. Based on the pre-2008 estimates, China was expected to become the world's largest economy by 2030. This could happen sooner because of the 2008–2009 recession in the U.S. and the slow growth thereafter, while China's economy continued its rapid growth. China is a one-party socialist country that, since the economic reforms in the 1980s, has been adopting more and more of a free market system.

India is the second most populous country in the world (after China), with a continuing but slightly slowing rate of population increase. Based on United Nations population statistics and forecasts, India is expected to surpass China as the most populous country around the year 2030, yet its area is only about one-third of that of China or the U.S. The population of India is increasing by approximately 17 million annually; however, the rate of growth is not geographically uniform. Today, India has the 10th- to 12th-largest global economy, based on the total GDP; however, in 2005 India's per capita GDP of U.S. $3300/person put India at the 158th place among the nations (GeoHive, 2007).

However, the population rise in the emerging economic powers has to be considered from the viewpoint of their per capita resource use and waste generation in comparison with the developed countries. Average per capita waste generation in the urban centers of developed countries can be as much as 20 times higher, and it is as high as 1000 kg/capita-year (McGranahan and Satterthwaite, 2003). Furthermore, both China and India import and reuse waste from abroad, mainly from the U.S., for recycling—from paper, cans, and scrap metal to decommissioned large ships that are converted to scrap metal and reused in India. McGranahan and Satterthwaite also point out that it is the upper- and middle-income groups in the affluent countries that account for most resource use, most generation of household waste, and highest per capita emissions of GHG. The authors also state that it would be highly appropriate to require consumers in countries with high income, consumption, and waste production to reduce their levels of natural resource use, including water, and/or halt the damaging ecological impact of their demands for fresh water and other resources.

Countries in Asia and some in Latin America (Mexico, Brazil, Venezuela) anticipate massive changes towards urbanization by 2050. In the next 30 years, China is planning to relocate 300 million people from rural to urban areas, where the jobs are needed for the growing economy. Some of these new cities will be ecocities, and China is spending funds on ecocity research and importing know-how from other countries, including Sweden, Singapore, and the U.S.

I.3 DRIVERS FOR CHANGE TOWARDS SUSTAINABILITY

Chapter II will define, describe the concepts of, and address the needs for the new sustainable (fifth) paradigm of water centric urbanism. This paradigm will balance social, environmental, and economic factors and the resolution of stresses. There are numerous social, economic, and environmental drivers for a change and a switch to the new paradigm of sustainability such as:

1. Population increase and the resulting migration of population into cities; the emergence of megacities
2. Increasing water scarcity due to overuse and pollution, impacting both population and economy
3. The necessity to reduce emissions of greenhouse gases and the need to adapt to global warming
4. The increased frequency and magnitude of extreme meteorological events, and the need for cities to become more resilient

5. The deteriorating water infrastructure and the need to rebuild and/or retrofit cities to accommodate current and future stresses

6. Attaining and maintaining the ecological integrity of urban water resources, as mandated by environmental legislation and desired by the public in most countries

7. The increasing living standard of people in cities and suburbs, and the desirability of living near surface water bodies

8. The deleterious effects of continuing the status quo and building cities using the rules and methods of the current paradigm

9. The new technologies that have been developed and are available:

   a. Wastewater can be treated and reclaimed with a quality commensurate with or better than that in the unpolluted receiving water bodies; even potable water quality can be reclaimed in small (subdivision, commercial area, large office building) as well as large (regional) water reclamation facilities (Chapter VII).

   b. Methods for reclaiming energy from wastewater supplemented by solar, wind, and geotechnical renewable sources are available and economical. New methods of reclaiming energy based on hydrogen gas rather than carbon will be available (see Chapter VIII).

   c. Best management practices mimicking nature and blending with the urban environment have been developed and are desired by the public. After capture and treatment of rainwater and stormwater, these BMPs can provide water for reuse that can also be blended with reclaimed wastewater effluents (Chapters III and IV).

   d. Green buildings and low-impact subdivisions are now being built on a large scale that provide substantial water reuse and energy savings (Chapters III and VIII).

   e. Vehicles fueled by hydrogen or electricity are being developed and will be available for mass market in one or two decades.

   f. Living in cities, not in distant suburbs, which has always been preferred in European and other cities outside of the U.S., is now becoming a popular alternative in the U.S.

   g. Restored or daylighted urban streams stimulate the economic revival of cities and provide recreational and leisure opportunities (Chapter IX).

10. Because rainwater and wastewater will be considered as a resource and not waste, significant economic benefits will become available that, under the best-case scenario, can pay for the sustainable urban water centric developments. Urban sewage can be converted to a clean effluent for reuse, and methane gas and hydrogen for energy (Chapters VVI, VII and VIII).

The building blocks of the Cities of the Future—the ecocities—are available, and the necessity of adaptation to the future's very serious stresses calls for the change.

I.3.1   Population Increases and Pressures

The magnitude and consequences of the expected population increases have been on people's minds for decades. Demographic experts coined the term "population explosion" to describe the population growth, reviving the predictions of British economist Thomas Malthus, who predicted in 1798 that the world's population would eventually outpace food production, which would lead to massive starvation and famine. At that time the world population was several hundred million people. It reached one billion in the late 1800s, and over the last century, the earth's population has increased from about one billion to six billion—"officially" reached on October 12, 1999 (United Nations, 1999). The world population more than doubled in the last 50 years. In 2009, the world population reached 6.9 billion. Most of the growth has occurred in developing and undeveloped countries. Malthus's predictions have been shown to be overly pessimistic; although large famines have occurred in the last 50 years in China, North Korea, and Africa, the main reasons were institutional and political mismanagement and faulty demographic and agricultural policies. For example, in China during the Great Leap Forward in the 1950s, massive relocation of people from cities to rural areas and faulty agricultural policies and methods resulted in a mismanaged agricultural economy and famine.

In 2010, China was the most populous country of the world, followed by India and the U.S. China and the conterminous U.S. have about the same area; hence, the population density of China is about 4.3 times greater than that of the U.S. The world population is expected to stabilize at around 9 to 10 billion after 2050. Figure 1.24 and Table 1.3 present the population numbers of the world and of several sample countries, including the U.S., the United Kingdom (Western Europe), the Czech Republic (Central Europe), Russia (Eastern Europe and Asia), and China and India.

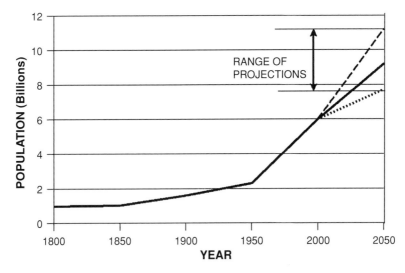

Figure 1.24   World population in billions according to United Nations statistics and projections.

Table 1.3   Population numbers in selected countries

			Population in Millions				
YEAR	World	U.S.	Czech Republic	United Kingdom	China	Russia	India
1960	3041	180.671	9.66	52.372	650.6	119.936	445.4
1980	4452	227.726	10.288	56.314	984.7	139.038	684.9
2000	6084	282.338	10.270	59.522	1268.8	146.709	1004.1
2010	6866	309.162	10.201	61.068	1347.6	139.390	1184.1
2030	8373	363.811	9.628	64.462	1461.5	124.094	1532.5
2050	9538	420.080	8.54	63.977	1424.2	109.187	1807.9

Source: U.S. Census Bureau, International Data Base, year 2008 data.

The demographic population models balance births, deaths, and migration and several factors affecting the model parameters. Traditionally in most countries, until a few decades ago, rural farming families desired and had a large number of children, who provided labor. The high death rate of children also encouraged large families. During the generational transition, the oldest son was given the farm, and the other siblings had to relocate, mostly to cities. During the industrial age there was always a movement of people from the villages to the cities in search of employment. The population explosion in the 20th century in developing countries occurred mainly because of the improvement in health care and a dramatic reduction in child mortality.

Table 1.3 shows interesting trends. First, most European countries (including Russia) and also Japan have already reached their peak plateau, and by 2050 the population is expected to stay steady or decline, which, oddly, could be caused by increased living standards and a cost of living that often requires both parents to be employed. The U.S. is expected to gain almost 40% in population numbers between 2007 and 2050, mainly due to immigration. China will reach its peak population around 2030, and then the population will remain steady or slightly decline. The biggest population increases are expected in India (60%) and also in other Asian countries such as Bangladesh, Indonesia, Thailand, and the Philippines, and in several Latin American and African countries such as Brazil, Mexico, Nigeria, Ethiopia, and Egypt.

Beyond the year 2050, the population numbers will be affected by global warming and accompanying increases in seawater levels and frequency of catastrophic events that may dramatically affect population in low-lying high-density countries such as Bangladesh.

Even under the current population numbers, some cities are already overpopulated and cannot provide even basic services to a large part of the rapidly increasing number of their citizens, especially in developing countries. The conflicts between the population numbers and the services provided by communities have reached a breaking point with the emergence of "megacities" (cities with a population of more than 10 million) in several rapidly growing countries in Asia, Africa, and Latin America. In these countries the population migration is different from that in the U.S. and,

Figure 1.25   In slums in India, the street is the drainage (photo courtesy Operation ASHA www.opasha.org).

partially, in some European countries. In developing countries the net flux of population is away from the rural areas that are unable to provide a livelihood to the increasing number of rural people. This flux is often undertaken for survival and, having moved from the rural areas into cities, entire families live in small crude cardboard or tin houses, or even under tents in large shantytowns abutting the city centers. The environment in the shantytowns of some developing countries is on the same level as that in the medieval cities of Europe centuries ago (Figure 1.25).

The most serious aspect of the population increase and migration is the fact that most of the new population will be residing in the cities. Hence the population growth of the cities will be more than twice that for entire countries (Table 1.4). To put these changes into perspective, China's population is expected to increase in the next 30 years by 100 million, but the urban population increase that the Chinese planners anticipate is about 300 million, which implies that each year between 2010 and

Table 1.4   Annual population growth (geometric increase)

	Growth/Year (%)
Total world population increase	
1970–1990	1.76
2020–2050	0.72
More developed regions	
1970–1990	0.64
2020–2050	~ 0
Less developed regions	
1970–1990	2.12
2020–2050	0 (China) to 0.95 (India)
Urban areas	
More developed regions 1970–1990	1.58
Less developed regions 1970–1990	5.27

Source: U.S. Census Bureau, International Data Base; Neimczynowicz, 1996; United Nations statistics.

2040 China will have to build new cities and expand existing cities to accommodate 10 million people. This situation is even more serious in India and other developing countries of Asia, Africa, and Latin America with large population increases.

In the 1950s about 30% of the total world population lived in urban areas; the corresponding estimate for 2000 by the experts from large urban areas during the 2004 Stockholm World Water Week (Biswas et al., 2004) was 50%, and it will increase in the future to more than 60%. In some countries, the urban population will represent more than 90% of the total population (UN Secretariat, 2005).

The major difference between urbanization in the developed and developing countries is the fact that in developed countries the largest rate of increase occurred during the Industrial Revolution in the 19th and 20th centuries, and the population movement—after involuntary serfdom and slavery were abolished—was commensurate with the economic development and lasted for more than a century. However, by the end of the 20th century, industries fueling the industrial expansion were in decay, and most were abandoned. This was the period when Milwaukee (Wisconsin) lost its breweries and major manufacturing industries; Manchester (U.K.), Youngstown (Ohio), Pittsburgh (Pennsylvania), and the Ruhr area of Germany lost their steel mills and deep mines; and many cities in New England, North Carolina, and other regions of the U.S. lost their textile manufacturing. St. Louis, Missouri, on the Mississippi River, was the fourth-largest U.S. city at the beginning of the 20th century, with a population of more than 700,000, and had several large industries. One hundred years later, at the beginning of the 21st century, most industries had left the city or were sold, and the city population was around 300,000.

Megalopoli-Megacities   The period after 1970 has witnessed the emergence of megalopoli/megacities—that is, cities with a population of more than 10 million. Before World War II, the largest city in the world was London, which has

Figure 1.26   One of the largest megacities, São Paulo, Brazil.

never achieved the dubious distinction of a megalopolis. New York, which was the second-largest city, became the first megalopolis in 1956 (Lewis, 2007). In 2008, the Tokyo–Yokohama conglomeration was the largest city of the world, with a population of 35 million, followed by Mexico City, São Paulo (Brazil) (Figure 1.26), New York, and Mumbai (India). However, Tokyo's rate of growth is being reduced; therefore, it is predicted that Tokyo will be overtaken by Mumbai, Shanghai (China), and Dhaka (Bangladesh) (Lewis, 2007). There were 20 megacities in 2008, and 15 of them were in developing countries. Lewis (2007) expects that even as the world's overall population will eventually level off somewhere between 9 to 10 billion, the megacities and smaller cities with populations in the millions will continue their expansion, as the rural population moves to the cities and—in most developing countries—become the urban poor.

In the U.S., the opposite trend has occurred. The net flux of population is away from the cities in the Northeast and Midwest that have been losing population for decades to the suburbs or to rapidly growing cities in the South and Southwest. Hence, the dichotomy of urban population migration is the formation of shantytowns and rapid urban growth in the developing megacities, and a loss of population and urban sprawl mainly in the U.S. Northeast and Midwest.

Of the world's megacities, the Los Angeles, California, metropolitan area has the greatest surface area ($27{,}800\,\text{km}^2$), resulting in the least dense population. Cairo's population (10.4 million) is reported as being confined in $214\,\text{km}^2$, making it the most densely populated megacity.

Nevertheless, in spite of the fact that historic cities in the U.S. have stagnant or dropping population numbers, the cities provide the employment, and the majority of suburban areas are not rural farming communities but low-density bedroom communities connected to the city by freeways and sometimes by public transportation. One can see now and foresee in the future the development of urban megalopolis agglomerations consisting of several cities with suburbs and satellite cities between them. The most notable urban conglomerations in the U.S. are:

Los Angeles–Riverside–Orange County, in California

Chicago, Illinois–Gary, Indiana–Kenosha, Wisconsin

Washington, D.C.–Baltimore, Maryland–Northern Virginia

San Francisco–Oakland– Berkeley–San José, in California

Dallas–Ft. Worth–Arlington, in Texas

Tokyo–Yokohama, in Japan, the largest urban conglomeration in the world

In the developing world, the growth of megacities and one-million-plus cities is rapid. For example, the population of the Mexico City metropolitan area increased from 3.1 million in 1950 to 13.4 in 1980, a 425% increase in only 30 years (Biswas et al., 2004). Consequently, the infrastructure may not be available to provide adequate drinking water service, stormwater management, and sanitation. Flooding is sometimes a major problem, since the burgeoning population often settles in floodplains, either by design or by necessity, to find space for living. The number of people exposed to flooding tripled from the 1970s to the 1990s and is currently about two billion (Biswas et al., 2004), and it is expected to grow even faster due to the effects of global warming on the frequency and magnitude of floods, as experienced already in the U.S. Midwest in 1993, 1998, and 2008.

Gurjar et al. (2008) reported that Tokyo, Beijing, Shanghai, and Los Angeles have the highest CO (carbon monoxide) emissions on an annual basis, and Kolkata, Dhaka, Mumbai, Cairo, and Rio de Janeiro have the lowest emissions. Cairo, Tokyo, and Moscow rank among the highest emitters of CO per unit of surface area, and Rio de Janeiro and Los Angeles rank among the lowest emitters per unit area. CO emissions can be correlated to CO_2 (carbon dioxide) GHG emissions.

I.3.2   Water Scarcity Problems and Flooding Challenges of Large Cities

In the U.S. some large cities—such as Los Angeles (CA), Tucson (AZ), Santa Barbara (CA)—have grown in dry arid climatic conditions and anticipate droughts and water scarcity as is also true of Beijing (China). However, water scarcity problems are not limited to cities located in arid zones. The rapid growth of some cities in more humid areas that rely on relatively small water resources, such as Atlanta (Georgia), or draw water from limited groundwater resources, such as many suburbs of Chicago (Illinois), Boston (Massachusetts), and Milwaukee (Wisconsin), results

Table 1.5   Freshwater withdrawals and water use distribution in some countries (from Gleick, 2003)

Country		Per Capita Withdrawals (Liters/ Capita-Day)	Municipal and Domestic Uses % (Liters/ Capita-Day)	Industrial Use (%)	Agricultural Use (%)
Africa	Libya	1972	11 (217)	2	87
	Uganda	24.6	32 (7.9)	8	60
Americas	Canada	3918	11 (431)	80	8
	U.S.	4625	12 (555)	46	42
	Mexico	2156	17 (366)	5	78
	Brazil	878	21 (184)	18	61
	Paraguay	214	15 (32)	7	78
Asia	Bangladesh	1060	12 (127)	2	86
	China	1181	11 (130)	21	69
	India	1361	5 (68)	3	92
	Iran	2501	6 (150)	2	92
	Israel	767	16 (123)	5	79
	Japan	1981	19 (376)	17	64
	Pakistan	2732	2 (54)	2	96
Europe	Austria	833	19 (158)	73	8
	Czech Republic	764	23 (175)	68	9
	France	1619	16 (257)	69	15
	Germany	1477	14 (206)	68	18
	Italy	2693	14 (377)	27	59
	Russian Fed.	1443	19 (274)	62	20
	Spain	2293	12 (275)	28	62
	Sweden	912	36 (328)	55	9
Australia	Australia	2589	15 (388)	10	75
	New Zealand	1457	46 (670)	10	44

in an inadequate water supply, mainly due to overuse and large losses from the water distribution systems.

Gleick (2003) compiled worldwide and country-by-country water use and withdrawal statistics. Some data for selected countries are included in Table 1.5, which contains data mostly reported in the 1990s. The municipal/domestic water use in the table includes household, municipal, commercial, and government uses. The industrial sector uses includes water used for cooling and production. Agricultural uses are for irrigation and livestock. The largest use is for irrigation, especially in Asia.

Regarding domestic/municipal use, there is great disparity in water use among countries and continents. The highest municipal/domestic use is in New Zealand, which has abundant water resources throughout the entire country. The second and third highest are the U.S. and Canada, respectively, but these numbers may be misleading in the U.S., where Southwestern urban areas have severe water short-ages. Lawn irrigation is the largest domestic water use in the U.S. suburbs, and often treated drinking water is used for lawn irrigation. Municipal water use in Europe is about 50% of that in the U.S., which is due to the fact that the lots on which houses are built are much smaller than those in the U.S. or Canada, and many people live in apartments. Municipal/domestic water use in large Asian countries is 25% (China) of that in the U.S., or less. Chapter V presents the concepts and data on water conservation.

Gleick (2003) also summarized the forecasts of future use and pointed out that extrapolating from past trends may be misleading. For example, in the U.S. the total water withdrawals of fresh and saline water between 1975 and 1990 increased from about 2000 liters/person-day to about 7500 liters/person-day, but between 1975 and 2000 the water withdrawals dropped to 5500 liters/person-day. The per capita water use rates have dropped in many cities also, due to mandatory or voluntary implementation of some water-saving devices and by plugging the leaks and minimizing the losses in the water distribution systems, as occurred in Chicago (Lanyon, 2007) and in Boston (Breckenridge, 2007), which realized about 20% savings on water demand, even when the population increased. Gleick (2003) surveyed the literature reporting the effect of water conservation in several municipalities throughout the world. See Chapter V for more details about water conservation.

The effect on water use of switching to sustainable water management can be seen in the prototype of a sustainable city, Hammarby Sjöstad, a district of Stockholm, Sweden (see also Chapter XI). Note that the average domestic (municipal) use in Sweden reported in Table 1.5 is 328 liters/person-day; typical water use in Stockholm in 2000 was about 200 liters/person-day. In the U.S. it is about 550 liters/person-day, as shown in Table 1.5. The municipal water use of Hammarby Sjöstad in 2008 was about 150 liters/person-day, and the goal, after the full implementation of water conservation, is to reduce the water use to 100 liters/person-day. The new ecocity (see Chapters V, VII, and XI) Masdar in the United Arab Emirates, and those planned elsewhere throughout the world can, by water conservation, reclamation, and reuse, reduce water demand from the grid or other freshwater sources or desalination to 50 liters/person-day, yet still maintain a comfortable water use commensurate with that of other cities practicing water conservation.

I.3.3 Greenhouse Emissions and Global Warming Effects

The recent report by the International Panel on Climatic Change (2007) outlined the challenges human beings are facing due to the effects of global climatic changes. These effects will be both global and regional. As a result of these changes, it is very likely that large and catastrophic storms will increase in magnitude and frequency, resulting in more frequent flooding. Droughts in dry zones will also be more frequent and more severe. This necessitates, on one side of the issue, the development of adaptation and risk management practices for the urban water sector and better human response management during extreme events and, on the other side, connecting water conservation and management with a reduction of greenhouse gases.

As pointed out in the preceding sections of this chapter, global warming is caused by the emission of greenhouse gases (GHGs). The principal greenhouse gases that enter the atmosphere because of human activities are:

- **Carbon dioxide (CO_2)**: Carbon dioxide enters the atmosphere through the burning of fossil fuels (oil, natural gas, and coal), solid waste, trees and wood products, and also as a result of other chemical reactions (e.g., manufacture of cement). Carbon dioxide is also removed from the atmosphere (or "sequestered") when it is absorbed by plants as part of the biological carbon cycle. The global warming potential (GWP) of CO_2 has been set as 1. Carbon dioxide (CO_2) is the most important greenhouse gas.
- **Methane (CH_4)**: Methane is emitted during the production and transport of coal, natural gas, and oil. Methane emissions also result from livestock raising and other agricultural practices, and from the decay of organic waste in municipal solid waste landfills and organic matter in wetlands. The GWP of methane is 25 over 100 years—that is, it is 25 times more potent as a greenhouse gas than carbon dioxide—but there's far less of it in the atmosphere, and it is measured in parts per billion. When related climate effects are taken into account, methane's overall climate impact is nearly half that of carbon dioxide.
- **Nitrous oxide (N_2O)**: Nitrous oxide is emitted during agricultural and industrial activities, as well as during combustion of fossil fuels and solid waste. GWP = 300 over 100 years. It is also emitted by natural and constricted wetlands.
- **Fluorinated gases**: Hydrofluorocarbons, perfluorocarbons, and sulfur hexafluoride are synthetic, powerful greenhouse gases that are emitted from a variety of industrial processes. Fluorinated gases are sometimes used as substitutes for ozone-depleting substances (i.e., CFCs, HCFCs, and halogens). These gases are typically emitted in smaller quantities, but because they are potent greenhouse gases with GWP in the thousands, they are sometimes referred to as "high global warming potential" (high GWP) gases.

Short-wave solar radiation penetrates the earth's atmosphere, and it is partly absorbed by the earth's surface and partly reflected as long-wave radiation back into space. The ratio of the reflected solar radiation to the total radiation is called albedo. Albedo depends on the color of the surface and the angle at which the radiation reaches the surface. White surfaces reflect most of the incoming short-wave radiation; dark surfaces absorb it and emit a portion of it back into the atmosphere as long-wave radiation. Greenhouse gases can prevent part of the reflected long-wave radiation from being sent back into space, which will warm up the atmosphere as a glass roof does in a greenhouse. Short-wave solar radiation can penetrate glass, but glass will keep the long-wave radiation (heat) in the greenhouse. Hence, the temperature of the atmosphere is related to the concentration of the greenhouse gases in the air. The natural concentration of carbon dioxide in the atmosphere during the

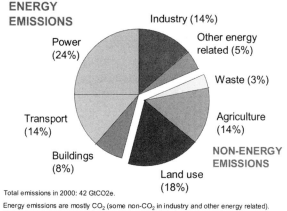

ENERGY
EMISSIONS

Industry (14%)

Power
(24%)

Other energy
related (5%)

Waste (3%)

Transport
(14%)

Agriculture
(14%)

NON-ENERGY
EMISSIONS

Buildings
(8%)

Land use
(18%)

Total emissions in 2000: 42 GtCO2e.

Energy emissions are mostly CO_2 (some non-CO_2 in industry and other energy related).
Non-energy emissions are CO_2 (land use) and non-CO_2 (agriculture and waste).

Figure 1.27 Main sources of greenhouse gases according to the IPCC (2007).

pre-industrial period was about 279 ppm (parts per million) in the volume of air, or 0.028%, but currently it is significantly higher and is increasing.

Sources of Greenhouse Gas Emissions Most of the natural CO_2 and methane (CH_4) in the atmosphere originates from natural sources such as decay of organic matter, the respiration of living organisms, and natural forest fires. Volcanic eruptions today account for about 1% of the natural emission. Methane is a product of anaerobic decomposition and is emitted naturally from wetlands by the anaerobic decomposition of organic matter and by living organisms. These sources of GHG emissions are counterbalanced by sinks that include photosynthesis or dissolution in oceans and conversion into bicarbonate and carbonate compounds.

The main anthropogenic source of carbon dioxide is combustion of coal and natural gas in power plants, homes, and industries, gasoline burning in vehicles, deforestation by slash-and-burn farming, and grassland fires ignited by human beings—and these additional sources are not counterbalanced by commensurate sinks. Wastewater disposal and treatment operations represent 3% (Figure 1.27). The result is the increase of greenhouse gases in the atmosphere (Figure 1.28) that trap heat and increase the temperature on earth.

In Section I.2.5, "The Impact of Automobile Use," we reported the U.S. EPA estimate of CO_2 emissions as being 2.33 kg of CO_2 per 1 liter of gasoline fuel consumed in driving. The conversion of energy production into carbon emissions in power plants takes into account the efficiency of the power plant to convert fuel energy into electric energy and the caloric (heat) content of the fuel. The efficiency of power plants is:

$$\varepsilon = \text{energy produced by the power plant/energy in fuel}$$

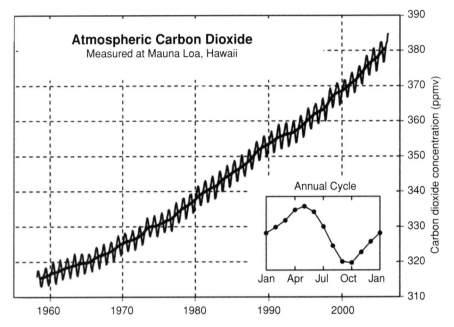

Figure 1.28   Carbon dioxide concentration in the atmosphere measured at the Mauna Loa observatory by the National Oceanic and Atmospheric Administration (ppmv = parts per million by volume = 10^{-6}).

The typical efficiency of fossil fuel power plants is 40 to 50%, which means that about 50 to 60% of the fuel energy is lost (wasted) as heat into the environment and can contribute to the thermal pollution of receiving waters that are a source of cooling water. The efficiency of nuclear power plants is less (they emit more waste heat), but these plants have a very small carbon footprint. Heat emitted from power plants may contribute slightly to global warming, but not as much as the effect of carbon dioxide emissions into the atmosphere from smokestacks.

Energy in fuel and produced by generators is today expressed in kilowatt-hours (kWh) (see also Chapter VIII). 1 kWh is 3.6 megajoules (MJ). The U.S. Department of Energy (2000) published estimates of the carbon equivalent of energy produced by fossil fuel power plants as:

> 0.96 kg of CO_2/kWh produced by coal-fired power plants
>
> 0.89 kg of CO_2/kWh produced by oil-fired power plants
>
> 0.60 kg of CO_2/kWh produced by natural gas power plants

Figure 1.29 shows that in the U.S., about 30% of energy is produced by processes that do not emit substantial quantities of GHGs (nuclear, renewables, and

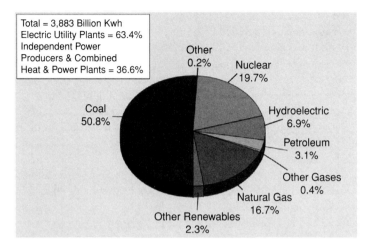

Total = 3,883 Billion Kwh
Electric Utility Plants = 63.4%
Independent Power
Producers & Combined
Heat & Power Plants = 36.6%

Other 0.2%

Nuclear 19.7%

Coal 50.8%

Hydroelectric 6.9%

Petroleum 3.1%

Other Gases 0.4%

Natural Gas 16.7%

Other Renewables 2.3%

Figure 1.29   U.S. electric power generation sources. The total energy generation is for 2003. In 2008, the energy consumption increased to 4,110 billion kWh. From U.S. Department of Energy – Energy Information Administration (2008).

hydropower). Because the power plants are interconnected in a grid for GHG emission estimations in this monograph, we will consider a weighted average, which is:

$$0.62 \text{ kg of } CO_2 \text{ emitted per kWh of energy produced}$$

The proportions between the sources of energy vary from country to country. For example, in France, relying heavily on nuclear power, or Austria, relying on hydropower and renewable energy, most energy production does not emit GHGs. Austria, for example, has embarked on large-scale development of wind energy, in addition to the hydropower that is abundant in this mountainous country. In the U.S., on the other hand, sources of hydropower are almost all already developed, and no major dams will be built. As a matter of fact, there is pressure by environmental and wildlife nongovernmental organizations (NGOs) to remove some dams (e.g., on the Columbia River) that impede the movement of anadromous fish (salmon). Hence, the future of the development of the new energy sources is mainly in renewable sources and a return to the development of nuclear power. China has just about finished building the world's largest hydropower dam, the Three Gorges Dam on the Yangtze River.

Effects of Greenhouse Gas Emissions   Since the beginning of the Industrial Revolution in the 19th century, the atmospheric content of CO_2 has increased by 30% and the temperature of the atmosphere by 0.8°C. However, if the anthropogenic emissions of greenhouse gases continue to increase or even remain at the present levels, including uncertainties in future greenhouse gas concentrations and climate modeling, the IPCC (2007) and Meehl et al. (2005) anticipate warming by another 0.6°C by the end of the 21st century and potentially an ultimate rise by 4.5°C (8.1°F) relative to 1990. It was noted in the IPCC report that during the last warm interglacial period, about 125,000 years ago, when the temperatures were 3–5°C warmer than during the 20th century (due to the change of the earth's rotation that exposed polar regions to more warming), sea levels were 4 to 5 meters higher.

Ocean water levels will rise due to the volume expansion caused by warming water and the corresponding decrease of volumetric density, and from melting of glaciers on dry land. The melting of ocean ice and icebergs does not contribute to the water level rise, but it is a part of the global thermal balance. Meehl et al. (2005) in *Science* estimated global temperature rise at 1–3°C by 2100 over that at the beginning of the Industrial Revolution, and corresponding sea level rises of about 15 to 25 cm due to sea volume expansion alone. Adding the effect of glacier melting, mainly from Greenland, the sea level rise will double to about 0.5 meters. On the basis of their models, the global warming scientists expect very little contribution from the ice over Antarctica, which will keep accumulating precipitation in the form of snow and ice. A similar seawater rise by the year 2100 was predicted by Gregory and Orlemans (1998) in their article in *Nature*. Under the "no action" scenario, the doomsday seawater rise of 4–5 meters, caused by the temperature rise of 5 to 8°C, could occur over centuries to millennia.

The IPCC (2007) report states that a temperature increase of 1–3°C may have both positive and negative effects, but that climatic change and a continuing increase in temperature would, in the long term, likely exceed the capacity of natural, managed human systems to adapt.

Implication of global warming for cities. At the end of the 20th century, the implications of global warming began to be felt on a wide scale. Arctic scientists and satellite observations noticed significant losses of Arctic ice, including Greenland; glaciers were disappearing from high mountains; polar bears were affected; and the summer and winter temperatures were noticeably higher, especially in polar regions.

All of these and other effects were widely reported by the media. On December 11, 1997, the Kyoto Protocol was adopted in Kyoto, Japan, and it entered into force on February 16, 2005. It set binding targets for 37 industrialized countries and the European community for reducing GHG by 5% measured against 1990 levels over the 5-year period 2008–2012. One hundred eighty nations have ratified the treaty to date, but not the United States and a small handful of other countries. The U.S. is by far the largest emitter of greenhouse gases per capita of the largest developed countries; larger per capita emissions can be found in small and wealthy Middle Eastern countries. Until recently, the U.S. was also the largest total emitter of greenhouse gases. Table 1.6 lists top countries with per capita and total annual emissions. The impact of global warming on cities will be large, especially on those located in low-lying coastal areas, where the effects will be exacerbated by increased coastal erosion due to more frequent extreme storms. US President Obama attended in 2009 the follow-up Copenhagen Global Warming Convention and more action of reducing GHG emissions in the US will ensue.

Table 1.6 Greenhouse emissions by selected countries

Per Capita Emissions in 2007 (tons person^{-1} year^{-1})		Countries' Total Annual GHG Emissions in 2007, in Tons, and Share of the Total Global Emissions, in %	
U.S.	19.1	Total GHG emissions	27,944
Australia	18.8		tons
Canada	17.4		
Saudi Arabia	15.8		% share
Czech Republic	11.8	China	21.5
Russian Federation	11.2	U.S.	21.1
Germany	9.7	Russian Federation	6.1
Japan	9.7	India	4.6
United Kingdom	8.6	Germany	3.1
Norway	7.9	Canada	2.2
Japan	9.4	United Kingdom	2.1
		Rest of the world	39.3
China	4.6		
India	1.2		
World average	3.8		

Source: "United Nations Millennium Goals Indicators" accessed by Wikipedia (2009) http://en.wikipedia.org/wiki/List_of_countries_by_carbon_dioxide_emissions_per_capita.

Because most of the world and U.S. population will be residing and working in cities, living and production processes and commuting will produce GHG emissions and will be responsible for most of the temperature increase. Large cities and (ultra-large) megacities require large amounts of energy, and the production of energy requires large amounts of water both for hydropower and for cooling. Fossil-coal- and oil-fueled power plants produce most of the anthropogenic GHGs.

Besides the rise of temperature, especially in polar regions, cities will face, and have to adapt to, two other major serious impacts related to global warming:

• Seawater level rise due to melting of Greenland glaciers and thermal expansion of the sea volume
• Increased frequency and magnitude of extreme events

Several large cities already have a portion of their area below the elevation of the high tide, or even below the mean sea level (e.g., New Orleans in Louisiana, several cities in Holland, and Venice, Italy). Some have built or are building tidal surge dams across the estuaries on which they are located. Such dams have been built, for example, in London (UK), Boston (Massachusetts), and across several estuaries in Holland. The Dutch situation is especially troublesome because most of the country has very low elevation, and a large portion of the country was actually reclaimed from the sea and is below the main seawater level. The sea tide effects are also a problem in the historic city of Venice in Italy (Figure 1.30). This more than 1300-year-old

Figure 1.30 Thirty years ago, Venice experienced only one or two floods a year, sometimes none. In the 2000s there were on average 70 significant floods of the historic city due to tidal surges, and the increase is partially attributed to global warming. The elevation of Piazza San Marco is less than 1 meter above the average high tide. This picture shows the flooding in December of 2008, one of the largest on record. Any significant increase of sea level elevation will have detrimental effects if flood prevention controls are not implemented. (Picture courtesy of Wikimedia-common and an annonymous photographer)

city, former metropolis and a naval power in the Mediterranean during the time of the Venetian Republic (697–1797 A.D.), is located on the archipelago of 118 small islands inside the brackish 500-km² Lagoon of Venice. We have already said that its palaces and churches were built on wood pile foundations, and the city, because of the groundwater overdraft, has been sinking for a long time, reaching a maximum subsidence of 2 cm/year in 1970s. After bringing drinking water by a pipeline from nearby mountains on the mainland, which eliminated groundwater mining, the subsidence and the central city have now stabilized. However, seawater rises measured in 2000 had increased to 1.3 mm/year (Carbognin et al., 2000).

The city responded to this threat with a massive project of increasing (repaving) the grade elevation of all streets and walkways and closing the gaps between the barrier islands and peninsulas that separate the lagoon from the Adriatic sea. Seventy-nine mobile gates were designed to hold the high tides and tidal surges from entering the lagoon in which the historic city is located.

The chair of the U.S. president's Office of Science and Technology Policy, John Holdren (2008), summarized the problems and goals related to global climatic changes as:

- Climatic disruptions and their impact are growing more rapidly than predicted just a few years ago.
- Harm from these impacts is already significant and much more is coming.
- It is too late to avoid "dangerous anthropogenic interferences" in earth climate. The question is: "Can we avoid catastrophic interferences?"
- To avoid severe consequences of climatic changes requires stabilizing human influences on the atmosphere below 450 parts per million CO_2 equivalent.
- Not exceeding 450 ppm CO_2e requires emissions to begin falling in industrial countries by 2015, elsewhere by 2025.
- Doing this will require much better technologies and much stronger policies all over the world.

Several countries (e.g., Great Britain), as well as the U.S. president's National Science and Technology Council (NSTC) (2008) are calling for the development and implementation of net zero CO_2e emission goals for the Cities of the Future, which would include building, transportation, and also water and wastewater (used water) management (see Chapters VI to VIII and X).

I.3.4   Aging Infrastructure and the Need to Rebuild and Retrofit

Water delivery, existing stormwater drainage, and wastewater disposal infrastructure systems are now aging to an extent that is leading to problems (older components may be more than 150 years old), and some parts of the cities' infrastructures are becoming obsolete. At the same time, as performance standards for pollution control continue to become more stringent, the required sewer maintenance or rehabilitation often does not keep pace with the system requirements. To a lesser degree, these concerns also apply to stormwater management facilities built 35 to 25 years ago that are approaching the age when they require major repairs and or upgrades to meet the current expectations (Marsalek et al., 2007; Ashley and Cashman, 2006). While some urban drainage service providers may be equipped for corrective action, in other cases, the financing of rehabilitation and upgrading of drainage systems must be planned in competition with other priorities—and drainage, being in most cases out of sight, is often rated as less important (Gaudreault and Lemire, 2006). There is also a lesson to be learned from this situation by developing countries without extensive centralized infrastructures: that is, distributed systems may offer better services.

The IPCC (2007) report asks cities and society to develop a portfolio of strategies to reduce the trend and cope with global warming. These should include:

- Mitigation of GHG emissions
- Adaptation to irreversible impacts and reducing vulnerability to existing and future disasters

- Technological developments to enhance both adaptation and mitigation
- Research on climate change, impacts, adaptation, and mitigation

The portfolio must be combined with policies, including incentive-based approaches, and actions at all levels from individual citizens through local and national governments.

I.3.5   The Impossibility of Maintaining the Status Quo and Business as Usual

The International Water Association (IWA) Specialist Group on Urban Drainage convened a panel of experts that analyzed several variants of the future developments of urban water, wastewater, and stormwater infrastructures (Marsalek et al., 2007). These alternative approaches included (1) business as usual, (2) privatization, (3) technocratic, and (4) green scenarios.

The business as usual (BAU) scenario has the fundamental weakness of the lack of any explicit consideration of the risks, opportunities and needs for action; inadequate and insufficient funding and financing; low-level involvement of stakeholders; small investment in R&D and hesitation to apply innovative approaches; and, in spite of the good progress in new developments, insufficient attention being paid to older areas that were originally designed for much lower environmental performance standards (e.g., without modern stormwater controls). Thus, the BAU approach is responsible for many of the current problems and is not sustainable. The persistence of this scenario is fueled by the inertia of technical, administrative, economic, and political systems. In the absence of acute problems (catastrophes), the incentive for change is low. The prevailing water, wastewater, and drainage system architecture is a centralized but not integrated system, with some experimental use of decentralized facilities (particularly in suburbs or satellite developments). The system is managed by cities or regional public utilities, which derive their payment from users' fees and taxes.

This scenario, however, neglects the trends of global warming, population increase, and increased water scarcity. The response to these threats is more infrastructure, more imperviousness, more freeways with more traffic lanes, and continuing worsening trends of the quantity and quality of urban runoff and receiving waters. Water scarcity will be increasing not only in the arid areas of the Southwestern U.S. but also in the humid areas relying on smaller streams and groundwater for the water supply, such as the suburbs of Boston, Massachusetts, or the entire metropolitan areas of Atlanta, Georgia, or Tucson, Arizona, or Southern California (see Chapter V). The risks and acute problems of this scenario might be severe in the developed world, but they are unbearable for developing countries, where neither inherited infrastructure, nor money, nor implementation capacity is available.

The business as usual scenario using the third or fourth paradigm tradition is governed by economic targets and goals (both paradigms) and under environmental legislative constraints (fourth paradigm) that are sometimes detached from the reality of social and ecological impacts. Until the end of the 20th century, environmental constraints (criteria and standards) focused only on the chemical parameters of air and water quality.

The privatization scenario has its foundation in the belief that private enterprises are more efficient than bureaucratic public agencies and utilities. The IWA panel stated that in this scenario the private sector is systematically involved in buying or assuming a license for the entire deteriorating water infrastructure and providing water services for contracted fees with a profit. In essence, these large operators are monopolistic entities because citizens and industries do not have a choice in selecting among several providers except for patronizing small-scale haulers of septic tank sullage or buying bottled water in supermarkets. The main purpose for these large monopolistic operations is to make profit, but the result in some cases is soaring increases in the cost of the services (for example, after privatization of British watershed management agencies in the 1980s) or failure to meet the expectations.

At the beginning of the third millennium, private companies provided approximately 15% of all water services in the U.S., and this proportion has been relatively constant since 1940 (Cech, 2005). In the United Kingdom, Berlin (Germany), Buenos Aires (Argentina), Johannesburg (South Africa), and Mexico City, all water supply is provided by private companies (National Research Council, 2002). The British water utility, Thames Water, which evolved from the public watershed management agency of the same name after the privatization decrees implemented by the Thatcher government in the 1980s, has over 11.5 million customers in England and abroad, from Chile to Turkey and Australia to China (Cech, 2005). They compete with other large private companies such as Suez Environment (a parent company of United Water) and Veolia Environmental Services. Recently Siemens Corporation entered the business of developing and implementing green cities in China and Singapore.

On the regional scale, with respect to the large centralized utilities, the main risk of the privatization scenario is that the price for water service may become unreasonable, especially for developing countries (this has already happened in South America, in the case of drinking water), and water could assume the same role as energy today: no longer a natural resource, but a tradable commodity like crude oil or electricity—a situation that led to well-known failures (e.g., ENRON) in early 2000. To avoid these pitfalls the contract with a public oversight agency must be well formulated with guarantees of compliance with environmental goals and standards and protections against excessive profits and futures trading.

The IWA panel identified four major driving forces that can make this scenario happen: (1) Selling public water infrastructure assets generates large one-time income that can be used by the cities for other high-priority purposes; (2) all urban dwellers need urban water services, and water service can be a profitable business and an attractive investment for private interests; (3) actual (or perceived) failures of the conventional technocratic approach will support the opinion "private is better than public"; and (4) where privatization is one of several options (e.g., in neighboring cities), private industry will attract the best engineers and offer them opportunities and resources to apply their technical talents. However, to maximize profits and satisfy investors, the private operators may resort to cutting labor costs and looking for potential voids in the regulations with which they have to comply. For example, the permits generally are expressed in terms of limits on BOD and suspended solids, and the private company may have no interest in addressing other issues such as nutrient removal or nitrification or stream restoration.

However, privatization can be attractive if the regional systems are decentralized into smaller semiautonomous clusters, or "ecoblocks," operating in a market situation. Private companies can be and have been effective in installing and operating smaller treatment plants, providing water for reuse, heat, and other water- and energy-related services, provided there are public and institutional oversights and well-defined limits and compliance. The future water/stormwater/used water utilities will also be commercially "producing" electricity, heat, hydrogen, biogas, soil conditioning solids, and nutrients (struvite), which could make the integrated resource recovery very attractive to private investors (see Chapters VIII and X).

The technocratic scenario was defined by the IWA panel (Marsalek et al., 2007) as a situation in which engineers are fully in charge and strive for technical excellence, with a minimum involvement of the public or politicians. In some way, this scenario resembles the way public utilities used to be operated in wealthy countries, 40 or more years ago, under the late third and early fourth paradigm, but with a greater emphasis on technical excellence, performance standards, and new developments in science and technology. The main operating principle of this scenario is technological excellence, based on the application of well-proven technology coupled with redundancy and adequate safety factors for a chosen design event return frequency. The system would be protected by fail-safe devices and fallback alternatives to keep operational risks small.

Under this scenario, while traditional cost/benefit analysis is undertaken, emphasis is placed on maximizing benefits and system performance, rather than on balancing such factors against costs. Long-term planning (development of master plans) is emphasized, and such plans are frequently updated. Furthermore, retrofitting and renewal of the central drainage systems and wastewater treatment are a top priority. Responsibility for system operation and maintenance is centralized and mostly public; the provision of water services remains a monopoly. In essence, this scenario assumes that: (1) it is unrealistic to expect essential changes in individual and corporate behavior with respect to environmental protection, and consequently such changes are not needed; (2) most urban citizens are not interested in urban water, wastewater, and drainage issues; and (3) politicians are satisfied with a low level of control, as long as there is no trouble. Thus, the system could be operated largely independently of the economic, social, and political context. Such a system would be quite expensive and might not lead to a balanced sustainable solution that would consider social, environmental, and economical factors equally and equitably. The technocratic scenario is feasible (affordable) in many developed countries, but is essentially irrelevant for developing countries, where the lack of funds, engineering expertise, and operation and maintenance capacities prevents the adoption of such solutions.

Under the third paradigm, technocratic solutions chose sometimes grandiose projects wherein the goals were noble and urgent, such as the elimination of terrible epidemics of the late 19th century, saving Venice from flooding, providing land reclamation (drainage of the Everglades in Florida), the diking and channelization of the Mississippi River, tide surge barriers in Holland or the United Kingdom, or compliance with mandatory point source control standards embedded in the Clean Water Act and similar laws in Europe and other industrialized developed countries. Often money was not the problem, goals were set with high priority, funds were appropriated by governments, and the only question was how to achieve the goals at the least cost. Some of these projects subsequently caused great harm to the environment (e.g., Everglades drainage), and some turned out to be ineffective (e.g., diking and constricting the Mississippi River and other Midwestern U.S. rivers to prevent flooding of cities and farmlands) and must be redone by renaturalization in this century at enormous economic and social cost.

In 1968, the U.S. Congress passed the National Environmental Policy, followed shortly by equivalent state legislative acts, requiring government-conducted or -funded projects to prepare a comprehensive Environmental Impact Statement (EIS) that introduced more social and environmental considerations into all government projects. It is interesting to note that the developers of Disney World transformed a large portion of central Florida into amusement parks and one of the most visited man-made semiurban areas in the world, but were not required to prepare an EIS because the entire project was financed by private funds.

The green scenario concepts considered by the IWA panel included Low Impact Development (LID), smart growth, Sustainable Urban Drainage Systems (SUDS), Water Sensitive Urban Design (WSUD), and others. The main characteristics of this scenario are the replacement of conventional central water service systems by distributed systems, with more accounting for sustainability, attention to environmental concerns, restoration and renaturalization of receiving waters, and so forth. The panel did not include consideration of the integrated resource recovery and energy issues. Also, the new concepts of ecocities that were concurrently evolving in China and the Middle East (Chapter XI) were not considered. The IWA panel correctly pointed out that while these concepts are currently being promoted as "new" and "sustainable," they were proposed and implemented in some places as early as the 1970s (e.g., The Woodland, Texas) as developments with natural drainage and best management practices conceived for diffuse pollution, efficient water use, and flood controls based on natural soft approaches.

The panel noted that the objectives and performance criteria of the green scenario are not well defined, as reflected, for example, by the existence of more than one hundred definitions of "sustainable development." The scenario is based on appealing principles, ideas, and visions, but its sustainability, when used on a large scale, was found to be unclear at the time of the IWA panel report. Many of the perceived or real risks of the green scenario arise from the fact that it is a new concept that has not been truly tested in the field, certainly not on sufficiently large scales and for sufficiently long periods for larger urban areas. The panel identified several weaknesses of the green concepts prevalent at the beginning of the 21st century:

- The notion of "no impact" development is unrealistic and physically impossible.
- Risks arise from the fact that green solutions transfer much of the maintenance responsibility to the property owners.
- The transition from the existing centralized systems to future decentralized systems is not clear.

The first stipulation and desire—no impact developments—led to serious requirements based on superficial observations that urban developments should be kept at less than 15% of the watershed and imperviousness to less than 10% because these were the thresholds at which biotic integrity of the watershed begins to deteriorate (Schueler, 1994). If this is taken literally, only low-density developments leading to urban sprawl would be acceptable. The second weakness may lead to litigations among property owners and between the property owners and some unidentified regulators who would oversee compliance that currently has no or only very poorly defined standards. The third problem can be simply stated as "What to do with the existing medium- to high-density urban areas?" Note that the new knowledge and state of the art of planning ecocities provide answers to these questions elaborated also through this treatise.

The IWA panel concluded that the main driving force supporting the green scenario is its positive political and economic appeal (at least in the short run) to those public utilities that struggle financially. The scenario's objectives are undisputed, both internationally and locally (see Chapter II), and it receives "green political support," particularly in relatively affluent countries where stakeholders are concerned about the overexploitation of nature and want "to do something good," and it is defended by a lot of enthusiastic supporters. From the sociological point of view, it appeals to the well-educated, well-to-do part of society, often living in upscale developments or ecovillages. Furthermore, the green scenario may even be more feasible in developing countries, where large central infrastructures are almost nonexistent and their construction is hardly feasible, and low-cost "green" solutions such as constructed wetlands, waste stabilization ponds, and reuse in agriculture are available and have even been implemented.

The panel equated "green" low-density developments with the image of sustainability. Mihelcic et al. (2003) pointed out the fact that "green" development and cities based on late-1990s ideas may not be necessarily sustainable. Only a balanced *triple bottom line – life cycle assessment* will be a testament of sustainability (see Chapters II and X); however, macroscale metrics and methods to derive balanced, societal environmental and economic assessment methods and criteria were not fully available in the first decade of this millennium. Chapter XI will document that the sustainable ecocities are not low-density developments.

The solution of choice for the panel was a mix of the four alternatives. The corner-stone of a realistic future vision for the panel was decentralized wastewater treatment and localized urban drainage networks comprising mostly surface, rather than un-derground, systems that could then be utilized as resources. The panel concentrated mostly on urban drainage and did not consider the imperatives of coping with the future effects of global warming on the cities, the effects of running out of (cheap) oil, nor the effects of population growth.

I.4 THE 21ST CENTURY AND BEYOND

U.S. cities such as Chicago (Illinois), Portland (Oregon), Seattle (Washington), Boston (Massachusetts), New York, Philadelphia (Pennsylvania), San Francisco (California), and Milwaukee (Wisconsin), and, on the international scale, cities in Sweden (Stockholm and others), England, Singapore, China (Tianjin, Harbin, Shenyang, Beijing, Chengdu, cities cluster in the Pearl River Delta), and Australia, and parts of Canada (British Columbia) are implementing sustainable (green) devel-opment policies requiring renewable energy and green buildings, added trees, green roofs, and parks to improve air quality and reduce stormwater runoff and create a more livable urban space. They have also added bike paths and walkways to encour-age biking and walking. Many of these same cities are leaders in "smart growth" development that is close to public transportation and built around commercial cen-ters, preserves open space, reuses land, and protects mixed uses. These efforts also have significant economic development potential—from fostering new technology-based industry clusters to creating well-paying jobs in housing and construction and manufacturing (Fitzgerald, 2007). However, outside of these notable cases, most de-velopments currently are still piecemeal efforts rather than an integrated effort of the entire community to introduce interconnected functioning ecotones into the urban area and watersheds (Hill, 2007). There is a need to expand the scope of the green development visions and plans to a metropolitan/regional scale at the intersection of urban aquatic and terrestrial ecology, society, and infrastructure.

The reality of the fourth paradigm is that after almost 40 years of extensive in-frastructure building programs and hundreds of billions spent, the goals of the CWA have not been met. We have systems that are functional under normal conditions but highly vulnerable during extreme events and unsustainable. The gravity of the fu-ture plight of the water resources in the world's cities and of their future under the "no action" or "proceed as usual" or "traditional" scenarios was recognized only less than two decades ago, and the first serious attempts to find the solutions appeared about the same time. However, there is now a consensus among experts that changes are needed and forthcoming—but what these changes will be is still uncertain on a worldwide scale. Resistance and inertia, as well as the tradition of the current urbanisms based on hard infrastructures and pavements, are persisting and will be difficult to overcome. Nevertheless, most experts agree that the water-impacted in-frastructure in some cities is at a breaking point and that it will take trillions of dollars (euros, etc.) just to fix it. But no matter how many billions will be spent under the current paradigm of building new hard water/wastewater infrastructure and/or fixing old ones the old way, the ecological goals of the Clean Water Act in the U.S., the Water Framework Directive in the EC countries, and similar goals in many other countries will still not be met. There is also a need to build many new cities, especially in Asia and Latin America, to accommodate anticipated population increases and the flux of people from rural areas to the cities. Then, instead of fixing the old infrastructure the old way at an enormous price, let us do it right and make the water and other infrastructure systems sustainable and energy efficient for future generations, reduce GHG emissions substantially, and save—or even make—money doing it.

Looking far ahead (considering the impact of the "business as usual" scenario and the continuation of the current trends), we can quote from the Abel Wolman Distinguished Lecture to the U.S. National Research Council by Peter Gleick (2008), who looked into the future to 2100. Regarding sustainability and carbon imprint in 2008, he prefaced his predictions by a reference to a study by the Pacific Institute (2007):

Things were so bad in the United State that people actually spent vast sums of money [100 billion U.S. dollars annually] to buy small quantities of water in plastic bottles, when they could get safe water from the tap at a thousandth the price. By some estimates, as much as 17 million barrels of oil equivalent were used annually just to make the plastic bottles used in the U.S., most of which were then thrown away.

Gleick then continued with his vision of 2100 (paraphrased):

- With very few exceptions of very high altitudes, all mountain glaciers will be gone, and the impact on local water supplies will be especially severe in China and parts of South America.
- Downhill skiing in resorts will be mostly gone.
- The Everglades, which were saved by restoring their natural flows and function during the early part of the 21st century, were ultimately lost to the rising seas, along with coastal aquatic ecosystems and some major cities (e.g., Venice) all over the world.
- Floods from increased precipitation and the increased intensity of storms will continue to be the leading killer of people worldwide.

Other gloom and doom predictions have been made, especially by physical scientists who projected the current trends. Engineers by their training have a tendency to "fix" problems, sometimes with unforeseen adverse effects. In general, Peter Gleick was an optimist in his presentation. He outlined several steps to avert the doom and gloom scenario, calling for a fundamental paradigm shift by rethinking water use, reducing waste and losses, and improving efficiency and productivity on the drinking water side. The new paradigm of integrated water/stormwater/wastewater and urban landscape management will enlarge Gleick's vision to other water and urban sectors. Changing the paradigm will provide immense opportunities for small and large businesses. Even today, spontaneous, localized, and limited "green" developments such as green business and government buildings (e.g., Chicago) require innovation and are beginning to generate employment opportunities (Fitzgerald, 2007). A "green" high-rise in New York (Battery Park, see Chapter VI) saves up to 50% of water through reuse (Engle, 2007). Politicians have been promising a bonanza of green benefits that would come from large- and small-scale government and private projects. The facts of the "business as usual" alternative that must be considered are:

1. Most of the water and wastewater infrastructure in cites is almost 150 years old, and is deteriorated or deteriorating rapidly and will have to replaced.
2. Combined sewer flows have been and will have to be separated, and both flows stored and treated.
3. Urban stormwater must be treated if discharged into storm sewers.
4. Building underground conduits for the conveyance for relatively clean water of buried streams and cleaner urban runoff does not make sense and does not provide protection against flooding.
5. Runoff from transportation systems will have to be captured and treated to avoid severe ecological damage.

Hence, it is clear that the first 50 years of this century will see massive investment in new urban water/stormwater/wastewater infrastructures, on the order of trillions of dollars or euros, or whatever currency the country is using. If it is not done right, if business as usual (in developed countries) or even doing nothing (in undeveloped countries) scenarios prevail this century will see increased severe inconveniences at best and human catastrophes at worst. Thus, it must be done right.

The concepts of sustainability and "Cities of the Future" have now been discussed and addressed by a number of research and outreach initiatives in Europe, Asia, and Australia, research conferences and congresses organized by the International Water Association (IWA), Stockholm Water Conferences, and NGOs, and initiatives funded by private foundations. The IWA has established an International Steering Committee and made the Cities of the Future one of its primary programmatic goals. UNESCO has funded an extensive international research and pilot implementation collaborative project, SWITCH – Managing Water for the City of the Future – (http://www.switchurbanwater.eu/about_mgmt.php), in several countries throughout the world. These initiatives are driven by the widespread public desire for "green" "sustainable" everything, from houses to urban landscapes, food and agriculture, manufacturing, cleaning products, and, finally, entire cities. The engineering and scientific communities now have the tremendous mission of responding comprehensively to these stresses and public desires for action by developing and implementing the new concepts of sustainable urbanisms that, with their water systems, would not only satisfy the present and future needs for water and sanitation but also be resilient to the stresses, demands, and extreme events of the future and have a positive impact on GHG emissions.

REFERENCES

Aldrich, H. P and J. R. Lambrechts (1986) "Man-made structures that permanently lower groundwater levels can have adverse effects on buildings with water table sensitive foundations," *Civil Engineering Practice: Journal of the Boston Society of Civil Engineers* 1(2)

American Public Works Association (1969) *Water Pollution Aspects of Urban Runoff*, WP-30-15, U.S. Department of Interior, Washington, DC

Ashley R. M. and A. Cashman (2006) *Assessing the Likely Impacts of Socio-Economic, Technological, Environmental and Political Change on the Long-Term Future Demand for Water Sector Infrastructure*, OECD report, Pennine Water Group, Sheffield, UK

Barboza, D. (2007) "China's growth accelerates to 11.9% and food prices spur inflation," New York Times, June 20

Barnard, J. L. (2007) "Elimination of eutrophication through resource recovery," *The 2007 Clarke Lecture*, National Water Research Institute, Fountain Valley, CA

Biswas, A. K., C. Tortajada, J. Luindqvist, and O. Varis (2004) *The water challenges of megacities*, Stockholm World Water Week Publication, Stockholm Water Institute

Breckenridge, L. P. (2007) "Ecosystem resilience and institutional change: the evolving role of public water suppliers," in *Cities of the Future: Towards Integrated Sustainable Water and Landscape Management* (V. Novotny and P. Brown, eds), IWA Publishing, London, UK, pp. 375–387

Carbognin, L., P. Tetini, and L. Tosi (2000) "Eustasy and land subsidence in the Venice Lagoon at the beginning of the new millennium," *Journal of Marine Systems* 51(1–4):345–353

Carson, R. (1962) *Silent Spring*, Houghton Mifflin, Boston, MA

Cech, T. V. (2005) *Principles of Water Resources – History, Development, Management, and Policy*, John Wiley & Sons, Hoboken, NJ

Chomat, P. R. (2008) *The Delusion of Progress – A Fallacy of Western Society*, Universal Publishers, Boca Raton, FL

Dilworth, R. (2008) "Measuring sustainability in infrastructure: The case of Philadelphia," *Drexel Engineering Cities Initiative (DECI) Working Paper Series, 2008-01*, Drexel University, Philadelphia, PA

Dittmar, H. (1995) "A broader context for transportation planning," *Journal of the American Planning Association* 61(1):7–13

Emanuel, K. A. (2005) "Increasing Destructiveness of Tropical Cyclones over the Past 30 Years," *Nature* 436, pp. 686–688, <ftp://texmex.mit.edu/pub/emanuel/PAPERS/NATURE03906.pdf>

Energy Information Center (2008) *Residential Energy Prices: A Consumer Guide*, Department of Energy, Washington, DC, http://www.eia.doe.gov/neic/brochure/electricity/electricity.html

Engle, D. (2007) "Green from top to bottom," *Water Efficiency* 2(2):10–15

Field, R., J. S. Heaney, and R. Pitt (2000) *Innovative Urban Wet-weather Flow Management Systems*, TECHNOMIC Publishing Co., Lancaster, PA

Fitzgerald, J. (2007) "Help wanted – Green: Green development could be a big generator of good jobs—if America will seize the opportunity," *The American Prospect* January-February, pp. A16–A19

Fujita S. (1984) "Experimental sewer system for reduction of urban storm runoff," Proceedings of the 3rd International Conference on Urban Storm Drainage, Gutenberg, Sweden

Furumai, H. (2007) "Reclaimed stormwater and wastewater and factors affecting their reuse," in *Cities of the Future: Towards Integrated Sustainable Water and Landscape Management* (V. Novotny and P. Brown, eds.), IWA Publishing, London, UK, pp. 218–235

Furumai, H. (2008) "Urban water use and multifunctional sewerage systems as urban infrastructure," in *Urban Environmental Management and Technology* (K. Hanaki, ed.), Springer Japan, Tokyo

Gaudreault V. and P. Lemire (2006). *The Age of Public Infrastructure in Canada.*, Analysis in brief, No. 11-612-MIE2006035, Statistics Canada, Ottawa, ON

GeoHive (2007) www.geohive.com/charts/

Gleick, P. H. (2003) "Water Use," *Annual Review of Environmental Resources* 28, pp. 275–314

Gleick, P. H. (2008) *The 15th Abel Wolman Distinguished Lecture*, The U.S. National Research Council, National Academy of Sciences, Washington, DC, April 23, 2008

Gregory, J. M. and J. Orlemans (1998) "Simulated future sea-level rise due to glacial melt based on regionally and seasonally resolved temperature changes," *Nature* 391, pp. 474–475

Gurjar, B. R., T. M. Butler, M. G. Lawrence, and J. Lelievels (2008) "Evaluation of emissions and air quality in megacities," *Atmospheric Environment* 42(7):1593–1606

Hardoy, J. E., D. Mitlin, and D. Satterthwaite (2001) *Environmental Problems in an Urbanizing World: Finding Solutions for Cities in Africa, Asia, and Latin America*, Earthscan, London

Heaney, J. P. (2000) "Principles of integrated urban water management," Chapter 2 in R. Field, J. S. Heaney, and R. Pitt, *Innovative Urban Wet-weather Flow Management Systems*, TECHNOMIC Publishing Co., Lancaster, PA

Hill, K. (2007) "Urban ecological design and urban ecology: an assessment of the state of current knowledge and a suggested research agenda," in *Cities of the Future: Towards Integrated Sustainable Water and Landscape Management* (V. Novotny and P. Brown, eds.), IWA Publishing, London, UK, pp. 251–266

Holdren, J. (2008) *Clean Energy and Climate Change – International Partnership and Path Forward, PowerPoint/Keynote Address at the U.S. – China – India Innovation Partnership Conference*, Boston, MA, December 11, 2008

Imhoff, K. (1931) "Possibilities and limits of the water-sewage-water-cycle," *Engineering News Report*, May 28

Imhoff, K. and K. R. Imhoff (1993) *Taschenbuch der Stadtentwässerung (Pocket Book of Urban Drainage)*, 28th ed., R. Oldenburg Verlag, Munich, Germany

IPCC (2007) "Summary for Policy Makers," in *Climate Change 2007: Impacts, Adaptation and Vulnerability*. Working Group II, Intergovernmental Panel on Climate Change, Cambridge University Press, Cambridge, UK

Karr, J. R., K. D. Fausch, P. L. Angermeier, P. R. Yant, and I. J. Schlosser (1986) *Assessing biological integrity of running waters. A method and its rationale*, Spec. Publ. #5, Illinois Natural History Survey, Champaign, IL

Krenkel, P. A. and V. Novotny (1980) *Water Quality Management*, Academic Press, Orlando, FL

Lanyon, R. (2007) "Developments towards urban water sustainability in the Chicago metropolitan area," in *Cities of the Future: Towards Integrated Sustainable Water and Landscape Management* (V. Novotny and P. Brown, eds.), IWA Publishing, London, UK, pp. 8–17

Lewis, M. (2007) "Megacities of the Future," Forbes.com, http://www.forbes.com/2007/06/11/megacities-population-urbanization-biz-cx_21cities_ml_0611megacities.html

Macaitis, B., S. J. Povilaitis, and E. B. Cameron (1977) "Lake Michigan diversion - stream quality planning," *Water Resources Bulletin* 13(4):795–805

Marsalek, J., R. Ashley, B. Chocat, M. R. Matos, W. Rauch, W. Schilling, and B. Urbonas (2007) "Urban drainage at cross-roads: four future scenarios ranging from business-as-usual to sustainability, in *Cities of the Future: Towards Integrated Sustainable Water and Landscape Management* (V. Novotny and P. Brown, eds.), IWA Publishing, London, UK, pp. 339–356

McGranahan, G. and D. Satterthwaite (2003) "Urban centers: An assessment of sustainability," *Annual Review of Environmental Resources* 28, pp. 243–274

Meehl, G. A., W. M. Washington, W. D. Collins, J. M. Arblaster, A. Hu, L. E. Buja, W. G. Strand, and H. Teng (2005) "How much more global warming and sea level rise?" *Science* 307(5716):1769–1772

Metcalf & Eddy (2003) *Wastewater Engineering: Treatment and Reuse*, 4th ed. (revised by G. Tchobanoglous, F. L. Burton, and H. D. Stensel), McGraw-Hill, New York

Mihelcic, J. B., J. C. Crittenden, M. J. Small, D. R. Shonnard, D. Hokanson, Q. Zhang, V. U. James, and J. L. Schnoor (2003) "Sustainability Science and Engineering: The emergence of a new metadiscipline," *Environmental Science & Technology* 37, pp. 5314–5324

Nappo. S. (1998) *POMPEII: A Guide to the Ancient City*, Barnes and Noble, NY

National Research Council (2002) *Privatization of Water Services in the United States*, National Academy Press, Washington, DC

National Science and Technology Council (2008) *Federal Research and Development Agenda for Net-Zero Energy, High Performance Green Buildings*, Committee on Technology, Office of the President of the United States, Washington, DC

Neimczynowicz, J. (1996) "Megacities from a water perspective," *Water International* 21(4):1998–205

Newman, P. (1996) "Reducing automobile dependence," *Environment and Urbanization* 8(1):67–92

Novotny, V. (2003) *WATER QUALITY: Diffuse Pollution and Watershed Management*, John Wiley & Sons, Hoboken, NJ

Novotny, V. (2007) "Effluent dominated water bodies, their reclamation and reuse to achieve sustainability," in *Cities of the Future: Towards Integrated Sustainable Water and Landscape Management* (V. Novotny and P. Brown, eds.), IWA Publishing, London, UK, pp. 191–215

Novotny, V., K. R. Imhoff, P. S. Krenkel, and M. Olthoff (1989) *Karl Imhoff's Handbook of Urban Drainage and Wastewater Disposal*, John Wiley & Sons, Hoboken, NJ

Novotny, V., N. O'Reilly, T. Ehlinger, T. Frevert, and S. Twait (2007) "A River is Reborn: The Use Attainability Analysis *for the Lower Des Plaines River, Illinois*," Water Environment Research 79(1):68–80

Novotny, V. *et al.* (1999) *Urban and Highway Snowmelt: Minimizing the Impact on Water Quality*, Water Environment Research Foundation, Alexandria, VA

Oregon State University, GeoSyntec Consultants, University of Florida, and The Low Impact Development Center (2006) *Evaluation of Best Management Practices for Highway Runoff Control*, National Cooperative Highway Research Program (NCHRP) Report 565, Transportation Research Board, Washington, DC

Pacific Institute (2007) *Bottled water and energy: Getting to 17 million barrels*, http://www.pacinst.org/topics/integrity_of_science/case_studies/bottled_water_energy.html

Ripley, A. (2008) "10 things you can like about $4 gas," *Time*, July 14

Rohlich, G. A. (1969) *Eutrophication: Causes, Consequences, Correctives*, National Academy of Sciences, Washington, DC

Sansalone, J. J. and D. W. Glenn (2002) "Accretion of pollutants in snow exposed to urban traffic and winter storm maintenance activities," *Journal of Environmental Engineering* 128(2):151–166

Schipper, L., F. Unander, S. Murtishaw, and M. Ting (2001) "Indicators of energy use and carbon emissions: Explaining the energy economy link," *Annual Review of Energy and the Environment* 26, pp. 49–81

Schipper, L. (2008) *Counting energy and carbs: Indicators for sustainable energy in an urban context*, Presentation to the Urban Sustainability Workshop, Drexel University, June , 2008

Schueler, T. (1994) "The importance of imperviousness," *Watershed Protection Technology* 1(3):100–111

Scott, C. A., A. J. Zarazua, and G. Levine (2000) *Urban-Wastewater Reuse for Crop Production in the Water-Short Guanajuato River Basin, Mexico*, International Water Management Institute, Colombo, Sri Lanka

Sedlak, R., ed. (1991) *Phosphorus and Nitrogen Removal from Municipal Wastewater – Principles and Practice*, Lewis Publishers, Boca Raton, FL

Shaheen, D. G. (1975) *Contribution of Urban Roadway Usage to Water Pollution*, EPA-600/2-75-004, Office of Research and Development, U.S. Environmental Protection Agency, Washington, DC

United Nations (1999) *World Population Prospects*, United Nations Department of International Economic and Social Affairs, New York, NY

United Nations Secretariat (2005) *World population prospects: The 2005 revision*, United Nations Department of Economic and Social Affairs, New York, NY

U.S. Department of Energy (2000) *Carbon Dioxide Emissions from the Generation of Plants in the United States*, also published by U.S. EPA, Washington, DC

U.S. Environmental Protection Agency (2008) *Emission facts: Greenhouse gas emissions from a typical passenger vehicle*, htt://www.epa.gov/oms/climate/420f5004.htm (accessed July 2008)

Zaitzevsky, C. (1982) *Frederick Law Olmsted and the Boston Park System*, The Belknap Press of Harvard University, Cambridge, MA

II

URBAN SUSTAINABILITY
CONCEPTS

II.1 THE VISION OF SUSTAINABILITY

The onset of the fourth paradigm period in the late 1960s and early 1970s was marked by revelations and controversies about the costs of the environmental degradation caused by unrestricted economic development and accelerated population growth. The societal goals had been to satisfy the needs of the population for necessities and amenities that required the continuous growth of industrial, agricultural, and commercial economies, and large defense expenditures fueled by the Cold War. However, it was becoming clear that the environmental costs of unrestricted growth and production were too high, that they had reached a point at which they posed a very serious threat to nature as well as to human health. Furthermore, it was realized that with the rapidly increasing population, future generations might be threatened by diminishing resources due to overuse, pollution, diminishing raw materials, and more people competing for the use of these finite resources. Meeting the fundamental needs of present and future generations, while preserving the life-supporting systems of nature, became a goal and a basic tenet of sustainability. At the end of the last century, it became apparent that, given a future world population at least 50% larger than it then was, it might be impossible to provide goods to all people at the level enjoyed then by the more affluent population in developed countries. If the 50% larger population demanded the same living standard as that in the developed countries, the amount of resources and gross national product would have to grow by three to five times (Daigger, 2009; Rees, 1997). Hence, sustainability is both intra- and intergenerational. It has also become clear that the engineering/technological community is locked in the fourth paradigm rules and only slowly adapting to the sustainability requirements for the future.

Cities, and the urbanized environment in general, are the fundamental foundations of civilization, places where civilization began almost 10,000 years ago, where society develops, persists, and provides conditions for advances of knowledge and rapid progress. In the future, most people on earth will be living in cities (see Chapter I). A majority of cities are built along the water, but even some desert cities—for example, Las Vegas (Nevada), Phoenix (Arizona), Dubai (United Arab Emirates), and Masdar (Abu Dhabi)—have water resources nearby (in the case of Dubai and Masdar, it is the sea) and man-made freshwater bodies within the city, including those fed by desalination facilities. Managing water wisely is a key prerequisite of the existence of cities.

In Chapter I, the existing fourth (current) paradigm of urban water/stormwater/ wastewater management was identified as one based on long-distance predominantly *subsurface transfer of water and wastewater, fast conveyance of stormwater, and end-of-pipe treatment.* The current paradigm also continues an unsustainable expansion of cities (urban sprawl), economic growth with increasing energy demands, and greenhouse gas (GHG) emissions. Unlike the third paradigm when economic growth was not restricted by environmental concerns, the fourth paradigm operates under environmental restrictions and new concepts, which, however, are incapable of reversing the past trends. Furthermore, there are now several emerging economic giants in Asia, to be followed soon by some in Africa and Latin America, still operating their societies and economies under third paradigm rules (or lack thereof), with the same disastrous impact on the environment typical of the pre-1970 period in developed countries.

II.2 THE SUSTAINABILITY CONCEPT AND DEFINITIONS

Economics, social development and quality of life, and the ecology/environment are intertwined as an interacting trinity (Figure 2.1). A change in one component affects the other two. An ideal situation occurs when the three components are in balance, which is the fundamental premise of sustainability. An imbalance in favor of economics, or caused by societal pressures (e.g., weak governments, social upheavals) leads to pollution or to impairment of the integrity of the environment's ecology, but in a worst-case scenario, it could also lead to social stresses such as starvation. The balance of the three components is achieved by discourse among the various groups that constitute society (Allan, 2005; Novotny, 2003).

Discussions among experts in the worldwide literature have concluded that the present paradigm of urban water and landscape management is neither sustainable nor resilient enough to accommodate climatic changes and the ensuing increase of extreme meteorological events.

II.2.1 A New (Fifth) Paradigm Is Needed

Figure 2.2 shows how the shifts in the weights of the trinity's components led to the changes of the paradigms. This concept of water resources paradigms was formulated

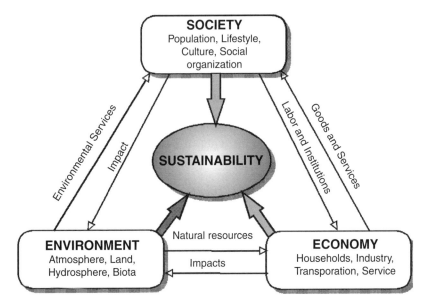

Figure 2.1   The trinity of factors and impacts determining sustainability. Adapted from Brundt-land (1987), Novotny (2003) and Allan (2005).

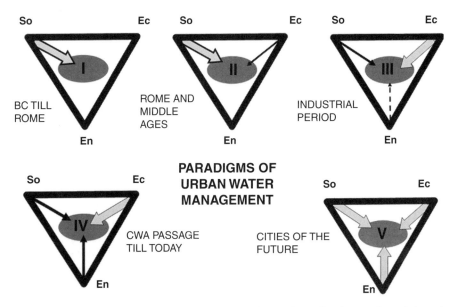

Figure 2.2   The impacts of three categories of policy factors determine the paradigms (adapted and modified from Allan, 2005, 2008). So – social drivers, Ec – economic drivers, and En – environmental drivers.

by Allan (2005, 2008) and was applied originally to use of water. The paradigms emerged from a discourse within society driven by social, economic, and, later, environmental protection interests. The first paradigm was solely driven by the desire of society for water for self-preservation and protection against enemies. Other aspects such as waste disposal were supplied by gravity—that is, used potable water and stormwater flowed downhill to the nearest watercourse or infiltrated the soil. During the second paradigm period, the economy and the skill of the workers allowed the engineering of water and drainage systems. From the onset of the Industrial Revolution in the third paradigm period, economic utilization of water accelerated. Urban rivers provided the power, water, and navigation that drove the industries. Later, as the streams became heavily polluted, some concerns about the environment emerged. Environmental concerns received much greater attention during the fourth paradigm period, but systems still operated on the premise of unlimited water availability, oil was plentiful, and global warming was not a concern. Consequently, the effect of large withdrawals on the ecologic status of receiving waters was not considered, carbon emissions were not an issue, and nonpoint pollution abatement—including the control of polluted stormwater—was deemed to be achieved by the voluntary efforts of dischargers.

Allan (2005) pointed out that the switch to achieve the fifth paradigm of *integrated resource management (IRM)* (originally "integrated water resources management"— the term "water" was dropped to recognize the fact that other resources, such as energy or transportation, must also be considered) requires a new holistic approach and a high level of political cooperation. The inclusive political process of the fifth paradigm requires the interests of civil society, hierarchy (government), social movements (NGOs), and the private sector to be included in the policy-making discourse. (Thompson et al., 1990). To attain the fifth paradigm will require an unprecedented level of political cooperation.

If accepted and implemented, the new fifth paradigm connecting green cities and the infrastructure with drainage, landscape, and receiving waters will be not only ecologically sound, but also acceptable to and desired by the public. In order to be sustainable, conserve water, and reduce carbon footprint, the new "ecocities" will most likely implement distributed and decentralized cluster-based water/stormwater/ wastewater management with water and energy reclamation that could also be supplemented by geothermal and solar energy extraction. Under the fifth paradigm concepts, the notion of "wastewater" becomes obsolete because waste and wastewater are resources from which water, nutrients, energy, and other by-products will be recovered. Hence, the new term "used water" has been introduced, and reclaimed water becomes "new water" (see Chapter VI). The developing and retrofitting of older cities under the fifth paradigm will incorporate surface and underground drainage infrastructure and landscape that will (1) store and convey water for reuse, provide ecological flow to urban flow-deprived rivers, and for safe downstream uses, (2) treat and reclaim polluted flows, and (3) integrate the urban hydrologic cycle with multiple urban uses and functions to make it more sustainable. More on the features of the fifth paradigm will be covered in the subsequent sections of this chapter and throughout the book.

II.2.2 Definition of Pollution

Before beginning the discussion of "sustainability," the term "pollution" has to be defined. For a more general and detailed discussion on this topic, see Novotny (2003). Generally, pollution has been considered as a perceptibly bad quality of the environment and, before the passage of the Clean Water Act (CWA) by the U.S. Congress in 1972, the definition of pollution most accepted by scientists was "unreasonable interference of water quality with the beneficial uses of water." "Air" or "soil" can be substituted or included to extend this definition to other domains. Pollution was generally expressed by comparing chemical or bacteriological measurements of the quality of the environment with numeric standards that in the U.S. were developed for each recognized use of the water body.

However, today the quality of the environment in general and water in particular is understood in a more comprehensive manner, expressed as *integrity*. Attaining and preserving integrity is the main goal of the CWA, expressed in Section 101. The statutory definition of pollution is then included in Section 502-119 as:

> The term 'pollution' means the man-made or man-induced alteration of physical, chemical, biological, and radiological integrity of water.

Based on the linguistic definition, integrity means "being unimpaired"; therefore, adverse alteration of integrity means impairment. *Biological integrity*, the most important integrity component, implies the ability of the water body's ecological system to support "a balanced, integrated, adaptive community of organisms comparable to that of natural biota of the region" (Karr and Dudley, 1981). *Physical integrity* implies habitat and riparian zone conditions conducive to the support of a balanced biological community. *Chemical integrity* would mean a chemical composition of water and sediment that would not be injurious to aquatic biota and human use of the water body for drinking and contact recreation. A composition of aquatic organisms resembling or approaching that of unaffected similar water bodies in the same ecoregion without invasive species represents the biological integrity (Novotny, 2003).

Sustainability, however, is not limited only to water. The serious threats to human beings and other species from global warming add another component of integrity—atmosphere—now impacted by air pollution by GHG emissions. They will adversely affect both nonhuman organisms and human beings. As a working hypothesis in this book, we make an assumption that the concept of integrity is universal and applies to the environment in general—that is, air, water, and soil—and also considers damage to human beings, in addition to nonhuman organisms. It would then be appropriate to substitute "environment" for "water" in the definition of pollution above and redefine environmental integrity as:

> the ability of the environment (air, water, and soil) to support, on a regional or global scale, balanced communities of organisms comparable to those of natural biota of the region and also provide conditions for the unimpaired well-being of present and future human generations

This definition recognizes that human beings are a part of the ecological system and, in addition to biota, they can also impose damage upon themselves. According to these definitions, pollution is caused by human beings or their actions and is differentiated from changes of the quality (integrity) of the environment due to natural causes such as the natural CO_2 content of the atmosphere, natural erosion, weathering of rocks, volcanic eruptions, or fly ash and CO_2 emissions from natural forest fires. GHG emissions are clearly pollution.

Pollution not only implies the addition of harmful substances into the environment, but includes also any human action or alteration of the environment that would impair its integrity. Under this definition, pollution includes channel alterations such as the straightening and lining of streams with concrete or riprap lining, or even converting them into underground conduits resulting in loss of species and impairment of integrity; cutting down the trees and vegetation abutting a stream, which deprives the biota of refuge and causes disappearance of species; as well as excessive withdrawals of water resulting in insufficient ecological flows or ephemeral flow conditions.

The Need for Regulations and Other Socioeconomic Tools to Solve the Pollution Problem Solving pollution problems cannot be done solely by market forces; that is, "let the market decide how much pollution the environment should receive" will not work alone. Events on the financial and stock markets in 2008 and thereafter revealed that the market sometimes does not function correctly even in its own domain—which is moving money around, looking for the best returns on investments—and needs regulation to control its excesses. Pollution has been defined as an economic *externality* or *diseconomics*, which can be best explained by the case of two cities or industries located on the same river. The upstream source is enjoying good water quality but polluting the river with its waste, and the pollution is transferred to the downstream city by the river, causing water quality degradation and economic damage exhibited by the increased cost of treatment of potable water and the loss of recreational opportunities, fishing, and aesthetical amenities of the water body. However, the downstream users have no economic market means to recover the cost of the damage from the upstream polluter.

A well-known political economist, R. M. Solow (1971), defined *externality* as follows: "One person's use of a natural resource can inflict damage on other people who have no way of securing compensation, and who may even not know that they are being damaged." Besides water and air pollution, other externalities affecting cities include crime that is exported from one area of the city to another or to suburbs, or airport noise (noise pollution). The most dreadful examples of externality are regional, such as excessive uses of fertilizers and nutrient discharges from urban areas (including effluents, overfertilized lawns and golf courses) in many upstream regions of the watersheds, causing massive developments of cyanobacteria (blue-green algae) in many water supply reservoirs or excessive algal development of coastal waters that results in oxygen-deprived dead zones (Niemi, 1979; Paerl,

1988; Rabalais et al., 1999). Notable examples are:

- The majority of all reservoirs in the Czech Republic, other countries of Europe, and China suffer from excessive toxin-producing algal blooms of cyanobacteria that create anoxic conditions in the deeper parts of the reservoirs, interfere severely with recreation, and dramatically affect the taste and odor of potable water (Czech Academy of Sciences, 2006). In the absence of regulations or some kind of court injunction, the users of affected waters damaged by upstream pollution sources do not have any economic means to stop the pollution-causing activities short of "bribing" the polluters to stop them.
- In the late 1980s, a large hypoxic "dead" zone, whose insufficient dissolved oxygen concentration is deadly to fish and shellfish, was found off the Louisiana coast in the Gulf of Mexico. This occurrence of a hypoxic zone has dramatically reduced fishing and shellfish harvesting (Figure 2.3). The

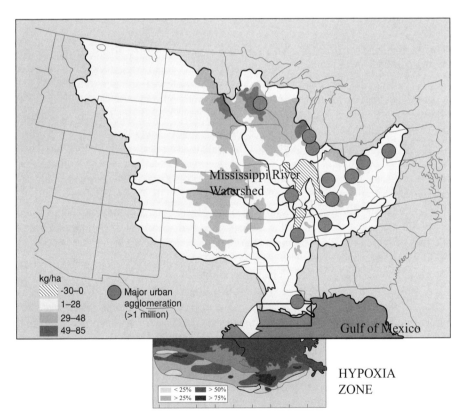

Figure 2.3 Agricultural and urban nitrogen sources in the Mississippi River watershed (Burkart and James, 1999; U.S. Census Bureau statistics) stimulating excessive algal growths in the Gulf of Mexico and the extent of the hypoxia zone of the Louisiana coast (lower box) at the end of the last millennium (Rabalais et al., 1999).

excessive growth of algae responsible for the dead zone was stimulated by nitrogen discharges mainly from agricultural operations in the Mississippi watershed and to a lesser degree by nitrogen loads from cities (Chicago, Illinois; Minneapolis, Minnesota; St. Louis, Missouri; Memphis and Nashville, Tennessee; New Orleans, Louisiana; Pittsburgh, Pennsylvania; Cincinnati and Columbus, Ohio; and many smaller cities). Subsequently, dead zones have been discovered in coastal waters of China, Japan, the Baltic and Black Seas of Europe, and many other locations throughout the world.

These examples imply that the externality problem could also have a transboundary dimension sometimes called *pollution export.* For example, decades ago, lakes in Scandinavia and in Switzerland suffered from the impact of acid rainfall caused by sulfur oxide emissions in Central Europe and the United Kingdom. Loss of the polar ice sheet is due to GHG emissions, and the largest polluters are again thousands of kilometers away. We have already showcased the effect of nutrient discharge on the formation of dead zones in seas.

Typically, the solutions of the externality problem are limited and evolve from a discourse among the involved parties, which can be individuals, groups, or countries. The main tools are:

- Regulations that restrict the polluting activities or ban them entirely; imposing the cost of remedying the situation on polluters
- Subsidies to polluters to implement pollution control practices, or reduce or eliminate the polluting activities that are imposing on the users of the resource or environment being damaged by polluting activities
- In developing countries, pressures exerted by the public and NGOs directly on polluters, which are often the most effective means to curtail pollution-causing activities because governments are typically ineffective
- Bringing the polluters to court and engaging in litigation

Is it possible to categorize GHG emissions as an externality? There are no distinct social differences between those who pollute and those who will be suffering from the consequences of global warming. Everyone, with the exception of some native tribes in the Amazon region, is contributing to the emissions, although there are several orders of magnitude of disparity between the emissions of slum dwellers in India and those of highly affluent and mobile suburban dwellers in the U.S. True, the consequences will be different in different regions of the world. The most affected will be low-lying coastal areas, increasingly affected by flooding, which includes many large cities or even large portions of some states or countries (Louisiana, Florida, Bangladesh, the Netherlands), many natural treasures, including the Everglades, and historic cities (New Orleans, Louisiana, Venice, Italy). Water shortages will also increase (see Chapter I). It is also true that countries contribute to global warming disproportionately at different levels, with the U.S. and China leading in total GHG emissions. But the global impact will be of such magnitude that the global society must react and consider itself as a major sufferer of global GHG emissions. Hence, the tools are the same: (1) worldwide regulations, (2) international aid to those countries that cannot afford to install significant GHG emission reducing or eliminating measures, and (3) continuing international pressure and vigilance.

II.2.3 Sustainability Definitions

The notion of "sustainability" has been deeply embedded in human nature and culture. In most civilizations and cultures, parents sacrificed and saved for the future of their children. Thomas Jefferson, the second U.S. president, wrote, "the earth belongs in usufruct to the living" (Howarth, 2007), where "usufruct" means "the right to use and enjoy the profits and advantages of something belonging to another as long as the property is not damaged or altered in any way" (*American Heritage Dictionary of the English Language*). Section 101 of the U.S. National Environmental Protection Policy Act of 1969 specifically highlights the government's duty to "fulfill the responsibilities of each generation as trustee of the environment for succeeding generations."

Parkin (2000) presented an excellent analysis of sustainability concepts, starting by pointing out the meaning of the word "sustainable," which means having a capacity to continue. Sustainability is a quality and objective, not a process. It is an entity's intrinsic capacity to keep itself going in perpetuity, and this applies to society, habitat, resources, and environment. Sustainable development is a process toward achieving and then maintaining sustainability. It concerns the people and their behavior towards nature and the environment, and it is a path for human progress. Nature without excessive stresses is sustainable and operates in a sustainable manner.

However, in the late 20th century and early 21st century, some governments generated huge debts, and budget deficits will have to be paid by future generations, which obviously is a gross violation of the cognitive sustainability principle. Essentially society has been living beyond its means, continuing to exhaust nonrenewable resources at a very high rate, and pushing the cost of these excesses onto their children. In the second half of the 20th century, arguments were made that the development and use of resources should be maximized irrespective of the impact on future generations, and that budget deficits were acceptable. It was argued (see Howarth, 2007) that unborn or young children cannot affect the welfare or enjoyment of the present generation, and that the unborn generation has no moral right to impose binding duties on the present generation. Voluntary restriction would constitute a sacrifice. Howarth (2007) quoted the arguments of others against the sustainability concept and then provided legal facts and arguments showing that future generations are morally entitled to enjoy opportunities at least as good as those enjoyed today. Not to accept the concept of sustainability would mean that it would be morally permissible for the present generation to inflict harm on young children, which is ethically and also legally unacceptable. Sustainability is based on the legal doctrine that equal opportunities should be provided to each member of society, including children, which is also a well-known concept of intergenerational (parents vs. children) and intragenerational (developed vs. developing nations, affluent vs. disadvantaged population groups) fairness. Howarth (2007), considering both inter- and intragenerational justice, introduced a "fair sharing principle":

> "Each member of present and future society is entitled to share fairly in the benefits derived from environmental resources and specific stocks of environmental resources should not be depleted without rendering just compensation to members of future generations."

There are hundreds of definitions of sustainability (Parkin, 2000; Marsalek et al., 2007), but the one that is quoted most is that of the World Commission on Environment and Development, chaired by then prime minister of Norway G. Brundtland. The definition is as follows (Brundtland et al., 1987):

> "Humanity has the ability to make development sustainable—to ensure that it meets the needs of the present without compromising the ability of future generations to meet their own needs."

The Brundtland Commission's definition, which has been generally accepted since the end of the last century, does not differentiate between sustainable development and sustainability. The report expresses the desire of society to use and manage resources on the bases of economic sustainability, social equity, intergenerational justice, and the intrinsic value of nature (Dilworth, 2008), and it treats intergenerational preservation as paramount. It is now generally agreed that these values are not mutually exclusive; they overlap to some degree, but according to Dilworth, each also can stand independently of the others. The term "sustainability" and adherence to its principles represent the historical shift from "a maximum economic use model," which considered resources to be merely raw materials for production and sinks for the disposal of waste (a purely anthropogenic view), to a more biocentric optimal model that recognizes the environment as a finite resource that needs to be conserved through public stakeholder involvement and governmental regulation, in order to create a long-term relationship between the economy and nature (Dilworth, 2008). "Sustainability" is also salient to the "land ethic" (including water and watershed) expressed by Leopold (2001), who emphasizes a balance between preserving nature and development.

The British government recognizes four objectives of sustainable development (DETR, 1999; Parkin, 2000):

- Social progress which recognizes the needs of everyone (intragenerational sustainability)
- Effective protection of the environment
- Prudent use of natural resources
- Maintenance of high and stable levels of economic growth and employment

The U.K. government's understanding of sustainability is only one of many found in the world literature. Nevertheless, it is now generally agreed that sustainability has three interacting dimensions: environmental, economic, and societal, and that sustainability can be achieved only when these components are balanced (Figures 2.1 and 2.2). A change in one component of the trinity of sustainability affects the other two components. If they are not balanced, the outcomes can be numerous—not only pollution and environmental degradation but also social injustice, or unsustainable development (Novotny, 2003). However, the Clean Water Act and federal court rulings following the passage of the CWA established that these components of sustainability are not equal; a highly beneficial economic development cannot result in degradation of the environment, even when standing ambient criteria are not violated. This is the principle of the *antidegradation* rules embedded now in most environmental regulations (U.S. EPA, 1994), and relaxation of ambient quality standards can only be allowed if meeting the standards would bring about "a wide spread social and economical hardship on the population" (U.S. EPA, 1994; Novotny et al., 1997). In the U.S., the "hardship" reasoning is extremely difficult to prove to the regulators and the courts.

II.2.4   Economic versus Resources Preservation Sustainability

Quoting Solow (1993), Howarth (2007) stated that sustainability could also be defined in terms of maintaining the utility or welfare of a typical member of society. Maintaining per capita utility may not require the conservation of a specific natural resource, as long as compensating natural or manufactured resources are provided to the future generation as a substitution for an exhausted nonrenewable resource. However, the substituted manufactured capital should not result in increase of the demand on other nonrenewable resources or increase of GHG emissions. For example, when oil is exhausted or becomes too expensive to extract, or its use will cause unbearable damages to the environment and society, automobiles fueled by electricity or biogas may be a satisfactory replacement for gasoline from oil, given the fact that such fuel would be cleaner with less carbon footprint than the current fossil fuels. Hence, an economic sustainability can be defined as one following the *net investment rule* (Pezzey, 2004; Howarth, 2007)

> A dynamic economy will maintain a constant or increasing level of per capita utility only if investment in manufactured capital exceeds the monetary value of natural resource depletion on an economy-wide basis.

This rule obviously is not applicable to water or water quality or ecology. Fish or shrimp that disappeared from a natural water resource as a result of pollution or economic overuse cannot be replaced by farmed seafood. Many authors have argued that investment in manufactured capital is necessary but not sufficient to guarantee sustainability (Howarth, 2007). Furthermore, Section 101 of the Clean Water Act in the U.S., tested and confirmed by court rulings, states that preserving and maintaining the integrity of the nation's waters is a supreme goal. This implies that, in an economic sense, maintaining healthy and balanced aquatic biota has an irreplaceable economic value that cannot be degraded because of economic benefits (antidegradation rule).

A new interdisciplinary field of sustainability science emerged in the 1980s and 1990s that seeks to understand the character of the intersection between nature, society, and economic development that is eminently pertinent to cities (Kates et al., 2001; Mihelcic et al., 2003) and their sustainability. Sustainability as an interdisciplinary science is just emerging; nevertheless, calls for sustainable development from the public, public officials, and the media have been heard since the 1990s and have intensified in the new millennium. People who are well informed can sympathize and align themselves with the movement towards sustainability, which, however, may not mean the same thing to different people. The cognitive values of sustainability can be related to:

- Preservation of the human race today and in the future
- Preservation of nature and restoration where nature is damaged
- Achieving good economic status for present and future generations
- Minimizing or eliminating risks to public health and providing healthy and green urban environments
- Integration of water resources management systems

Sustainability is evolving from the gradual merging of and discourse between population groups subscribing to two social views:

1. The anthropogenic view regards nature as a resource that should be used and developed for economic gains.
2. The biocentric view regards preserving and restoring nature as the ultimate goal for human beings.

Novotny (2003) pointed out that most people subscribe to both views; that is, they want to increase their living standard, yet they do not want to live in a polluted or severely damaged environment. Protection of the environment and public health is a cognitive value. They also want to preserve nature for future generations (Novotny et al., 2001). This adherence to two apparently contradicting principles could have been one of the reasons for urban sprawl under the fourth paradigm. Those who could afford to leave deteriorating cities with polluted air and water, to be closer to nature, might have participated in the economic activities in the cities that created the problem. This process of the affluent leaving degraded urban zones—and the poor—behind is a social problem known as *environmental injustice* which is rampant in shantytowns of some developing countries (see Figure 1.25) but can also still be seen in the U.S. and other developed countries (McGranahan and Satterthwaite, 2003). There may be other reasons besides pollution for such intragenerational injustice, such as crime or quality of education.

Mihelcic et al. (2003) also said that sustainability is not merely a preference for an economic development with some environmental protection (an anthropogenic development view) nor for preserving nature with "green" development (a biocentric view). It is a megascience defined as a design of human and industrial systems to ensure that the development and use of natural resources and cycles do not lead to diminished quality of life due either to losses in future economic opportunities or adverse impact on society, human health, and the environment. The time point of reference must be added to these concepts. It was shown in Chapter I and in Section 2.1 that each one of the previous four paradigms adhered to one or two values of the trinity of sustainability (Figure 2.2) in an unbalanced way that shifted the paradigm towards resource overuse and disregard of, or less emphasis on, nature. This imbalance can work in several ways. Adherence to anthropogenic economic principles without considering the environmental consequences may obviously lead to pollution, degradation, and ultimately loss of resources. On the other hand, strict adherence to purely biocentric views, not permitting the use of the resources and leading to strictly green "no impact" developments, will perpetuate the problem of urban sprawl by the affluent and urban environmental injustice for the urban poor or less fortunate. Sustainability implies that all values are in balance and symbiotically considered, such that the urban economic development will be carried out within the environmental and resource limitations given by ecological and GHG emission footprints (Rees, 1997).

Resilience, the Fourth Dimension of Sustainability   As referred to previously, since the publication of the Brundtland (1987) Report of the World Commission on Environment and Development, many authors have attempted to refine and improve the definition of sustainability. Mays (2007) presented several definitions of water resources sustainability that comply with Brundtland. A more general definition, for example:

> "Water resources sustainability is the ability to use water in sufficient quantities and quality from the local to the global scale to meet the needs of humans and ecosystems for the present and the future to sustain life, and to protect humans from the damages brought about by natural and human-caused disasters that affect sustaining life."

The above definition by Mays (2007) introduces a concern about water systems, and urban systems in general, and vulnerability to extreme events that may be magnified by global warming. This definition is more pertinent to the issues of the Cities of the Future because no matter how much social equity and sustainable economic development are considered, one extreme event of the proportion of the Hurricane Katrina in 2005 in New Orleans can devastate the urban area to a point that the entire viability of the city is disrupted or destroyed, and the intergenerational sustainability is irreversibly lost. Hence, urban areas in vulnerable zones must be resilient enough to survive extreme events. Resilience is the fourth dimension of sustainability. However, resilience of urban design is always related to the risk of failure, which, in turn, is related to the cost of protection. The magnitude of an acceptable risk has never been satisfactorily established.

II.2.5 Sustainability Components

Society The attitudes of people towards cities and water are linked first to self-preservation. People built fortified settlements to protect their water and their existence from enemies and to exchange goods in the settlement market, Without water there is no life, and if the availability of good water disappears through a change of hydrologic conditions or pollution or upstream overuse, civilization vanishes. The attitudes of people towards water and other natural resources are also impacted by their living standard and working hours. As more leisure time becomes available, people gravitate towards water resources for enjoyment, recreation, and living. In recent years, people have also become aware of the threats from global warming.

The societal criteria are based on environmental ethics and standards that are built on common sense, self-preservation, and the common desire to protect human health, the environment, and natural resources, in order not to harm the present and future generations and their natural surroundings. Since human beings are not always perfect and often suffer from weaknesses or are affected by wrong information, ethical (societal) norms must be incorporated into policies, legal rules, and laws (Leopold, 2001). Novotny (2003), following Rogers and Rosenthal (1988), identified 10 social policy imperatives related to the policies of diffuse pollution abatements. Similar imperatives will be presented throughout the subsequent sections of this chapter and throughout the book. The social/political imperatives on which the urban sustainability projects will be judged are:

- Preservation of the well-being of human beings
  - *Reduction of and adaptation to global warming.* Scientific observation and modeling of trends show that warming might be a serious threat to human beings in the short run (one or two generations) because of rising temperatures, more frequent extreme events with flooding, sea levels rising, and the disappearance of species. In the long run, the survival of the human race may be threatened in some parts of the world. Consequently, policies must be enacted that would lead to a rapid reduction of GHG emissions and adaptation to future anticipated global warming on local, regional, and global scales.
  - *Public health protection.* More than a century ago, society made a commitment and set a goal to eliminate epidemics, including those that are water- or airborne. This self-preservation cognitive rule has been a part of human nature since the beginning of civilization; however, it was not until the 19th century that people recognized the connection between epidemics and human health and water and air quality. The waterborne health threats include cholera (Figure 2.4), typhoid, cryptosporidium, diarrhea and other

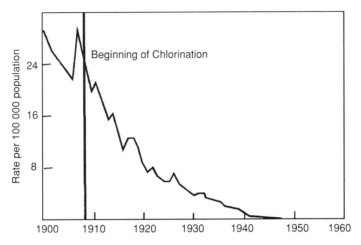

Figure 2.4   Death rate for typhoid fever in the United States 1900–1960 (*Source:* U.S. Centers for Disease Control and Prevention – 1999).

gastrointestinal diseases, and a plethora of diseases caused by water contamination from toxic and carcinogenic priority pollutants. Hence, no action by human beings should increase the risk of disease or death to present and future populations. The risks should be kept at the lowest level possible.

- *Sustainability cognitive values.* Society has made a commitment that the well-being of future generations must not be infringed by the current and past overuse of resources, or by exceeding the capacity of the environment to safely absorb and attenuate the emission and discharge of pollutants, or by inappropriate land use and modification. No actions should be permitted that would irreversibly harm the environment and natural resources or society. Society recognizes the right of both present and future generations to undiminished use of the environment, a balanced ecology and the enjoyment of nature.
- Social Policies
  - *Regulations and statutes.* Due to the failure of the general market, caused by externalities, to control the quality of the environment and reduce GHG emissions, there is a need for international and national regulations that must be clear, understandable, fair, and implementable. Such regulations will be derived by discourse and agreement of all parties involved in the process of protecting the environment and reducing the impact of global warming. Standards and criteria commensurate with the societal goals of sustainability must be developed, on which current and future actions in the environmental and economic domains will be judged.
  - *Acceptance.* There must be concurrence on the part of the people and groups being regulated that they will obey the regulations.

- *Reasonableness and/or avoidance of widespread social and economic impact.* The U.S. Clean Water Act, and regulations derived from it, recognize the fact that implementation of pollution control should not bring about undue widespread economic and social hardship. In the context of sustainability, this imperative should be extended to land use, traffic, and other human activities causing pollution, not just apply to discharges into receiving waters. However, preservation of the human race takes precedence over reasonableness.

II.2.6 The Environment and Ecology

Throughout most of human history, the environment was not recognized as a factor of importance to society; this did not happen until the middle of the 20th century. However, in the 19th century, during the Industrial Revolution, streams became heavily polluted to the point of nuisance and danger to public health. The first environmental attitudes were directed towards public health protection and the elimination of waterborne diseases, which often led to converting the polluted and unsightly urban rivers to combined sewers and culverts buried out of sight. In the second half of the 20th century, protection and propagation of aquatic life, addressing toxic contamination of water and land, and protection of recreational swimmers became the center of attention of the public and their representatives. Today, new concerns affecting policies have emerged, including the global warming issues, endocrine disruptors and pharmaceutical residuals, and cyanobacteria blooms. Scientists and the public realize the vulnerability and finite nature of energy (oil) and water resources, which led to discussions and calls for sustainability at the end of the 20th century.

The increasing concerns about the state of the environment in the second half of the 20th century and the forecasted impact of global warming have shifted the trinity of factors more towards the biocentric views and approaches. These views were first formulated during the 1940s by Aldo Leopold (1887–1949), a professor at the University of Wisconsin, who formulated the ethical standard for environmental protection and conservation (Leopold, 2001), as follows: "A thing is right when it tends to preserve the integrity, stability, and beauty of the biotic community. It is wrong when it tends to do otherwise." His ethical standard in a modified form was made a key rule of the Clean Water Act, which has a goal of restoring and maintaining the integrity of the nation's waters. Leopold, in his *Sand County Almanac*, extended the rule of environmental ethics as follows: "The land ethic simply enlarges the boundaries of the community to include soils, water, plants, and animals, or collectively: the land." Thus, the notion of land extends to general ecological terrestrial and aquatic systems that are intertwined, and both must be considered together and protected. Therefore, this books covers urban landscape and waters as one system. Leopold realized that "ethics of course cannot prevent alteration, management, and use of these 'resources' [i.e., environment] but it does affirm their right to continued existence, and, at least in spots, their continued existence in a natural state."

This philosophy was put into reality more than 50 years before Aldo Leopold's writing by the well-known urban landscape architect/civil engineer Frederick Law

Olmsted (1822–1903), who, in the second half of the 19th century, designed many urban parks, including Central Park in New York City, and lake parks in Chicago, Illinois, and Milwaukee, Wisconsin. He was given a commission in Boston to restore urban land around a small body of water, the Muddy River (Zaitzevsky, 1982). Olmsted approached the design by considering the river and the land surrounding it as one functioning system, called the Emerald Necklace (Figures 2.5 and 2.6). Olmsted's commission was to convert the polluted Muddy River marshlands into a linear river park bordering the cities of Boston and Brookline. He sculpted the landscape and river shorelines, increased velocity in the river by straightening the channel, and developed a new functioning ecosystem by means of a combination of ponds and channels (Figure 2.5). The park is a desirable place for recreation. The philosophy of park creation during the 19th century was to improve health and provide recreation to the urban population. The Emerald Necklace is not preserved nature; such park creation could be called "ecomimicking" (Novotny and Hill, 2007). However, the 20th century's third-paradigm urban development severely degraded the water and especially the sediment quality of the river, and the river and the park are now being restored at great cost.

Similar urban stream corridors with natural ecotones (a transition between the river and the urban built-in habitat) can be found in many other cities, notably the Menomonee and Milwaukee Rivers in the Milwaukee, Wisconsin, metropolitan area, and streams in Vancouver, British Columbia. More detailed discussion on renaturalization of urban streams will be presented in Chapters III and IX.

Olmsted created an ecological system within the urban environment. In some cities, patches of the original natural systems have been preserved (see also Chapter III). Thus, cities are a mix of the built infrastructure environment occupied by human beings and domesticated animals with open land that includes other organisms and wildlife. In this system, environment and ecology are interconnected. Ecology generally describes a system of living organisms (flora and fauna), including human beings, and their environments. Some ecologists (see Rees, 1999) have described urban ecosystems as assemblages of nonhuman species adapted to the structural and chemical characteristics of the area. In this concept people are implicitly involved, but only as agents potentially causing damage to which the nonhuman species are adapting. *Ecotones* in urban areas are transitional areas between the nonhuman ecosystem, such as streams, and the built environment. In a natural system, the transitions between two different ecosystems are gradual, and the ecotones buffer diffuse pollution, but in an urban system the transitions are sharp boundaries.

Bolund and Hunhammar (1999) have identified the following urban ecosystems: (1) street trees and green street drainage (ditches and swales), (2) lawns and parks, (3) vacant land, (4) urban forests, (5) cultivated land, (6) lakes and wetlands, (7) coastal areas of seas, and (8) streams and floodplains (riparian zones). Development has often eliminated some or all ecosystems and replaced them with impervious surfaces, including drainage of wetlands and putting streams underground. In order for the urban ecosystems to be functional and sustainable, the individual ecosystems must be interconnected to allow passage of animals over land and fish and other water organisms in streams, so they can populate and repopulate the area.

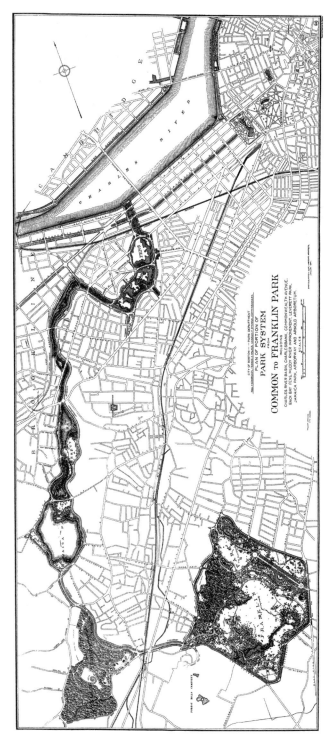

Figure 2.5 The plan of the Emerald Necklace. Olmsted designed a linear river and ecotone park that was ecologically functional. The park includes floodplain.

Figure 2.6 A scene in the Emerald Necklace in Boston/Brookline, Massachusetts.

II.2.7 Living within the Limits in the Urban Landscape

If pollution is to be controlled and then eliminated, people will first have to adapt to living with the effects of global warming and try to reduce them. Society has to switch to sustainability, and limits must be defined on human activities, production, and other processes that generate pollution, including GHG emissions from industries, commerce, and household processes and buildings, land use development, traffic, product safety failures, and so on. Many limits are cognitive, such as not throwing trash on a neighbor's land, some are included in legal doctrines (e.g., trespassing) or are religious (e.g., the Ten Commandments), but most are included in civic regulations and statutory laws. Limits have been developed from past bad experiences, but today they are based on science.

Kates et al. (2001) have expressed the need for limits in several questions aimed at the scientists working on defining sustainability:

- What determines the vulnerability or resilience of the nature–society system in particular places and for particular ecosystems and human livelihoods?
- Can scientifically meaningful "limits" or "boundaries" be defined that would provide effective warning of conditions beyond which the nature–society systems incur a significantly increased risk of serious degradation?

- What system of incentive structures—including markets, rules, norms, and scientific information—can most effectively improve social capacity to guide interactions between nature and society towards sustainability?

Other more technical questions can also be added such as:

- What are the margins of safety society needs, given the considerable uncertainty of current predictions of trends fifty to one hundred years in the future, when the brunt of global warming, the exhaustion of some current resources (oil), and population increases will be much worse than it is today?
- How to change century-long traditions of doing things as usual in an unsustainable way?

Urban areas may range from almost completely impervious zones with all water/stormwater/wastewater infrastructure underground to open "garden" cities and subdivisions that constitute ecologically and hydrologically functioning urban ecosystems. There are many cities throughout the world, especially in less-developed countries, that derive food and resources from nearby lands, and their impact on water and air is not damaging. Hence, one can entertain a notion of limited human and infrastructure capacity that can be supported permanently (Rees, 1999) without imposing damage on the environment. Percent imperviousness or percent build (urbanization) or urban population density and total numbers are the most obvious parameters to which the biotic, chemical, and physical integrity of urban areas can be correlated. For example, Schueler (1994) and others related the Index of Biotic Integrity of aquatic macroinvertebrates (see Chapter IX) to percent urbanization (Figure 2.7). To arrive

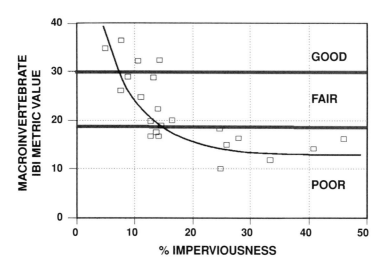

Figure 2.7 The simplified relationship of the macroinvertebrate Index of Biotic Integrity to the imperviousness of the area (Schueler, 1994).

at the value of this popular biotic index, biologists collect benthic macroinvertebrates in a predefined stretch of the stream and classify them into metrics. Each metric is then assigned a numeric ranking, and the total sum of the rankings then constitutes the index (see Barbour et al., 1999). Schueler has come to the conclusion that aquatic integrity is impaired when the imperviousness of the watershed is greater than 15%. Percent of imperviousness is a surrogate for many adverse stresses caused by urbanization and development (Field et al., 2000). Unfortunately, the percent imperviousness parameter is irreversible in most cases and, typically, can only increase with time. It would not be logical to argue that every watershed with more than 12% to 15% imperviousness was degraded, and that urban development should consist of low-density scattered subdivisions. This is an oversimplified conclusion, and it will be shown that this would not adhere to the reality of the Cities of the Future (see Chapter XI and discussions throughout this treatise). Furthermore, investigations of Yoder et al. (2000) in Ohio demonstrated that, due to many other stressors, a simple relationship of the biotic indices to a simple surrogate stressor—such as imperviousness or percent of some other land use parameter—may not exist. That was also emphasized in Karr and Chu (1999). Later research (Bedoya et al., 2009) revealed that the quality of the 30-meter riparian buffer along the streams and connectivity are far better indicators of habitat quality for aquatic organisms than percent imperviousness (Chapter IX). Better parameters will have to be developed (Novotny et al., 2005).

Rees (1997; 1999) suggested there is a need to enter the human component, especially when considering that the stresses generated by human beings also adversely affect the human beings themselves. It has been established beyond any doubt that society is under threat, that development as usual is not sustainable or acceptable in the long run. Experts must develop, and society must accept, ambitious yet reasonable/acceptable sustainability goals and the ensuing criteria. Since sustainability is integrated over several domains, limiting criteria may have to be multidimensional and include:

I. *Global climatic change* is a new (discovered only few decades ago) serious threat to the global ecosystem.
   a. Limits: Emission standards/criteria are needed to dramatically reduce GHG emissions; otherwise, scientists predict, severe social and ecological adverse effects will occur in the next fifty to one hundred years. Some serious warning signs have already materialized, such as the gradual disappearance of the Arctic ice sheet, melting of Greenland and mountain glaciers, increasing frequency of flooding, and increasing temperatures. The criteria (or goals) will be related to carbon footprint and may apply, among others, to:
      i. Urban infrastructure performance, e.g., reduction of energy use and GHG emissions, increasing green development
      ii. Transportation, e.g., fuel efficiency (kilometer per liter or miles per gallon) and GHG emission restrictions in the short run, and GHG elimination in the long run

      iii. Switching to non-fossil fuel energy production and manufacturing, such as developing wind, solar, geothermal, and nuclear energy

    b. Resilience of urban lands against the effects of extreme events, including coastal flooding. Limits are needed on development in vulnerable flood zones, restricting the impacts of extreme meteorological events fueled in the future by global warming.

II. *Protection of nature* has already become a goal under the fourth paradigm in the second half of the last century and includes:

    a. Protection of aquatic and terrestrial species as formulated in the U.S. in the Clean Water Act and the Endangered Species Act

    b. Restricting or eliminating emissions of harmful toxic substances into water and air, some of which, such as fluorocarbons, have global consequences (ozone hole, greenhouse effects)

III. *Public health protection* standards and criteria were also defined during the last decades of the last century. As the new threats related to global warming and population increases emerge, the criteria have to be periodically updated and upgraded:

    a. Criteria for the protection of the public against water- and airborne diseases

    b. Recreation protection criteria

IV. *Urban and suburban land use.* These traditional limits currently cover restrictions on minimum lot sizes, square-meter size of the houses, height of the buildings, septic tank installation, etc., which either have low sustainability relevance or are even counterproductive. Serious effort will have to be made by the communities and state regulators towards incorporating "smart," "green," and "low impact" development concepts that would cover both retrofitting of existing urban areas and new developments. These concepts may also be redefined:

    a. As pointed out above, highly impervious cities with little natural drainage and no green interconnected zones (for recreation and propagation of urban wildlife) are not sustainable, while some other "garden cities" might be. Urban planners and landscape architects must address and develop guidelines and quantitative criteria for sustainable urban zones that might include the ratio of green to built impervious areas, public transportation, buffer riparian zones along the urban streams, and connectivity.

    b. Drainage guidelines and criteria that would encourage renaturalization of urban drainage systems, and stormwater and used water reuse need to be developed.

Stresses that impair the integrity of aquatic and terrestrial ecosystems have been defined as pollution. Cities generate pollution in numerous forms, but almost all forms affect urban water resources and land in one way or another, within and outside of the city. In general, urban ecosystems have a limited resiliency towards pollution and other ecological stresses. Two examples of existing methodology and use of criteria are introduced herein and will be further discussed in pertinent chapters of this book. They are:

- *Total Maximum Daily Load (TMDL) planning process* (U.S. EPA, 2007) has been included in Section 303 of the Clean Water Act. This process is mandatory for water bodies that do not meet the standing water quality standards even after the mandatory effluent limitations are imposed on and implemented for point sources of pollutants. Maximum *loading capacity* of urban and downstream water bodies—the capacity to accept potential pollutant loads without impairing the integrity of the biota and the opportunities for primary and secondary recreation—is calculated using established numeric standards. The total maximum daily load is then estimated by applying a margin of safety to the calculated loading capacity, and the TMDL is then allocated to the pertinent dischargers causing the exceedance of the water quality standard. The TMDL process is mostly ineffective to deal with pollution that is not derived from pollutant discharges, such as channel straightening or burying, development in the riparian and flood zones, or excessive flow withdrawals. TMDL can handle pollutants for which numeric standards are available, or standards are available for surrogate cross-correlated pollutants (Novotny, 2003).
- Similar concepts have been implemented for emissions of pollutants into the air, such as those that result in acid rain (sulfur and nitric oxides). At the time of writing this book, restrictions on emissions of GHGs included in the Kyoto protocol were adopted by a majority of countries, but not by the largest emitters, the U.S., China, and India (see Chapter 1). "Cap and trade" legislation for GHG regulation by the beginning of 2010 was passed by the U.S. Congress but stalled in the U.S. Senate. The impact of cities on global warming is numerically calculated by *carbon footprint* (see Chapter VIII).

Environmental Policy Imperatives
- *Urban waters are the lifeline of the cities and when damaged or converted to underground storm and combined sewers they should be restored to a state that would be ecologically and hydrologically functional. The stream restoration includes the stream corridor.*
- *The concept of achieving and preserving the integrity of environmental systems, first included by the U.S. Congress in the Clean Water Act of 1972 to address the issue of the nation's water, should also include the integrity of the landscape and global atmosphere.*
- *Antidegradation: No action should bring about worsening of the quality of air, water, or soil environments that presently meet environmental standards.*

II.2.8 The Economy

Water is a commodity, but it is also public goods, and everyone must have a right to use water in a sustainable manner. At the Hague World Water Forum, two opposing views emerged and were discussed by the delegates regarding the economic value

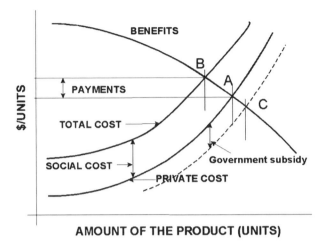

Figure 2.8 The classic economic concept of the benefits and private and social costs in a production process. An optimum (points A, B, and C) is reached at the intercept of the benefit and cost curves. Social cost should be included in the unit costs. Subsidies will alter the relationship.

of water (Allan, 2008). One political-economical attitude regards water, all water, as an economic resource; the other view regards it as a social resource. The former leads to the conclusion that water resources can be developed and engineered and economically managed by water-pricing instruments and even by privatization. During the centuries water resources have been developed for economic benefits by building reservoirs, making withdrawals for various uses, pumping, building dikes to reduce floodplain for development, burying surface streams to get land for development, and ultimately, disposing of waste. This attitude and the paradigm that nature and its water resources can be engineered and economically developed was shaped and driven by the economic development of societies, especially after the Industrial Revolution, and it lasted, in the developed countries, till the 1970s and still persists in the rapidly developing countries of Asia (China, India, the Middle East) and Latin America (Brazil).

Figure 2.8 shows the classic economic concept of the price, demand, benefits, and cost. The benefit curve shows that, as the availability of a commodity on the market increases, the unit market price of the commodity or the willingness to pay for it decreases. On the other hand, the cost of the commodity (product) increases with the volume, as the producer may be forced to look for more expensive raw materials and pay higher wages after the availability of cheap materials is exhausted and the cost of labor is increased. The optimum quantity of the product on the market is then determined by the intercept of the unit (marginal) benefits and costs. If the producer only includes his/her private cost, this optimum will be represented by the point A. However, making a product generate other costs such as pollution, which causes damage to the downstream uses of the receiving water body and additional cost to the downstream users. This causes the social cost (externality). Without regulation enforcement, the producers have no incentive to include the social cost in their reasoning, and the product will appear to be cheaper because the costs of those who have to cope with the pollution impact are not considered. With the cost of damage imposed on the producer (by a regulation and or tax on the product), the cost of the product will be increased, and a new optimum will be reached at point B. The amount of taxation should theoretically equal the amount of damage or recovery cost to society. This is the Pigouvian tax, so called to commemorate the economist who first proposed this concept. An alternative to the tax is a regulation that would force the polluters to implement pollution reduction on their premises, at their cost. Implementing this concept has another effect: Pollution is reduced because there is less demand for the more expensive product.

Ignoring externality has serious consequences, as it did in the Central and Eastern European (CEE) countries before the political transformation in 1989. The previous totalitarian regimes of CEE countries disregarded the economic principles of externality and its costs. For instance, acid rain caused by emissions from fossil fuel burning operations (power plants and traffic) damaged lakes and soils to the point that large areas of forest in Central Europe were lost. This also resulted in increased erosion and severe health effects on the population. Because this damage was not included in tariffs of electricity, electric power was "cheap" for consumers, which led to more pollution damage. In this way societies subsidized those using a lot of electricity and caused increased environmental and health problems that, when enumerated, were larger than the cost differential in the price of electricity. The cost of externality then becomes a *social cost* (Figure 2.8). The same reasoning applies to industrial and agricultural production.

The problem with sustainability is that it covers the entire society and not just large industrial operations or big water utilities. It reaches even to such odd consequences as those of the production of charcoal for heating and cooking in Congo in Africa, which not only generates small amounts of GHG emissions but also impacts the mountain gorilla's endangered population. There are obviously economic tools that can be used to spur the behavior of people and industries towards sustainability, involving both subsidies to encourage conservation and penalties for not meeting emission standards.

Marginal cost, benefits, and economy of scale. The classic economic theory of a free market states that the unit value of a benefit (product) decreases with the amount of the product on the market, while the unit cost of production may be actually increasing because the cost of raw materials, transport, and labor may be increasing due to exhaustion of nearby cheap resources and reliance on more distant sources, requiring higher costs of transportation and energy with a concurrent increase of GHG emissions. The marginal cost and benefit are the unit cost or the benefit to produce one additional unit of the product. For example, if in Figure 2.8 the current (existing) volume of production is A, then the marginal benefit and cost are the unit benefit and cost of producing one additional unit of the product, over the current level, A. The marginal pricing works well for private business, which has maximizing profit as its main goal.

The benefits can be both *tangible* (i.e., they can be expressed as monetary, such as lower expenses for treatment or increased fees for recreation and fishing) and *intangible* (such as enhancement of aesthetical values, improved recreation, or saving of nature). Costs are for capital expenditures and for operation, maintenance, and replacement (OMR) of the facilities and best management practices. Both benefits and costs are affected by the cost of borrowing the funds to implement the facilities and the opportunity cost, which is the value of benefits that would have accrued from spending the money elsewhere, instead of using the borrowed funds on building and operating the facility.

However, current large utilities operating and managing water/stormwater/ wastewater facilities and other urban-water-related infrastructure are designed and operate under mandated environmental goals and standards that must be achieved. According to the National Pollutant Discharge Elimination System (NPDES) and TMDL regulation, and sometimes court injunctions, if the mandated limits are not met, stiff penalties and fines may be imposed whose costs may exceed the cost of meeting the mandate. Consequently, the economics of operating these facilities is reduced to finding the design that would meet the environmental limitations at the least cost. Typically, the financing of large utilities is accomplished by issuing a bond offered to investors who receive interest payments, and the cost of borrowing must also be considered.

When only capital operation and OMR costs are considered, to some degree, the larger the facility becomes, the lower cost is. This is called the *economy of scale*, which in the 1970s led to the abandonment of many smaller treatment plants, which were replaced with long large interceptors and one or very few large regional treatment plants. Also when increased pollution of municipal wells and local water sources caused their decommissioning, some municipalities were looking for distant large sources and regional water treatment facilities. The result was a regionalization of the urban water and wastewater systems resulting in large costs of water and wastewater transfers (including clean water inputs). For example, in 2002, the large regional wastewater treatment plant in Fusina near Venice, Italy, was oversized but was operating near capacity because it was receiving infiltrated clean water and hence treating 25% dry-weather sewage flows and 75% clean water inflows that infiltrated or entered the sewer and interceptor system during rainfall and also due to a high groundwater. Because of the flat topography, the cost of pumping was very large.

In many older cities with legacy infrastructures, the switch to the new paradigm will be gradual. In this case marginal pricing will be employed because the replacement of the old infrastructure will start with the existing most costly component; for example, new nutrient, heat, and energy recovery facilities will replace old and very expensive to operate and maintain secondary (activated sludge) and tertiary treatments with high energy demand and chemical cost.

II.3 TOWARDS THE FIFTH PARADIGM OF SUSTAINABILITY

It is now generally agreed that present cities, their landscapes, and their water/stormwater/wastewater systems are not sustainable; many cities, especially in the developing world, cannot provide an adequate amount of water, and the water that is provided for a few hours a day is contaminated by cross-connections with sewers due to low pressure and damaged pipelines and severe environmental pollution. In the poorest nations, even the basic necessities such as sanitation are not provided. Even in developed countries, infrastructure built decades or even centuries ago is crumbling and will require massive investments for repairs. Traditional subsurface stormwater drainage can only handle smaller storms, with the recurrence intervals ranging from less than two years (e.g., Tokyo) to five years (standards for storm sewer designs in many U.S. and European communities); CSO overflows are allowed even with abatement ten or more times per year, which again is unacceptable to the public because after each storm beaches and swimming areas are closed, or are unsuitable for swimming at all times. Hence, these systems are not resilient to extreme events. In Tokyo and Osaka, Japan, large portions of the metropolitan areas are actually located in the floodplain, and city engineers cope with this fact by turning the yards of apartment buildings into flood storage basins, with a forebay and a warning system providing 15 minutes to the tenants to evacuate the yard when floods occur (personal observation by the primary author). The current unsustainable situation will be further exacerbated by:

- Population increases (urban population is expected to increase by 50% in the next 20–30 years; many new cities will be built in Asia and other parts of the world)
- Increasing living standards (more demand on food and, consequently, water resources)
- Global warming (increasing sea levels, changes in drought and water availability patterns) (ICPP, 2007)
- Emerging new pollutants (endocrine disruptors, pharmaceuticals residuals, more frequent massive cyanobacteria bloom outbreaks)
- Increasing water scarcity; currently about 0.7 billion people experience true water scarcity (they live on less than 25 liters per person per day), which is expected to grow to more than 3 billion people by 2025 if nothing is done (Zhang, 2007; Colwell, 2002)
- Conversion of urban waters into effluent dominated water, which will require management of the total urban water hydrological cycle and decentralization of the urban sewerage
- Increased flooding due to global warming effects, increased imperviousness, and other land use changes in the watershed
- Energy shortages because the world is running out of oil; production of biofuel from corn and other crops is driving food prices up

The new fifth paradigm of urban water/stormwater/used water management is a variant of the fifth paradigm of integrated water resources management (IWRM) described by Allan (2008) for water resources development. This paradigm is derived from the premise that water is an economic and social resource. Adoption or rejection of IWRM is closely related to the political system in which water users and policy makers operate, and poverty is the main impediment to the adoption of economically and environmentally sound water management (Allan, 2008). Treating water resources as an abundant social free good is a feature of the third paradigm, or current developed countries, and still is a paramount requirement of developing countries trying to catch up with the developed countries. However, today, it can be argued that even some developing countries, such as China and Singapore, have adopted IWRM concepts that have revolutionized their integrated urban water management and brought them to the forefront of world development, which may save these societies from destruction by pollution and bring them great societal and economic benefits (China). Singapore today must be considered a highly developed country and leader in the area of integrated water management and sustainability.

II.3.1 Emerging Sustainable Urban Water/Stormwater/ Used Water Systems

The concepts of the new paradigm of sustainable water centric ecocities have been emerging for the last 15 years in environmental research and landscape design laboratories in Europe(Sweden, Germany, the United Kingdom), Asia (Singapore, China, Japan, and Korea), Australia, the U.S. (Chicago, Portland, Seattle, Philadelphia, San Francisco), and Canada (British Columbia, the Great Lakes region). This paradigm is based on the premise that urban waters are the lifeline of cities and the focus of the movement towards more sustainable cities (Novotny, 2008), and its evolution ranges from the microscale "green" building, subdivision, or "ecoblock" to macroscale ecocities and ecologically reengineered urban watersheds, incorporating transportation, and neighborhood urban living as well. The new paradigm must include consideration of energy and greenhouse gas emission reductions, and must treat stormwater and reclaimed used water as a resource to be reused, rather than wasted (with high disposal costs). Therefore, the Cities of the Future will combine concepts of "smart/green" development and natural landscape systems with control of diffuse pollution and stormwater flows from the landscape. They will reuse highly treated effluents and urban stormwater for various purposes, including landscape and agricultural irrigation; groundwater recharge to enhance groundwater resources and minimize subsidence of historic infrastructure; environmental flow enhancement of effluent dominated and flow-deprived streams; and, ultimately, for water supply. The organic content and energy in used water will be treated as a recoverable resource along with reclamation and reuse of urban stormwater (Rittmann, Love, and Siegrist, 2008). The most obvious differences between the current and future paradigms are summarized in Table 2.1.

Mihelcic et al. (2003) focused on developing sustainability science in the water/wastewater field and emphasized that just focusing on green engineering, even with pollution prevention and industrial ecology, may not be sufficient to achieve sustainability because the material flow from these systems may still overwhelm the limiting carrying capacity of the ecosystem or may lead to unbalanced situations such as urban sprawl. Mihelcic et al. outlined the evolution from the environmental

Table 2.1   Comparison of traditional cities and Cities of the Future (fifth paradigm) concepts[1]

Traditional	Cities of the Future
Drainage: **Rapid conveyance of stormwater from premises by underground concrete pipes or culverts, curb and gutter street drainage**	Storage oriented: **Keep, store, reuse, and infiltrate rainwater on-site or locally, extensive use of rain gardens, drainage mostly on surface**
Wastewater: **Conveyance to distant downstream large treatment plants far from the points of reuse**	Local reuse: **Treat, reclaim, and keep a significant portion of wastewater locally for local reuse in large buildings, irrigation, and providing ecological low flow to streams**
Urban habitat infrastructure: **No reuse, energy inefficient, excessive use of water**	Green buildings (LEED certified): **Water-saving plumbing fixtures, energy efficient, larger buildings with green roofs**
Water, stormwater/wastewater infrastructure: **Hard structural, independently managed**	Local cluster decentralized management: **Soft approaches, best management practices as a part of landscape, mimic nature**
Transportation, roads: **Overloaded with vehicular traffic and polluting**	Emphasis on **less polluting fuel, urban renewal to bring living closer to cities, good public transportation, bike paths, best management practices to reduce water pollution by traffic**
Energy for heating and cooling, carbon emissions: **Energy (electricity, gas, oil) brought from large distances, no on-site energy recovery, high carbon emissions**	Energy recovery and reduction of use: **Part of the heat in wastewater will be recovered and used locally without carbon emissions, biogas production from organics in waste, fuel saving by people traveling shorter distances, use of geothermal, solar, and wind energy that reduces carbon emissions**
Overuse of potable water: **Drinking water is used for all uses (household, irrigation, street washing, fire protection), large losses in the distribution system**	Use of treated drinking water **from distant sources should be limited to potable uses only, reused water or water from local sources for other uses, reduced losses in distribution**
Economies of scale **in treatment cost and delivery are driving the systems—the bigger the better**	Triple bottom line pricing and life cycle assessment **of the total economic, social, and environmental impact**
Community expectation of water quality **distorted by hard infrastructure and past abuses such as buried urban streams, fenced-off streams converted to flood conveyance and/or effluent dominated**	Daylighting and/or renaturalization **of the water bodies with ecotones (parks) connecting them with the built areas enhances the value of the surrounding neighborhoods and brings enjoyment**
Low watershed resilience to extreme events, **underground stormwater conveyance can handle only smaller storms, infiltration is low or nil, fast conveyance results in large peak flows**	Surface drainage with floodplain ecotones, **in addition to storage and infiltration, dramatically increases the resilience of the watersheds to handle extreme flows and provide water during times of shortages**

[1] *Source:* Adapted from Valerie Nelson, unpublished document.

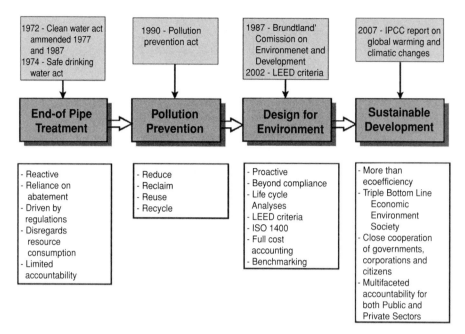

Figure 2.9 Adaptive progression of water management and urban pollution control from the end-of-pipe control to sustainable systems. Adapted and modified from Mihelcic (2003).

issues typical of the fourth end-of-pipe treatment paradigm to sustainable development (Figure 2.9). It could be noted that even in the countries that are making the fastest progress toward sustainability of some of their urban areas (Singapore, Sweden, China), the state of sustainable development is still more in design studios than a full reality.

Mihelcic et al. linked progress to the legislative acts that stimulated the activities. This indirectly says that society, by means of a discourse among the major groups of stakeholders and legislators, should decide on the goals. In a parallel process, science must provide the knowledge and support for these societal decisions. The process towards sustainability in a democratic society is stepwise and adaptive. For example, the focus of water pollution abatement after the passage of the Clean Water Act in 1972 was almost exclusively on point sources of pollution—that is, on polluted discharges from sewer effluents of cities and industries. The CWA included requirements for point sources to apply for discharge permits under NPDES permitting regulations, which also contained penalties for noncompliance. However, in less than a decade or so after the passage of the CWA and fast implementation of point source control goals, it became evident that these actions will not be enough to attain the goals of the Act, and it was realized that other sources must be included, such as urban stormwater. In 1983, the U.S. Environmental Protection Agency (EPA) conducted a scientific study of the pollution of urban runoff, the Nationwide Urban Runoff Project (NURP) (U.S. EPA, 1983), that gave the impetus to Congress to ask the U.S. EPA to issue regulations for the control of the pollution from urban runoff. In 1990 the Pollution Prevention Act was passed by the U.S. Congress, providing impetus for conservation, reclamation, and reuse of reclaimed stormwater and wastewater. In 1987, the Brundtland Commission report was published, making sustainability a global goal, and in 2007 the International Panel on Climatic Change made the need to drastically reduce carbon footprint and GHG emissions another global goal. The drivers for change towards sustainability have been presented in Chapter I.

The *fifth paradigm*, discussed at the Wingspread Workshop (Novotny and Brown, 2007; Novotny, 2008), offers a promise of adequate amounts of clean water for all beneficial uses. The new paradigm of sustainable urban waters and watersheds is based on the premise that urban waters are the lifeline of cities and the focus of the movement towards more sustainable and "green" cities. Summarizing the discussions at the Wingspread Workshop and their literature, the concepts of the new sustainable urban water management system and the criteria on which its performance will be judged include:

- Replacing the linear flow-through systems by the integration of water conservation, stormwater management, and wastewater disposal into one system managed on the principle of a closed-loop hydrologic balance concept (Figure 2.10) (Heaney, 2007)

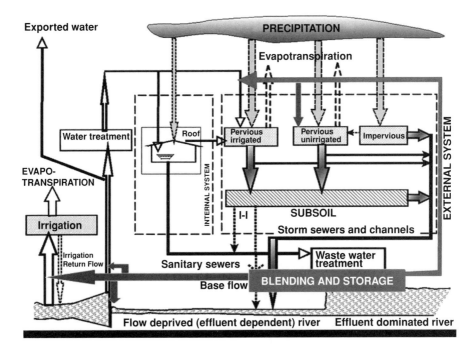

Figure 2.10 Total urban hydrologic cycle concept (adapted from Mitchell et al., 1996, and Heaney, 2007). The traditional fourth paradigm linear flow-through system will be replaced or retrofitted by a closed-loop system promoting and enabling conservation, water and energy reclamation, and reuse.

- Considering designs that reduce risks of failure and of catastrophes due to the effects of extreme events and that are adaptable to future anticipated increases of temperature and associated weather and sea level changes (IPCC, 2007);
- Incorporating **green buildings** (LEED certified) that will reduce water use by water conservation, and reduce storm runoff with best management practices (BMPs), including **green roofs, rain gardens, and infiltration**
- Incorporating heat energy and cooling water recovery from sewage in cluster water reclamation and energy recovery facilities (Engle, 2007)
- Implementing new innovative and integrated infrastructure for reclamation and reuse of highly treated effluents and urban stormwater for various purposes, including landscape irrigation and aquifer replenishment (Hill, 2007; Ahern, 2007; Novotny, 2007; LEED criteria (U.S. GBC, 2005, 2007))
- Minimization or even elimination of long-distance subsurface transfers of stormwater, wastewater, and mixtures (Heaney, 2007; Anon, 2008)
- Energy recovery from used water (wastewater); environmental flow enhancement of effluent dominated and flow-deprived streams; and ultimately a source for safe water supply (Anon, 2008)
- Striving for net zero GHG footprint by incorporating renewable energy sources into the system of the water–energy nexus
- Implementing **surface stormwater drainage** and **hydrologically and ecologically functioning landscape**, making the combined structural and natural drainage infrastructure and the landscape far more resilient to extreme meteorological events than the current underground infrastructure; the landscape design will emphasize interconnected ecotones connected ecologically with a viable interconnected surface water systems; surface stormwater drainage is also less costly than subsurface systems and enhances the aesthetic and recreational amenities of the area (Hill, 2007; Ahern, 2007)
- Considering the pollution-loading capacity of the receiving waters as the limit for residual pollution loads (Rees, 1992; Novotny, 2007), as also defined in the TMDL guidelines and documents (U.S. EPA, 2007); striving for zero pollution load systems (Metcalf & Eddy, Inc. 2007)
- Adopting and developing new green urban designs through new or reengineered resilient drainage infrastructure and retrofitted old underground systems interlinked with the daylighted or existing surface streams (Novotny, 2007)
- Reclaiming and restoring floodplains as ecotones buffering the diffuse (nonpoint) pollution loads from the surrounding human habitats, and incorporating best management practices that increase attenuation of pollution, such as ponds and wetlands (Novotny, 2007)
- Connecting green cities, transportation needs, and infrastructure with drainage and receiving waters that would be ecologically based, protec aquatic life, provide recreation, and by doing so be acceptable to and desired by the public
- Decentralizing water conservation, stormwater management, and used water treatment to minimize or eliminate long-distance transfer, enable water reclamation near the use, and recover energy (Heaney, 2007; Anon, 2008)

- Developing surface and underground drainage infrastructure and landscape that will:
  1. Store and convey water for reuse and provide ecological flow to urban flow-deprived rivers, and safe downstream uses
  2. Treat and reclaim polluted flows
  3. Integrate the urban hydrologic cycle with multiple urban uses and functions to make it more sustainable

II.3.2 Triple Bottom Line—Life Cycle Assessment (TBL—LCA)

Coined in 1994 by John Elkington, the expression "triple bottom line" (TBL) was introduced to expand the notion of sustainability from a largely environmental agenda to include social and economic dimensions. Elkington suggested that companies should be using three bottom lines: "One is the traditional measure of corporate profit—the 'bottom line' of the profit and loss account. The second is the bottom line of a company's 'people account'—a measure in some shape or form of how socially responsible an organization has been throughout its operations. The third is the bottom line of the company's 'planet' account—a measure of how environmentally responsible it has been" (Economist.com, 2009). In the simplest terms, the triple bottom line agenda encourages corporations to focus not just on the economic value that they can enhance, but also on the environmental and social values that they can enhance—or degrade (Elkington, 1997, 2001).

Hence, sustainability should also be evaluated using the "triple bottom line" (TBL) criteria which include (1) Environmental/ecological protection and enhancement, (2) Social equity, and (3) Economics (Anon, 2008; Brown, 2007; Novotny, 2008). Figure 2.11 is an illustration of the TBL concept. Using the TBL approach over the life cycle of the systems (40 or more years), it should be logically expected that ecocities built according to the fifth paradigm will outperform the current urban developments. To evaluate resiliency to extreme events in a TBL-LCA, consider: (1) flood-causing precipitation, (2) water shortages, and (3) extreme pollution. All three are affected by global warming; therefore, the TBL-LCA must consider emissions of GHGs. Research should increase understanding of how the new integrated urban drainage, water management, transportation, and resource systems work during times of stress, between stresses, and after stresses, and of how they impact population and respond to population increases and other socioeconomic stresses. Urban drainage systems must be clearly reengineered to become more sustainable and resilient to increased stresses.

The TBL-LCA methodology has been used by industries, but its use in the ecological domain is still evolving. While the economic and to some degree environmental sides of the assessment can build on the traditional analyses, the societal and ecological components are being researched. For one thing, societies in developed countries have already decided through discourse in the political and societal arena (e.g., the Clean Water Act in the U.S. and the Water Framework Directive in E.U. countries) that the components of the TBL are not equal—that is, the integrity of the

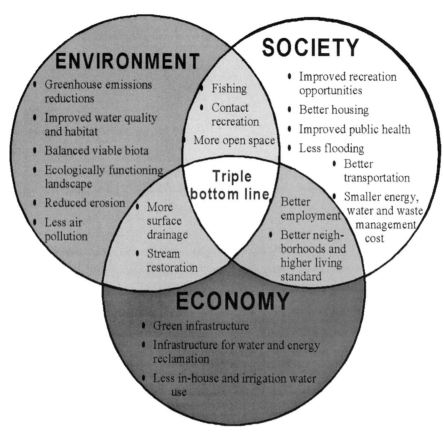

Figure 2.11 Triple bottom line assessment concept. Adapted and modified from Elkington (1997).

water resources cannot be compromised to increase economic interests (antidegradation rule). All aspects of society are dependent on underlying ecosystem services. Nevertheless, to consider the tangible and intangible benefits and costs in the TBL we have to convert them into a common denominator, which is monetary (see Chapter X). Ecology and economics are thus inextricably linked. Very often it also means restoring the ecological function of the system that was damaged by the past third and fourth paradigm developments that focused only on economics. If either system is unhealthy, social systems fail. Both private and government sectors must be assessed and, obviously, must cooperate in a well-defined regulatory framework (Mihelcic et al., 2003). To some degree, in water resource development, the concept of the triple bottom line is not new, and the trinity of criteria have been incorporated (in a slightly different form) in the water resources development guidelines (Maas et al., 1962). Hence, TBL concepts and criteria are built on the previous economic concepts by incorporating social and environmental costs into the total cost of producing goods, and accounting for all social and ecological tangible and intangible benefits of preserving and improving the environment/ecology and providing aesthetic and recreational amenities to the population in the intergenerational context, which is sustainability.

II.3.3 Water Reclamation and Reuse

Integrated resource management concepts view urban treated effluents as a resource, not as waste. As a matter of fact, the trend today has been to take "waste" from "wastewater" and replace it by "used," creating a "used water" resource which can be reclaimed and used for various purposes: phosphorus needed for fertilizers can be recovered, and—because effluents, especially in sanitary separate sewers, maintain a relatively constant temperature—heat and cooling energy can also be extracted (Barnard, 2007). Effluent reuse for irrigation, even in an incompletely treated (or untreated) form, has been practiced in some countries for decades, sometimes for more than one hundred years. Examples include large-scale effluent irrigation in water-poor regions of China (e.g., the Beijing region), Mexico, India, the U.S. (e.g., Tucson), and Israel. Some irrigation systems in developing countries use untreated effluents, and the irrigation practice is a substitute for treatment. For example, 34 m^3/s of untreated wastewater from Mexico City irrigates more than 100,000 ha of agricultural land up to 60 kilometers away from the city, with some serious groundwater and surface water contamination consequences and public health concerns because farmers live near the irrigated fields and the watercourses bringing the raw sewage to them (Scott, Zarazua, and Levine, 2000). Treated effluent from Tucson, Arizona, irrigates golf courses and city parks. Use of a highly treated effluent for potable uses was attempted several decades ago in Namibia, in Africa, and is being seriously considered in Los Angeles and other Southern Californian cities (see Chapter VI). An important part of water reclamation, reuse, and energy recovery is flow separation into:

- *Black water* containing fecal matter that contains most of the biodegradable organic matter that can be converted to biogas; pathogens; and also water from kitchen sinks with grinders (comminutors)
- *Urine (yellow) water* that contains most of the unoxidized nitrogen and about half of the phosphorus from human-used water (wastewater) in less than 1% of the total flow (without flushing)
- *Gray water* containing discharges from laundry, bath, and kitchen containing nonfecal organic solids from kitchen dishwasher, soap, and detergents, and some pathogens from showers and baths
- *White water* which consists of surface street and highway runoff containing most of the toxic, sometimes carcinogenic, compounds such as metals, PAHs, petroleum hydrocarbons, oil and grease, salt, cyanides, and nonhuman pathogens

- *Blue water* which is clean water that may enter the drainage systems from clean infiltration-illicit inflows (I-I) and rainwater from illegally connected roof downspouts

Each stream contains different reusable resources such as that providing clean water for recycling (gray, white, and blue); fertilizer recovery (yellow and black); biogas and organic fertilizer/soil conditioner (black); irrigation (treated black, yellow, white, and blue); and raw water for water supply (white and blue). The highly concentrated supernatant from digestion of sludge, other organic solids, and leachate from landfills is a highly valuable resource from which fertilizer struvite (magnesium ammonium phosphate hexahydrate) can be extracted. The advent of smaller often packaged and automated treatment (water reclamation plants, or WRPs), providing high-quality effluent based on membrane bioreactors and filters (Chapter VII) enables the implementation of a distributed and safe water-reclamation-near-the-points reuse, which can be a high-rise building, commercial area, subdivision, one or several city blocks, or a small suburban or even rural community. Reclaimed water for irrigation could retain most of its nutrient content, but the remaining pollutant should be at a level that would not impair the integrity of surface and groundwater resources, and be safe for human contact. In addition to irrigation, the reclaimed high-quality effluent can be used for:

- Toilet flushing in buildings
- Street flushing and washing of infrastructure
- Flow augmentation to provide ecological flow to streams that have lost their base flow due to excessive upstream withdrawals and hydrologic modification of the watershed by urbanization
- Cooling
- Groundwater recharge for indirect potable and nonpotable reuse

The present technologies that are being developed, tested, and put on the market also provide (Rittmann, Love and Siegrist, 2008) (see Chapters VII, VIII and X):

- Biogas produced by anaerobic decomposition of organic solid wastes
- Hydrogen (H_2) gas produced by fermentation of organic materials in special microbial fuel cells or as an intermediate product of digestion in hydrogen fuel cells
- H_2 gas can be used as fuel for a conventional chemical fuel cell, which produces combustionless, pollution-free electricity
- Direct electricity production in microbial fuel cells
- Heat and cooling energy recovered by heat pumps from warmer effluents
- Energy supplements by tapping into geothermal sources, and wind and solar energy

Compact treatments providing high BOD, suspended solids, nutrients, and pathogen removals are available, ranging from serving a few houses to populations of up to 20,000. These units provide effluents that could be as clean as the receiving waters into which they may be directed (Furumai, 2007; Barnard, 2007). Ultimately, potable water quality is achievable (Barnard, 2007); however, direct potable reuse is still not recommended (Chapter V). Today, distributed small-scale treatment plants can be installed in neighbourhoods, in the basements of shopping centers buildings (see Figure 10.3 in Chapter X) or large high-rise . Energy can be recovered from both wastewater solid residues (sludge) and organic solid wastes (Anon, 2008). Currently, methane produced by landfills or anaerobic wastewater treatment plants is often flamed out without reuse.

Chapters V, VI, and VII present the most advanced and efficient water, fertilizer, and energy reclamation schemes.

II.3.4 Restoring Urban Streams

Urban streams and lakes in many cases spurred city development by providing hydropower for mills, navigation, water supply, flood conveyance, fishing, and recreation. Because of excessive pollution and demand for development land, at the end of the 19th century, pollution of urban streams became unbearable and urban surface water bodies began to disappear from the surface by being converted into underground storm and combined sewers or placed in culverts (Figure 1.13. Those water bodies that stayed on the surface lost the floodplain and riparian habitat through development. Because of the changed hydrology from imperviousness due to urbanization, floods increased, and the capacity of the streams was no longer sufficient to handle them. Cities responded to this by lining the surface streams with concrete or masonry and converting them into lifeless flood conveyance channels, often fenced off to prevent public access (see Figures 1.16, 1.17, and 2.12 left).

Today, along with the cleanup of urban runoff and separation of combined sewers, stream restoration and daylighting (bringing a buried stream to the surface) projects

Figure 2.12 Lincoln Creek in Milwaukee, Wisconsin. Left, channelized before restoration; right, after restoration. Photos by V. Novotny.

are being carried out in many cities. It does not make sense to use sewers to carry cleaner treated runoff. Restoration of urban streams is only possible after the major point sources of pollution, including CSOs and SSOs, have been eliminated. It is a complex process that begins with the identification of the cause of impairment (impaired habitat, insufficient base flow, and erosive high flows), followed by implementation of best management practices to control the stormwater flow and pollution inputs, removal of lining, restoration of natural sinuosity, pool and riffle sequence and habitat restoration, removal of stream fragmentation (bridges, culverts, channel drops, and small dams impassable to fish and other aquatic organisms), and riparian (flood) zone restoration (Novotny, 2003). Habitat degradation is the primary cause of the impairment of the integrity of urban streams (Manolakos et al., 2007; Novotny et al., 2008).

The major reasons for and benefits of restoration and daylighting are:

- Water, fertilizer, and energy reclamation from municipal used water (wastewater) appears to be inefficient if the reclamation unit is located many kilometers downstream, and the reclaimed water with or without fertilizers would have to pumped back to the city for reuse. It may be more efficient to install smaller compact reclamation units closer to the points of reuse of both reclaimed water and energy. It would also make a lot of sense to use some of the reclaimed flow to improve base flow conditions in the existing restored or daylighted streams.
- Bringing the streams to the surface provides larger capacity to handle flows (Chapter IX).
- Reclaimed and renaturalized floodplain with storage ponds, wetlands, and buffers provides treatment and attenuation of runoff from surrounding areas and storage of excess flows, and it has many other uses such as wildlife habitat, parks, and recreation (Chapters III and IV).
- The stream corridor is an ecotone that provides ecological and hydrological connectivity needed to sustain aquatic and terrestrial wildlife and provide public recreation and enjoyment.
- Restored streams are universally known to bring great economic and revitalization benefits (Lee, 2004).

Figure 2.12 shows the stream restoration in Milwaukee, Wisconsin. Restoration is still more or less an art, rather than a science. Restoration of streams damaged by urbanization—often to the point of conversion into underground sewers—should be a key component of green development. Today, raw sewage inputs into surface streams, or underground culverts carrying the buried streams have been or are being eliminated, and the buried streams are becoming storm sewers (with insufficient capacity to handle flows from extreme storms, in most cases). The restored and daylighted streams will become technically a part of the surface drainage system, but they should be ecologically viable and functioning, pleasing to the public, and able to provide recreation as well as enjoyment. Surface drainage is also more resilient to flooding, as documented in the case of the buried Stony Brook branch under the

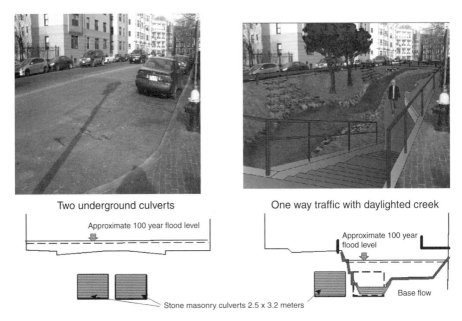

Figure 2.13 Proposal for daylighting of the historic Stony Book buried under the streets of Boston about one hundred years ago on the Northeastern University campus. The photo on the left shows the current situation; on the right is the daylighting proposal by the Capstone Design student project. The left culvert will carry sewage flows with a portion of stormwater runoff. The channel on the right will carry a portion of the "clean" Stony Brook.

campus of Northeastern University in Boston (Figure 2.13) (see also Chapter 1 for a brief history of Stony Brook). Today, most of the buried Stony Brook is not a combined sewer anymore; it carries relatively clean water from an upstream nature reserve and stormwater from the city. One of the key requirements of daylighting and urban stream restoration is to provide and recreate good base flow that can be from natural sources (springs, wetland), if available, or created or supplemented by highly treated effluent from nearby high-efficiency treatment plants or stormwater runoff stored in ponds, wetlands, and recharged shallow aquifers. Base flow of urban streams has been lost because of the high imperviousness of the surrounding water-shed and shallow groundwater infiltration into sanitary sewers, basement dewatering into sanitary sewers, and leaks into other underground infrastructure (underground garages, subway and freeway tunnels). More discussion on urban stream restoration and daylighting will be covered in Chapter IX.

II.3.5 Stormwater Pollution and Flood Abatement

Since the late 1970s scientists and urban planners have been developing and implementing best management practices (BMPs) for controlling pollution and peak flow of urban runoff. Prior to 1970, urban runoff was considered clean and a "diluter" of more concentrated, often untreated, point source pollution. Sewer separation or building underground storage basins and tunnels were general solutions to the problem. An extensive U.S. Environmental Protection Agency (1983) study, the National Urban Runoff Project (NURP), disputed this policy and found that urban runoff contains unacceptable concentrations of pollutants, including extreme concentrations of pollutants from de-icing chemicals in winter flows, such as salinity, sodium, chlorides, metals, and cyanides (Novotny et al., 1999), and, in the nonwinter runoff, suspended solids, oil and grease, COD, pathogens, toxic metals, and organics. BMPs to control diffuse pollution developed and implemented in the last 30 years can be categorized as (Novotny, 2003):

1. Source control measures (control of atmospheric deposition, reduction of urban erosion, especially from construction; street sweeping; switching from irrigated lawns using large quantities of fertilizers to nonirrigated xeriscape)
2. Hydrologic modification focusing on infiltration (porous pavements, landscape infiltration, infiltration trenches)
3. Reduction of delivery (silt fences at construction sites, buffer strips, grass swales, in-line solids separation in sewers)
4. Storage and treatment (wetlands, ponds, underground storage basins with a follow-up treatment)

The BMPs listed above can be divided into structural (hard) and nonstructural (soft) (Chapter IV). Most structural BMPs implemented until the end of the last century were "engineered" and did not blend with the natural environment, nor did they try to mimic nature. Since one of the requirements of sustainable development is to restore and protect nature, most of the structural BMPs are not considered sustainable, nor are they appealing.

Landscape architects (Ahern, 2007; Hill, 2007) have proposed that the BMPs listed above also be divided into:

• Those that remedy landscape disturbance and emission of pollutants
• Those modifying the landscape and the hydrologic cycle to make them more ecologically and hydrologically sustainable
• Those that remove pollutants from the flow

Developers and landscape architects at the end of the last century realized that BMPs can be an architectural asset that can blend with nature and mimic natural systems. Almost every structural engineered BMP has a natural-looking, hydrologically and ecologically functioning, and nature-mimicking equivalent (Figure 2.14).

With the exception of the source control measures mentioned above, in the past, BMPs were designed and implemented a posteriori—that is, after pollution had been generated from the land. BMPs provided treatment, and use as drainage was secondary. The typical drainage design preference of the fourth paradigm was to divert urban runoff and snowmelt collected by street gutters and catch basins from impervious road and parking surfaces into underground conduits (storm sewers).

Figure 2.14 Landscaped swale providing infiltration and pollutant removal (photo from Marriott, 2007).

Subsequently, the sewer outlets were connected to a pond or a wetland or—directly, without any treatment—to a receiving water body. Traditionally designed geometric ponds were intended to attenuate the peak flows and provide some removal of pollutants, but their ecological worth was minimal.

At the end of the last millennium, the "green movement" began to change BMPs from relatively unappealing in appearance, with little or no ecologic value, to attractive and desirable assets of the urban landscape. Hence, mowed grass ditches, swales, and dry detention ponds were converted to rain gardens and bioretention facilities (Chapters III and IV). Now it is being realized that BMPs are not only additions to the drainage, but can be the drainage itself, in a modified more attractive form (Novotny, 2007). Best management practices can:

- Mimic nature
- Provide and enhance surface drainage
- Repair unsustainable hydrology by reducing flooding and providing enhanced infiltration, and provide some ecological base flow to sustain aquatic life, as well
- Remove pollutants from the ecological flow
- Provide water conservation and enable water reuse

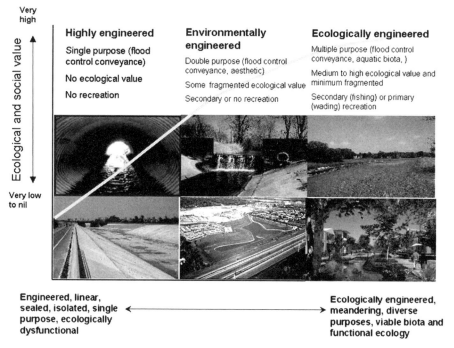

Figure 2.15 Engineering approaches to urban drainage from traditional to eco-engineering (adapted from Ahern, 2007).

- Buffer and filter pollutants and flow for restored/daylighted streams
- Enhance recreation and the aesthetic quality of the urban area
- Save money and energy (expensive underground conduits and pumping may not be needed; swale-type rain gardens combined with green roofs and permeable pavements for parking lots and some streets may dramatically reduce the need for underground storm sewer capacity and reduce energy use)

Ahern (2007) and Lucey and Barraclough (2007) have pointed out the differences between the traditional (civil) engineered and ecologically engineered components (Figure 2.15). Ecological engineering is becoming a new engineering discipline needed for the paradigm shift towards sustainable ecocities.

II.3.6 Urban Landscape

Landscape ecologists (e.g., Forman, 1995; Forman et al., 2003; Ahern, 2007; Hill, 2007) have proposed an ecologically balanced urban landscape with a river or a chain of urban lakes as a centerpiece. Based on these concepts, the urban landscape of the future will be made of interconnected ecotones preserving or imitating nature, threaded through the inhabited space with the river corridor. In addition to supporting biota and preserving nature and hydrology, the ecotones will also attenuate pollution coming from surrounding urban areas. In most cases they will contain the floodplain and provide storage of floodwater during extreme events.

Connectivity refers to the degree to which a landscape facilitates or impedes the flow of energy, materials, nutrients, species, and people across the landscape, and it is an emergent property that results from the interaction of landscape structure and functions, including flow, nutrient cycling and maintenance of biotic diversity (Ahern, 2007). Connectivity of urban ecotones and water systems is needed to provide conditions for sustainability of the aquatic biota and terrestrial ecology. If the biota is disturbed or lethally impacted by a stress (e.g., toxic spill), the biotic system can be repopulated by migration from neighboring unaffected ecotones. In urban systems, fragmentation of ecosystems—that is, separation of the ecology into isolated landscape elements—is a common feature of the landscape and aquatic systems (Figure 2.15). Water flow connectivity and water systems are the primary examples where connectivity is important to maintain sustainable and balanced aquatic biota. Connectivity must be considered on a watershed scale and must also include flood plains—that is, the entire water body corridor and contributing watershed areas. See Chapter III for a more detailed discussion.

II.4   CITIES OF THE FUTURE—WATER CENTRIC ECOCITIES

II.4.1   Drainage and Water Management

The time has come to critically evaluate what the developments during the last 25 years in the field of urban drainage and diffuse pollution abatement, green city concepts, and come up with a new approach to drainage that would mimic nature and the predevelopment hydrology. Other trends can also be considered such as reduced or eliminated GHG emissions from vehicles and improved public transportation, which also reduces urban/highway pollution. Table 2.2 presents the components

Table 2.2   Components of 21st-century urban water/stormwater/used water management (adapted from Daigger, 2008)

- Water conservation
- Distributed stormwater management
  - Low-impact development
  - Rainwater harvesting and rain gardens
  - Mostly surface drainage
- Distributed water treatment
- Water reclamation and reuse in buildings, for irrigation, and for ecologic stream flow
- Heat and energy recovery from used water and waste organic solids
- Organic management for energy recovery
- Source separation
- Nutrient recovery

and features of the new 21st-century urban water/stormwater/used water sustainable systems. Some concepts also consider organic farms surrounding the cities and significant reduction of nonpoint pollution from farms supplying food to the cities.

The new drainage will make a switch from strictly engineered systems (sewers) to ecologic systems (rain gardens, new and restored surface wetlands and ponds, and daylighted water bodies). Most municipal sewage management is expected to be decentralized rather than linear and regionalized. At some point surface drainage containing BMP elements must become a sequence of ecotones connected to the major receiving water body (Hill, 2007; Ahern, 2007; Novotny and Hill, 2007). Chapters III and IV cover the concepts of sustainable landscape and BMPs. The emerging urban landscape incorporating aesthetical and desirable BMPs (rain gardens, grass filters, wetlands, ponds, and also green roofs) will:

- Dramatically reduce the needs for hard infrastructure pollution controls and combined and storm sewers
- Mitigate pollution by urban runoff and reduce flooding
- Enhance the aesthetic value of the landscape

How to mimic nature in the Cities of the Future? The natural drainage systems begin with ephemeral small vegetated channels and gullies. At some point several of these channels will form a first-order perennial stream. A second-order stream is formed when several first-order streams join together. Springs and wetlands feed and provide perennial flows to natural streams. It is possible to do the same in urban areas, but then it would be called integrated best management practices. Table 2.3 presents a comparison of natural systems and their BMP equivalents. In urban areas perennial

Table 2.3   Natural systems and their equivalent BMPs (Novotny, 2007)

Natural Systems	Nature Mimicking Best Management Practices
Watershed with infiltration	Pervious pavements, green roofs with French well or rain garden infiltration of downspout excess water
Ephemeral pre-stream channels	Rain gardens, buffers, sand filters connected to landscaped swales or dry storage ponds for floodwater
First-order perennial streams with base water flow from:	Daylighted, restored, or created streams with base flow from:
	Groundwater infiltration, including dewatering basements
Springs	Decentralized high-efficiency treatment plant effluents
Headwater wetlands	Restored or created wetlands
Headwater lakes	Wet detention ponds with stored stormwater
Second-order streams	Restored original streams with reclaimed floodplains and riparian wetlands; floodplain converted to recreational park and buffer zones; storage in lakes and ponds in the reclaimed floodplains
Third- and higher-order streams	Removal of channelization and impoundments wherever possible, providing flood storage; significant portion of flow may originate from upstream nonurbanized areas

base flow can be provided by high-quality effluents from the cluster treatment plants, as has already been done in Tokyo and elsewhere. This sequence is also covered in Chapter X.

II.4.2 Microscale Measures and Macroscale Watershed Goals

LEED Criteria The U.S. Green Building Council has proposed and is developing standards for "green" buildings and neighborhoods (U.S. GBC 2005, 2007) that are becoming standard for building and development. For example, each federal-, state- and city-owned building in Chicago, Illinois, is expected to comply as close a possible with the LEED (Leadership in Energy and Environmental Design) standards, install a green roof wherever possible, and implement water conservation. Green roofs reduce runoff and provide substantial savings on energy use, which again reduces greenhouse emissions. New green tall buildings are being showcased in New York and elsewhere. Most consultants and city planners try as best as they can to adhere to LEED concepts and standards (U.S. GBC, 2005, 2007). "Green" subdivisions and satellite cities are now sprouting up throughout the world and in the design studios of landscape architects. The concept and designs of "Ecocities" with up to several hundred thousand inhabitants are now being implemented in, Sweden, Singapore, China, Australia, United Arab Emirates (Masdar), and elsewhere.

The U.S. GBC standards for "green" certification were formulated for homes, neighborhood development, and commercial interiors (http://www.usgbc.org). The new construction and reconstruction standards (U.S. GBC, 2005) include the following categories:

- *Sustainability of the sites* such as site selection and development, brownfield development, transportation, and stormwater design
- *Water efficiency* in landscape irrigation, innovation in wastewater technologies and reuse, and water use reduction
- *Energy and atmosphere*
- *Material and resources* such as construction materials and waste reuse and recycling
- *Indoor environmental quality*
- *Innovation and design*

Under the pilot LEED Neighborhood Rating System (U.S. GBC, 2007), additional categories are:

- *Smart location and linkage*, which include, among others, required indices of proximity to water and wastewater infrastructure, floodplain avoidance, endangered species protection, wetland and water body conservation, and agricultural land conservation

- *Neighborhood pattern and design* such as compact development, diversity and affordability of housing, walkable streets, transit facilities, access to public spaces, and local food production
- *Green construction and technology*, essentially LEED building certification
- *Innovation & design process*

This comprehensive list of standards is a potpourri of many "good sense" ideas. The LEED standards are aimed at buildings and small neighborhoods. They are not a priori related to natural resources, and the value (total number of points) for natural resource protection and water resources conservation is relatively small; only about 15% of the points are credited for reducing water use and for potential contribution to improving the integrity of waters and natural resources. There are no credits for restoration of water bodies or wetlands as a part of the neighborhoods. A maximum of two points is available for implementing sound stormwater management strategies and diffuse pollution controls. The standards were developed by volunteers of various nongovernmental organizations and developers. It is becoming clear that the scientific basis of ecological sustainability has not been sufficiently incorporated into the LEED standards.

ISO Environmental Performance Criteria ISO 14001:2004 provides the generic requirements for an environmental management system for industries and communities. The system follows the 1992 Rio protocols and is used in EC countries. It does not lay down levels of environmental performance; hence, the standard can be implemented by a wide variety of organizations, whatever their current level of environmental maturity. Meeting ISO criteria provides evidence that the system is environmentally sound and complies with environmental regulations. At this time ISO criteria have not been applied to watershed management but could provide another measure of the impact on environment, in addition to the traditional environmental risk assessment used in the U.S.

Low Impact Development (LID) LID concepts—covered extensively in Chapters III and IV—are used in and restricted to subdivision-size developments that practice mostly on-site stormwater containment, storage, infiltration, and conveyance. The LID approach selects "intergraded management practices" which are distributed small-scale controls that closely replicate the predevelopment hydrology. The goal is to achieve the highest efficiency or effectiveness at approximating the predevelopment conditions (Oregon State University et al., 2006). LID goals are aimed primarily at control of urban runoff, and water conservation and other aspects of "green" development contained in LEED or ISO criteria are not a priori considered. However, there is no such thing as "no impact" development. LID developments often are situated in rural settings with very high open/built space ratios, which could imply long-distance travel and urban sprawl.

Best Management Practices LID practices can be considered as a subgroup of a more general category of best management practices (BMPs), known by this term in North America or as Sustainable Urban Drainage Systems (SUDS) in the United Kingdom. Many manuals have been developed and published on the BMPs and SUDS, their design and implementation. It should be noted that the BMPs category is broad; BMPs deal with diffuse pollution caused by precipitation and other causes and are not focused only on drainage (Novotny, 2003; Oregon State University et al., 2006). Chapter IV will cover the new outlook, description, and designs of the BMPs.

Missing Links – Macroscale (Watershed-Wide) Goals   Architects, builders, developers, local governments, and consultants are pushing for implementing "sustainable" and "green" infrastructure, land, and resources development. The LEED index with its metric is a well-meant step towards better developments and more sustainable urbanization. These microscale LEED criteria are aimed at individual buildings and small subdivisions and commercial developments; ISO standards and LID criteria also are applicable only to small neighborhoods and commercial establishments. However; what is the impact of LEED certified and similar developments and infrastructure? Their impact on the sustainability of water resources, water quality, increasing resilience against extreme events (such as floods or catastrophic storms), and the protection and enhancement of natural terrestrial resources has been fuzzy at best, and some developments could be found irrelevant at worst, when macroscale (for example, watershed-scale) hydrological and ecological goals and impacts are considered. The development of the Cities of the Future, the ecocities, requires a comprehensive and hierarchical macroscale approach to the microscale and often fragmented piecemeal transformation (Hill, 2007) of the current unsustainable urbanization into the new eco-friendly and sustainable urban areas, and finally entire cities. There is a strong rationale for integrating urban water management concepts into the ecocities concepts, and vice versa. The convergence of efforts to improve the quality of life in urban communities and the campaign to improve our water quality offer potential synergies that could overcome the often confrontational encounters that can occur between environmental regulation and economic development.

The macroscale goal of the fifth paradigm is to develop an urban watershed and its landscape that mimics, but not necessarily reproduces, the processes and structures present in the predevelopment natural system, and to reduce GHG emissions. The goal should also include protection of the existing natural systems. Eco-mimicry includes hydrological mimicry, whereby urban watershed hydrology imitates the predevelopment hydrology, relying on reduction of imperviousness, increased infiltration, surface storage, and use of plants that retain water (e.g., coniferous trees).

The macroscale goals of the fifth paradigm for water centric communities are:

- Developing an urban watershed and its landscape that is sustainable and resilient over the long run and mimics, but not necessarily reproduces, the hydrologic processes and ecological structures present in the predevelopment natural system

- Protection of the natural systems and restoration of the natural drainage (daylighting)
- Mimicking predevelopment ecology and hydrology, relying on reduction of imperviousness, increased infiltration, surface storage, and use of plants that retain water (e.g., coniferous trees)
- Developing or restoring interconnected green ecotones, especially those connected to water bodies, that provide habitat to flora and fauna, while providing storage and infiltration of excess flows and buffering pollutant loads from the surrounding inhabited, commercialized, and urban areas with heavy traffic. (Hill, 2007; Ahern, 2007)
- Adaptation to the trends of global warming and the stresses caused by increasing population—for which it is not enough to be carbon neutral, i.e., keep the emission at the present level; the new development must reduce carbon emissions and increase resources to accommodate anticipated urban population increases
- Retrofitting and reconnecting old underground systems interlinked with the daylighted or existing surface streams

The macroscale goals should be evaluated by a watershed or citywide summation of the TBL-LCA impacts of the clusters and intracluster components (infrastructure, BMP, water, green developments and retrofitting, stream restoration and daylighting, and water reclamation).

One Planet Living (OPL) Principles   The World Wildlife Fund (WWF, 2008) has developed and is promoting ecocity principles that include social and technological metrics under the name of *One Planet Living*. Some ecocity developments are now aiming at OPL certification (e.g., Masdar, Sonoma Mountain Village—see Section II.5). These criteria for ecocities are far more broad and stringent than LEED or LID criteria. OPL criteria are as follows:

- Zero carbon emissions with 100% of the energy coming from renewable resources
- Zero solid waste with the diversion of 99% of the solid waste from landfills
- Sustainable transportation with zero carbon emissions coming from transportation inside of the city
- Local and sustainable materials used throughout the construction
- Sustainable foods with retail outlets providing organic and or/fair trade products
- Sustainable water with a 50% reduction in water use from the national average
- Natural habitat and wildlife protection and preservation
- Preservation of local culture and heritage with architecture to integrate local values

- Equity and fair trade, with wages and working conditions following the international labor standards
- Promotion of health and happiness with facilities and events for every demographic group

II.4.3   Integrated Resource Management Clusters—Ecoblocks of the Cities of the Future

At the onset of the fourth paradigm in the 1970s, in some regions of the U.S., many medium-sized suburban communities had or built local small treatment plants and also had local private and/or municipal wells, including water treatment facilities. The wastewater treatment consisted mainly of trickling filters or aerated (facultative) lagoons, requiring minimum supervision, less energy, and almost no chemicals. Some small as well as large communities in Europe had primary treatment only (e.g., Prague) or no treatment at all (Vienna, Brussels, Milan).

When the mandatory effluent standards based on the Best Available Treatment Economically Achievable (BATEA) were issued and the National Pollution Discharge Elimination System (NPDES) permits were enacted to implement and enforce the standards, communities realized that the old treatment would not meet the standards and that implementing new treatment required relatively rapid action. Because the cost was the only issue and optimization parameter, the economy of scale led the communities to opt for large regional facilities consisting of long and deep interceptors, deep tunnels, and large regional treatment plants. The water/stormwater/wastewater systems became linear and separated—that is, water brought to the city from larger distances is used within the city, converted to wastewater collected by sewers, and, with rainwater, discharged into interceptor tunnels and conveyed to treatment plants and untreated or partially treated overflows. The management of these systems is typically done by separate agencies managing and operating water treatment and distribution, wastewater conveyance and treatment utilities, and, in communities with separated sewer systems, local stormwater management districts.

The integration of the complete water management system under the fifth paradigm—which includes water conservation and reclamation, storage of reclaimed water and stormwater for reuse, used water treatment, and energy from waste recovery—cannot be achieved in a linear system that incorporates long-distance transfer, underground subsurface and deep tunnels, and distant wastewater treatment plants. The concept of clustered distributed and decentralized complete water management has been evolving (Lucey and Barraclough, 2007; Heaney, 2007; Daigger, 2009). An integrated resource management cluster (IRMC) system should be developed with reuse of the reclaimed water and energy reclamation, which would also minimize GHG emissions when compared to the traditional linear long-distance transfer systems.

An IRMC is a semiautonomous water management/drainage unit that receives water, implements water conservation inside the structural components of the cluster

Figure 2.16 Architect images and realities of the water centric urban landscape (Courtesy Patrick Lucey, Aqua-Tex Scientific Consulting, Victoria, BC).

and throughout the cluster, reclaims sewage for reuse (such as flushing, irrigation, and providing ecological flow to restored existing or daylighted streams), recovers heat energy from used water, and possibly recovers biogas from organic solids (Figure 2.17). The energy reclaimed from wastewater, other organic solid residues (food waste, vegetation), and possibly from stored stormwater can be supplemented by wind, solar, and geothermal energy. The concept enables privatization (Rahaman and Varis, 2005) and commercialization (e.g., selling reclaimed water, energy and biogas), although privatization and commercialization are not key prerequisites. Clusters may range from a large high-rise building, larger shopping center, or a subdivision, to a portion of a city (Furumai, 2007; Lucey and Barraclough, 2007). *For these reasons, the term"wastewater" has become obsolete, and it is being generally replaced by "used water"—water that is available for reuse and resource recovery.*

The size of the cluster and the number of people it serves must be optimized. Even today, many cities have multiple wastewater management and water reclamation districts within their borders. For example, Chicago (Illinois), Minneapolis–St. Paul (Minnesota), and Los Angeles (California) have four regional treatment plants; Beijing (China) had in 2010 sixteen and is building more. However, in most cases the cost of reclamation is not favorable. The cost is represented by the cost of transporting wastewater and stormwater mixture towards a treatment plant, its treatment and water reclamation, and transporting the reclaimed water back to the city for reuse on landscape, for toilet and street flushing, and for recovering energy. Benefits include fees for the recovered water, nutrients and energy, fees for accepting organic solids, savings on the size and length of sewers, savings on energy due to installation of

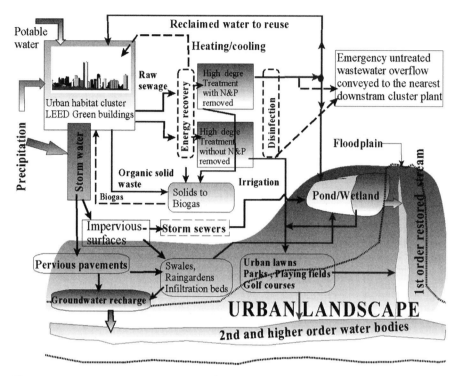

Figure 2.17   A concept of the integrated resource management cluster (IRMC) of sustainable water/stormwater and wastewater management with water reclamation and reuse.

green roofs and less pumping, benefits related to the recreational use of restored and daylighted streams, and so on. The size and distance of transfer matter. The longer the distance, the more costly water and used water transfers are, and the less revenue can be derived from water, nutrient, energy, and biogas recovery. It is quite possible that cluster stormwater/wastewater management based on LID concepts, water conservation, and reuse can make the deep and large interceptor sewers and tunnels obsolete. Furthermore, bringing stormwater conveyance to the surface can make existing sewers oversized, and the freed space can be used for other underground conduits such as fiber-optic cables and phone cables, for which the water management utility can charge a fee, as is being done in Tokyo and other cities.

China's, Singapore's, UK's and Sweden's programs of developing the concepts, planning, and building several ecocities have brought into the forefront concepts developed by landscape architects with international cooperation. Indeed, China s may be the places where these concepts could be implemented on the largest scale, and soon. For example, the College of Environmental Design of the University of California, working on the concepts for the new cities of Qingdao and Tianjin in China, developed the concept of an ecoblock (see Chapter XI), which is a self-contained and self-sufficient (in terms of energy), area of the city that is carbon-footprint-neutral. Most or all used water would be recycled on-site; energy generation would also be on-site, and any energy generated on-site from waste, sun, and wind would be used to treat rainwater and gray water and provide residents with high-quality potable drinking water. Even food waste and landscaping waste are proposed to be converted into energy to power residents' homes. The Qingdao ecoblock envisiones a 600-unit habitat with the water and energy recovery infrastructure encompassing an area of about 3.5 hectares (Fraker, 2008). China established joint ventures with Sweden to cooperate on the development of the new city Cao Fei Dian, and with Singapore for the development of Tianjin (see chapter XI). Singapore is a worldwide center for the development and testing of the most advanced technologies and infrastructure for used water and stormwater capture, reclamation, and reuse.

In the ecoblock, constructed wetlands and swales would collect and treat water for reuse, serving the dual purpose of enhancing the aesthetic value of each neighborhood and creating green waste that can be transformed into energy within an on-site anaerobic digester. Ecoblocks are expected to use less energy than standard developments of the same size, and in well-designed cases would be energy self-sufficient and carbon neutral (Fraker, 2008).

II.4.4 Interconnectivity of Clusters—Spatial Integration

Each integrated resource management cluster or ecoblock may provide complete management of water delivered from a regional water supply system or reclaimed within the cluster from stormwater and used water. In this way, the urban hydrological water cycle will be closed. However, landscape ecologists and planners (e.g., Forman, 1995; Forman et al., 2003; Ahern, 2007; Hill, 2007) have emphasized the need for an ecologically balanced urban landscape with a river or a chain of urban lakes as a centerpiece. Based on these concepts, as pointed in out in Section II.3.6 and in Chapter III, the urban landscape of the future will be made of interconnected landscape ecotones in the river corridor threaded through the inhabited space and imitating or preserving nature. Connectivity of urban ecotones and water systems is needed to provide conditions for sustainability of the aquatic biota and terrestrial ecology. If the biota is disturbed or lethally impacted by a stress (e.g., toxic spill), the biotic system can be repopulated by migration from neighboring unaffected ecotones. Fragmentation is the opposite of connectivity. Fragmentation in urban environments is caused by roads (Forman et al., 2003), culverts and drops impassable by fish and other larger organisms, zones of poor water and sediment quality, and/or by high temperature due to cooling water discharges.

Another need for interconnectivity is for the safety of the water reclamation facilities and the prevention of raw sewage overflows into the surface flow conveyance of preserved and restored or daylighted streams. Although the IRMCs are semiautonomous and integrated in water, sewage, and energy recovery management, they should be interconnected to increase resiliency against the failure of the cluster operating system, namely its WRP. In the case of failure, there should be an option to store and send the untreated used water to the nearest cluster WRP that has available capacity. Consequently, an online real-time optimization and control cyber infrastructure will have to be developed. Then a regional real time control (RTC) center will monitor online the flow and mass of water and its quality through the system and may also monitor other flows, such as people and traffic movement through the system, or meteorology. RTC will make short-time (hours to days) predictions of the status of the system and observe and resolve disturbances such as unwanted mass accumulation of water and pollutants at bottlenecks, potential water quality standards violations, or the need for de-icing operations on roads that would not infringe on water quality. If a bottleneck situation occurs anywhere in the system, RTC will resolve the overload by sending the excess to the nearest cluster with excess capacity, or finding a storage capacity within the system.

II.5   ECOCITY/ECOVILLAGE CONCEPTS

The first definition of an ecocity was coined by Richard Register (1987) and could be paraphrased and expanded as:

> A sustainable city or ecocity is a city or a part thereof that balances social, economic, and environmental factors (triple bottom line) to achieve sustainable development. A sustainable city or ecocity is a city designed with consideration of environmental impact, inhabited by people dedicated to minimization of required inputs of energy, water, and food, and waste output of heat, air pollution (CO_2, methane), and water pollution. Ideally, a sustainable city powers itself with renewable sources of energy, creates the smallest possible ecological footprint, and produces the lowest quantity of pollution. It also uses land efficiently, composts and recycles used materials, and converts waste to energy. If such practices are adopted, overall adverse contributions of the city to climate change will be none or minimal below the the ecosystem safe assimilatiob capacity. The cities of the future will contain energy and water frugal (green) infrastructure, resilient and hydrologically functioning landscape, protected and restored interconnected natural features and water bodies within their zones of influence. Urban (green) infrastructure, resilient and hydrologically and ecologically functioning landscape, and water resources will constitute one system.

Table 2.4 shows the degree of decentralization and cluster management of various components of the future hydrologically and ecologically functioning ecocities. Summarizing the concepts as they are emerging, it may be expected that the ecocities paradigm will have the following components:

I. Sanitary sewage conveyance mostly underground but decentralized

High-efficiency treatment (water reclamation) plants located so that they can provide reclaimed flow for (1) reuse in buildings (toilet flushing, on-site energy recovery, cooling, etc.) and/or (2) ecological base flow to perennial streams, and/or (3) park and golf course irrigation. Hence, decentralized urban used water management could be organized into (1) clusters of one or several large (high-rise) buildings, (2) one or more subdivisions, or (3) smaller urban districts (Figure 2.17). The quality of the effluent should be commensurate with

Table 2.4 Centralized and decentralized components of the future ecocities (Adapted from Daigger, 2009)

Component	Centralized	Distributed/Decentralized in Clusters
Stormwater management, rainwater harvesting		BMPs – pervious pavements, rain gardens, green roofs, surface and subsurface storage, infiltration basins and trenches
Water conservation	Reducing or replacing leaking pipes, systemwide education of citizens about water conservation, dual water distribution (potable and nonpotable)	Wide variety of commercial water-saving plumbing fixtures and technologies for potable and nonpotable use; changing from lawns to xeriscape
Treatment	Treatment for potable use and nonpotable reuse	Treatment for local potable use (from local wells and surface sources) and nonpotable reuse (from used water) in small cluster size water and energy reclamation units; stormwater treatment in biofilters, ponds, and wetlands; effluent post-treatment in ponds and wetlands
Energy recovery	Methane from anaerobic treatment and digestion of residual organic solids, thermal microbial fuel cells, electricity from methane by hydrogen fuel cells	Capture and distribution of heat and cooling energy (heat pumps); geothermal, wind, and solar energy; small-scale biogas production by digestion
Nutrient recovery	Land application of biosolids, struvite (ammonium magnesium phosphate) precipitation and recovery	Irrigation with reclaimed water with nutrients left in it; reclaimed irrigation water distribution to parks, golf courses, and homeowners
Source separation	Treatment of black wastewater and organic solids with energy (biogas) production	Supply potable and nonpotable water; treatment of black, gray (laundry and kitchen), and yellow water for nonpotable reuse (irrigation, toilet flushing)
Landscape management	Habitat restoration; fish management and restocking, wildlife management in ecotones	Stream and ecotones maintenance, installation and maintenance of BMPs, including ponds and wetlands; on and off water recreation

the purpose of reuse. In the effluent reuse for irrigation, nutrients (nitrogen and phosphorus) should not be removed. For effluent used to provide base flow, a high-quality effluent with removed nutrients and pathogens is desirable. Removed nutrients can be converted to biofertilizer and reused, and heat can be extracted from the effluent. In this way, treated effluent and extracted nutrient fertilizer (see Chapter VII) become a commodity that can be commercially distributed (e.g., Milorganite fertilizer produced by the Milwaukee Metropolitan Sewerage District, Wisconsin). New and better methods of water, energy, and fertilizer reclamation will be developed by new "green" industries and utilities.

Energy recovery from used water (see Chapter VIII). The temperature of urban sewage/used water is warmer than that of water supply, due to the addition of warm water from households and cooling water from industrial operations. Depending on geographical location, the mean annual temperature of urban sewage/used water varies between 10 and 20°C (Metcalf & Eddy, Inc., 2003). Both cooling and heating energy can be recovered by heat pumps and other similar energy recovery units, still to be developed, without emitting carbon dioxide. In winter, the energy needs could be supplemented by geothermal energy sources in groundwater. Groundwater typically has a stable temperature around 12°C.

Electric energy can be recovered from methane biogas in digestion of sludge and other organic solid residues by combustion of the biogas or by converting it into hydrogen and electricity in hydrogen fuel cells, or it can be recovered directly by microbial fuel cells (Chapter VIII) .

Valuable nutrients (nitrogen and phosphorus) can be retrieved in the integrated resource recovery facility from the concentrated supernatant in the form of struvite or by separating urine (yellow water), which contains 75% of the phosphorus and 50% of the nitrogen in 1% of the total used water flow (Chapters VII and VIII).

II. Surface drainage for stormwater and treated effluent discharges

1. *Ephemeral swales landscaped as rain gardens (see also Chapter III).* On side streets, low- to medium-density urban zones, less frequently traveled urban highways, and parking lots, in combination with pervious pavement, no storm sewers would be needed. The swale/rain gardens will be designed to have minimum (to prevent standing water and development of unwanted cattails and other vegetation) and maximum (to prevent erosion and gullying) slopes and engineered flow capacities. Stormwater runoff from impervious roads and streets would be filtered by grass or sand filters. Rainwater from downspouts would be directed to French wells or other infiltration devices for infiltration and/or to rain gardens.

   Flow from storm sewers, if installed, should be treated by various best management practices available for treatment (filters, ponds, etc).

2. *First-order perennial drainage channels – streams.* In older cities the original first-order streams disappeared and were converted into sewers. In new

planned communities, they should be preserved. As soon as perennial flow becomes available from reclaimed effluents, from stored rainwater (in subsurface man-made basins), from groundwater pumped from basements, or from wetlands, smaller natural or natural-looking channels (sinusoidal, with pools and riffles) should be created (see Chapter IX), or the original streams should be preserved or restored. Hydrologically, the channels and landscape could be designed with the channel capacity to hold a two-year flow, considering also flood storage capacity, and the extended channel with vegetated banks to hold flows with a large recurrence interval. Landscape should be resilient to floods with a one-hundred-year recurrence interval. Storage ponds and/or wetlands may be included to create water parks and enhance the landscape. The purpose of the ponds and wetlands in the first-order stream systems is to store excess peak flows for longer times (not 24 hours or less, as in conventional designs) so that the stored water can be used for irrigation, supplementing base flow, and other purposes, and also to provide post-treatment of effluents discharged into them. Created wetlands are the best place for receiving treated effluents; however, the problem with methane emissions from wetlands should be considered. Most first-order streams may not have natural base flow unless they originate in a nature reserve within the city.

Some ponds on the first-order streams may be stocked with fish but may not sustain a large quality of less-tolerant fish species. Surface urban runoff not infiltrated through the pervious surface (vegetated areas and porous pavement) will be filtered by grass or sand filters or, if storm sewers are used in dense settlements, by storm separators, filters installed in sewers, and other stormwater treatment units.

3. *Second- and higher-order streams*. These larger streams should sustain balanced viable fish population. Since these streams will consist mostly of preserved original or daylighted and restored streams, the pollution control laws in many countries will call for attaining and maintaining "a balanced indigenous aquatic biota" (in the U.S.) or achieving and preserving the "best ecologic potential" (In EU countries) of the water body. The streams should be surrounded by buffer zones encompassing the flood zone. The buffer and flood storage zones should be landscaped as interconnected parks, nature reserves, with bike and walking trails, and picnic areas. Recent research on the integrity of receiving water has been discovering the beneficial role of ecological green riparian zones surrounding the water bodies (Novotny et al., 2007).

The differences between the second-, third- and higher-order streams are primarily in the origin of the flow they receive. Second-order streams receive flows primarily from the first-order water bodies located within the urban area. Third- and higher-order streams carry a significant proportion of flow originating from outside nonurbanized areas.

Streams, straightened and/or channelized with lining, may have to be restored, with the lining removed and the channel renaturalized. Lakes on these streams would be part of the park and the overall urban ecosystem.

Long-distance wastewater transfers and large effluent discharges into second- and third-order streams should be minimized or avoided completely. The most preferable discharge location of effluents from cluster water reclamation plants is into the first-order wetlands and/or polishing ponds.

Ecocities or ecovillages are now emerging on subdivision or urban levels in reality, and on large city levels (up to 500,000 people) in planning, with some already in construction (see Chapter XI). China is looking for urban housing for up to 300 million people in the next 30 years because of intensification of agriculture (loss of jobs of indigenous population) and a large increase of GNP being derived from industries in the cities. Essentially, it is a planned attempt to manage the migration from rural to urban areas that has been so devastating in several other fast-developing countries, including Brazil, Mexico, India, and the like. An ecocity on a large scale is still a vision, but the realities are fast emerging in Sweden, the U.S., the United Kingdom, China, Masdar (UAE), Saudi Arabia, Canada, Germany, Australia, Japan, and elsewhere. Chapter XI presents several current ecocities already built or under development. Figure 2.18 shows Hammarby Sjöstad, a part of Stockholm in Sweden built on ecocity concepts.

In developed countries, the movement towards ecocities is based on the realization that the limits of the current paradigm have been reached, population will be

Figure 2.18 Hammarby Sjöstad in Stockholm (Photo courtesy Malena Karlsson, GlashusEtt, Stockholm).

increasing, the technology (e.g., high-level treatment) is available, new architectural diffuse pollution controls are functioning and desired by the public, the intensity and frequency of catastrophic storms will be increasing and must be mitigated, and the population desires these). On the other hand, in spite of a lot of interest and work being done in academia, until the first decade of the new millennium, the progress in the US was a piecemeal approach mostly by individual developers or some agencies trying to use technologies that had not yet been developed and scientifically tested. In the U.S., progressive developers focused on LID concepts. However, now that the progress is moving at a rapid pace in some countries, soon, hopefully, it will be seen on a worldwide scale, including the megapololi of the developing world and the retrofitting of the older cities.

REFERENCES

Ahern, J. (2007) "Green infrastructure for cities: The spatial dimension," in *Cities of the Future: Towards Integrated Sustainable Water and Landscape Management* (V. Novotny and P. Brown, eds.), IWA Publishing, London, UK

Allan, J. A. (2005) "Water in the environmental/socio-economic development discourse: Sustainability, changing management paradigms and policy responses in a global system," *Government and Opposition* 40(2):181–199

Allan, J. A. (2008) "Millennial water management paradigms: Making Integrated Water Resources Management [IWRM] work," http://www.mafhoum.com/press/53aE1.htm (accessed June 2008)

Anon. (2008) *Resource from Waste – Integrated Resource Management Phase I Study Report*, Prepared for B.C. Ministry of Community Services, Victoria, BC, Canada

Asano, T, F.L. Burton, H.L. Leverenz, R. Tsuchihashi, and G. Tchobanoglous (2007) *Water Reuse – Issues, Technologies, and Applications, Metcalf & Eddy/WECOM*, McGraw Hill, New York

Barbour, M. T., J. Gerritsen, B. D. Snyder, and J. B. Stribling (1999) *Rapid Bioassessment Protocols for Use in Streams and Wadeable Rivers: Periphyton, Benthic Macroinvertebrates, and Fish*, 2nd ed., EPA-841-B-99/002, U.S. Environmental Protection Agency, Washington, DC

Barnard, J. L. (2007) *Elimination of eutrophication through resource recovery*, The 2007 Clarke Lecture, National Water Research Institute, Fountain Valley, CA

Bedoya, D., E. Manolakos, and V. Novotny (2009) *Biological Response to Environmental Stress. Environmental Similarity and Hierarchical, Scale-dependent Segregation of Biotcv Signatures for Prediction Purposes*, Technical Report #16, Center for Urban Environmental Studies, Northeastern University, Boston, MA, www.coe.neu.edu/environment

Bolund, P. and S. Hunhammar (1999) "Ecosystems services in urban areas," EcologicalEconomics 29(20):293–301

Brown, P. (2007) "The importance of water infrastructure and the environment in tomorrow's cities," in *Cities of the Future: Towards Integrated Sustainable Water and Landscape Management* (V. Novotny and P. Brown, eds.), IWA Publishing, London, UK

Brundtland, G., ed. (1987) *Our Common Future: The World Commission on Environment and Development*, Oxford University Press, Oxford, UK

Burkart, M. R. and D. E. James (1999) "Agricultural-nitrogen contributions to hypoxia in the Gulf of Mexico," *Journal of Environmental Quality* 28(3):85, 0–859

Centers for Disease Control (1999) "Achievement in Public Health, 1900–1999, Safer and Healthier Foods," *MMWR - Weekly* 48(40):906–913, http://www.cdc.gov/mmwr/preview/mmwrhtml/mm4840a1.htm

Colwell, R. (2002) "A Global Thirst for Safe Water: The Case of Cholera," The 2002 Abel Wolman Distinguished Lecture, The National Academy of Sciences, http://www.nsf.gov/news/speeches/colwell/rc020125abel.htm

Czech Academy of Sciences (2006) *Proceedings of the 5th International Conference on Reservoir Limnology and water Quality*, Institute of Hydrobiology, Ceske Budejovice and Centre for Cyanobacteria and their Toxins, Masaryk University, Brno, Czech Republic

Daigger, G. (2009) "Evolving Urban Water and Residuals Management Paradigms: Water Reclamation and Reuse, Decentralization, Resource Recovery," *Water Environment Research* 81(8):809–823

DETR (1999) *A Better Quality of Life: A Strategy for Sustainable Development of the UK*. CM 4345, The Stationery Office, London

Dilworth, R. (2008) "Measuring sustainability in infrastructure: The case of Philadelphia," *Drexel Engineering Cities Initiative (DECI) Working Paper Series, 2008-01*, Drexel University, Philadelphia, PA

Economist.com (2009) "Management idea: Triple Bottom Line," November 17, 2009, http://www.economist.com/businessfinance/management/displaystory.cfm?story_id=14301663 (retrieved on December 16, 2009)

Elkington, J. (1997) *Cannibals with Forks: The Triple Bottom Line of 21st Century Business*, Capstone Publishing Ltd, Oxford, UK

Elkington, J. (2001) *The Chrysalis Economy: How Citizen CEOS and Corporations Can Fuse Values and Value Creation*, Capstone Publishing Ltd, Oxford, UK, pp. 1–16

Engle, D. (2007) "Green from top to bottom," *Water Efficiency* 2(2):10–15

Field, R., J. P. Heaney, and R. Pitt (2000) *Innovative Urban Wet-weather Flow Management Systems*, TECHNOMIC Publishing Co., Lancaster, PA

Forman, R. T. T. (1995) *Land Mosaics*, Cambridge University Press, Cambridge, UK

Forman, R. T. T. *et al.* (2003) *Road Ecology: Science and Solutions*, Island Press, Washington, DC

Furumai, H. (2007) "Reclaimed stormwater and wastewater and factors affecting their reuse," in *Cities of the Future: Towards Integrated Sustainable Water and Landscape Management* (V. Novotny and P. Brown, eds.), IWA Publishing, London, UK

Heaney, J. (2007) "Centralized and decentralized urban water, wastewater & stormwater systems," in *Cities of the Future: Towards Integrated Sustainable Water and Landscape Management* (V. Novotny and P. Brown, eds.), IWA Publishing, London, UK

Hill, K. (2007) "Urban ecological design and urban ecology: an assessment of the state of current knowledge and a suggested research agenda," in *Cities of the Future: Towards Integrated Sustainable Water and Landscape Management* (V. Novotny and P. Brown, eds.), IWA Publishing, London, UK

Howarth, R. B. (2007) "Toward an operational sustainability criterion," *Ecological Economics* 63, pp. 656–663

IPCC (2007). *Summary for Policy Makers, Climate Change 2007: The Physical Scientific Basis*, Fourth Assessment Report, Intergovernmental Panel on Climatic Change, Working Group (WG 1), Geneva

Jordan, J. L. (1999) "Externalities, water prices, and water transfers," *Journal of the American Water Resources Association* 35(5):1007–1014

Karr, J. R. and D. R. Dudley (1981) "Ecological perspectives on water quality goals," *Environmental Management* 5, pp. 55–68

Karr, J. R. and E. W. Chu (1999) *Restoring Life in Running Waters*, Island Press, Washington, DC

Kates, R. W. *et al.* (2001) "Environment and development: Sustainability science," *Science* 292(5517): 641–642

Lee, T. S. (2004) "Buried treasure: Cheonggyecheon restoration project," *Civil Engineering*, the magazine of the American Society of Civil Engineers, 74(1):31–41

Leopold, A. (2001) *Sand County Almanac: With Essays on Conservation*, Oxford University Press, Oxford, UK

Lucey, P. and C. Barraclough (2007) "Accelerating adoption of integrated planning & design: A water-centric approach, green value and restoration economy," PowerPoint presentation at Northeastern University (Boston, Massachusetts), Aqua-Tex, Victoria, BC

Maas, Arthur, Maynard M. Hufschmidt, Robert Dorfman, Harold A. Thomas, Stephen A. Marglin, and Gordon M. Fair (1962) *Design of Water Resource Systems*, Harvard University Press, Cambridge, MA

Manolakos, E., H. Virani, and V. Novotny (2007) "Extracting Knowledge on the Links between Water Body Stressors and Biotic Integrity," *Water Research* 41, pp. 4041–4050

Marriott, D. (2007) "What does green mean in Portland, Oregon?" PowerPoint presentation at the 2007 National Association of Clean Water Agencies Summer Conference, July 18, Washington, DC

Marsalek, J., R Ashley, B. Chocat, M. R. Matos, W. Rauch, W. Schilling, and B. Urbonas (2007) "Urban drainage at cross-roads: four future scenarios ranging from business-as-usual to sustainability," In *Cities of the Future: Towards Integrated Sustainable Water and Landscape Management* (V. Novotny and P. Brown, eds.), IWA Publishing, London, UK, pp. 339–356

Mays, L. W. (2007) *Water Resources Sustainability*, McGraw-Hill, New York, and WEF Press, Alexandria, VA

McGranahan, G. and D. Satterthwaite (2003) "Urban centers: An assessment of sustainability," *Annual Review of Environmental Resources* 28, pp. 243–74

Metcalf & Eddy (2007) *Water Reuse: Issues, Technologies, and Applications*, McGraw-Hill, New York

Mihelcic, J. R. *et al.* (2003) "Sustainability science and engineering: The emergence of a new discipline," *Environmental Science & Technology* 37, pp. 5314–5324

Mitchell, V. G., R. G. Mein, and T. A. McMahon (1996) "Evaluating the resource potential of stormwater and wastewater: An Australian perspective," *Proceedings of the 7th International Conference on Urban Storm Drainage*, Hanover, Germany, pp. 1293–1298

Niemi, A. (1979) "Blue-green algal blooms and N:P ratio in the Baltic Sea," *Acta Botanica Fennica* 110, pp. 57–61

Novotny, V. (2003) *WATER QUALITY: Diffuse Pollution and Watershed Management*, John Wiley & Sons, Hoboken, NJ

Novotny, V. (2008) "Sustainable urban water management," in *Water & Urban Development Paradigms* (J. Feyen, K. Shannon, and M. Neville, eds.), CRC Press, Boca Raton, FL, pp. 19–31

Novotny, V. and H. Olem (1993) *WATER QUALITY: Prevention, Identification and Management of Diffuse Pollution*, Van Nostrand-Reinhold Publishing, New York, NY, reprinted by John Wiley & Sons, Hoboken, NJ 1997

Novotny, V., J. Braden, D. White, A. Capodaglio, R. Schonter, R. Larson, and K. Algozin (1997) *Use Attainability Analysis: A Comprehensive UAA Technical Reference*, Proj. 91-NPS-1, Water Environment Research Foundation, Alexandria, VA (republished in 2005)

Novotny, V. *et al.* (1999) *Urban and Highway Snowmelt: Minimizing the Impact on Water Quality*, Water Environment Research Foundation, Alexandria, VA

Novotny, V., A. Bartošová, N. O'Reilly, and T. Ehlinger (2005) "Unlocking the Relationships of Biotic Integrity to Anthropogenic Stresses," *Water Research* 39, pp. 184–198

Novotny, V. *et al.* (2001) *Risk Based Urban Watershed Management - Integration of Water Quality and Flood Control Objectives*, Final Report, U.S. EPA/NSF/USDA STAR Watershed program, Institute of Urban Environmental Risk Management, Marquette University, Milwaukee, WI

Novotny, V. and K. Hill (2007) "Diffuse pollution abatement – A key component in the integrated effort towards sustainable urban basins," *Water Science and Technology* 56(1): 1–9

Novotny, V. *et al.* (2007) *Developing Risk Propagation Model for Estimating Ecological Responses of Streams to Anthropogenic Watershed Stresses and Stream Modifications*, Final Report, STAR R 83-0885-010, submitted to U.S. Environmental Protection Agency, Washington, DC, www.coe.neu.edu/environment

Novotny, V., D. Bedoya, H. Virani, and E. Manolakos (2008) "Linking Indices of Biotic Integrity to Environmental and Land Use Variables - Multimetric Clustering and Predictive Models," Paper accepted for platform presentation at the International Water Association World Water Congress, Vienna, September 2008

Oregon State University, GeoSyntec Consultants, University of Florida, and The Low Impact Development Center (2006) *Evaluation of Best Management Practices for Highway Runoff Control*, NCHRP Report 565, Transportation Research Board, Washington, DC

Paerl H. W. (1988) "Nuisance phytoplankton blooms in coastal, estuarine, and inland waters," *Limnology and Oceanography* 33(4):823–847

Parkin, S. (2000) "Sustainable development: the concept and the practical challenge," Paper 12398, *Proceedings of the Institution of Civil Engineers, Civil Engineering* 138, pp. 3–8

Pezzey, J. (2004) "One-side sustainability tests with amenities, and change in technology, trade and population," *Journal of Environmental Economics and Management* 48, pp. 614–631

Rabalais, N. N., R. E. Turner, J. Dunravko, Q. Dortsch, and W. J. Wisman, Jr. (1999) *Characterization of Hypoxia*, Topic 1 Report for the Integrated Assessment on Hypoxia in the Gulf of Mexico, NOAA Coastal Ocean Program Decision Analysis, Series No. 17, Silver Spring, MD

Rahaman, M., and O. Varis (2005) "Integrated Water Resources Management: Evolution, prospects, and future challenges," *Sustainability Science, Practice & Policy* 1(1):15

Rees, W. E. (1992) "Ecological footprints and appropriate carrying capacity: What urban economist leaves out," *Environment and Urbanization* 4(2):121–130

Rees, W. E. (1997) "Urban ecosystems: The human dimension," *Urban Ecosystems* 1(1):63–75

Rees, W. E. and M. Wackernagel (1996) "Urban ecological footprint: Why cities cannot be sustainable and why they are a key to sustainability," *Environmental Impact Assessment Review* 16, pp. 223–248

Rees, W. E (1999) "Understanding urban ecological systems: An ecological economic perspective," Chapter 8 in *Understanding Urban Ecosystems: A New Frontier for Science and Eduaction* (A. R. Berkowitz, C. H. Nilon, and K. S. Hollweg, eds.), Springer Verlag, New York, pp. 115–136

Register, R. (1987) *Ecocity Berkeley: Building cities for a healthy future*, North Atlantic Books, Berkeley, California

Rittmann, B. E., N. Love, and H. Siegrist (2008) "Making wastewater a sustainable resource," *Water* 21, pp. 22–23

Rogers, P. and A. Rosenthal (1988) "The imperatives of nonpoint source pollution control," in *Political, Institutional and Fiscal Alternatives for Nonpoint Pollution Abatement Programs* (V. Novotny, ed.), Marquette University Press, Milwaukee, WI

Schueler, T. (1994) "The importance of imperviousness," *Watershed Protection Techniques* 1(3):100–111

Scott, C. A., A. J. Zarazua, and G. Levine (2000) *Urban-Wastewater Reuse for Crop Production in the Water-Short Guanajuato River Basin, Mexico*, International Water Management Institute, Colombo, Sri Lanka

Solow, R. M. (1971) "The economic approach to pollution and its control," *Science* 173, pp. 497–503

Solow, R. M. (1993) "Sustainability: An economist's perspective," in *Economics of the Environment* (R. Dorfman and N. S. Dorfman, eds.), Norton, New York

Thompson, M., R. Ellis, and A. Wildavsky (1990) *Cultural Theory*, Westview Press, Boulder, CO

United Nations (1996) An Urbanizing World: Global Report on Human Settlements, *UN Center on Human Settlements*, Oxford University Press, Oxford, UK

U.S. Environmental Protection Agency (1983) Results of the Nationwide Urban Runoff Program, Volume 1, Final Report, Water Planning Division, Washington, DC

U.S. Environmental Protection Agency (1994) *Water Quality Standards Handbook*, 2nd ed., EPA 823-B-94-005A, Office of Water, USEPA, Washington, DC

U.S. Environmental Protection Agency (2007) *Total Maximum Daily Loads with Stormwater Sources: A Summary of 17 TMDLs*, EPA 841-R-07-002, Office of Wetlands, Oceans and Watersheds, Washington, DC, www.epa.gov/owow/tmdl/techsupp.html

U.S. GBC (2005) *Green Building Rating System for New Construction & Major Renovations, Version 2.2*, U.S. Green Building Council, Washington, DC., http://www.usgbc.org

U.S. GBC (2007) *LEED for Neighborhood Development Rating System, Pilot Version*, U.S. Green Building Council, Washington, DC, http://www.usgbc.org

World Wildlife Fund (2008) "One Planet Living," http://www.oneplanetliving.org/index.html

Yoder, C. O., R. J. Miltner, and D. White (2000) "Using biological criteria to assess and classify urban streams and develop improved landscape indicators," In *Proceedings of the*

National Conference on Tools for Urban Water Resource Management and Protection, U.S. Environmental Protection Agency, Washington, DC and Region V, Chicago, Illinois

Zaitzevsky, C. (1982) *Frederick Law Olmsted and the Boston Park System*, The Belknap Press of Harvard University, Cambridge, MA

Zhang, Y. (2007) "'Coping with Water Scarcity': UN Marks World Water Day," *UN Chronicle Online Edition*, http://www.un.org/Pubs/chronicle/2007/webArticles/032907_waterscarcity.htm

PLANNING AND DESIGN FOR SUSTAINABLE AND RESILIENT CITIES: THEORIES, STRATEGIES, AND BEST PRACTICES FOR GREEN INFRASTRUCTURE[1]

III.1 INTRODUCTION

III.1.1 Achieving Sustainability

This chapter reviews the issues, challenges, and best practices that are being conceived and applied by planners and designers to bring sustainability and resilience
to urban environments. Here "planning and design" is understood broadly to include
civil engineering, architecture, landscape architecture, and urban planning. The theory behind new initiatives on sustainability and resilience is discussed and illustrated
with international applications to urban planning and design. In this context, cities
are understood as urban ecosystems—with distinct ecological processes—driven significantly by human activities (Pickett et al., 2004). Prominent among these urban
ecological processes is hydrology, which is understood here as both an indicator
of urban conditions and as a driving factor that influences—or controls—a suite of
biophysical, cultural, and ecological processes. Although many issues, drivers, resources, and processes are considered, water is understood as the "tail that wags the
dog" of urban sustainability and resilience.

   According to the United Nations and other sources (see Chapter I), the world's
population has recently become predominantly urban, and the total world urban population is projected to double by 2050 (United Nations Habitat, 2006). In response to
this urbanization trend, the U.N. Habitat has reasoned that "the millennium ecosystem goals will be won or lost in cities!" (United Nations Habitat, 2006). In response
to this new urban reality, the "design and planning" disciplines of civil engineering,
urban planning, architecture, and landscape architecture are increasingly focused on
developing and testing new theories, strategies, and best practices to enhance the

---

[1]Written by J. Ahern, University of Massachusetts Amherst.

135

sustainability of cities. In addition, the design and planning disciplines are increasingly engaged in interdisciplinary practice with other engineers, biologists, ecologists, social scientists, and economists to address sustainability in an integrated way, to address the "trinity of factors," also known as the "triple bottom line" of sustainability (economic, social, environmental) (see Chapter II). Beyond interdisciplinarity, transdisciplinarity is arguably becoming the modus operandi for sustainability research and practice, because transdisciplinarity involves not only the professional and academic specialists, but also engages the stakeholders and decision makers, in a genuine and meaningful manner, throughout a continuous, interactive, and iterative process of urban planning and design (Tress, Tress, and Fry, 2005). Transdisciplinarity answers the call for a new, more holistic level of involvement in policy development, including public, private, and not-for-profit interests in developing and implementing a fifth paradigm of water management for cities of the future.

This great, global, transdisciplinary experiment in urban sustainability planning and design is evolving rapidly through theoretical research, innovative urban policies, and pilot or demonstration projects intended to test new approaches, especially in the development of ecocities—where the sustainability challenge meets the real urban world, at least at the scale of pilot or demonstration projects and communities. Also prominent among these pilot initiatives are Low Impact Development (LID), brownfield redevelopment, green urbanism, and ecological urbanism. Each urban sustainability initiative can be characterized by its: (1) goals (e.g., water management, net zero energy use, enhanced public and pedestrian transportation, mixed use urban form, recycling of waste, housing affordability), (2) scale (building/site, neighborhood, sub-watershed, city, metropolis), and (3) urban context (urban core/ CBD, established neighborhood, infill development, brownfield, periurban, suburban, or rural).

This monograph argues for a new understanding of urban change in the context of resilience—based on a growing theory of sustainable urbanism—in pursuit of urban sustainability. A working method for transdisciplinary planning and design is presented to organize and integrate diverse perspectives and fields of knowledge, to explicitly address sustainability goals and to integrate strategies to build resilience capacity. The method is focused on the challenge of planning for the uncertain future of cities that are growing rapidly in area, population, and complexity, in a context of global climate change. We offer theoretical foundations for the planning method as well as examples and case studies of successful plans and programs that have applied similar methods. These examples will include urban retrofit projects, brownfield restoration projects, low impact new development, and ecocities. While sustainability has three interrelated and interdependent dimensions, or pillars—environmental, economic, and social—this chapter primarily addresses the environmental/physical/spatial dimension of sustainability in cities, specifically as it can be understood and addressed through urban/landscape planning and design (Pickett et al., 2004).

Achieving sustainability of water resources must be accepted as a central, or perhaps the central, challenge for sustainable urbanism, because water is essential for all life—the universal solvent that transports and redistributes nutrients and pollutants across entire watersheds—and because future water quality and management for urban uses are threatened by urbanization itself in most cities of the world. Urban planning and design can play a key role to preserve, protect, restore and reuse the full spectrum of water uses that cities depend on, including: drinking water provision; wastewater collection, treatment, disposal, and reuse; stormwater management; and innovative, more holistic systems to create a new "urban hydrological cycle" (National Resources Defense Council, 2001). Because water is the essential and primary integrating resource, planning for water affects—and is affected by—most other sectors of physical urban planning, including land use, transportation, infrastructure, open space, waste processing, and energy generation and transmission.

To achieve sustainability and resilience in cities, urban infrastructure must be reconceived and understood as a means to improve and contribute to sustainability. If planners and designers only think defensively about avoiding or minimizing impacts related to infrastructure (re)development, the "target is lowered," actions become conservative, and the possibility to innovate is greatly diminished. Arguably, achieving sustainability will depend on significant innovations. In the 21st century, much of the infrastructure of the developed world will be replaced or rebuilt, and even more infrastructure will be needed to service the rapidly expanding cities of the developing world (Nelson, 2004). Ironically, when viewed as an opportunity, the magnitude of global infrastructure (re)development represents an unprecedented opportunity to redirect and reconceive the process of urbanization from one that is inherently destructive to one that is sustainable in specific terms.

III.1.2 Sustainability through Urban Planning and Design

Urban planning and design offers a particular and important perspective on water resource planning design and management. The physical form of any city is the result of its history of planning and design decisions and actions—or inactions. The physical urban form of a city is a major determinant of hydrological processes, directly affecting interception, infiltration, runoff, waste generation, and processing—and the overall urban water budget and many attributes of water quality (Paul and Meyer, 2001). Urban planning is understood here as a process for discussing and deciding on collective goals and priorities, in order to accommodate present and future human needs, and to make decisions and institute policies that reduce uncertainty and risk. Planning has been described as "the process of choice based on knowledge about people and land" (Steiner, 1991, p. 520). Contemporary urban planning integrates scientific knowledge and theories into an inclusive, public process regarding present and future use of resources, including the physical organization and use of urban space.

To the extent that the existing physical/built form of a city is limiting or problematic, urban planning and design can articulate policy and physical/spatial solutions to address the problems. More innovative urban planning and design is possible in large-scale ecocity projects where ambitious goals for multiple aspects of sustainability can be adopted to guide and focus the design process, including: a sustainable urban hydrology model, zero net energy use, a mix of urban uses, inclusion of biodiversity, and providing a healthy environment for people. The planning and design disciplines have developed a significant body of knowledge with respect to sustainable water resources, and more recently have begun to use cities as laboratories to test new practices and thereby promote innovation (Kato and Ahern, 2008; Lister, 2007). This chapter looks at the issue of sustainable urban water resources through the lens of urban planning and design, to understand the theories, issues, trends, emerging strategies, and current best practices from an international perspective.

In the contemporary urban planning and design literature, there is a convergence of research and case applications addressing sustainable cities and sustainable urbanism (Newman et al., 2009; Farr, 2008; Birch and Wachter, 2008; Newman and Jennings, 2008; Novotny and Brown, 2007; Girling and Kellett, 2005; Low et al., 2005; Moughtin and Shirley, 2005; Steiner, 2002; Beatley, 2000; Hough, 1995; among others). Because of cities' inherent density and tendency towards efficient compact form, they can arguably be planned and designed to provide ecosystem services while accommodating a range of human needs in a sustainable manner (Newman and Jennings 2008).

In earlier environmental planning discourse, a presumed, or de facto, polarity or tension existed between city and country, built and unbuilt, and perhaps even between human beings and nature (McHarg, 1969). This tension focused environmental planning on protecting intact ecosystems for conservation and on identifying the most suitable undeveloped "greenfields" for new development. After the global adoption of sustainability in the late 1980s, a more nuanced, or balanced environmental planning discourse evolved in the context of the trinity of sustainability principles: environmental sensitivity, economic opportunity, and social equity. This discourse led to new international policies—particularly the 1992 Earth Summit's Rio Declaration and Agenda 21—to actively promote and practice sustainability principles, including the design and environmental management of cities and communities (United Nations Department of Economic and Social Affairs, www.un.org/esa/dsd/agenda21;Johnson, 1993).

III.2   ECOSYSTEM SERVICES

III.2.1   Concepts

The United Nations' Millennium Ecosystem Assessment (2005) focuses on the ecosystem services concept in the context of sustainability, arguing that the protection of landscapes that provide ecosystem services can be justified on economic terms—and conversely, that their absence or degradation can have negative economic, as well as ecological, effects. Ecosystem services are classified as provisioning, regulatory, and cultural services (United Nations Millennium Ecosystem Assessment, 2005). Examples of ecosystem services related to water resources include: drinking water (provisioning), flood protection (regulatory), and recreational and aesthetic benefits (cultural). The concept of ecosystem services now provides a powerful, broadly accepted, logical argument for the protection and responsible

A Transdisciplinary Method for Spatial Planning of Resilient - Sustainable Cities

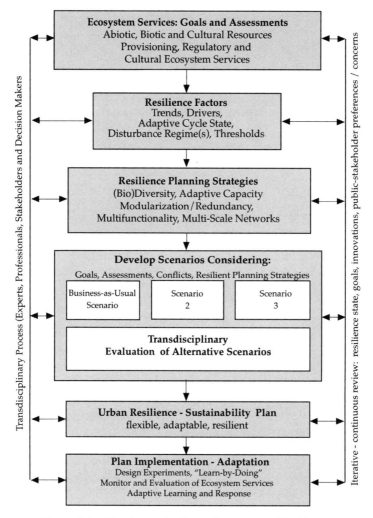

Figure 3.1 Planning method for resilient-sustainable cities.

development of landscapes—justified by the specific functions that landscapes provide, often with direct and measurable economic benefits for human beings (Figure 3.1).

While the concept of ecosystem services, as defined, includes cultural services, in popular literature and understanding it primarily addresses biophysical resources and processes, including water resources, biodiversity, and climatic regulation. Interdisciplinary environmental thinkers and leaders expand this concept to more explicitly include resources and services that directly benefit human beings as well as

Table 3.1   Selected examples of ecosystem services organized in abiotic, biotic, and cultural categories (ABC functions)

Abiotic Services	Biotic Services	Cultural/Landscape Services
Maintain hydrological regime(s)	Habitat and movement routes for generalist and specialist species	Opportunity for active and passive outdoor recreation
Accommodate disturbance and adaptive response	Support metapopulation dynamics in fragmented landscapes	Context for social interaction
Support nutrient cycling, buffering, sequestration	Bioremediation of wastes and toxics	Stimulus for aesthetic expression
Protection from floods	Maintenance of disturbance and successional regimes	Opportunity for environmental education
Stabilizing climate fluctuations	Biomass and food production	Reduce human stress
Filtering and improving air quality	Reservoir of genetic diversity	Supports economic activity (e.g., tourism)
Waste processing, digestion	Support flora:fauna interactions	Access to quiet/solitude

biotic and abiotic resources and processes (Table 3.1). These are often described as cultural, or landscape, services, and complement and expand biophysical ecosystem services to include: recreation, aesthetic inspiration, and opportunities for social interaction, stress reduction, and environmental education, among others (Temorhuizen and Opdam, 2009; Ahern, 2007; Ndubisi, 2002). Landscape architects Meyer (2008) and Dreiseitl and Grau (2009) argue that sustainable landscapes must do more than provide biophysical functions and services, they can and must perform socially and culturally, intersecting with social routines and spatial practices. This expanded and holistic conception of ecosystem/landscape services is closely aligned with a broad, multidimensional, and balanced conception of sustainability in general. The holistic view of ecosystem services provides specific, often quantitative, answers to the question: "What do sustainable landscapes do?"

Through the interdisciplinary field of landscape ecology, the limitations of some previous planning theories and methodologies have been revealed, and new methods have been proposed to apply the knowledge generated from landscape ecology to planning (Musacchio, 2009; Ndubisi, 2002; Leitão and Ahern, 2002; Hersperger 1994). The pattern and process principle from landscape ecology is particularly relevant to planning. The principle articulates the fundamental causal relationships among landscape pattern, process, and scale (Wu and Hobbs, 2002; Farina, 2000; Forman, 1995; Turner, 1990). The principle explains how flows of species, information, resources, and energy are influenced by the spatial composition and configuration of the unbuilt and built environment of cities, and how urban planning and design, in turn, influences urban landscape pattern. By making the links explicit between spatial pattern (form) and landscape process (landscape and ecosystem services), the "pattern-process" principle provides a key scientific basis for planning and designing urban form to provide ecosystem services under a resilient and sustainable model. The form and process principle applies particularly well to linking urban spatial form and configuration with urban water resources and hydrological processes (Marsalek et al., 2008).

Since the sustainability principle was adopted globally in the late 20th century, theorists increasingly appreciate the profound role that change, dynamics, and uncertainty play in sustainability. Sustainability is now understood as an "inherently moving target." This new understanding of environmental change and dynamics has pointed to the concept of resilience, and it has already influenced the global discussion of sustainability, geographically and demographically, in an interdisciplinary manner. To further examine the concept of urban resilience, we discuss non-equilibrium dynamics and resilience from the perspective of urban planning and design for water resources.

We pose an urban planning method and apply the concept of resilience to derive strategies intended to build urban resilience capacity—in support of sustainability in a non-equilibrium world. We focus on planning and design of urban form for sustainable water resources. The key argument is that urban environments, properly configured and managed, are key to providing sustainable water resources and uses to meet the needs of expanding urban populations.

Case studies of planning policies and pilot demonstration projects examine and demonstrate the effectiveness of interdisciplinary innovations and design experiments, and point to the need for continuing evolution and expansion of the knowledge base by practicing "learning by doing" (Gunderson and Holling, 2002; Kato and Ahern, 2008).

III.2.2 The Non-Equilibrium Paradigm

"Expect the best, plan for the worst, and prepare to be surprised"

Denis Waitley

The fourth paradigm of the 20th century was arguably developed around an equilibrium conception of natural, landscape, biological, and technological systems. Certainly many of the great technological achievements of the 20th century support and benefited from this equilibrium, or deterministic conception of nature. Advances in scientific knowledge, medicine, technology, and manufacturing supported a growing confidence that nature functions according to known rules, or laws, and that by understanding these laws and rules, human beings could manage or control nature, and consequently would prosper and thrive. The motto of modernism was to design machines for living. The new fifth paradigm of sustainability is to design living machines.

In the later 20th century, many thinkers argued for an alternative, non-equilibrium paradigm of science, of systems, and of the understanding of the natural and built environment (Rohde, 2005; Botkin, 1990; Steiner, 2002; Gleick, 1987). This view, known as chaos or non-equilibrium theory, argued that nature and natural-cultural systems are inherently variable, uncertain, and prone to unpredictable change.

Table 3.2 Paired concepts and terms relating to
equilibrium and non-equilibrium paradigms

Equilibrium	Non-Equilibrium
Modern	Postmodern
Linear	Networked
Rational	Chaotic
Closed, One way	Open, Circular
Predictable	Uncertain
Hierarchy	Panarchy
Deterministic	Stochastic
Reductionistic	Holistic
Tactical	Strategic
Disciplinary	Transdisciplinary
Terra firma	Terra fluxus

Non-equilibrium theory is manifest in a fundamental principle of landscape ecology: that landscapes are by definition heterogeneous, and that every landscape has an inherent disturbance regime in terms of type, frequency, and intensity of disturbance(s) (Turner, 1990). Important in this concept is an acknowledgment of stochastic processes, by which systems (i.e., cities) change according to known and unknown causes—often with explicit allowances for uncertainty—while framing the change or disturbance within some reasonable boundaries of time, space, and physical and biological processes. With this acknowledgment comes the professional responsibility for urban planners and designers to identify what the stochastic processes are that particular cities will face, the frequency and intensity of these events, and how cities can build the adaptive capacity to respond to these disturbances while remaining in a functional state of resilience.

Table 3.2 presents a series of paired, or complementary, concepts and terms that characterize and distinguish equilibrium and non-equilibrium theory. Arguably, the equilibrium view is largely responsible for the physical form of today's cities, which were built on the deterministic principles of modernism and have resulted in the current non-sustainable world. A new way of thinking about urban stability and change is arguably needed. As Albert Einstein said, "We cannot solve problems by using the same kind of thinking we used when we created them." While the equilibrium/non-equilibrium question remains appropriately unresolved, here we accept the non-equilibrium view of uncertainty and unexpected change among the fundamental characteristics of complex systems, including cities, and therefore of conditions that sustainability and resilience must address.

Under a non-equilibrium view, change and disturbance become accepted, even expected, characteristics of the system or process being planned—in this case, planning for urban sustainability. This raises the importance of resilience, the ability of a system to respond to change and disturbance without changing its state. The real challenge for urban planning and design for sustainability and resilience is to plan for the infrequent and the unexpected, while simultaneously planning for the routine, the familiar, and the very real requirements and processes that define and operate 21st-century cities. The planning method proposed here, and the supporting strategies and examples provided, address resilience explicitly as a necessary condition of sustainability.

III.3 PLANNING FOR RESILIENT AND SUSTAINABLE CITIES

An original method for planning resilient and sustainable cities is presented here (Figure 3.1). The method builds on established planning methods and models (Ndubisi, 2002; Leitão and Ahern, 2002; Steiner, 1991; Ahern, 1995; Steinitz 1990). The method has five themes: (1) goal-oriented and ecosystem-services-based, (2) strategic, (3) scenario-driven, (4) transdisciplinary, and (5) adaptive. Each of these five themes is discussed in the following sections.

The planning process begins by determining, or reviewing, ecosystem service goals, defined in the context of resilience factors—which are the trends and drivers of change. In planning to meet specific ecosystem service goals, resilience planning strategies are considered in the context of the public will, the economic climate, and existing urban conditions. Spatial concepts are used to design scenarios to explore possible futures, including the means to their realization. With expert and stakeholder participation, the scenarios are evaluated and ultimately revised or modified as an urban resilience sustainability plan. The plan is adaptively implemented, with monitoring of key indicators recommended to yield new knowledge and to continuously inform and (re)direct the planning process. While the method in Figure 3.1 is graphically represented as a linear process, in application it is cyclical, iterative, and may be entered or initiated at any point. For example, the planning process may start with an evaluation of a pre-existing plan, followed by goal re-determination, and development of new scenarios to explore new alternative strategies.

III.3.1 Ecosystem Service Goals and Assessments

The ecosystem services concept was developed as an integral part of the United Nation's Millennium Ecosystem Assessment (2005) to explicitly articulate the full complement of provisioning, regulatory, and cultural services provided by ecosystems by which humankind meets its needs. Ecosystem services, broadly defined to include cultural services (Figures 3.1 and 3.2), are appropriate as goals for sustainability planning because they are explicit and can be scientifically measured and analyzed and discussed in a transdisciplinary process. Such discussion may lead to the definition or appreciation of new ecosystem services, which can, in turn, also be discussed in the planning process.

Because ecosystem services represent the "process" side of the "pattern-process" dynamic; they can be explicitly "mapped" with geographic information system (GIS) models and algorithms. In other words, alternative spatial patterns can be modeled or tested for their effectiveness in providing specific ecosystem services, such as providing coastal flooding protection, or corridors for wildlife movement through a

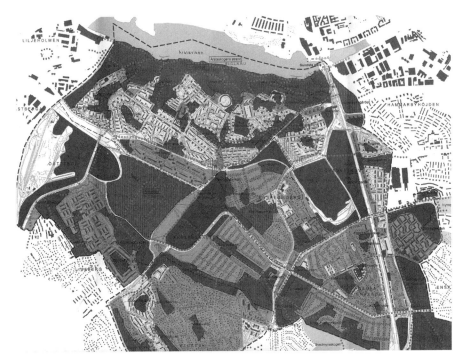

Figure 3.2   The City of Stockholm, Sweden, conducted an extensive survey exercise to identify the specific outdoor activities people engage in, and the locations where these activities occur. The resulting "Sociotope Map" defines spatially explicit patterns of use of all the public open spaces in the city. The map explicitly links patterns and process, for the important social functions of urban green spaces, informing and supporting green space planning and management decisions (Courtesy City of Stockholm).

city. When ecosystem service patterns are mapped with GIS, they can be combined with other ecosystem service "maps" to identify spatial patterns of compatibility or conflict, and then be adjusted or modified to resolve the spatial conflicts.

Ecosystem services therefore can serve well as assessment metrics linking urban form (pattern) with urban process (ecosystem services), to support an informed discussion of goals and their associated spatial requirements and consequences. Once articulated, quantified, and mapped, ecosystem services logically become the goals and benchmarks of progressive urban sustainability planning, for example, to improve water quality standards, or to provide a diversity of open spaces suited for particular outdoor activities (Figure 3.2).

III.3.2   Resilience Strategies

Urban planning is inherently a strategic process in that it attempts to understand and proactively manage the elements and forces that are the causes of change, rather than employing tactics to respond to the changes themselves (Ahern, 1995; Sijmons,

1990). Planning is, by definition, proactive—but not all planning is strategic. For urban planning to be strategic, it requires integration of interdisciplinary knowledge to define strategic goals consistent with political expectations, economic factors, and the reality of the existing landscape condition. Strategic urban planning requires a particular blending and integration of knowledge, vision, creativity, and political skills.

A proposed suite of five urban planning and design strategies for building urban resilience includes: multifunctionality, (bio)diversity, multiscale networks, redundancy and modularization, and adaptive capacity. These strategies represent new ways of thinking and acting that address the inherent uncertainty of cities. They also represent a somewhat radical rethinking about sustainability and change. The paradox of sustainability relates to the intrinsic need for stability and security while simultaneously accepting the existence of and the need for change in all systems. To resolve or confront the paradox of sustainability requires strategic thinking, which understands the forces and drivers of change, and seeks opportunities to influence these forces proactively, rather than reactively responding to the inevitable unexpected "surprises" characteristic of any urban environment over time.

Resilience is defined as the ability of a system to experience disturbance and still retain its basic function and structure (Walker and Salt, 2006). Understanding resilience is central to understanding sustainability, since sustainability addresses the need for a long-term, multigenerational view, and under a non-equilibrium view all systems will change in unpredictable ways, especially over the long term. Resilience theory is at the frontier of contemporary urban planning and design, serving as a robust platform for shaping and articulating the regenerative work of landscape architects, planners, and architects in volatile times (Vale et al., 2005).

Resilience can be better understood in the context of the paradox of efficiency and optimization—two pillars of the modern, equilibrium paradigm (Walker and Salt, 2006). Optimization assumes that change will be incremental and linear; it also tends to ignore changes that occur at higher or lower scales of organization. Optimization doesn't always work because the world is often configured and reconfigured by extreme events, rather than by average, day-to-day events and incremental change. Efficiency leads to the elimination of redundancies, and keeping only those elements that are immediately beneficial. The concepts of optimization and efficiency diminish the importance of unquantifiable or unmarketed values (e.g., ecosystem services) and reduce time horizons below those at which important changes occur (e.g., climate change). Optimization is a large part of the sustainability problem, not the solution—because in a non-equilibrium view, there is no optimal state for a dynamic system. Therefore, embracing change is the essence of resilience. The resilience of any system depends on its current state, cross-scale connections, and its context (Walker and Salt, 2006).

Resilience is a new way of thinking about sustainability, rather than a specific set of guidelines, instructions, or checklists. Resilience is more strategic than normative, because, to be effective, resilience must be explicitly based on, and informed by, the environmental, ecological, social, and economic drivers and dynamics of any particular place, and it must be integrated across a range of linked scales (Pickett

Table 3.3   Strategies for building urban resilience capacity

Strategies	Attributes/Characteristics	Examples
A) Practice Multifunctionality	Spatially efficient Economically efficient Builds a constituency of social/political support	Green Streets, Portland Oregon Stormwater wetlands
B) Practice redundancy and modularization	Risk-spreading Backup functionality Metasystems Decentralized, adaptable Can "contain" disturbance Flexibility, adaptability Spatial segregation	Created wetlands in Green Wedges, Green Infrastructure Watersheds and "neighbor-sheds" Gray water recycling systems
C) Promote (bio)diversity and heterogeneity	Differential response to disturbance, stress, and opportunity Bio-library of memory/knowledge Complementarity of resource requirements	Urban bioreserves Conventional, ecosystem-based, and hybrid functional types
D) Build and restore networks and connectivity	Metasystems Circuitry and redundancy, risk-spreading Design for functions and flows	Bluebelt, Staten Island, New York City Ecological Networks
E) Build adaptive capacity	Actions as opportunities for experimentation and innovation "Learn-by-doing," "Safe-to-fail" design experiments	SEA Street, Seattle

et al., 2004). In addition, by definition, resilience depends on being able to adapt to unprecedented and unexpected changes. Many such changes will affect, and will be affected by, the work of urban planners and designers, including: the effect of sea level change on coastal cities; the changing intensity of precipitation and runoff; and the changing composition and dynamics of urban forests, with implications for biomass production, carbon storage, and biodiversity. Urban planners and designers also affect the changing demographics of cities that define cultural identity and social resilience. By focusing explicitly on ecosystem services, the design fields are poised for leadership in building resilience capacity in cities at multiple scales and in multiple contexts. Table 3.3 summarizes five key strategies for building urban resilience capacity. These strategies will be discussed further.

Practice Multifunctionality   It is incumbent on planners and designers to think strategically to find new ways to provide for sustainable ecosystem services in the increasingly limited spaces within compact cities. This can be accomplished by intertwining/combining functions (Tjallingii, 2000). Another design strategy is vertical integration, where multiple functions can be "stacked" in one location, as with

Figure 3.3 A proposal for a multifunctional wildlife and pedestrian crossing of a state highway at Walden Pond near Boston, Massachusetts. The highway overpass supports wildlife and pedestrian movement, with a major highway passing underneath—an example of "stacking functions" (Image: Jinglan Wang).

wildlife crossings located under or over roads (Figure 3.3), stormwater infiltration systems located beneath buildings or parking lots, or green roofs used on top of buildings (see Chapter IV). Innovative and real-time scheduling can also be employed to coordinate the time dimension of ecosystem services/functions to achieve multiple functions in the same location but at different times of day, or over the course of a year. Examples of infrastructure real-time scheduling include limited human use of hydrological systems during periods of high flows (e.g., floodplain parks), restrictions of recreational use of habitat areas during sensitive breeding periods, and the closing of roads at night when nocturnal species movement is concentrated (Kato and Ahern, 2009).

Planning for multifunctionality is also an effective strategy for cost effectiveness and for building a broad constituency of public support (Ahern, 1995). Doing more than one thing in one place is an obvious and important strategy for urban sustainability. Monitoring is necessary to verify that the actual benefits expected in multifunctional projects are realized, and measured under dynamic urban conditions.

Portland Oregon's Green Streets Program demonstrates the multifunctional strategy for urban resilience (Figure 3.4). Green Streets are conceived as extensions of the natural drainage system, as part of an integrated approach to urban development. Street design alternatives are literally assessed from the perspective of a spawning fish. The street designs that arise from this policy contribute to water quality improvements, as they have no artificial outfalls, have naturally irregular banks, and receive input from clean cold groundwater to the greatest extent possible. In addition, the Green Streets support the vehicular and pedestrian transportation needs of

Figure 3.4   Portland Oregon's Green Streets Program: Urban road with conveyance swale and detention basin (Courtesy Metro 2002).

the city, contribute to urban biodiversity, and mitigate the urban climate. Portland's Green Streets Program has been recognized for its leadership and culture of innovation (Metro, 2002).

Redundancy and Modularization   If resilience is largely about expecting the unexpected and planning for recovery after failure, then redundancy and modularization are important supporting strategies. Redundancy is defined as multiple elements or components providing the same, similar, or backup functions. Modularization is defined as the construction or use of standard units allowing for flexibility and variety in use. Modularization also refers to design and operation of discrete, subsystems rather than centralized integrated systems. Systems composed of redundant and modular components may also be integrated at higher levels of organization or scale. Both terms relate directly to the paradox of optimization and redundancy, as discussed above.

Ironically, in the context of resilience thinking, redundancy and modularization are valued specifically for the same reasons for which they would be criticized in modernist or industrial planning for optimization and efficiency. This is because redundancy and modularization spread risks—across time, across geographical areas, and across multiple systems. When a major urban function or service is provided by a single entity or infrastructure, it is more vulnerable to failure from a disturbance or extreme event. When the same function is provided by a "meta" or distributed modular system (a system of subsystems), it has an inherent insurance against failure, and if combined with a diversity of functional types, it would even more directly support resilience. Distributed wastewater treatment and reclamation plants are another example of redundancy and modularization (See Chapter 2). Redundancy and modularization are strategies to avoid putting "all your eggs in one basket," and for preparing and pre-planning for when (not if) the system fails.

Landscape-based stormwater or gray water recycling systems in urban environments represent a modular approach to urban wastewater management. By capturing, treating, and reusing stormwater locally in housing projects, for example, peak flows of centralized drainage systems are reduced, and combined sewer overflows can be managed more efficiently and economically. Recent research in a Chicago area urban watershed found that upstream infiltration-based stormwater management can significantly reduce downstream drainage infrastructure construction and maintenance/replacement costs (Johnson et al., 2006). In Berlin, Germany, Bureau Kraft has developed several successful modular, landscape-based stormwater treatment systems for housing projects in different contexts and densities. The systems include green roofs, cisterns, porous paving, swales, and small ponds to intercept, retain, and convey stormwater to larger ponds which become attractive and healthy landscape features of the projects (see Chapter IV). The systems are designed for the five-year storm. Monitored water quality is high in the receiving pond, as evidenced by healthy fish populations in some projects. The systems are completely modular and operate with solar power, thus demonstrating resilience on a neighborhood or project basis (Figure 3.5).

(Bio)Diversity   Here we define diversity to include biodiversity with social, physical, and economic diversity as important and effective strategies to support urban

Figure 3.5   The Landseberger Tor Community in Berlin features a modular landscape-based system to collect, convey, treat, and recycle stormwater, while providing a self-regulating system and amenity for the community; designed by Bureau Kraft, Berlin (Photo, Jack Ahern).

resilience. Biodiversity has been described metaphorically as a "library of knowledge," some of which is familiar and valued, while some remains "unread, but on the library shelves" waiting for its value or function to be discovered (Lister, 2007). Some obscure microbe living in the city could hold the information to cure a human disease, stabilize or remediate a terrible toxin, or provide a critical function or service that we can't even imagine being of value at this time. Maintaining or enhancing biodiversity is undisputedly an essential component of resilience.

All forms of diversity can be understood in two categories: functional diversity and response diversity (Walker and Salt, 2006). Functional diversity refers to the different functions that collectively define and operate an ecosystem, or a socioeconomic system. Biological communities live as integrated systems with discrete functions, including primary production, respiration, growth and reproduction, decomposition, and nutrient cycling. Urban communities, likewise, depend on a large group of functions, including energy generation and transmission, housing, manufacturing, water supply, waste removal, transportation, and communication, among others. Functional diversity is more a measure of system complexity than an operational strategy for resilience per se.

Response diversity in biological systems refers to the diversity of species within functional groups that collectively have different responses to disturbance and stress (e.g., temperature, pollution, disease). Thus with a greater number of species performing a similar function, the ecosystem services provided by any functional group—for example, the decomposers—are more likely to be sustained over a wider range of conditions, and the system will have a greater capacity to recover from disturbances, over time (=sustainability) (Walker and Salt, 2006, p. 69). But do functional and response diversity apply to urban systems?

Urban ecosystems support a suite of basic "urban" functions, including hydrological drainage and flood control, nutrient and biogeochemical processes, biomass production, climatic regulation, waste disposal, and wildlife habitat, among many others. Table 3.4 shows alternative modes of providing selected urban ecosystem functions (e.g., hydrology, nutrient cycling) with conventional infrastructure, ecosystem-based infrastructure, and hybrid systems. Under a fourth-paradigm optimization and efficiency model, these functions are provided by a centralized, engineered system—for example, a citywide sanitary waste-processing system. This system is susceptible to failure resulting from unexpected disturbances. In a resilient urban model, a diversity of response types, including ecosystem-based and hybrid infrastructures, contributes to each function and will respond differently to disturbance, strengthening it's resilience capacity.

Expanding urban tree canopy is a particularly effective strategy to build response diversity. Research in Davis, California, found that open-grown trees intercept 15% (*Pyrus calleryana* "Bradford") to 27% (*Quercus suber*) of rainfall. While interception rates varied with species, antecedent moisture conditions, rainfall duration and intensity, and windspeed, the research is conclusive: Trees intercept significant amounts of rainfall that will never reach the ground, thereby adding to the response diversity of the urban stormwater system, and reducing the amount of storm drainage infrastructure that a city needs to build and maintain. Urban trees are also

Table 3.4   Selected urban ecosystem functions and functional response infrastructure types, including conventional engineered systems, natural systems, and hybrids

Selected Urban Ecosystem Functions	Functional Response Infrastructure: C = conventional/engineered E = ecosystem-based, H = hybrid
Hydrology (interception, infiltration, storage, discharge, flow)	(C) Conventional drainage system (H) Stormwater best management practices (BMPs) (E) Created wetlands (E) Extensive urban forest
Nutrient cycling/ bio-geo-chemical processes (nitrification, denitrification, sequestration, ionization, ...)	(C) Sanitary treatment systems (H) Biological waste treatment/polishing (E) Created wetlands (E) Bioremediation
Biomass/food production	(C) National/global food supply system (E) Urban agriculture (E) Urban forestry (H) Transportation corridor plantings
Biomass/waste disposal and decomposition	(C) Sanitary landfill and incineration (H) Biodigestion, methane generation (E) Composting
Climatic regulation	(C) Engineered HVAC systems (C) green, well-insulated buildings (E) Green wedges/fingers @ metro scale (E) Extensive urban forest (canopy 30%+) (E) Urban greening
Wildlife habitat	(E) Habitat reserves (E) Naturalized areas (E) Abandoned areas (H) Formal planted areas

valuable for multiple functions, including reducing the urban heat island effect, reducing air pollution, storing carbon, reducing noise, and providing aesthetic benefits (McPherson et al., 2005; Xiao et al., 2000). It should be no surprise that expanding urban tree canopy has become a central goal of many contemporary urban greening projects, for example, New York City (City of New York, 2007; Hsu, 2006) and Chicago (Attarian, 2007).

With respect to urban stormwater drainage systems, response diversity can be increased with decentralized, landscape-based solutions that attempt to "start at the source" in intercepting, storing, slowing, and treating stormwater with a large—and growing—diversity of Low Impact Development (LID) systems and best management practices, including green roofs, rain barrels, rain gardens, street trees, porous paving, green surfaces, bio-infiltration swales, cisterns, created wetlands, and detention/retention ponds (Low Impact Development Center, www.lowimpact development.org; Richman et al., 1999). By employing a greater number of such

Green Alley Pilot Approach #2:
Full Alley Infiltration Using
Permeable Pavement

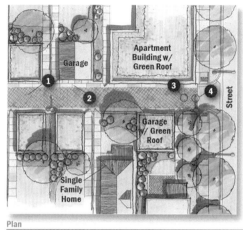

1 Permeable
pavement material
(permeable
asphalt, permeable
concrete, or
permeable pavers)

2 High albedo
concrete paving
with recycled
aggregate and slag

3 Optional inlet
structure with pipe
under drain

4 Energy efficient
dark sky compliant
light fixture

Plan

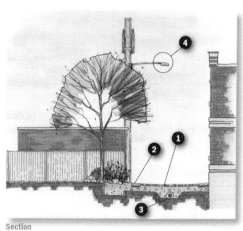

1 Permeable
pavement material
(permeable
asphalt, permeable
concrete, or
permeable pavers)

2 High albedo
concrete paving
with recycled
aggregate and slag

3 Optional pipe
under drain

4 Energy efficient
dark sky compliant
light fixture

Section

Figure 3.6　Chicago's Green Alleys Program employs a suite of green infrastructure/stormwater best practices to intercept, store, and infiltrate stormwater, significantly keeping runoff from entering the city's overloaded combined sewer system (Courtesy Attarian 2007).

systems across an urban landscape, the response diversity of the urban drainage function is increased, and the drainage function becomes more resilient to disturbance. And these landscape-based solutions all provide numerous important collateral benefits. Figure 3.6 shows the Chicago Green Alley system suitable for retrofit of densely populated urban zones, which combines green roofs with permeable pavements.

Multiscale Networks and Connectivity　In general, connectivity refers to the degree to which a landscape facilitates or impedes the flow of energy, materials, nutrients, species, and people across a landscape (Forman and Godron, 1986). Connectivity is also a property of cities that links urban form and function. Connectivity is an emergent property of systems, like cities, that results from the interaction of structure and function over time, for example: water flow, nutrient cycling, and the maintenance of biological diversity (Leitão et al., 2006). In urban environments, connectivity of natural systems is typically greatly reduced, often resulting in fragmentation—the separation and isolation of urban landscape elements with significant impacts on specific ecological processes that require connectivity (e.g., species dispersal and movement). And cities are also connected in other respects, for example transportation and communication systems, that also require high levels of connectivity to function properly.

Networks, by definition, are systems that support a range of functions by way of connectivity. Ecological networks are broad-scale landscape systems that include core wildlife habitat and connecting corridors, generally planned at the continental, national, or regional scale (Jongman and Pungetti, 2004; Bennett, 1999). Greenways are networks that provide multiple functions, including wildlife habitat and movement, but also recreation, water resource management, and cultural resource connections (Ahern, 1995). Urban greenways and urban networks are gaining popularity to provide stormwater, recreation, and alternative transportation functions—perhaps owing to the inherent benefits of connectivity.

The concept of connectivity applies directly to water flow, arguably the most important flow in any urban environment. Disruption of the hydrologic connectivity of nonchannelized stream reaches is a major concern when planning for sustainability. Because human culture relies on water in many respects, maintaining a connected and healthy hydrological system supports multiple ecosystem services and functions. Water is, arguably, the "tail that wags the dog of sustainability."

In urban or built environments, roads represent the greatest barrier to connectivity and are the primary contributor to habitat fragmentation (Forman et al., 2003). Connectivity is arguably a primary generator of sustainable urban form around blue-green networks that support biodiversity, hydrological processes, pedestrian transportation, climatic modification, and aesthetic enhancements—as illustrated in a site design proposal for Brentwood, British Columbia, in Figure 3.7 (Condon and Proft, 1999). The strategy of multifunctionality is also well illustrated in this example. Cross-scale connectivity is equally important, in this case to link the neighborhood with its larger urban context by ecotones—that is, supporting functions and processes that operate over larger scales of space and time, for example walking trails that link with bus stops, or urban drainage swales that connect to nonchannelized low-order streams, that, in turn, link with higher-order streams (Chapters IX and X).

When an urban landscape is understood as a system that regularly and continuously performs functions, connectivity is often the parameter that is responsible or critical. Such urban landscape systems are organized hierarchically, with some functions operating at broader or finer scales. Again, connectivity is the means through which many functions operate simultaneously at multiple scales, and conversely, the lack of connectivity is often a prime cause of malfunction or failure of particular functions.

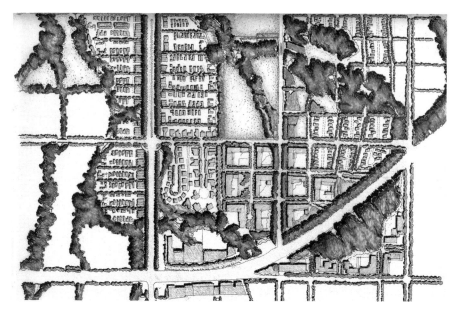

Figure 3.7 Concept design for Surrey, British Columbia, as a model sustainable urban community, clearly showing the organizing function and connectivity of the blue-green network (Courtesy Condon, 1996).

Ironically, connectivity of engineered drainage systems is understood as a hydrological problem. When impervious surfaces and areas are linked with engineered drainage systems, they contribute to effective impervious surfaces. When these surfaces are systematically and incrementally disconnected, and runoff directed to infiltration areas, effective impervious surfaces can be reduced, allowing stormwater to infiltrate rather than remain in the centralized engineered drainage system (Villareal and Bengtsson, 2004). Connectivity combined with imperviousness of urban watersheds increases runoff peak flow (by a factor of 4 to 10) and volume and, hence, the magnitude and frequency of flooding (Novotny, 2007).

Adaptive Capacity A fundamental challenge and impediment to applying sustainability principles to real projects is the common lack of empirical evidence of the effectiveness of a given intervention in a specific location. Wildlife corridors provide an example. While corridors have been implemented across the world to move species across agricultural and suburban locations (Bennett, 1999), the recommendations for corridor width, length, or structure are specific to the particular species and the landscape context involved. Thus, a wildlife corridor system for koala bears, in Australia, has questionable transferability for planning a moose corridor in the Northeastern U.S. The dilemma faced by planners is that the specific recommendations needed to implement a corridor system cannot be proven by applications or guidelines developed in other landscapes for different species. Unfortunately, the result is often inaction because of the lack of definitive recommendations.

Adaptive planning provides an alternative strategy. Under an adaptive approach, plans and policies can be developed in the face of uncertainty and incomplete knowledge. An adaptive plan is based on the best available knowledge, structured as experiments and monitored to learn how the actions result in specific goals, in terms of measurable ABC functions (Table 3.1). For example, to monitor cultural functions, surveys and observations of green corridor users can be kept over time to track the numbers of users, their motivations, their expectations, and their impressions of the resource or experience. Implicit in the adaptive approach is the potential to fail, but also the possibility to succeed. An adaptive approach requires a transdisciplinary effort involving scientists, stakeholders, decision makers, and planning and design professionals (Kato and Ahern, 2008).

The adaptive approach is promising for sustainability and resilience because the knowledge to plan and implement sustainable and resilient systems is, by definition, continuously evolving. In fact, if one accepts the non-equilibrium concept of urban and ecological systems, then there can never be complete knowledge, because every process is subject to disturbance and unpredictable change over time. However, if experimental applications and designs can be routinely practiced, the potential to build empirical knowledge, while exploring sustainability in the built environment, is quite profound. This would require a change in the culture of innovation, a support for experiments, and a willingness to accept (small) failures as a consequence of experiments that can also yield great innovations and new knowledge.

Resilience for urban systems can be understood as a complex and multidimensional challenge. The five strategies discussed above are offered to stimulate critical thinking, testing, and further refinement and development. In the true spirit of resilience, these strategies are offered as a starting point for a sustained discourse, not an a priori conclusion.

III.3.3 Scenario Planning

Scenarios are important tools for urban planning and are integral with the resilient-sustainable planning method presented above. Scenarios provide a perspective that is not constrained by the present situation. Scenarios have been used in corporate and governmental decision making since the 1970s because of their inherent advantages over expert judgments and other planning approaches. In urban planning, future scenarios are particularly well suited to linking goals and assumptions with potential future spatial changes in cities. A complete scenario should include a description of the current situation, a potential future state, and a means of implementation. Without all three of these elements, scenarios can be faulted as utopian. Scenarios are different from forecasts, which attempt to predict the expected future. In contrast, scenarios pose and answer a series of "what if" or "if then" questions. What if the city population changes? If the city builds a demonstration ecocity neighborhood, will it influence other future plans? Scenarios may be based on mathematical or other models, or they may be normative. The scenario approach is more appropriate when there is a great deal of uncertainty concerning the future, or when there is a general dissatisfaction with the present. Trend breaks are one reason that scenarios may be more useful than forecasts. Changes in technology or global economics can cause a trend break or paradigm shift that can alter the most fundamental assumptions in a planning activity (Schooenboom, 1995; Veenenklaas and Van den Berg, 1995).

Two fundamental types of scenarios can be defined, neither of which is predictions; instead, scenarios are vignettes of possible futures. A "forecasted" scenario projects current trends and control practices to produce a trajectory upon which a possible future may be conceived. A common forecasted scenario in urban planning is the "build-out" or "business-as-usual" scenario, in which current land use controls are used to determine a theoretical, or maximum, level of future development. Build-outs should not be represented as predictions, but used to establish a theoretical maximum level of development as a "conservative" base line for comparison of other alternatives. A "backcasted" approach, in contrast, is based on an idealized spatial concept—or vision—of what the future could be. Backcasted scenarios are often designed to articulate, and to visualize, the spatial consequences of planning goals or assumptions and the steps necessary to realize them (Schoonenboom, 1995).

In many instances, alternative scenarios are intentionally generated with the explicit purpose of demonstrating a range of alternatives. This has been described as identifying the four corners of an abstract frame within which a more balanced or compromised alternative may be selected (Harms et al., 1993). Or these alternative scenarios can be considered as the four points of a tetrahedron, indicating a more dynamic third-dimensional aspect to the relationships among the alternatives (Forman, 1995; Steinitz et al., 2002).

Scenarios, and innovative plans, are often based on spatial concepts, which acknowledge the centrality of the spatial dimension of sustainable landscape planning (Forman, 1990). A spatial concept expresses through words and images an understanding of a planning issue and the actions considered necessary to address the issue. Spatial concepts often manifest basic goals or assumptions upon which more specific decisions can be based. Spatial concepts are often carefully selected metaphors—for example, "compact city" or "green heart"—which communicate the essence of the concept clearly, build consensus, and serve as a basis for specific planning decisions (Ahern, 2002; Van Langevelde, 1994; Steiner, 1991). Spatial concepts are related to the proactive, or anticipatory, nature of planning, in that they express solutions to bridge the gap between the present and the desired future situation.

Although scientific input from many disciplines can contribute spatial concepts, its potential is limited. Many scientists are reluctant make the "leaps of faith" that are essential to conceive spatial concepts. There is an essential element of creativity in the design of spatial concepts. They represent an interface of empirical and intuitive knowledge (Zonneveld, 1995; Lyle, 1994). Through spatial concepts, rational knowledge is complemented with creative insights. Spatial concepts in urban planning can be thought of as design concepts—essential ideas that transcend basic knowledge and that result in successful solutions. In site-scale urban design, concepts are the basis for giving physical form in response to goals, resource assessments, and the designer's creative insight.

The Netherlands has a rich tradition of landscape planning and has long employed spatial concepts in the planning process. The "green heart" is a good example. It is a spatial strategy to maintain a "green core" of agriculture, forests, and recreation within the densely populated western Netherlands. The core is surrounded by the Randstad (Ring City), which is a reciprocal strategic spatial concept. The "green heart" concept has significantly guided Dutch planning and development strategies since the 1950s, during a major period of population growth and land use change.

The appropriate use of scenarios and spatial concepts moves the sustainability discussion away from abstract theory and policy towards specific mappable-spatial solutions, by integrating ecological, engineering, and planning principles and knowledge with creative solutions appropriate to a specific urban spatial context. Scenarios and spatial concepts represent the fundamental concepts with clear, memorable images to both inform and stimulate the discussion of planning goals and strategies.

III.3.4 Transdisciplinary Process

An urban plan is fundamentally different from a research project. It offers recommendations that may be implemented and will influence residents and stakeholders. Therefore, a participatory process involving nonexpert public officials, local inhabitants, and special interest representatives is essential. This type of planning process promotes "mutual learning" (Friedman, 1973) through which experts and participants are jointly involved in the process, leading to: goal determination; integration of local knowledge, perceptions, and values; evaluation of alternatives; and ultimately, implementation, monitoring, and management. When the participatory planning process achieves a higher level of integration, it becomes genuinely transdisciplinary. Tress et al. (2005) define transdisciplinarity as "innovative because, unlike other models of urban planning, it involves systematic, significant and reciprocal involvement of experts, professionals, decision makers and stakeholders." Transdisciplinarity is arguably the defining and distinguishing characteristic of planning in the era of sustainability (Ahern, 2004).

III.3.5 Adaptive Planning

As discussed earlier, there are multiple dimensions to urban sustainability, including economic, social, ethical, and spatial. Here, urban planning is most closely linked with the spatial dimension, predominantly at the neighborhood or municipal scale. Urban plans can be understood as hypotheses of how a policy will influence landscape processes. If the planning policies are implemented, the plan, as a hypothesis, becomes an experiment from which experts, professionals, and decision makers may gain new knowledge through monitoring and analysis (Kato and Ahern, 2008).

As ecological knowledge has become more routinely integrated in planning, a common dilemma regarding the accuracy and certainty of the knowledge recurs. The planners ask legitimate questions, such as, "How much impervious surface can a city have and retain water quality and manageable streamflows?" to which the hydrologist replies, "It is impossible to generalize this type of information. Detailed, site-specific research is the only path to the answer." Adaptive management addresses this common dilemma by reconceptualizing the "problem" (of making specific planning decisions with imperfect knowledge) as an "opportunity" (Holling, 1978). In addition to contributing the best current knowledge available for making the initial planning decision, the hydrologist provides guidelines for implementation and monitoring, through which the planning decision may become a field experiment from which new knowledge may be generated. While adaptive management has been practiced successfully in natural resource management for decades, its application to urban planning and design is rare (Kato and Ahern, 2008).

The adaptive aspect of the planning method is an intellectual strategy to address uncertainty, and to "learn by doing." Applying the strategies for resilience in urban planning and design can result in innovations. While many such innovations remain to be validated over time, they collectively provide examples of building adaptive capacity to: (1) understand the dynamic state of the system, (2) seek opportunities for effective intervention—often as "design experiments" (Lister, 2007), and (3) promote "learning-by-doing" (Kato and Ahern, 2008).

The adaptive approach is inherently well suited to address the challenge for urban resilience: maintaining urban structure-function and form in the face of increasingly frequent and unpredictable disturbances and disruptions—without urban collapse. Resilience is a powerful concept that addresses the stochastic, or seemingly random, occurrence of disturbance and change. If cities are to be sustainable over significant periods of time, they will also need to be resilient because, over time, cities will be affected by unexpected change and disturbance. Many argue that cities are already being affected by more extreme weather attributable to climate change (Van Heerden et al., 2007). While resilience has been researched and discussed in natural resource management for some decades now (Holling, 1978), it has only recently been addressed in the context of the resilience of cities (Walker and Salt, 2006).

In the resilience-sustainability method presented above (Figure 3.1), the discussion and evaluation of goals, resilience strategies, and alternative scenarios lead to a Resilience/Sustainability plan, which, unlike conventional plans, is conceived as a framework based on goals and objectives, but a flexible and adaptive framework for continuous monitoring, goal reevaluation, and plan adjustment—all conducted in a transparent and transdisciplinary process.

If urban planning and design is truly innovative and adaptive in its pursuit of sustainability and resilience, it has an inherent potential to fail. This is the nature of true innovation: applying new ideas, testing new procedures and applications. Urban planners and designers can innovate by implementing practices that are untested while acting with professional responsibility. To further reduce the risk of failure, innovations can be "piloted" as "safe-to-fail" design experiments (Lister, 2007).

III.4   BEST PRACTICES FOR GREEN INFRASTRUCTURE

Infrastructure is commonly used to describe an administrative or management system that supports a specific function or service. Infrastructure also refers to the physical systems that support large-scale public functions, such as transportation, communication, or energy generation and distribution. Common to these two definitions is the general understanding of infrastructure as an underlying foundation, or structural system, that supports specific social or physical functions and services. Infrastructure is widely understood as the fundamental and essential support organized and managed to make the modern world function—and if the 21st-century world is to function in a more sustainable manner, green infrastructure will arguably play a central role.

Recently, the concept of green infrastructure has been used in relation to landscapes and physical systems that support the provision \of ecological and environmental processes and services that contribute to a sustainable landscape conditon. This usage adds another meaning to the administrative and functional meanings of infrastructure described above. Benedict and McMahon (2006, p. 1) define green infrastructure as "an interconnected network of natural areas and other open spaces that conserves natural ecosystem values and functions, sustains clean air and water, and provides a wide array of benefits to people and wildlife." This commonly referenced definition emphasizes the idea of infrastructure as a networked system that provides multiple functions and services.

Here, green infrastructure is defined as "spatially and functionally integrated systems and networks of protected landscapes supported with protected, artificial and hybrid infrastructures of built landscapes that provide multiple, complementary ecosystem and landscape functions to a broad public, in support of sustainability" (Ahern, 2007). Green infrastructure directly addresses the urgent social need to make built/urban environments more sustainable and resilient in new developments and in rebuilding or adapting existing developments to new, more sustainable uses. In addition to supporting core urban functions (transportation, drainage, communication, waste disposal), green infrastructure delivers measurable ecosystem services and benefits that are fundamental to the concept of the sustainable city.

While green infrastructure is a new, and still evolving, urban planning and design concept, a significant body of completed built projects exists. The following green infrastructure projects have been selected to provide instructive case examples of sustainable and resilient urbanism at a range of scales and contexts: Seattle SEA Street; Westergasfabriek Park, Amsterdam; Staten Island Blue Belt, New York City; and Augustenborg Neighborhood and Western Harbor, Malmö, Sweden. These case studies will also be discussed in the context of the five resilience strategies presented earlier (multifunctionality, redundancy-modularization, (bio)diversity, networks-connectivity, adaptive capacity), which can be used also in high-density neighborhoods (Figure 3.8).

III.4.1 SEA Street Seattle

The Street Edge Alternative (SEA Street) project in Seattle, Washington, is a good example of innovative green infrastructure and of "learning by doing" to build adaptive capacity (Figure 3.9). Loss of tree canopy and increasing stormwater runoff motivated the Seattle Public Utilities Department to establish a natural drainage system (NDS) program that mimics natural processes to slow stormwater runoff, increase infiltration, and improve water quality. The first pilot project of the NDS, the

Figure 3.8 The Bremen Street Park in East Boston, Massachusetts, an example of green infrastructure that integrates multiple modes of transportation and recreation in a dense urban environment. Photo: Jack Ahern.

Street Edge Alternative (SEA Street), was established in 2001 in the Pipers Creek Watershed. The project was coordinated with the University of Washington, where engineers collaborated on design and monitoring to rigorously verify the project's hydrological effectiveness. Seattle Public Utilities engaged the local community throughout the planning, design, and implementation of the project as a model socially inclusive process.

The primary goal of the SEA Street project was to reduce the runoff from the 2-year, 24-hour storm (4.25 cm) to predevelopment levels, and to convey 100% of the 25-year, 24-hour storm, in accordance with city drainage requirements. To achieve these hydrologic goals, the project designers narrowed the street to 5.5 meters wide (4.25 m plus 2 shoulders of 0.6 m each) representing 11% less impervious surface than a conventional street. The SEA Street includes planted swales and basins to create a long flow path with high surface roughness to increase the hydrologic time of concentration and promote infiltration. Swales and basins are planted with native and noninvasive ornamental species of trees, grasses, sedges, and rushes. Plantings were designed in collaboration with residents, resulting in a sense of ownership and stewardship; neighbors are now actively involved in voluntary care and maintenance of the SEA Street plantings.

Figure 3.9 The Street Edge Alternative (SEA Street) in Seattle, Washington, is a natural drainage system (NDS) designed to direct street runoff from a narrowed street pavement to planted infiltration swales. Note the adjacent street in the background showing the pre-SEA Street condition with wider road pavement and conventional stormdrains. Photo: Jack Ahern.

   Three years of hydrologic monitoring by Seattle Public Utilities and the University of Washington shows 98% of wet-season and 100% of dry-season stormwater runoff has been eliminated by the project (Seattle Public Utilities, 2009; Horner et al., 2002). The success of SEA Street has helped inspire and inform subsequent natural drainage system designs in Seattle and in other cities (Vogel, 2006).

   The SEA Street is also notable for providing multiple functions and benefits and a rare demonstration of "learning by doing." In addition to improving the hydrological performance of the street, the SEA Street has slowed vehicle traffic, while providing adequate space for emergency and delivery vehicles. The slower street has encouraged more pedestrians (+ 400%) and has encouraged more neighborly social interactions. The project has also raised awareness of the urban watershed issues that it was intended to address. Because the project was proven to be effective through scientific monitoring, it has served as a model for other projects in Seattle (110th Street Cascade, Broadview Greengrid, and Highpoint) and other cities worldwide (Girling and Kellett, 2005). The project has received numerous awards and has been published internationally, demonstrating the potential of a small project to serve as a defensible, effective, and visible model. The SEA Street illustrates the resilience strategies of modularization, multifunctionality, increased response diversity, and building adaptive capacity.

III.4.2   Westergasfabriek Park, Amsterdam

The design of public green infrastructure projects is challenged to make beautiful, visible new urban environments that provide ecosystem services that people see, understand, and value (Meyer, 2008). The Westergasfabriek Park in Amsterdam was developed on the site of a 19th-century coal-to-gas plant, a classic example of an urban "brownfield" (contaminated site). The plant was closed in the 1960s but left behind a highly toxic post-industrial landscape. In the 1990s, the city of Amsterdam decided that the 14.5-hectare site (36 acres), located very close to the city center, should be developed as a public park. The park, completed in 2003, has been recognized as a model for other cities addressing the post-industrial legacy of 19th- and 20th-century industrialization, representing "the passage of brownfields from dereliction and pollution to culturally energetic and socially sustainable creative centers that thrusts these formerly discounted lands into becoming vital agents of change" (Kirkwood, 2003).

At Westergasfabriekpark the toxic materials were largely capped and buried to prevent public contact. Many industrial buildings were retained and restored for artistic and cultural activities and organizations. The park was completed in 2003 and features an innovative water treatment system through which stormwater is circulated over multiple cascades and treated sequentially with biofilter plantings of baldcypress, reeds, sedges, and shrub dogwoods (Figure 3.10). Park visitors can enjoy these water gardens from walkways and boardwalks in the lower "Theatre Pool" for their aesthetic effect, and then gain additional experience when they learn of the bioremediation functions of the plantings. Here the infrastructure for filtering and "polishing" the stormwater before discharging into an adjacent canal has become an attractive water feature—a kind of post-industrial 21st-century fountain, but one with an ecological function for the public to experience and to learn from.

The Westergaspark makes numerous contributions to resilience, including: the multifunctionality of the park, including artists studios, performance spaces, and large areas for public events; contributions to urban biodiversity with its numerous wetland and water gardens; and adaptive capacity by applying new techniques for brownfield remediation.

III.4.3   Staten Island Blue Belt, New York

The Staten Island Bluebelt is an example of green infrastructure that provides multiple functions and ecosystem services. Staten Island is the least populated borough of New York City and the last part of the city to provide storm and sanitary sewer service. Parts of Staten Island have a history of drainage problems and septic system failures due to low topographic relief, high water table, and soils with low permeability. Staten Island also has the largest and last concentration of freshwater wetlands in New York City, a motivation for considering an alternative to an engineered stormwater system. Since 1997 the New York City Department of Environmental

Figure 3.10   Stormwater biofilter and cascade, Westergasfabriek Park, Amsterdam. Here, green infrastructure goes beyond functionality; it creates a designed experience for park visitors that informs about sustainability, the role of plants, and the beauty of the simple process of circulating and cleaning stormwater (Koekebakker, 2003). Photo: Jack Ahern.

Protection has been building an alternative stormwater management system that uses sewers to convey stormwater to detention areas employing created wetlands, settling ponds, and sand filters (New York City DEP, 2005; Wu, 2007). The effluent from this treatment is discharged into natural wetlands and watercourses to provide conveyance, storage, and filtration of stormwater. The overall system, known as the Staten Island Bluebelt, services 11 watersheds with an area of some 5000 hectares. The system was built at a cost savings of over $50 million in comparison with a conventional separated stormwater system, including the cost of land acquisition (Eisenman, 2005). The Bluebelt was planned to protect, salvage, and maintain the native flora to sustain ecological and hydrological functions, making a significant contribution to local urban biodiversity (Figures 3.11 and 3.12).

The Bluebelt system has been proven to be effective to reduce peak stormwater flows, to increase groundwater recharge, and to remove contaminants from stormwater. Importantly, the Bluebelt is recognized for providing additional functions, including recreation, wildlife habit, historic preservation, and neighborhood beautification. The Bluebelt has been integrated with public parks and trails in Staten Island. Anecdotal evidence shows that proximity to the Bluebelt adds to property value. By providing functional ecosystems as well as urban drainage systems, the Bluebelt builds resilience capacity and contributes to the sustainability of multiple urban watersheds.

Figure 3.11 The Staten Island Bluebelt is a green infrastructure system that provides stormwater drainage for many watersheds in Staten Island, part of New York City. The system employs stormwater best management practices and created wetlands to collect, slow, treat, and release stormwater. Source USGS, Based on NYC, Department of Environmental Protection.

III.4.4 Ecostaden (Ecocities): Augustenborg Neighborhood and Western Harbor, Malmö, Sweden

Malmö is the third-largest city in Sweden and capital of the southern province of Skåne. Since 2000, Malmö has been directly linked with Copenhagen, Denmark, via the Øresland bridge—important infrastructure to support the city's planned transition from an industrial to a service-based economy known as the City of Knowledge. Malmö implemented the Ekostaden (Ecocity) program in response to the global Agenda 21 (Johnson, 1993) and in coordination with the Swedish National Environmental Program. The Malmö Ekostaden program focused on two very different and distinct projects that illustrate sustainability, resilience, and innovative green infrastructure. Augustenborg, a 1950s medium-density residential neighborhood, has been retrofitted with an open drainage system to address chronic combined sewer overflow problems, among other environmental objectives. The Western Harbor (Västra Hamnen) is a new state-of-the art ecocity built on an industrial brownfield on Malmö's former industrial waterfront.

The Augustenborg neighborhood (Figures 3.13 and 3.14) is a 1950s-era moderate-density housing project in Malmö, Sweden (ca 100 units/ha). Augustenborg was

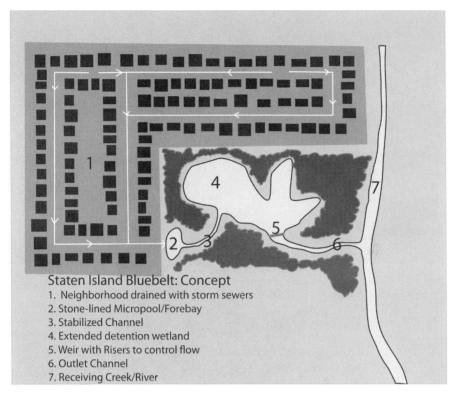

Staten Island Bluebelt: Concept
1. Neighborhood drained with storm sewers
2. Stone-lined Micropool/Forebay
3. Stabilized Channel
4. Extended detention wetland
5. Weir with Risers to control flow
6. Outlet Channel
7. Receiving Creek/River

Figure 3.12 Staten Island Bluebelt Concept, a multifunctional green infrastructure for stormwater management, wildlife habitat, recreation, and neighborhood beautification in New York City.

built circa 1950 as a model future urban community with residences, schools, and businesses in an integrated plan with internal pedestrian circulation, generous open spaces, and access to public transportation (Scania Ecoguide, www.eco-guide. net/skane/Ekostaden_Augustenborg_project.php?db_id=). Over time the district declined physically and economically, and developed social problems. Augustenborg was selected as a focal project for Malmö's Ekostaden program in the late 1990s. The program included an eco-cycle waste management program that reduced waste by 65% and practices organic waste recycling within the district.

Older urban neighborhoods like Augustenborg are often drained with combined drainage systems. At Augustenborg, the combined sanitary and storm sewer system experienced frequent combined sewer overflows (CSOs) and basement flooding. In 2001 the Malmö Water and Wastewater Works and Malmö Municipal Housing completed implementation of a sustainable open drainage system, as part of the larger Ekostaden sustainability initiative. The plan involved the construction of a series of structural best management practices to drain the site's stormwater with an open drainage system consisting of green roofs, swales, channels, ponds, and small wetlands (Villareal and Bengtsson, 2004). The goal was to reduce flooding by 70% and

Figure 3.13 The Augustenborg neighborhood was retrofitted to manage stormwater on-site through a series of best management practices, including this attractive planted wet retention pond to store, infiltrate, and clean stormwater. Photo: Jack Ahern.

Figure 3.14 The Augustenborg neighborhood has 1450 square meters of green roofs for stormwater management and to express the project's ecological aesthetic and commitment to sustainability. Here, the green roof is added to a "resource house" to support organic composting and material recycling. Photo: Jack Ahern.

to eliminate CSO events. The open drainage system improves stormwater management in the project area, but also improves the sanitary sewer that serves a larger area outside of the Augustenborg district. The amount of storm drainage that enters the combined system is now negligible (Villareal and Bengtsson, 2004).

The open drainage system demonstrates multiple strategies for sustainability and resilience, including modularization and decentralization by taking storm drainage "offline" from the existing system. The series of structural BMPs also provides multiple functions and ecosystem services, including biodiversity enhancement and aesthetic improvement. As the project has been monitored, it contributes to adaptive capacity to learn how districts with older housing stock can be retrofitted to provide open, surface-based drainage systems that build resilience capacity and provide multiple ecosystem services.

Bo01 - Malmö's Western Harbor (Västra Hamnen)
Western Harbor is an ecocity built on a former shipyard and industrial site on Malmö's waterfront (Figures 3.15 to 3.17). Malmö's shipbuilding industries suffered economically in the

Figure 3.15   Wetlands are part of the stormwater system of Bo01 Western Harbor, as well as being integral to the aesthetic landscape character of the neighborhood. Photo: Jack Ahern.

Figure 3.16 A courtyard canal in Bo01/Western Harbor, part of the surface drainage system and an important element for organizing and structuring the urban form of the project. Photo: Jack Ahern.

Figure 3.17 Landscape spaces in Bo01 are integrated with the stormwater management system and also provide wildlife habitat through diverse plantings of native species. Photo: Jack Ahern.

1970s and were abandoned, leaving a contaminated post-industrial landscape behind. The Western Harbor is part of Malmö's strategic transition from an industrial to a knowledge-based service economy. The project is also known as "The City of Knowledge." The Western Harbor is planned for a total area of 160 ha, eventually to support 10,000 residents and 20,000 workers and students. Chapters X and XI present a summary of ecocities throughout the world, including another Swedish ecocity, Hammarby Sjöstad, an area of Stockholm. Western Harbor was left in this chapter to introduce the ecocity concepts, which were also outlined in the One Planet Living criteria in Chapter II.

The first phase of the project, known as Bo01, was built in 2001 as Sweden's first international housing exhibition. The goal of Bo01 was to create a model ecologically sustainable city, combining aesthetics, ecology, and high technology as part of Malmö's Ekostaden program. Bo01 has 1000 housing units on 25 ha (40 units/ha). To prepare the contaminated site for development, 6000 m^3 of contaminated soil was removed for treatment and replaced with 2 m of clean soil.

Bo01 has a goal of renewable energy generation. The project's energy is provided 100% by locally produced renewable energy, including: 1400 m^2 solar photovoltaic collectors, solar thermal panels, 2-megawatt wind turbines, and a geothermal heat system. Biogas is produced via collection of organic waste with a vacuum collection system and used to heat homes and power vehicles. Other waste is also collected, sorted, and recycled or incinerated for energy to heat buildings. District heating supports heating and cooling distribution throughout the project. The project also uses an integrated electric grid to manage energy generation and use efficiently. Buildings are designed to minimize energy use through efficient insulation and natural daylight.

The urban design of the Bo01 neighborhood is modeled after the nearby medieval Swedish town of Lund, with small interior streets and taller buildings on the waterfront to enclose the space and block the very consistent and strong wind. The signature building of the project is the renowned 45-story residential tower, the "Turning Torso," designed by Spanish architect Santiago Calatrava. To promote aesthetic diversity, many architects were involved with building designs in the district.

The Bo01 neighborhood has an extensive sustainable urban drainage system including green roofs, open channels and swales, courtyard ponds, canals, and a large stormwater pond. The drainage system is fully integrated with the neighborhood design at multiple scales—from community squares, blue-green open space canals and corridors to fine-scaled drainage details (Figures 3.16 and 3.17). Over all, the drainage system gives the project a distinct and attractive "sustainable design" identity. As part of the project's open space network, the drainage canals and corridors provide recreational opportunities and support biodiversity with green roofs and created wetlands.

Bo01 employed a "Biotope Area Factor" (BAF) to ensure that the neighborhood has a minimum amount of "green" associated with each building/ building block. This incentive-based tool has been used effectively to promote "greening" in Berlin, Germany (Keeley, 2007). The BAF requires a specific percentage of ecologically effective land area that contributes to ecosystem functions by storing and infiltrating stormwater, and by creating wildlife habitat in all development parcels. Each plot needs to have a minimum green factor of 0.5. Developers have the choice of different "green" elements from a menu that can be combined in variable combinations to reach the minimum factor of 0.5 for the plot—for example: impervious surfaces rate 0.0, trees rate 0.4, and green roofs 0.8. The BAF also promotes wildlife habitat with native plantings and gardens (Figure 3.17).

Bo01 is designed for sustainable transportation. All housing units are within 300 m of a bus stop, with regular service. Public transportation will run on renewable biogas, generated in part from recycled organic waste from the district. Only 0.7 parking spaces per unit are provided. The center of the project is a pedestrian car-free zone, enhanced with well-designed promenades, bicycle paths, alleys, and squares.

The Western Harbor can be considered as a model of early 21st-century sustainable urban living. The project is comprehensive in its commitment to sustainability in terms of: energy use, transportation, waste recycling, water (re)use, and ecological integrity. The quality of the environment is evident in the design of buildings and landscape. The project clearly demonstrates the application of numerous resilience strategies and has succeeded in reaching its sustainability goals.

III.5 DISCUSSION

The urban planning and design disciplines are now engaged in a fundamental re-alignment of working methods, practices, and goals to address the challenge for sustainability and resilience. This new planning and design paradigm accepts the 21st-century global urban demographic and the non-equilibrium view as axioms and prerequisites for urban sustainability. While recognizing that sustainability has multiple dimensions, or pillars, planners and designers address sustainability primarily through the spatial form of the built environment. And this focus on spatial form applies across a broad range of projects from the *de novo* urbanism of ecocities to the redesign and retrofitting of established neighborhoods and the reconception of the structure and function of urban infrastructure. Consistent with the theme of the book, there has been an emphasis and a focus on hydrology, water, and wastewater systems in this discussion.

The new planning and design reality needs new methods and practices to address the profound challenges to sustainable and resilient urbanism. The method proposed here addresses these challenges through:

A focus on ecosystem services—articulating specific abiotic, biotic, and cultural functions and services that, in the aggregate, define sustainability

A suite of planning and design strategies to build resilience (multifunctionality, modularization, (bio)diversity, networks, adaptive design)

The use of scenarios to effect communication and exchange knowledge reciprocally between stakeholders, decision makers and experts

An adaptive approach, in which planning and design actions are understood as "design experiments" to support "learning by doing" and promote innovation

Samples of best sustainable urbanism practices were selected to illustrate a range of issues and intervention types in diverse international contexts. Pilot projects receive great attention and critique in the planning and design disciplines. They push the familiar and the status quo, employing new models, technologies, and goals to demonstrate and test new systems, in pursuit of ever-more-ambitious sustainability goals. The ecocities discussed in this chapter, and elsewhere in the book, stand as innovative models and experiments in sustainable urbanism. Time will select the best ideas and practices from these innovations for broader application and adoption in other locales. Other projects presented and cited address the challenge to improve the sustainability of established cities and neighborhoods. Here renovations, adaptations, and retrofits are applied to rethink familiar aspects of urbanism with new materials, design strategies, and models of transdisciplinarity.

All concepts of ecocities have water and water bodies as a centerpiece of the design. Hydrology dictates, among other factors, the layout of the city, drainage, aesthetics, and ecology. It is noted in this chapter for the first time, and will be also pointed out throughout the book, that being frugal with water use and water bodies enables planners to locate new green cities in areas of brownfields and in areas with severe sewage overflow and pollution problems. In the following chapters, ecocity examples from water-short areas (e.g., desert environments) will also be introduced. Hence, green developments/ecocities are not a luxury; instead, they are part of a solution to the current severe problems of the urban areas, and may become accepted models for solving current urban problems and implementing future urban renewal.

The best practices for sustainable urbanism presented can be classified and evaluated against any of the recent and emerging sustainability guidelines, programs, and benchmarks. These programs include: The U.S. Green Building Council's Leadership in Energy and Environmental Design (LEED) Program, Low Impact Development (LID) Standards, One Planet Living Criteria (WWF), The Sustainable Sites Initiative (ASLA, 2009), and the "Cities of the Future – Fifth Paradigm" offered in this book. Each of these programs and benchmarks has distinct goals and criteria, with modest to very ambitious sustainability goals, and is aimed at a particular sector or activity for sustainability (green buildings, sustainable communities, sustainable landscape practices, and sustainable living in a broad, multidimensional context). More important than debating the relative merits or value of a particular sustainability guideline/system/program is the realization that these programs and benchmarks exist, that they are increasingly being applied and followed, and that they are increasingly recognized and supported through public policy. In this spirit, principles, methods, and practices are offered here to challenge, inform, and guide planning and design professionals towards the creation of more sustainable urban environments.

REFERENCES

Ahern, J. (2007) "Green Infrastructure, a spatial solution for cities," in *Cities of the Future: Towards Integrated Sustainable Water and Landscape Management* (V. Novotny and P. Brown, eds.), IWA Publishing, London, UK, pp. 267–283

Ahern, J. (2002) *Greenways as Strategic Landscape Planning: Theory and Application*, Doctoral Dissertation, Wageningen Agricultural University, Netherlands

Ahern, J. (2004) "Greenways in the USA: theory, trends, prospects," in: *Ecological Networks and Greenways: Concept, Design, Implementation* (R. H. G. Jongman and G. Pungetti, eds.), Cambridge University Press, Cambridge, UK, pp. 34–55

Ahern, J. (1995) "Greenways as a Planning Strategy," *Landscape and Urban Planning* 33(1–3):131–155

ASLA (2009) "The Case for Sustainable Landscapes," *The Sustainable Sites Initiative*, ASLA, LBJ Wildflower Center, U.S. Botanical Garden, www.sustainablesites.org (accessed November 14, 2009)

Attarian, J. (2007) *The Chicago Green Alley Handbook: An Action Guide to Create a Greener, Environmentally Sustainable Chicago*, Department of Transportation, Chicago, IL

Beatley, T. (2000) *Green Urbanism: Learning from European Cities*, Island Press, Washington, DC

Benedict, M. A. and E. T. McMahon (2006) *Green Infrastructure: Linking Landscapes and Communities*, Island Press, Washington, DC

Bennett, A. F. (1999) *Linkages in the Landscape: The Role of Corridors and Connectivity in Wildlife Conservation*. IUCN, Gland, Switzerland, and Cambridge, UK

Birch, E. L. and S. M. Wachter (2008) *Growing Greener Cities: Urban Sustainability in the Twenty-first Century*, University of Pennsylvania Press, Philadelphia, PA

Botkin, D. (1990) *Discordant Harmonies: A New Ecology for the Twenty-First Century*, Oxford University Press, New York

City of Malmö, "Västra Hamnen, The Bo01 area: a city for people and the environment," www.malmo.se/vastrahamnen (accessed November 12, 2009)

City of New York (2007) "PlanYC: A greener, greater New York," http://www.nyc.gov/html/planyc2030/downloads/pdf/full_report.pdf (accessed October 12, 2009)

Condon, P. (1996) *Sustainable Urban Landscapes: The Surrey Design Charrette*, University of British Columbia Press, Vancouver, BC

Condon, P. and J. Proft (1999) *Sustainable Urban Landscapes: The Brentwood Design Charrette*, University of British Columbia Press, Vancouver, BC

Dreiseitl, H. and D. Grau (2009) *Recent Waterscapes: Planning, Building and Designing with Water*, Birkhäuser, Basel, Switzerland

Eisenman, T. (2005) "A Watershed Moment in Green Infrastructure," Landscape Architecture Magazine (November 2005)

Farina, A. (2000) *Landscape Ecology in Action*, Kluwer Academic Publishers, Dordrecht, Netherlands

Farr, D. (2008) *Sustainable Urbanism: Urban Design with Nature*, John Wiley & Sons, Hoboken, NJ

Forman R T T, Sperling D, Bissonette J, Clevenger A P, Cutshall C D, Dale V H, Fahrig L, France R, Goldman C R, Heanue K, Jones J A, Swanson F J, Turrentine T, Winter T C. (2003). *Road Ecology: Science and Solutions*, Island Press, Washington.

Forman, R. T. T. (1995) *Land Mosaics*, Cambridge University Press, Cambridge, UK

Forman, R. T. T. (1990) "Ecologically sustainable landscapes: the role of spatial configuration," in: *Changing Landscapes: An Ecological Perspective* (I. S. Zonneveld and R. T. T. Forman, eds.), Springer-Verlag, New York, pp. 261–278

Forman, R. T. T. and M. Godron (1986) *Landscape Ecology*, John Wiley & Sons, Hoboken, NJ

Friedman, John. (1973) *Retracking America: A Theory of Transactive Planning*, Anchor Press, Garden City, NY

Girling, C. and R. Kellett (2005) *Skinny Streets and Green Neighborhoods: Design for Environment and Community*, Island Press, Washington, DC

Gleick, J. (1987) *Chaos: Making a New Science*, Penguin Books, New York

Gunderson, L. H. and C. S. Holling, eds. (2002) *Panarchy: Understanding Transformations in Human and Natural Systems*, Island Press, Washington, DC

Harms, Bert; Knaapen, Jan P.; Rademakers, Jos G. 1993. Landscape Planning for Nature Restoration: Comparing Regional Scenarios. In: *Landscape Ecology of a Stressed Environment*, Vos, Claire C. and Opdam, Paul, Eds. Chapman and Hall, London. pp. 197–218.

Helphand, K. (2002) "Housing the future: Swedish housing exposition aims to marry sustainability and urban form," ASLA Online, http://www.asla.org/nonmembers/lam/lamarticles02/march02/malmo.html (accessed November 12, 2009)

Hersperger, A. M. (1994) "Landscape ecology and its potential application to planning," *Journal of Planning Literature* 9(1):14–29

Holling C. S. (1978) *Adaptive Environmental Assessment and Management*, Wiley International Series on Applied Systems Analysis, vol. 3, Chichester, UK

Horner, R. R., H. K. Lim, and S. J. Burges (2002) *Hydrologic Monitoring of the Seattle Ultra-Urban Stormwater Management Projects*, Water Resources Series, Technical Report No. 170, November 2002, Department of Civil and Environmental Engineering, University of Washington, Seattle, WA

Hough, M. (1995) *Cities and Natural Process: A Basis for Sustainability*, Routledge, New York

Hsu, D. (2006) *Sustainable New York City*, Design Trust for Public Space and the New York City Office of Environmental Coordination, (accessed April 20, 2006) http://blog.americanrivers.org/wordpress/index.php/2007/05/29/the-staten-island-bluebelt-natural-stormwater-management-and-flood-control/

Johnson, D. M., J. B. Braden, and T. Price (2006) "Downstream Economic Benefits of Conservation Development," *Journal of Water Resources Planning and Management* 132(1):35–43

Johnson, S. P. (1993) *The Earth Summit: The United Nations Conference on Environment and Development (UNCED)*, Graham and Trotman, London

Jongman, R. and G. Pungetti, eds. (2004) *Ecological Networks and Greenways: Concept, Design, Implementation*, Cambridge University Press, Cambridge, UK

Kato, S. and J. Ahern (2009). "Multifunctional Landscapes as a Basis for Sustainable Landscape Development," *Landscape Research Japan* 72(5):799–804

Kato, S. and J. Ahern (2008) "Learning by doing: Adaptive planning as a strategy to address uncertainty in planning," *Environment and Planning* 51(4):543–559

Keeley, M. (2007) "Using Individual Parcel Assessments to Improve Stormwater Management," *Journal of the American Planning Association* 73(2):149–160

Kirkwood, N. (2003) "Brownfield Passages (Forward[AU: Foreword?])," in *Olof Koekebakker, Westergasfabriek Culture Park: Transformation of a Former Industrial Site in Amsterdam*, NAI Publishers, Rotterdam, Netherlands

Koekebakker, O. (2003) *Westergasfabriek Culture Park: Transformation of a Former Industrial Site in Amsterdam*, NAI Publishers, Rotterdam, Netherlands

Leitão, A. B., J. Miller, J. Ahern, and K, McGarigal (2006) *Measuring Landscapes: A Planner's Handbook*, Island Press, Washington, DC

Leitão, A. B. and J. Ahern (2002) "Applying landscape ecological concepts and metrics in sustainable land planning," *Landscape and Urban Planning* 59(2):65–93

Lister, N. M. (2007) "Sustainable Large Parks: Ecological Design or Designer Ecology?" in *Large Parks* (G. Hargreaves and J. Czerniak, eds.), Architectural Press, New York, Princeton, pp. 35–54

Low Impact Development Center, Inc., http://www.lowimpactdevelopment.org/ (accessed October 12, 2009)

Low, N., B. Gleeson, R. Green, and D. Radovic (2005) *The Green City: Sustainable Homes Sustainable Suburbs*, Taylor and Francis, New York

Lyle, J. T. (1994) *Regenerative Design for Sustainable Development*, John Wiley & Sons, Hoboken, NJ

Marsalek, J., B. Jiménex-Cisneros, M. Karamoutz, P. A. Malmquist, J. Goldenfum, and B. Chocat (2008) *Urban Water Cycle Processes and Interactions*, UNESCO Publishing, Taylor and Francis, Leiden, Netherlands

McHarg, I. L. (1969) *Design with Nature*, Natural History Press, Garden City, NY

McPherson, G., J. R. Simpson, P. J. Peper, S. E. Maco, and X. F. Xiao (2005) "Municipal Forest Benefits and Costs in Five US Cities," *Journal of Forestry* 103(8):411–416

Metro (2002) *Green Streets: Innovative Solutions for Stormwater and Stream Crossings*, Metro, Portland, Oregon

Meyer, E. K. (2008) "Sustaining Beauty: The Performance of Appearance," *Landscape Architecture* 98(10):92–131

Moughtin, C. and P. Shirley (2005) *Urban Design: Green Dimensions*, 2nd ed., Architectural Press, Oxford, UK

Musacchio, L. R. (2009) "The scientific basis for the design of landscape sustainability: A conceptual framework for translational landscape research and practice of designed landscapes and the six E's of landscape sustainability," *Landscape Ecology* 24, pp. 993–1013

Natural Resources Defense Council (2001) *Urban stormwater solutions: These case studies show new ways cities, developers and others are reducing stormwater pollution*, http://www.nrdc.org/water/pollution/nstorm.asp (accessed October 19, 2009)

Ndubisi, F. (2002) *Ecological Planning: A Historical and Comparative Synthesis*, Johns Hopkins University Press, Baltimore, MD

Nelson, A. C. (2004) *Toward a New Metropolis: the Opportunity to Rebuild America*, Brookings Institute, Washington, DC

New York City DEP (2005) Staten Island Bluebelt: Seminar presentation for U.S. EPA Urban Watershed Management Branch, Edison, N J., http://www.epa.gov/ednnrmrl/events/bluebeltseminar1.pdf (accessed November 1, 2009)

Newman, P. and I. Jennings (2008) *Cities as Sustainable Ecosystems: Principles and Practices*, Island Press, Washington, DC

Newman, P., T. Beatley, and H. Boyer (2009) *Resilient Cities: Responding to Peak Oil and Climate Change*, Island Press, Washington, DC

Novotny, V. and P. Brown, eds. (2007) *Cities Of The Future: Towards Integrated Sustainable Water And Landscape Management*, IWA Publishing, London, UK

Paul, M. J. and J. L. Meyer (2001) "Streams in the Urban Landscape," *Annual Review of Ecological Systematics* 32, pp. 333–365

Pickett S. T. A., M. L. Cadenassso, and J. M. Grove (2004) "Resilient cities: meaning, models, and metaphor for integrating the ecological, socio-economic, and planning realms," *Landscape and Urban Planning* 69(4):369–384

Richman, T., Camp Dresser McKee, and Ferguson Bruce (1999) *Start at the Source: Design Guidance Manual for Stormwater Quality Protection*, Bay Area Stormwater Management Agencies Association, San Jose, CA

Rohde, K. (2005) *Non-equilibrium Ecology*, Cambridge University Press, Cambridge, UK

Scania Ecoguide, "Ekostaden Augustenborg," http://www.eco-guide.net/skane/Ekostaden_Augustenborg_project.php?db_id= (accessed November 12, 2009)

Schooenboom, I. J. (1995) "Overview and State of the Art of Scenario Studies for the Rural Environment," in *Scenario Studies for the Rural Environment* (J. F. Schoute, P. A. Finke, F. R. Veenenklaas, and H. P. Wolfert, eds.), Kluwer Academic Publishers, Dordrecht, Netherlands, pp. 15–24

Seattle Public Utilities, "SEA Street and Natural Drainage Systems," http://www2.cityofseattle.net/util/tours/seastreet/slide1.htm (accessed Jan. 7, 2009)

Sijmons, D. (1990) "Regional Planning as a Strategy," *Landscape and Urban Planning*. 18 (3-4):265–273

Steiner, F. (2002) *Human Ecology: Following Nature's Lead*, Island Press, Washington, DC

Steiner, F. (1991) "Landscape Planning: A Method Applied to a Growth Management Example," *Environmental Management* 4(15):519–529

Steinitz, C. (1990) "A framework for theory applicable to the education of landscape architects (and other environmental design professionals)," *Landscape Journal* 9(2):136–144

Steinitz, C., H. M. A. Rojo, S. Bassett, M. Flaxman, T. Goode, T. Maddock III, D. Mouat, R. Peiser, and A. Shearer (2002) *Alternative Futures for Changing Landscapes*, Island Press, Washington, DC

Temorhuizen, J. W. and P. Opdam (2009) "Landscape Services as a bridge between landscape ecology and sustainable development," *Landscape Ecology* 24, pp. 1037–1052

Tjallingii, S. P. (2000) "Ecology on the edge: Landscape and ecology between town and country," *Landscape and Urban Planning* 48(3):103–119

Tress, B., G. Tress, and G. Fry (2005) "Integrative studies on rural landscapes: Policy expectations and research practices," *Landscape and Urban Planning* 70, pp. 177–191

Turner, M. G. (1990) "Landscape Ecology: the effect of pattern on process," *Annual Review of Ecological Systematics* 20, pp. 171–197

United Nations Department of Economic and Social Affairs, Agenda 21, http://www.un.org/esa/dsd/agenda21/ (accessed September 6, 2009)

United Nations Habitat (2006) *State of the World's Cities 2006/07*, Earthscan, London, UK

United Nations Millennium Ecosystem Assessment (2005) *Ecosystems and Human Well Being: Synthesis*, World Resources Institute, Island Press, Washington, DC

Vale, L., J. Campanella, and J. Thomas (2005) *The Resilient City: How Modern Cities Recover from Disaster*, Oxford University Press, Oxford, UK

Van Heerden, I. L., G. P. Kemp, and H. Mashriqui (2007) "Hurricane realities, models. Levees and wetlands," in *Cities of the Future: Towards Integrated Sustainable Water and Landscape Management* (V. Novotny and P. Brown, eds.), IWA Publishing, London, UK

Van Langevelde, F. (1994) *Habitat connectivity and fragmented nuthatch populations in agricultural landscapes*, Doctoral Dissertation, Wageningen Agricultural University, Netherlands

Veenenklaas, F. R and L. M. van den Berg (1995) "Scenario Building: art, craft or just a fashionable whim?" in: *Scenario Studies for the Rural Environment* (J. F. Schoute, P. A. Finke, F. R. Veenenklaas, and H. P. Wolfert, eds.), Kluwer Academic Publishers, Dordrecht, Netherlands, pp. 11–13

Villareal, E., Semadeni-Davies Lars Bengtsson, A. (2004) "Inner city stormwater control using a combination of best management practices," *Ecological Engineering* 22, pp. 279–298

Vogel, M. (2006) "Moving Towards High-Performance Infrastructure," *Urban Land*, October, 73–79

Walker, B. and D. Salt (2006) *Resilience Thinking: Sustaining Ecosystems and People in a Changing World*, Island Press, Washington, DC

Wu, J. and Hobbs, R. (2002) "Key issues and research priorities on landscape ecology: an idiosyncratic synthesis," *Landscape Ecology* 17(4):355–365

Wu, J. (2007) "The Staten Island Bluebelt: natural stormwater management and flood control," American Rivers Web site, http://www.americanrivers.org/ (accessed January 7, 2009)

Xiao, Q. F, E. G. McPherson, S. L. Ustin, M. E. Grismer, and J. R. Simpson (2000) "Winter rainfall interception by two mature open-grown trees in Davis, California," *Hydrological Processes.* 14(4):763–784

Zonneveld, I. S. (1995) *Land Ecology*, SBP Academic Publishing, Amsterdam

IV

STORMWATER POLLUTION ABATEMENT AND FLOOD CONTROL—STORMWATER AS A RESOURCE[1]

IV.1 URBAN STORMWATER—A PROBLEM OR AN ASSET?

IV.1.1 Problems with Urban Stormwater

Water quality of urban water bodies. Many urban waters have not met the goals of the U.S. Clean Water Act, Section 101, to attain and preserve the physical, chemical, and biological integrity of the nation's waters and the ability of the receiving waters to provide conditions for contact and noncontact recreation. However, in the biennial reports to Congress, the U.S. Environmental Protection Agency consistently cites urban sources of diffuse pollution (stormwater and erosion by urbanization) and modifications of urban streams as the leading causes of impairment of coastal waters and the second cause for urban streams and impoundments.

Urban watersheds are impacted by stormwater runoff from storm sewer outflows that contain pollutants washed from the city's many impervious surfaces and combined sewer overflows. Stormwater outflows also may discharge polluted dry-weather flow. Combined sewer overflows carry a mixture of stormwater with sewage. The flushing of pollutant particulates deposited on the surface or developed in the sewers (e.g., slime development in humid combined sewers) may contribute to toxic "first flush" discharges. The intermittent high flows from sewer overflows erode the stream bank habitat and threaten the well-being of aquatic organisms (Novotny, 2003). However, after treatment, stored rainwater and urban runoff can be a resource for providing ecological and even potable water flows and, after blending with the treated effluents and clean flows, can be used for rehabilitation of streams and reuse.

Hydrologic problems. The natural hydrologic and ecologic (habitat) status of many urban water bodies has been compromised by imperviousness and other

[1]This chapter was co-authored by Eric V. Novotny.

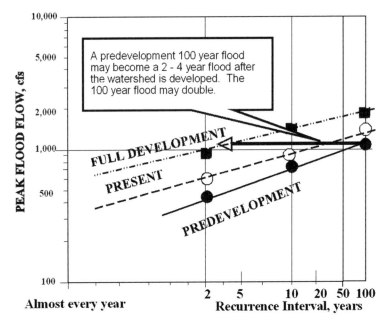

Figure 4.1   Effects of the increased magnitude and peak flows on a stream located in a rapidly urbanizing storm-sewered suburb of Milwaukee, Wisconsin. This phenomenon causes floodplain encroaching to previously flood-safe zones. To convert from cubic feet per second to m³/sec, divide by 35.3.

modifications of the watershed, by fast conveyance drainage, and by stream modifications. These hydraulic changes have resulted in (Novotny, 2003):

- Flow peak and flow volume increase by a factor of 2 to 10, as shown in Figure 4.1 (Hammer, 1972)
- Increased bank erosion caused by increased flow rates and cutting down the stream bank vegetation (Figure 4.2)
- Less base flow, sometimes to the point of disappearance of base flow, which impairs water body integrity
- Higher variability and increase of temperature of urban streams, which, in combination with less base flow, may lead to thermal shocks to the biota

Aesthetic problems. Streams carrying urban runoff have been converted to concrete channels or buried in underground culverts, and, in the ultimate demise, converted to combined sewers. As the streams were lost or became open sewers and lined and fenced-off flood conveyance channels, neighborhoods dilapidated.

Fragmentation of the ecological habitat. Chapter III introduced the concept of biodiversity that is compromised by urban development. Channelization and other modifications prevent the biota from developing and repopulating the water bodies

Figure 4.2 Stream bank erosion of a suburban creek located in a nature preserve near Milwaukee, as a result of suburban development in the watershed. Photo V. Novotny.

and the entire stream corridor (Figure 4.3). Other damages to the habitat include increased embeddedness (silting of the channels by fine sediments surrounding boulders and gravel) and loss of the riparian habitat. Chapter IX covers restoration of urban water bodies.

Pollution by urban runoff. Sources of urban runoff pollution are numerous and have been covered extensively by Novotny (2003) and many other stormwater pollution abatement manuals. The sources of urban stormwater pollution are:

- *Pollution contained in wet precipitation* – Urban rainfall is polluted by washoff of atmospheric pollutants originating from smokestacks and traffic.
- *Elutriation of toxic metals and PAHs* – Urban precipitation is acidic due to nitrous and nitric oxide emissions from traffic and sulphur acidity from power plants and dissolved metals on metallic roofs and downspouts (zinc or copper) and PAH (carcinogenic polyaromatic hydrocarbons) from asphalt shingles and pavements.
- *Erosion of pervious lands and construction sites* – Urban erosion is much higher than natural erosion, reaching magnitudes of more than 50 tons/ha-year.
- *Dry atmospheric deposition* – Urban dust, infrastructure deterioration, and pollen are the main sources of particles in dry atmospheric deposition. Some of these particles can come from large distances. For example, the atmosphere in Beijing, China, often contains high concentrations of solids originating from the Gobi desert, several hundred kilometers away.

Figure 4.3 Stream lining with a drop structure was the third and fourth paradigm engineering solution to reduce erosivity of urban streams. Biota cannot develop in such channels and cannot pass the drop structure when migrating from downstream to upstream. Photo V. Novotny.

- *Street refuse accumulation and washoff* – Street refuse that accumulates near the curb, in addition to atmospheric dry deposition in the form of street dust, also contains litter, dirt from deteriorating roads, organic solids from vegetation (fallen leaves, grass clippings), and animal (pets and birds) feces (Figure 4.4).

- *Traffic emissions* – Vehicles deteriorate and deposit solids on the street surface, including metals, asbestos, rubber, and oils.

- *Industrial pollution deposits* – Heavy industries located in or near urban areas are a source of pollution far beyond their premises.

- *Application of fertilizers and pesticides on urban lawns* – To maintain lush lawns and eradicate all "weeds," lawn-care companies and homeowners tend to use much larger quantities of industrial chemicals per area than farmers.

- *Application, storage and washoff of de-icing chemicals* – Keeping roads in winter free of ice and snow is done at great cost to receiving waters, soil, and ecology. The salt content in urban snowmelt after salt applications is extremely high and toxic (Novotny et al., 1999; Novotny et al., 2008; Novotny and Stefan, 2008).

- *Oil and fuel drips on parking lots*

- *Dry-weather infiltration-inflow (I-I) contributions* – These contributions from various sources are troublesome in most drainage systems. They can be divided into the following two categories:

Figure 4.4 Curb and gutter installation on urban streets and roads results in accumulation of pollutants near the curb and subsequent washoff into catch basins. With such practices in highly impervious urban neighborhoods, there is no attenuation of pollutants or reduction of urban runoff rates. Photo V. Novotny.

Polluted

- Flushing and cleaning impervious areas (dust, vegetation residues, litter, pet and bird fecal matter with pathogens)
- Construction site dewatering (sediment)
- Irrigation return flows from watering private lawns, public parks, and golf courses (nitrogen and phosphorus, pesticides)
- Snow and ice melt induced by de-icing chemicals applied on roads and other pavements (very high salinity, toxic metals, cyanides)
- Cross-connections from sanitary into storm sewers (organic matter, nitrogen, and pathogens)
- Filter backwash from swimming pools

Clean

- Underground springs and groundwater leaking into sewers
- Basement dewatering sumps
- Swimming pool drawdown and overflows
- *Illegal entry of pollutants* – Homeowners and small car repair shops sometimes allow car-washing detergents and oil to enter stormwater inlets.

- *Cross-connection of sewage and solids build up in sewers* – Leaking sanitary sewers can contaminate underground storm sewers and culverts and, vice versa, clean water inflows from leaking stormwater and illegal entries of roof drains into sanitary sewers can overload sanitary sewers and treatment plants, causing illegal sanitary sewer overflows (SSOs).

IV.1.2 Current Urban Drainage

Urban drainage determines the degree and type of pollution and also has an impact on flooding. The purpose of urban drainage is primarily to convey urban runoff and snowmelt from the urban area without causing flood damages, and also to transport wastes to the receiving water bodies with or without treatment. There are three types of urban drainage, each with a different pollution impact. In developed countries most cities have either combined or separated sewer systems. In some developing countries, surface channels collect most of the runoff and also gray water. As shown in Chapter I, the default drainage is on urban streets. Subsurface drainage of runoff in storm and combined sewers, or on the surface in lined channels, is not compatible with the concepts of sustainable cities; however, these systems may persist for some time.

Combined Sewers These conduits carry both dry-weather wastewater flows and, during wet weather, a mixture of rain (snowmelt) and dry-weather flows. As pointed out previously, I-I inputs into the systems may occur during both dry-weather and wet-weather conditions. Dry-weather flows mostly contain sewage, but the I-I inputs can sometimes be overwhelming.

When the capacity of the downstream sewer interceptor is exceeded during wet-weather conditions, the excess water overflows from the sewers into the nearest surface watercourse, such as a river or seashore and sometimes floods basements. The flow capacity of combined sewers is typically about six times the dry-weather flow. The pollution effect of *combined sewer overflows* (CSOs) is of great concern because:

- The overflow contains untreated sewage.
- Highly objectionable solids and slimes develop in the sewers because of insufficient flushing velocity during dry-weather conditions, causing these solids to be then flushed into the overflow during subsequent larger and higher-velocity wet-weather flows as first flush.
- During medium and large storms, the CSOs can bring far more pollution into the receiving water than the load discharged from treatment plants.

Pollution by CSOs must be controlled. In the United States and other advanced countries, CSOs are regulated, requiring municipalities and industries to apply for a discharge permit. Such permits may specify the allowable frequency of overflows and/or the necessary controls and penalties for noncompliance.

Typically, the ordinances used to control pollution by CSOs require a certain volume of the overflows to be captured and treated. The typical number of overflows from combined sewer systems with some excess sewer capacity ranges from 30 to 70 per year. In some EU countries (e.g., Germany) and some U.S. urban areas, the requirement of the ordinances is to capture and treat 90% of the CSO volume, which corresponds roughly to about 10 allowed overflows in an average year. Under the fourth paradigm, oversized interceptors ("deep tunnels") (Figure 1.18) were excavated or drilled in several communities worldwide, such as Milwaukee (Wisconsin), Chicago (Illinois), Tokyo (Japan), and Singapore, to store and convey wastewater flows and pump them into the treatment plants.

Separated Sewer Systems   Two conduits are required: the *sanitary sewer* carries more concentrated dry-weather flows, and *storm sewers* carry less-polluted urban runoff. I-I inputs can enter either system. The capacity of storm sewers in developed countries is a high flow that has a frequency of being equaled or exceeded once in five to ten years. Concrete-lined man-made surface channels and channelized lined streams carry larger flows (see Chapter I).

The level of pollution in urban runoff and the magnitude of the loads carried by storm sewers are related to the level of imperviousness of the drainage area. As the population density increases, higher emissions from cars, more pet (dog) waste and litter deposited on impervious surfaces, pesticides, oils from parked vehicle drippings and maintenance, and other wastes are washed directly into storm sewers or dumped directly into storm drains. Having more people in less space results in greater concentrations of pollutants that can be mobilized and discharged into the municipal storm sewer systems.

Many studies throughout the world have indicated that urban stormwater runoff carries higher annual loadings of chemical oxygen demand (COD), total lead, and total copper than effluent from secondary treatment plants, and much higher concentrations of oil and grease and polyaromatic hydrocarbons (PAH). The findings showed that fecal coliform counts in urban runoff typically range from thousands to millions of counts per 100 ml. Urban runoff also carries the highest loads of toxic micropollutants.

Sanitary sewer overflows (SSOs) and undesirable high-frequency CSOs in suburban areas can also be caused by rapid (uncontrolled) development that is not matched by increasing the capacity of sewers. Also, the point where the separate sanitary sewer from an outlying urban or suburban community is connected to an older combined sewer system of the urban center is a possible cause of SSOs. In some cases, if the downstream system is overloaded with combined sewage and cannot accept the upstream input of the sanitary flow, SSO will occur.

Figure 4.5 shows the origins and pathways of runoff and sewage pollution in communities drained by the fourth paradigm separate sewer systems.

Natural Drainage   "Natural" drainage may be a misnomer because even in low density less impervious urban areas most of the original natural drainage pathways were modified or lost by building roads, streets and houses. In low

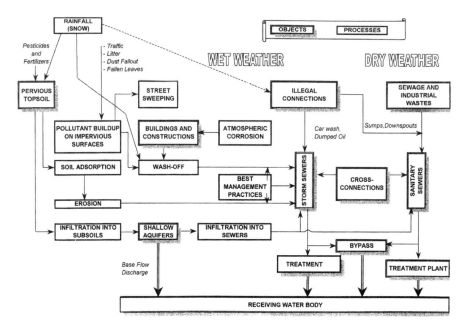

Figure 4.5 Schematic diagram of sources of overall pollution from urban areas (reprinted from Novotny, 2003 with permission).

impact development (LID) communities storm sewers are not needed, and the storm drainage can be accomplished by roadside swales, engineered grassed waterways, rain gardens, creeks and canals, and so forth. The pollution control benefits and cost savings are significant (see subsequent sections); measurements in Wisconsin have documented that pollution loads from urban and suburban low-density zones, even without best management practices, are only about 10 to 30% of the loads that would occur if the same area had storm sewers.

As pointed out above, the I-I problem and illicit connections with clean water require attention and the education of the public and plumbers. This problem may be magnified in communities that have sanitary sewers.

Natural drainage today is not just ditches and creeks. In the LID communities, and even more so in the Cities of the Future, natural drainage replaces storm and combined sewers, leaving only sanitary sewers in place, as has been done in Malmö, in Sweden (Stahre, 2008). Most of the remainder of this chapter focuses on implementation of natural storm drainage in the new and retrofitted sustainable communities.

IV.1.3 Urban Stormwater Is an Asset and a Resource

It is erroneous to look at urban runoff as pollution and a flood problem, and to deal with it using the same concepts as are applied to sewage—namely, putting it out of sight. That practice was a consequence of the urban drainage practice paradigm dating back to the ancient Greeks and Romans, which can be paraphrased as:

- Pave the city.
- Collect dirt, trash, and fecal matter from animals (in old times from horses and donkeys, and in India from cows) and sometimes from people on the pavement and in the gutters.
- Elevate the walkways by installing curbs so that people would not step into the solid waste, and also provide stepping stones (in ancient cities) so that streets can be crossed by pedestrians without stepping into the polluted street runoff.
- Allow rain and street flushing to push the putrescible solids and other pollution (e.g., animal urine) into underground conduits or surface masonry conduits.
- Do not worry about the consequences of pollution of urban streams; they can be covered and put out of sight and converted to combined sewers.
- Clean water can be brought from a large distance at a large cost.

The irony is that people in the pre-20th-century cities liked water and with great efforts brought water to elaborate fountains, to the gardens of the rich and the aristocracy, and to the public baths of ancient Rome. They enjoyed picnicking and boating on rivers, as beautifully illustrated by Monet and other 19th-century (and earlier) painters. In Rome, the public baths were the last place to lose water during shortages.

Water and waterscapes are a great asset to a city and to the new water centric architecture (Dreiseitl and Grau, 2009), and clean urban runoff, stored and reused, is as good a source of water for potable and nonpotable uses as the raw water brought from distant rivers (e.g., Colorado River aqueducts to California and Arizona) or drawn from deep saline groundwater aquifers and treated at great expense. The illustrations and discussions throughout this chapter document the rediscovered value of rainfall and urban runoff as:

- A valuable and easy-to-treat source of potable water
- A source for recharging depleted groundwater aquifers
- A good traditional source of water for irrigation
- A source of base flow of restored/daylighted urban streams, ponds, and lakes
- A source of water for the decorative purposes of urban waterscapes
- A source of water for cooling homes (e.g., by irrigating green roofs)
- A source for waterscapes in urban parks and recreational areas, for the enjoyment of people, and especially children
- A source of water for private courtyards and swimming pools
- A source for water arts and fountains, in city plazas and water festivals

What is needed are source controls, collection, storage, and passive soft treatment before rainwater and stormwater become heavily polluted.

This century is a period of rediscovering urban water resources, of which rainwater and urban runoff are the main component. Greater public awareness has been accelerated by the notable examples recently realized in many cities. This transition from runoff being a menace and hazard to runoff as an asset and social and ecological benefit will require a total change from

- *fast conveyance systems* characterized by high imperviousness, curb and gutter runoff collection, and conveyance by underground sewers, culverts, and surface concrete-lined channels and hard treatment by treatment plants

to

- *storage-oriented, slow-release systems* characterized by *storage* in ponds, on flat roofs, in underground cisterns, ponds, lakes, etc.; *infiltration* into shallow aquifers; *soft treatment* (rain garden biofilters, earth filters, wetlands, ponds); *slow conveyance* in grassed swales (raingardens) and natural or nature-mimicking surface channels

Fast conveyance has no social benefit except getting rid of water as quickly as possible. Storage-oriented management has many social as well as economical benefits.

IV.1.4   Low Impact Development (LID)

Starting in the 1970s (for example, The Woodlands near Houston in Texas), best management practices have been developed to control stormwater runoff pollutant levels and quantity. Originally the philosophy behind the development of the urban best management practices was to collect, convey, and treat urban runoff. Only very recently, in the last 15 to 20 years, has an increased emphasis on conservation and use of on-site natural features been considered. This type of water management is called **Low Impact Development** in the United States, **Sustainable Urban Drainage System (SUDS)** in Europe, and **Water Sensitive Urban Design (WSUD)** in Australia. Throughout this chapter all three of these systems will be referred to as Low Impact Development, or LID.

LID was created to decrease the impact of urbanization and increased impervious surfaces. It was first developed and demonstrated in Maryland (Prince George County, 1999) and was used to try to maintain the predevelopment hydrology of an area in order to control runoff and the transport of pollutants. Traditional stormwater management practices concentrated only on reducing peak flow rates to prevent flooding, whereas LID techniques also try to reduce runoff volumes by utilizing storage and infiltration systems to more effectively mimic the predevelopment hydrology (Dietz, 2007). LID techniques also promote the use of natural systems, which can effectively remove nutrients, pathogens, and metals from stormwater. A widely publicized example of LID is the Street Edge Alternative (SEA) developed by the

Seattle (Washington) Public Utilities, which was already introduced in Chapter III. Aspects of LID fall under the following categories (U.S. EPA, 2007):

Conservation design practices reduce the disturbances to the predevelopment environment. These practices can reduce the need for large structural runoff controls.

- Reduction of impervious surfaces through design, i.e., reducing road widths, parking areas, and sidewalks and using shorter or shared driveways
- Preserving existing wetland and riparian areas, forest tracts, and areas of porous soils, i.e., cluster development, open space preservation
- The use of site fingerprinting to minimize disturbances and compaction of soils during construction, i.e., delineating smallest possible areas, restricting ground disturbances to areas where structures, roads, and rights of way will exist after construction; elimination of complete clearing and grading of sites before construction

Infiltration practices are designed to capture and infiltrate runoff, reducing runoff volume and the amount discharged to receiving water bodies. These practices also recharge groundwater systems and help maintain stream temperatures.

- Infiltration basins, porous pavements, disconnected downspouts, rain gardens

Runoff storage is designed to capture runoff from impervious surfaces for infiltration and/or storage. The strategy lowers peak flows and reduces discharge to surface waters.

- Void space below parking lots, streets, and sidewalks used for placing storage tanks and rain cisterns; increasing depression storage in landscape islands and in tree shrub and green roofs
- Use of swales, as in the SEA development in Seattle, Washington, where swales are hydraulically connected into three groups, with each group controlled by a flow control structure; detention volume achieved by swales was 37% less than would have been required by the drainage ordinance

Runoff conveyance systems in LID are used to slow flow velocities, lengthen the runoff time of concentration, and delay peak flows discharge.

- Use of alternatives to curb and gutter systems: rough surfaces that slow runoff and increase evaporation and settling of solids, permeable and vegetated surfaces promoting infiltration, filtration, and some biological uptake of pollutants
- Use of grassed swales and/or grass-lined channels, roughening of surfaces, low flow paths over landscaped areas, smaller culvert pipes and inlets, terraces, and check dam

Filtration systems filter runoff through media that capture pollutants using the processes of physical filtration of solids and/or cation exchange of dissolved pollutants. These systems have the same benefits as infiltration systems, with the added benefit of pollutant removal.

- Bioretention/rain gardens, vegetated swales, vegetated filter strips/buffers

Low impact landscaping involves the use of vegetation to improve the hydrologic impact of the development.

- Planting native, drought-tolerant plants, converting turf areas to shrubs and trees, reforestation, encouraging longer grass length, planning wildflowers meadows rather than turf along medians and in open spaces, amending soil to improve infiltration

The use of LID can have a number of environmental benefits, including the following listed by the U.S. EPA (2007):

- Reduction of pollutants in runoff through settling, filtration, adsorption, and biological uptake
- Protection of downstream water resources by reducing the amount of pollutants and volume of water reaching the water bodies, which reduces stream channel degradation from erosion and sedimentation, and improves water quality as well
- Increased groundwater recharge through increased infiltration used to offset increases in impervious surfaces in a watershed
- Reduction in frequency and severity of CSOs by limiting the volume of water entering these systems
- Habitat improvement through increased vegetation and the use of indigenous plants

In addition to the environmental benefits of these systems are land value benefits. These include the following (U.S. EPA, 2007):

- Reduction in downstream flooding and property damage by reducing surface runoff
- Reduction in pollutant treatment costs
- Increases in real estate value
- Increased lot yield because LID practices typically do not require as much land area as traditional stormwater controls with larger ponds and wetlands
- Increased aesthetic value and improvement of quality of life by providing open space for recreation

The Street Edge Alternative LID system retains 98% of water in a two-year design storm. Because most pollution is carried by storms of lesser magnitude, the pollution load is also significantly reduced, which reduces the impacts of urban environments on receiving streams inhabited by salmon. The system is designed to contain, without excessive ponding or flooding, a 25-year storm (U.S. EPA, 2001).

IV.2   BEST MANAGEMENT PRACTICES TO CONTROL URBAN RUNOFF FOR REUSE

Stormwater management in the Cities of the Future will use LID alternatives to the greatest degree. Hence, the goal of best management practices will be changed from *collect, convey, store and treat* to *retain, attenuate, repair hydrology, provide water for ecology, and enjoy aesthetics.*

Hill (2007) adopted and investigated several urban BMPs that can be used to enhance urban landscape, such as woody buffers and grassed waterways (swales) along streets and in medians, infiltration areas, green roofs, and rain gardens, and the like. She outlined the necessary steps and a research agenda leading to more ecologically sustainable urban landscapes with urban water in focus:

- *Eco-mimicry.* The goal is to develop an urban landscape that would mimic, but not necessarily reproduce, the processes and structures present in a pre-development natural system. Eco-mimicry includes hydrological mimicry, i.e., relying on reduction of imperviousness, increased infiltration, surface storage, and use of plants that retain water (e.g., coniferous trees). Hill emphasizes that it is also necessary to ascertain that water flowing towards the (restored) water bodies would contain the necessary components and temperature to support aquatic life, and be free of pollutants impairing it. However, Hill realized there are limits to what can be mimicked, and urban systems will need active human participation to remove pollutants, pathogens, and/or invasive exotic species.

- *Green space for recreation and biodiversity conservation.* The health and recreational needs of an increasing human population living in new or retrofitted cities can be satisfied by interconnected green spaces inside the city and along water bodies. Pedestrian bicycling and walking paths in and to these interconnected spaces can support transportation alternatives as well. Riparian green zones, if properly designed, can also serve as successful habitat zones for urban biodiversity conservation, and as buffers to pollution inputs from the surrounding urbanized watershed and for flood storage.

- *Urban brownfield remediation and development.* Heavy industries and railroad yards that were central to urban industrialization in the past left behind a legacy of pollution. Serious public health and environmental justice issues must be addressed. Disadvantaged population groups often live on or near these polluted but remediable sites. Currently undeveloped sites with contaminated soil are a significant source of urban diffuse pollution. These sites can become public parks and conservation areas, provide new industrial or commercial land, contain ponds and wetlands for storage and treatment of runoff, and in some cases even be used for Cities of the Future (COTF) sustainable housing (Chapter XI). A large-scale (and costly) brownfield landscape and stream restoration is being conducted in the Emscher River watershed in the Ruhr district in Germany (U.S. EPA, 2006).

IV.2.1 Soft Surface Approaches

A number of approaches are used to reduce the impact a development has on the overall hydrology and/or pollutant transfer of a watershed. The key systems discussed in this section are LID soft approaches. Each of these systems can be incorporated into the existing landscape without the use of hard infrastructure. Each of the systems is designed to reduce the amount of runoff exported from a site and reduce the peak flow rates created by the addition of impervious surfaces to a watershed. An ecological landscape architect's design is crucial. The main systems discussed herein are green roofs, pervious pavements, rain gardens, rainwater harvesting, vegetation filter strips, biofilter strips, ponds, and wetlands. All of the systems are localized treatment or runoff reduction systems that can be used individually or as a part of a portfolio that will renaturalize the hydrology and ecology of the landscape without extensive storm sewer systems.

Green Roofs Green roofs are a LID system designed to limit the impervious surfaces of the city. They are the first step for collecting rainwater, using it for building cooling and insulation, and providing water for rainwater harvesting. Green roofs are a type of environmental system designed to help mimic the predevelopment hydrology of an urban development by incorporating vegetation and soils on top of the impervious rooftop (Figure 4.6). The systems consist of a vegetation layer, a substrate layer used to retain water and anchor the vegetation, a drainage layer to transport excess water off the roof, and specialized waterproofing and root-resistant material between the green roof system and the structural support of the building (Mentens et al. 2003).

These systems are divided into two main types, *extensive* and *intensive* (Figure 4.7), described also in Table 4.1. In extensive systems the substrate layer has a maximum depth of 15 cm. The shallow soil depth only allows for certain types of vegetation to grow, such as herbs, grasses, mosses, and drought-tolerant succulents such as *Sedum* (Getter and Rowe, 2006). The most widely used plant in extensive green roofs is *Sedum* because of its superior survival in substrate layers as thin as 2 to 3 cm (VanWoert et al., 2005; Heinze, 1985). This type of green roof can be installed on sloped surfaces up to 45° and is suitable for placement on existing buildings. Examples of extensive green roofs are shown in the bottom two pictures in Figure 4.6.

Two studies examined the effect that increased storage of runoff through the installation of green roofs could have on a city. In Toronto it was estimated that if 6% of the roof surface area were green, the impact on stormwater retention would be

Figure 4.6   Examples of green roofs. Top left: The Solaire, New York, NY; top right: Schwab Rehabilitation Hospital, Chicago, IL; bottom left: building in New York; bottom right: Chicago City Hall (photo credits: American Hydrotech, Inc., http://www.hydrotechusa.com; Columbia University Center for Climatic Research).

Figure 4.7   Illustrations of the two types of green roofs: (left) intensive, (right) extensive (American Hydrotech, Inc., http://www.hydrotechusa.com).

Table 4.1 Comparison of extensive and intensive green roofs

Characteristics	Extensive Roof	Intensive Roof
Purpose	Functional, stormwater management, thermal insulation, fireproofing	Functional and aesthetic, increased living space
Structural requirements	Typically within standard weight-bearing parameters, additional 70–170 kg/m^2	Planning required in design phase or structural improvements necessary; additional 290 to 970 kg/m^2
Substrate type	Lightweight high porosity, low organic matter	Lightweight to heavy, high porosity, low organic matter
Average substrate depth	2 to 15 cm	15 cm or more
Plant communities	Low-growing stress-tolerant plants and mosses (e.g., *Sedum spp.*, *Sempervivum spp.*)	No restrictions other than those imposed by substrate depth, climate, building height and exposure, and irrigation facilities
Irrigation	None required	Often required
Maintenance	Little or no maintenance required, some weeding and mowing as necessary	Same maintenance required as for similar garden at ground level
Cost (above waterproof membrane)	$100 to $300 per m^2	$200 per m^2 or more
Accessibility	Generally functional rather than accessible	Typically accessible
Percent runoff from rainfall[+]	20–75	15–35

Source: Oberndorfer et al. (2007), Mentens et al. (2005)
[+] Depending on the slope of the roof and vegetation

equal to building a $60 million storage tunnel (Peck, 2005). Another study calculated that if 20% of all buildings in Washington, D.C., capable of supporting a green roof had one, more than 71,000 m^3 would be added to the city's stormwater storage capacity, resulting in 958,000 m^3 of rainwater retained in an average year (Deutsch et al., 2005).

In addition to the main benefit of reducing runoff from the roof's surface, there are other benefits to the installation of a green roof. These include the ability to mitigate the urban heat island effect (U.S. EPA, 2003), improve building insulation and energy efficiency, increase biodiversity and aesthetic appeal, and reduce runoff water temperature. In Toronto, Canada, the roof membrane temperature for a typical nongreen roof structure reached as high as 70°C, while temperature on a green roof on the same day only reached up to 25°C (Liu and Baskaran, 2003). Covering the waterproofing membrane also eliminates degradation due to UV light exposure. The reduction in these two damaging processes can extend the life of a waterproofing membrane by 20 years (U.S. EPA, 2000).

The addition of a green roof can limit the transfer of heat through the roof, reducing the energy demands of a building (Del Barrio, 1998; Theodosiou, 2003). Greens roofs can also reduce the urban heat island effect. Because in cities with high imperviousness water is not retained in the soils as a result of runoff from impervious surfaces, the amount of water available for evapotranspiration is reduced. This allows more of the incoming solar radiation to be transformed into sensible heat instead of functioning to evaporate water (Barnes et al., 2001). This combined with other factors can result in urban air temperatures 5.6°C warmer than those of the surrounding countryside (U.S. EPA, 2003). Green roofs can be one tool towards the restoration of the heat balance. In Toronto 50% of green roof coverage distributed evenly throughout the city was found by modeling analysis to reduce temperatures by up to 2°C (Bass et al., 2003).

Porous (Pervious) Pavements Porous pavement is an alternative to conventional pavement whereby rainfall is allowed to percolate through the pavement into the subbase. The water stored in the subbase then gradually infiltrates into the subsoil or is drained through a drainage pipe.

Porous pavement provides storage and retention, and enhances soil infiltration that can be used to reduce runoff and combined sewer overflows. Porous pavements are made either from asphalt or concrete in which fine filler fractions are missing, or are modular. Typical construction of a porous asphalt system includes several layers (Figure 4.8) (U. of NH Stormwater Center [UNHSC], 2009). The top layer is the pervious pavement layer, which can range between 10 and 15 cm thick. In this layer

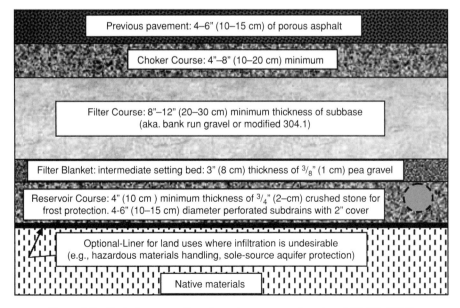

Figure 4.8 Typical cross section of pervious pavement system (UNHSC, 2009).

the sand particles below 2 mm are removed, creating a percent void space in the pavement of 18–20%. Below the surface layer is a choker course layer created with 2 cm crushed stone, followed by the filter course and filter blanket layers consisting of finer filter materials such as sand and gravel. Below the filter blanket material is a reservoir course layer, which can include a subdrain if needed. Below this layer would be the native materials. An optional impervious layer could be installed between the reservoir course layer and the native materials if infiltration is undesirable in the area, allowing for the system to be a reservoir with the runoff exiting through the drainage pipe.

There are a number of benefits of the use of porous pavements, the primary benefit being a significant reduction or even complete elimination of surface runoff rate and volume from an otherwise impervious area. If the pavement is designed properly, all or most of the runoff can be stored and subsequently allowed to infiltrate into the natural ground. Aquifer recharge by infiltrated water is the second important benefit. The third benefit is the reduced need for storm drainage, or even its elimination. A final benefit is the removal of contaminants from the runoff both through the filtration of the water through the soils and in the reduction of the amount of salt needed for de-icing purposes.

Field (1986) summarized the results of several U.S. EPA studies on the experimental applications of porous pavements. Results from a study in Rochester, New York, indicated that peak runoff rates were reduced by as much as 83% where porous pavement was used. In a recent study at the University of New Hampshire, no surface runoff was observed during the four years of operating a parking lot constructed with pervious pavements, even during 100-year storm events (Porous Asphalt Fact Sheet – UNHSC,2009). Typically, hydraulic conductivity (permeability) of porous pavements is much greater than runoff rates. Hydraulic conductivity measured by Jackson and Ragan (1974) was about 25 cm/hr, which is an order of magnitude higher than a typical catastrophic design storm. This means that infiltration into the base should occur without ponding. Infiltration rates remain high even during winter months when frost settles in the pavements. Winter infiltration remained constant in the New Hampshire study with some of the highest infiltration rates. Even during frost penetration, pores remained open and infiltration was able to occur. Hence, in addition to runoff attenuation and water quality, installation of porous pavements greatly improves traffic safety by reducing skidding.

The use of porous pavements can also increase the water quality of the water infiltrating through the soils. Water quality treatment exceeded U.S. EPA recommended levels of removal of suspended solids, and met regional ambient water quality criteria for petroleum hydrocarbons and zinc (UNHSC, 2009). Near 100% annual removal of total suspended solids (TSS), total petroleum hydrocarbons, and zinc was obtained with the pervious pavement systems in a parking lot in New Hampshire. However, due to the lack of vegetation in the system, there was limited phosphorus treatment (about 40%) and no removal of nitrate nitrogen.

While chlorides, due to their conservative characteristics, cannot be removed through the filter systems used in the porous pavements, less salt is needed. Due to the removal of ponding on the parking lots surfaces and the increased traction porous pavements have over standard impervious surfaces, chloride applications needed to remove ice were almost completely eliminated. Winter maintenance of porous asphalt required between zero and 25% of the salt routinely applied to impervious asphalt to achieve equivalent or better de-icing traction (UNHSC, 2009). This was a result of no black ice development due to reduction in ponding on the surface of the pavement.

Costs to construct a porous asphalt system are comparable to, but slightly higher than, a conventional impervious pavement with an additional 20–25% added material cost (UNHSC, 2009). However, real savings are observed when stormwater management practices are taken into account, to store and treat the runoff from a standard impervious pavement surface, especially when considering construction and land costs for the runoff treatment systems. Additional stormwater management practices are typically not needed in a pervious pavement system. However, maintenance costs will be higher for a pervious pavements system, requiring two to four vacuum sweepings (vacuum-assisted dry sweeper only) in a year. Researchers at the University of New Hampshire Stormwater Center have only experienced moderate clogging after two winters with no maintenance. Sand brought in by cars from other areas has been shown to cause clogging in the pervious pavements in high-trafficked areas of a parking lot.

The use of porous pavements has not been found to reduce the load-bearing strength or structural integrity of the system if properly installed. In fact the longevity of porous pavement is longer than that of traditional pavement systems due to the well-drained base of the system and the reduction in freeze thaw inside the pavement. The use of these systems is best in low-use roadways, parking lots, and alleys without a modified asphalt binder.

Contamination of shallow aquifers by toxic materials attributed to asphalt, vehicular traffic, and road usage, including salt application for de-icing, represents a slight to moderate environmental risk that depends on soil conditions and aquifer susceptibility. In order to minimize the impact on shallow groundwater, 1-meter vertical separation is needed between the pervious pavement and the seasonal high groundwater (UNHSC, 2009).

Rainwater Harvesting   Rainwater harvesting is another way to reduce the amount of runoff from an impervious area. By collecting runoff from a roof or other impervious surfaces and using it after the rainfall event for irrigation or even drinking water, the peak flows from a development can be reduced. Further examples and a more detailed explanation of rainwater harvesting are given in Chapter V. A well-known large concrete apron on the east side of the Gibraltar mountain connected to a cistern provided all the water for this British city at the southernmost tip of the Peloponnesian (Spanish) area. Today water is obtained by a pipeline from Spain.

Rain Gardens   Rain gardens, also known as bioretention, are depressed areas planted with shrubs, trees, or perennials that receive and infiltrate runoff. Infiltrating the water through a soil medium decreased surface runoff, increased groundwater recharge, and achieved removal of some pollutants (Galli, 1992). A more extensive explanation of rain gardens was presented in Chapter III.

Figure 4.9   Type of swale found in a Seattle SEA neighborhood. The swale conveys, treats, and infiltrates runoff Courtesy Seattle Public Utilities.

Grass Swales   Grass swales are grassed earth channels used to collect stormwater runoff while directing it to other stormwater management systems or conveyance elements (Marsalek and Chocat, 2003). Today, in contrast to old roadside highway engineering ditches, grass swales have milder slopes and are planted with native or even decorative flowers (Figure 4.9). Grass swales provide a number of advantages, including slower flow velocities than a standard stormwater pipe system. This results in a longer time of concentration and a reduction of peak discharges. Grass swales have the ability to disconnect impervious surfaces such as driveways and roadways, thus reducing the overall runoff curve number and the peak discharges of the National Resources Conservation Service's hydrologic model of rainfall-runoff transformation. They can also filter pollutants through the grass medium, soil filtration, and uptake of pollutants by plant roots, also known as phytoremediation. These systems promote infiltration, further reducing peak discharges (Clark et al., 2004). A typical cross section of a grass swale is shown in Figure 4.10. This type of system also could have a drainage pipe to remove excess runoff and reduce growth of cattails.

Filter Strips   Filter strips are vegetated sections of land designed to accept runoff as overland sheet flow from an upstream development, or flow from a highway or a parking lot (Figure 4.11). Filter strips remove the pollutants from runoff by filtering,

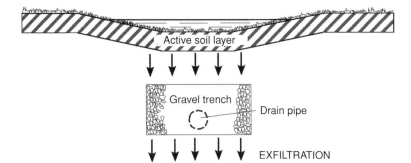

Active soil layer

Gravel trench

Drain pipe

EXFILTRATION

Figure 4.10   Typical cross section of a grass swale.

Figure 4.11   Example of a simple filter strip along a parking lot, filtering runoff before entering storm drain.

Table 4.2   Guidelines for grass filter strip design

Design Parameter	Design Criteria
Filter width	Minimum width 15 to 23 meters (50 to 75 feet), plus additional 1.2 meters (4 feet) for each 1% slope
Flow depth	5–10 centimeters (2–4 inch)
Filter slope	Maximum slope of 5%
Flow velocity	Maximum flow velocity of 0.75 m/s (2.5 fps)
Grass height	Optimum grass height of 15 to 30 cm (6 to 12 inches)
Flow distribution	Should include a flow spreader at the upstream end to facilitate sheet flow across the filter

Source: Minnesota Pollution Control Agency, 1989; Novotny and Olem, 1994; Novotny, 2003

provide some infiltration, and slow down the runoff flow. The dense vegetative cover facilitates pollutant removal. Filter strips cannot treat high-velocity flows. Therefore, they are generally used for small drainage areas. Grass filter strips provide higher pollutant removal rates than grass swales. The difference is in the type of flow. Flow depth in the filter strip is less than the height of the grasses, creating laminar flow conditions that enhance settling and filtering. Flow in swales is concentrated and flow depth is greater than the height of grasses, which results in turbulent flow. Vegetated filter strips are feasible in low-density developments with small drainage area and areas bordering roads and parking lots. General guidelines for grass filter strips are outlined in Table 4.2.

Strips are effective in removing sediment and sediment-associated pollutants such as bacteria, particulate nutrients, pesticides, and metals. Infiltration is an important removal mechanism in filter strips. Many pollutants, including phosphorus, are dissolved or associated with very fine particles that move into the soil with infiltrating water. Once in the soil profile, a combination of physical, chemical, and biological processes traps the pollutants. Infiltration is also important because it decreases surface runoff, which in turn reduces the ability of runoff to transport pollutants.

In a study of grass filter strips conducted by Lee et al. (1989), it was concluded that removal of sediment and nutrients from surface water by filter strips is primarily the result of the infiltration of dissolved nutrients and reductions in sediment transport capacity caused by decreased water volume and increased resistance to overland flow. Sediment deposition was determined to be the major trapping mechanism for phosphorus removal by buffer strips. Increased infiltration in buffer areas is ascribed to flow retardation due to the increased surface roughness caused by vegetation, and good soil aggregation due to increased soil organic matter. It was determined that biological uptake of nutrients during single events was negligible.

The distance at which 100% of the sediment is removed by a filter strip is called the "critical distance" (Novotny and Olem, 1994). In a study using Bermuda grass, >99% of the sand was removed in a distance of 3 meters, silt in 15 meters, and clay in 120 meters, respectively (Wilson, 1967).

Environmental Corridors and Buffer Zones   The urban environmental corridors—usually a park or open vegetated land along a stream or lake, or adjacent to the drainage system—act as an ecotone buffer between the polluting urban area and the receiving water body. They also contain walking and bike paths, picnicking areas, playgrounds, and the like (Figure 4.12). Ideally the corridor should include most of the floodplain. In most cases the buffer zones also provide storage for flood and pollution control (Wiesner, Kassem, and Cheung, 1982) and generally are a part of a floodplain and/or major drainage system. Chapter IX covers stream restoration, including preservation and restoration of corridors also serving as stream buffers. Stream buffers are used in both urban and agricultural watersheds. The corridors will lose their efficiency if the storm drainage outlet bypasses the grassed and vegetated areas and discharges directly into the receiving water body, or into a channel with concentrated flows that is directly connected with the water body. The "treatment" processes for storm runoff—such as vegetated filters, infiltration basins,

Figure 4.12   Environmental corridor of the Trinity River in the Dallas–Fort Worth area (Courtesy CDM, Dallas).

detention-retention ponds (dry or wet), and wetlands—are incorporated into the landscape of the corridor.

Buffer strips made of uneven dense shoreline vegetation may also be used to attenuate runoff pollutants, which otherwise would reach the water body. Woodard (1989) measured the efficiency of buffer strips that had vegetation typical of Maine lakeshore (mixed growth, uneven age stand, moderate ground cover of shrubs, ferns, etc.). Similar measurements were made by Potts and Bai (1989) in Florida. They found that the critical distance of a grass strip that was used for control of suspended sediment and phosphates from residential developments was 22.5 meters; however, it was concluded that the efficiency of the buffer strips is highly dependent upon a sufficient cover of organic matter (natural vegetation) and on the initial concentration of the pollutants or density of shoreline development. A 30-meter-wide buffer strip is recommended for protection of surface water reservoirs from which water is used for drinking water supply in states where residential development is permitted in such watersheds (Nieswand et al., 1990). In some countries, urban and agricultural land use practices are greatly restricted or not permitted at all in watersheds of water supply reservoirs.

Buffer strips are ineffective on steep slopes with loose soils. Also, their effectiveness is reduced by exposed soil on any part of the buffer strip, which can actually erode and contribute suspended solids and other pollutants, instead of attenuating them. Woodard (1989) recommends that a porous organic "duff" layer and/or a dense growth of underbrush must cover the mineral soil if buffer strips are to be effective.

Biofilters   Any system that uses vegetation to help remove pollutants from stormwater runoff is considered a biofiltration system. This includes rain gardens, swales, environmental corridors and buffer zones, grass filter strips, and to an extent green roofs. Incorporating plants into the filtration system instead of using only a soil-based structure, increases removal of nitrogen and phosphorus, along with heavy metals such as Pb, Zn, Cu, and Cd (Breen, 1990; Song et al., 2001; Read et al., 2008, among others). Other forms of biofiltration systems simply consist of an excavated trench or basin containing vegetated filter media. Below the filter media is a perforated pipe that is used to collect the treated water and deliver it to a stormwater drainage network or directly to a waterway (Davis et al., 2001a; Hatt et al., 2007; Henderson et al., 2007). Through the filtration process fine particles are trapped at the surface, while dissolved particles are removed as they travel through the soil. Dissolved particles are either removed through adsorption to the media or are taken in by the plant or microbiological community (Hatt et al., 2007).

Plants can contribute to the removal of pollutants through many processes. These include the degradation of organic pollutants and the uptake of macronutrients such as nitrogen and phosphorus used in plant cycles, and even the uptake of heavy metals (Breen, 1990; Schnoor et al., 1995; Cunningham and Ow, 1996). Plants can also influence the microbiological community present in the soil media. Microbes can be found both within and along the root structure or the rhizosphere. Plants can influence these communities by providing different organic substrates, modifying the water retention between storm events, and even through changing or controlling the soil pH (Schnoor et al., 1995; Salt et al., 1998).

The type of plants can influence the removal of phosphorus and nitrogen from the stormwater. Metals are generally removed in any soiled media. In a study in Australia, removal of nitrogen and phosphorus was found to differ up to 20-fold between plant species (Read et al., 2008).

According to the University of New Hampshire Stormwater Center, the efficiencies of vegetated swales to remove pollutants on an average annual basis were 60% for total suspended solids and 88% for zinc, respectively. Most of the removal was due to infiltration and grass filtration of solids.

IV.2.2 Ponds and Wetlands

Ponds   Ponds and storage basins are still the backbone of urban stormwater quantity-quality management. A combination of a pond (storage and pretreatment), wetland (treatment), and infiltration or irrigation can result in sustainable stormwater disposal and reuse. Very often ponds and wetlands are constructed together, or a constructed pond is combined with a restored wetland. As a matter of fact, the distinction between shallow ponds with a lush littoral zone, dry ponds with temporary pools of water and some wetland vegetation, and wetlands is fuzzy. They are a great asset to the urban landscape, if designed properly by landscape architecture concepts (Chapter III).

Today and in the future, wet ponds and retention areas are and will be designed by or in a close cooperation with landscape architects. These new designs are treated by the public as urban lakes that increase the value of riparian land and provide recreation and enjoyment, which contrasts with the past engineered, often lined rectangular retention basins with signs to keep the public away. Well-designed ponds can be stocked with fish. Nutrient control (e.g., by educating homeowners in the watershed about the overuse of fertilizers) is necessary to prevent eutrophication.

Two types of detention basins are used for quality control of urban runoff. The first type includes *wet detention ponds*, which maintain a permanent pool of water with an additional storage area designated to capture transient storm runoff. The second type is *extended or modified dry ponds*, which provide a part of their storage capacity for enhanced settling of solids and auxiliary removal of pollutants by filtering.

A *dry pond* is a stormwater detention facility that is designed to temporarily hold stormwater during high peak flow runoff events. The outlet is restricted to activate the storage during flows that exceed the outlet capacity. A safety overflow spillway is also a part of the pond, for conveyance of very high flows when the storage capacity of the pond is exhausted. Since the outlet of dry ponds used for flood control is typically sized for large storms, smaller but polluting runoff events will pass through such ponds, mostly without appreciable attenuation of the pollution load. Hence, such dry ponds are ineffective for urban runoff quality control.

Ponds can be either in-line or off-line (Figure 4.13). Overflows into an off-line pond are activated when the capacity of the stream (sewer) is exceeded by high flow, and the excess flow overflows into the storage. In in-line ponds the storage is a part of

STORAGE BASINS (PONDS)

IN-LINE

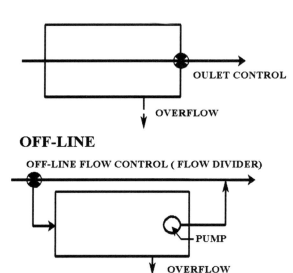

OULET CONTROL

OVERFLOW

OFF-LINE

OFF-LINE FLOW CONTROL (FLOW DIVIDER)

PUMP

OVERFLOW

BASIN, WETLAND OR COMBINATION

Figure 4.13   In-line and off-line pond schematics.

the conveyance channel. The in-line pond is controlled by the restricted capacity of the outflow. Water is pumped or released by gravity after the event from the off-line storage. Off-line storage does not have significant water quality impact.

By combining the dry detention pond with an infiltration system located at the bottom of the pond, the pollution control capability, which otherwise is minimal for a typical dry-weather pond, is enhanced and can be effective for pollution removal (Figures 4.14 and 4.15). These ponds have two outlets and a safety overflow for high flows. The smaller flows (up to two-year runoff event) discharge through a perforated pipe and orifice on the bottom; the higher outlet retains larger flows. Very large rare flows must safely overflow over the dam.

A *wet detention pond* has a permanent pool of water. A simple wet pond acts as a settling facility with medium removal efficiency. After years of operation accumulated solids must be removed (dredged) in order to maintain the removal efficiency and aesthetics of the pond. Improper design and maintenance can make such facilities an eyesore and a mosquito-breeding mudhole. A schematic of a wet pond is shown in Figure 4.16. The removal efficiency of wet ponds for constituents, obtained by a statistical analysis of National Urban Runoff Project (NURP) study sites, is shown on Figure 4.17.

Figure 4.14 A simple extended dry/bioretention cell used for treatment of smaller flows. Courtesy Delaware Department of Transportation

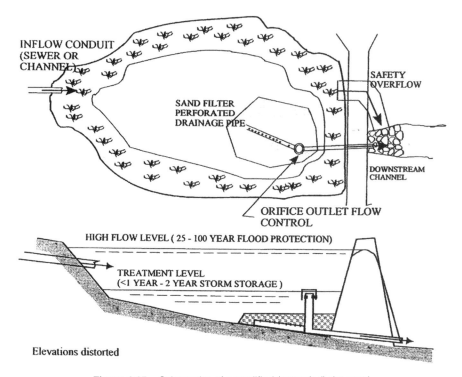

Figure 4.15 Schematics of a modified (extended) dry pond.

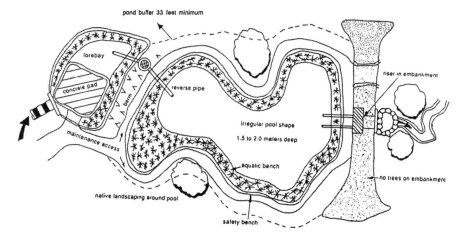

Figure 4.16   Enhanced high-efficiency wet pond system (after Schueler et al., 1991).

A well-designed engineered wet pond consists of (1) a permanent water pool, (2) an overlying zone in which the design runoff volume temporarily increases the depth of the pool while it is stored and released at the allowed peak discharge rate, and (3) a shallow littoral zone acting as a biological filter (Figures 4.16 and 4.18). Urbonas and Ruzzo (1986) stated that in order to achieve a 50% removal of phosphorus, a properly designed wet detention pond must be followed by filtration or infiltration.

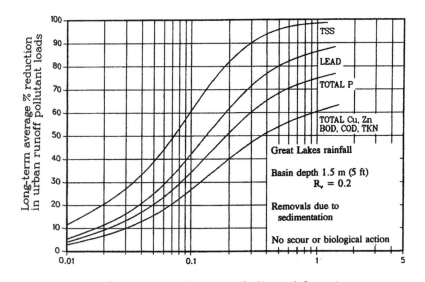

Basin area as % of contributing catchment area

Figure 4.17   Approximate removal efficiencies of conventional wet detention ponds (after Driscoll, 1988).

Figure 4.18   Ecologically designed wet pond. (Courtesy W.P. Lucey, Aqua-Tex Scientific Con-suting, Ltd., Victoria, BC).

Schueler and Helfrich (1988) described an improved design of detention ponds that includes a permanent wet pool, extended detention storage, and stormwater storage. The perimeter wetland area created by the extended detention and stormwater storage provides additional water quality improvement (Schueler et al., 1991). The depth of wet ponds should range from 1 to 3 meters. It is important that the side slope of the basin be mild (5 to 10 horizontal to 1 vertical) to minimize the danger of drowning.

In snowbelt zones of North America and Europe, modifications of the extended pond or wet pond may be necessary to secure its functioning during winter. These ponds receive chemically induced flows with extremely high concentrations of salt. Consequently, as demonstrated in Novotny et al. (1999), high salinity increases the solubility of metals in sediments and water; the sediment accumulated in ponds can actually become a source of metals rather than a sink, and the removal efficiencies in winter are much less than in nonwinter periods. Thus, the ponds should be used for storage of the small volumes of concentrated chemically induced flows, which can then be diluted by subsequent less-polluted snowmelt and/or rainfall. The ponds should be empty before the winter freeze period begins (Figure 4.19).

Hartigan (1989) compared removal efficiencies of modified (extended) dry ponds and wet ponds. He reported that removal of total phosphorus in wet ponds is 2 to 3 times greater than in modified dry ponds (50% to 60% vs. 20% to 30%) and 1.3 to 2 times greater for total nitrogen (30% to 40% vs. 20% to 30%). For other pollutants,

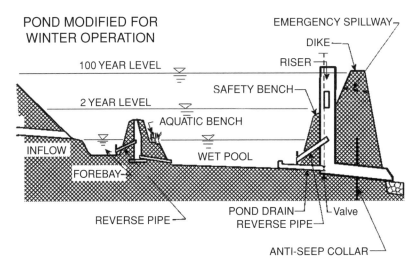

Figure 4.19   Extended engineered dry pond modified for winter operation (adapted from the Center for Watershed Protection and Novotny, 2003).

the average removal rates for wet detention basins and extended dry detention basins were very similar: 80% to 90% for total dissolved solids, 70% to 80% for lead, 45% to 50% for zinc and 20% to 40% for BOD and COD.

Dual use of detention basins for flood and quality control. From the foregoing discussion, it follows that the design of storage facilities for flood and for quality controls uses different objectives and design criteria for each. Hence, facilities designed solely for flood control, using statistically rare design storms, may be ineffective for quality control. It is, however, possible to retrofit existing flood control storage facilities, for example, by installing an additional small orifice and implementing a dual-level control strategy. This was shown in Figures 4.15 and 4.19.

Legacy pollution of urban stormwater ponds. Ponds accumulate suspended solids from runoff; therefore, the pond volume must be increased to accommodate the accumulated sediments. Managers and designers must take into account the fact that at some point the pond will be filled with sediment to a level at which it stops functioning and must be dredged or abandoned.

Ponds with accumulated sediments can become a source of pollution. Beside the need for dredging the sediments, the accumulated sediments can release some pollutants, such as metals or phosphates, especially when they receive high salinity flow from de-icing. This can be one of the reasons for the low or even negative pond efficiencies, especially during winter, for these hard-to-control pollutants. The release of pollutants from the accumulated sediments can occur in several ways:

1. By salt-laden snowmelt entering the pond. High salinity reduces the partitioning coefficient of metals and converts particulate metal into soluble ionic form (Novotny et al., 1999).

2. By exposing the sediments to the air when the water level is low. The sulfide forms of particulate metals in the anoxic sediments are converted by oxidation of sulfides to soluble sulphate forms. Phosphates can also be released by changing the redox status of the sediments (from oxidized to reduced).

3. The pond simply stops removing suspended solids when filled with sediments.

4. The majority of ponds were built in the last 10 years, after the release of the EPA stormwater control rules. The permits to the communities for installation of the ponds must include provisions for maintenance and final solution to the problem when the pond stops functioning. In some cases, abandoning and filling the pond may be the only solution. The problem is that those who built the pond may have no responsibility for the pond maintenance and final disposition of the accumulated sediment.

The planning and design of stormwater runoff retention/detention basins were extensively covered in the following publications:

American Association of State Highway and Transportation Officials, *Design of Sedimentation Basins*, 1980

P. Stahre and B. Urbonas, Stormwater Detention- For Drainage, Water Quality and CSO Management, 1990

V. Novotny WATER QUALITY: Diffuse Pollution and Watershed Management, 2003

Wetlands   Wetlands are attractive features of the urban and suburban landscape, providing citizens with an opportunity to enjoy diverse natural surroundings, similar in quality to urban forests. They can be restored and maintained even in urban centers, as in Portland, Oregon (Dreiseitl and Grau, 2009). But also "wetlands are the kidneys of the nature" (Mitsch and Gosselink, 2000). Both natural and man-made wetlands have been used for runoff pollution control. However, natural wetlands are considered as receiving waters and are subjected to water quality standards and restrictions. Constructed wetlands could be considered as treatment facilities, and the standards mostly apply to the outlet from the wetland (Stockdale and Horner, 1987; Linker, 1989; Hammer, 1989; Kadlec and Knight, 1996). Therefore, at present only constructed or restored wetlands may be used for treatment of runoff and CSOs. Wetlands combine both sedimentation and biological utilization effects to remove pollutants from runoff. The largest pollutant reduction can be achieved during the "active" wetland growing season—that is, during May to September in northern climatic conditions. During the dormant (winter) condition, wetlands may become a source of pollution that is leached from dead vegetation. In southern (Florida) climatic conditions, wetlands efficiency remains more or less constant throughout the year.

Wetland construction for runoff and wastewater pollution control is different from wetland restoration. Wetland restoration efforts are aimed at restoring nature's cleansing capability and creating habitat in places where former wetlands were drained and lost. Such created wetlands are located mostly in riparian zones of water bodies and serve as a buffer against pollution. Wetlands have been constructed for (Mitsch and Gosselink, 2000):

1. Flood control
2. Wastewater treatment
3. Stormwater or nonpoint source pollution control
4. Ambient water quality improvement (e.g., riparian and instream systems)
5. Wildlife enhancement
6. Fisheries enhancement
7. Replacement of similar habitat (wetland loss mitigation)
8. Research wetland

The following references are recommended design manuals for constructed wetland:

D. A. Hammer, *Constructed Wetlands for Wastewater Treatment: Municipal, Industrial, and Agricultural,* 1989

R. H. Kadlec and R. L. Knight, *Treatment Wetlands,* 1996

R. H. Kadlec and S. Wallace, *Constructed Wetlands and Aquatic Plant Systems for Municipal Wastewater Treatment,* 2008

U.S. Environmental Protection Agency, *Constructed Wetlands and Aquatic Plant Systems for Municipal Wastewater Treatment,* 1988

J. Vymazal and I. Kröpfelová, *Wastewater Treatment in Constructed Wetlands with Horizontal Sub-Surface Flow,* 2008

Water Environment Federation, *Natural Systems for Wastewater Treatment,* 1990 and 2000

Stormwater wetlands are shallow pools and/or saturated soils that create growing conditions suitable for the growth of marsh plants, used primarily for storage and treatment or storm runoff. There are two types of constructed wetlands: (1) free water surface (FWS) systems (Figure 4.20) and (2) subsurface flow systems (SFS). Wetlands are typically designed as several pools in a series, with or without a recycle (Kadlec and Knight, 1996; Kadlec and Wallace, 2008; Steiner and Freeman, 1989; Novotny and Olem, 1994). The wetland provides for shallow stormwater storage and enhanced pollutant removal (Figure 4.17). Visually attractive, they provide habitats and recreational areas. However, unmanaged they can create a habitat for mosquitoes, noxious weeds, and nuisance odors from stagnating materials. In this chapter we will cover free water systems as they are used mostly for stormwater and runoff treatment. Submerged flow wetlands are used primarily for wastewater (used water) treatment and will be covered in Chapter VII.

Originally, wetland treatment was considered as a universally applicable, almost miraculous, treatment of polluted waters and enhancement of the landscape and aquatic ecology. However, while wetlands cleanse and detoxify urban and highway

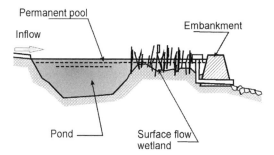

Figure 4.20 A simple schematic of the pond-wetland (preferable) configuration. TKN (total Kjeldahl nitrogen = organic N + ammonium) is oxidized in the pond to nitrate, which is converted to N_2 gas in the wetland. The pond also removes particulate pollutants and toxic compounds by sedimentation. Courtesy New South Wales (Australia) EPA.

runoff, concerns about global warming and GHG emissions should be considered. Wetlands perform both; they sequester carbon dioxide that is converted by photosynthesis into wetland flora and organic peat soils, but also emit methane and nitrous oxide that are 25 times more potent GHGs than carbon dioxide. The balance between sequestering carbon dioxide from the atmosphere by photosynthesis and emitting methane into the atmosphere has not been satisfactorily researched. Selection of plants may solve this dilemma.

An FWS system typically consists of basins or channels with a natural or constructed subsurface barrier of clay or impervious geotechnical material (lining) to prevent seepage. The basins are then filled with soils to support emergent vegetation. Water flows slowly over the soil surface through the basins with a shallow depth. Figure 4.21 shows an experimental wetland with multiple compartments at the University of New Hampshire Stormwater Center. A constructed wetland in Scotland treating industrial runoff is shown in Figure 4.22. This wetland has three compartments in series. The first is a shallow pond for settling and pretreatment, the second compartment is for treatment, and the third for polishing.

If the soil is brought from an existing wetland, wetland vegetation may emerge without seeding; however, seeding and planting of vegetation is a part of the construction process. Mitsch (1990) referred to the former type of constructed wetland as a *self designed wetland* and to the latter as *designer wetland*. To develop a wetland that is ultimately a low-maintenance one, the natural succession process needs to be allowed to proceed. Often this may mean some initial period of invasion by undesirable species, but if proper hydrologic and nutrient loads are maintained, this invasion is usually temporary. Table 4.3 has design parameters for surface flow wetlands.

The most important hydrological wetland design parameters are hydroperiods and hydraulic loadings.

Hydroperiod and depth. Hydroperiod is defined as the depth of water over time. This parameter is most important for natural wetlands. Wetlands with variable depths have the most potential for developing a diversity of plant and animal species. Alternate flooding and aeration of solids promote nitrification-denitrification. Deepwater

Figure 4.21   Experimental research gravel wetland treating runoff from a parking lot at the Stormwater Center on the campus of the University of New Hampshire. The compartmentalized arrangement works best for urban stormwater. Photo V. Novotny.

areas, devoid of emerging vegetation, offer habitats for fish (for example, *Gambusia affinis*, the mosquito-eating fish). Water levels can be controlled by inflow and outflow structures, including weirs and feed pumps.

Hydraulic loading. Hydraulic loading, HLR is defined as

$$HLR = Q/A$$

where

Q = flow rate (m^3 day^{-1} or m^3 $year^{-1}$), and A = wetland surface area (m^2). Note that the unit of HLR is m/day or m/year, which is the equivalent daily or annual flooding depth of the wetland.

An inverse of HLR is the area requirement per unit flow. Most of what is known about the hydraulic loading rates has been gathered from observations of wetlands

Figure 4.22 Stormwater wetland in Scotland treating runoff from an industrial site (photo V. Novotny).

receiving wastewater. Mitsch (1990) pointed out that the hydraulic loading rates used for wetlands treating wastewater would be too low for riparian wetlands used for runoff and stream quality control.

Retention time. Table 4.3 also provides the optimum retention time of water in the wetland. The retention time for free water surface systems can be calculated from a simple formula considering the water volume of the wetland and average flow, or

$$\text{HRT} = pV/Q$$

Table 4.3   Design parameters of free water surface flow wetlands (after Water Environment Federation, 1990; Vymazal and Kröpfelová, 2008)

Hydraulic loading rate (HLR)	0.03–0.05 m/day
Maximum water depth	50 cm
Hydraulic retention time (HRT)	5–7 days
Aspect ratio (length/width)	2:1
Configuration	Multiple beds in parallel series
Substrate gravel size	8–16 mm
BOD_5 load	100–110 kg/ha-day
Suspended solids loads	up to 175 kg/ha-day
Total nitrogen load	7.5 kg/ha-day
Total phosphorus load	0.12–1.5 kg/ha-day

where

HRT = hydraulic residence time (days); $V = W * H * L$ = active volume of the wetland (m^3); p = porosity or (water volume)/(total volume) ratio [$p = 0.9$ to 1 for free water surface (FWS) wetlands depending on the growth density of vegetation, and p = void fraction of the substrate for subsurface flow systems (SFS)]; W = width of the system (meters); H = average depth of the system (meters); L = length of the wetland (meters) = $W * AR$; AR = aspect ratio (length/width)

Alternatively, the hydraulic residence time for SFS systems may be calculated from Darçy's law as

$$HRT = pV/Q = L/(KpS) \quad 9.10$$

where

L = length of the bed (meters); K_p = hydraulic (saturated) permeability of the wetland (m/day); S = bed slope (m/m)

The treatment efficiencies of the gravel wetland shown in Figure 4.21 are given in Table 4.4 below.

IV.2.3 Winter Limitations on Stormwater Management and Use

Management of winter snowmelt runoff is different from that for nonwinter urban and highway runoff. Both urban and highway snowmelt and nonwinter runoff are polluted, laden with solids and nutrients, and contain concentrations of many pollutants that exceed accepted water quality criteria. The biggest single issue regarding winter snowmelt pollution is the use of de-icing chemicals (rock salt, calcium or magnesium chlorides, calcium magnesium acetate [CMA], glycols, etc.) or abrasives/salt mixtures. The use of these chemicals and abrasives along with the transport of heavy metals in snowmelt runoff can be a limiting factor in the use of LID in urban environments located in the snowbelt regions.

De-icing salts also limit the ability to use the stormwater. Water with high concentrations of sodium and chloride can be damaging to soils and vegetation, limiting the reuse of the runoff. They can also limit the effectiveness of rain gardens, swales, or other bioremediation techniques by damaging or even killing the vegetation in the structure. Sodium can also change soil properties by reducing the infiltration rates. Finally, the addition of road salt can limit the effectiveness these structures have in removing heavy metals from stormwater runoff.

Table 4.4 Removal average annual efficiencies of the gravel wetland on the campus of the University of New Hampshire treating runoff (UNH Stormwater Center)

Total suspended solids	100%
Dissolved inorganic nitrogen	100%
Zinc	100%
Total phosphorus	55%

If LID is to be used in an area using de-icing salts, all of these factors will need to be taken into consideration. The strategy for winter urban pollution management can be broken down into two approaches that should be implemented simultaneously (Novotny et al., 1999):

1. *Reduction of de-icing compounds use, selection of more environmentally safe de-icing and anti-icing chemicals, and better snow removal practices*
2. *Using BMPs to control residual pollution*

It has been reported that 21 million metric tons of road salt were used in the United States in 2005 to improve driving safety in the winter (U.S. Geological Survey, 2007). The accumulation of sodium and chloride ions in the environment degrades the water quality in a watershed (Jones and Jeffrey, 1992; Environment Canada Health Canada (ECHC), 1999; Ramakrishna and Viraraghocan, 2005). Increased chloride concentrations decrease the biodiversity of waterways and roadside vegetation (ECHC, 1999). If chloride reaches the groundwater, it can contaminate drinking water supplies. Elevated and/or increasing chloride concentrations attributable to road salt applications are present in groundwater and surface waters in urban environments in northern climate regions (Godwin et al., 2003; Kaushal et al., 2005; Lofgren, 2001; Marsalek and Chocat, 2003; Novotny et al., 2008; Thunqvist, 2004).

Drainage type is the primary factor affecting strategy selection. Areas with storm sewers are most susceptible to snowmelt pollution. Also, management and reduction of winter diffuse pollution are part of an overall stormwater management plan. In this strategy, "retrofitting" the stormwater BMPs and watershed management to function both in winter and summer is necessary (see Figure 4.19).

In urban environments, specifically on roadways, car traffic results in the deposition of heavy metals, asbestos, PAHs, oils, and other pollutants, which are transported by runoff from rainfall or snowmelt into receiving water bodies, resulting in contamination of soils, lakes, streams, wetlands, and groundwater. Wetlands and saturated soils detoxify metals by converting the toxic metallic divalent cations into insoluble sulfide minerals. Metal concentrations in stormwater runoff can reach levels high enough to be toxic or even severely toxic to biota, especially in runoff from major highways that may reach traffic densities (Kaushal et al., 2005). Copper, zinc, lead, cadmium, sediments, PAHs, and de-icing salts represent the main source of pollution found in runoff from roads (Backstrom et al., 2003). Sources of the metal pollutants are tire wear, engine and break parts, fluid leakage, vehicular component wear, atmospheric depositions, and road surface abrasion (Davis et al., 2001b; Marsalek et al., 1999). These metals are the most consistent pollutants in roadside runoff (Makepeace et al., 1995), with the highest concentrations of metals coming from major highways (Sansalone et al., 1996).

Metals accumulate in the snowpack and on roadways during the winter months (Figure 4.23). The accumulation, combined with the increased corrosion of metals and wear of the roadways, results in both higher concentrations and an increase in the total load of metals during snowmelt, when compared to rainfall (Davis et al., 2001b; Gobel et al., 2007; Kaushal et al., 2005; Brezonik and Stadelmann, 2002; Mitton and Payne, 1996). Concentration of metals in snowmelt can be up to two to four times

Figure 4.23 Highway 9 in Newton, Massachusetts. Heavy pollution of urban snow and snowmelt is acerbated in storm sewered urban areas where no attenuation or removal of high concentrations of pollutants is provided (Photo V. Novotny).

higher than during rainfall, with the highest concentration occurring during rainfall on snowpack (Mitton and Payne, 1996).

The higher concentrations of metals and the presence of de-icing salts in snowmelt runoff result in a larger number of toxic and severely toxic events compared to rainfall events (Kaushal et al., 2005). This increase in toxicity can reduce the diversity in the benthic and plant communities (Westerlund et al., 2003). The presence of de-icing salts, mostly in the form of NaCl, and drops in temperature can also have an impact on the transport of heavy metals through a watershed. These two parameters increase the mobility of metals in soil environments (Marsalek et al., 2003). The increased mobility is caused by a change in the partitioning coefficients between the dissolved and particulate phases of the metals, especially cadmium, zinc, and copper. Partitioning between the particulate (non toxic) and dissociated (toxic) metals and other pollutants is expressed by a simple linear equation

$$C_p = \Pi C_d$$

where C_p is the particulate (adsorbed) pollutants in the sediment or soil in mg of pollutant/kg of adsorbing sediment, C_d is the dissolved (dissociated) pollutant concentration in water (or pore water of sediment) in mg/L, and Π is the partitioning coefficient in L/kg which is essentially dimensionless. Π is related to the type of the compound and the concentration of the sediment. As stated above it is also affected by the salinity (for metals) and concentration of particulate organics (tocic chemicals and metals).

When the salinity is increased or the temperature drops in the water, the partitioning coefficients decrease, resulting in a higher portion of the metals in the bioavailable dissolved phase (Warren and Zimmerman, 1994; Marsalek et al., 2003; Novotny et al., 1998).

Examination of roadside soils in Sweden found that Pb, Cu, and Zn were susceptible to increased mobilization and leaching into the groundwater where high concentrations of NaCl, reducing conditions, or lowered pH levels were present (Grolimund and Borkovec, 2005). In Germany concentrations of cadmium and zinc in roadside soils were observed to be extremely different during the winter and fall than in summer and spring, due to leaching from application of de-icing salts (NaCl) (Novotny et al., 1998). In a small pond near a major highway in Ontario, Canada, chloride and sodium concentrations were measured at higher than 3 and 2 g/L respectively at the sediment-water interface. These concentrations decreased gradually with depth into the sediments, but chloride concentrations were still as high as 1.5 g/L 40 cm into the sediments. These high concentrations of chloride in the sediment pore waters were found to increase the dissolved concentrations of Cd, resulting in increased toxicity of the water to benthic organisms (Norrstrom and Jacks, 1998).

Snowplowing operations move snow from the road surface to the side while concurrently applying salt to the road (street) surface. Sometimes, prewetting of the surface with brine is done before the snowfall. Movement of contaminated snowmelt from the road (street) surfaces occurs in two waves (Novotny et al., 1999):

(1) *Chemically induced snowmelt* has extremely high concentrations of salt (in tens of grams/liter) and associated pollutants, but very small flows.

(2) *Radiation snowmelt* is a less polluted snowmelt generated by solar radiation and/or by melting at above-freezing temperatures. Snowmelt water percolates through the roadside snow piles and snowpack. This percolation removes the particulates from the snowmelt by filtering, leaving large quantities of particulates near the curb after the winter period. Radiation-induced snowmelt rate is very low and does not cause flooding. Consequently, the first spring rainfall washes large quantities of pollutants accumulated near the curb into storm drainage.

These characteristics of urban pollutant snowmelt support the argument for abandoning the curb-and-gutter street and road practices and replacing them with permeable road edges and swales with biofilters connected to modified dry ponds, such as that shown in Figure 4.19. This pond should be designed to retain highly concentrated and polluted chemically induced snowmelt and keep it in storage until the spring rainfall runoff will provide enough dilution and carry the salts safely into a receiving water body. It is quite likely that the highly concentrated and saline chemically induced snowmelt will be separated by density stratification in the pond from less-diluted smaller freshwater flows from rainfalls, until there is a larger freshwater flow that has enough energy to break the stratification (Novotny et al., 1999; Novotny et al., 2008).

With the addition of increased infiltration and the use of bioremediation for the treatment of runoff from impervious surfaces in urban environments, the use of de-icing chemicals has to be addressed. Sodium and chloride in the most common de-icing chemical (NaCl) are both considered conservative materials and cannot be removed from runoff even through typical practices used in LID. Sodium can be removed in small amounts through ion exchange in soils, but for the most part both sodium and chloride ions stay in solution and move with the water. The only known way to remove these ions is by reverse osmosis or distillation of the water, each requiring high energy amounts. This raises a dilemma for LID in cities that use high amounts of road salt during the winter. LID promotes infiltration, but if the water being infiltrated is contaminated with large concentrations of sodium, chloride, and other dissolved pollutants, then the recharged groundwater will have to be treated by reverse osmosis before it can be used by human beings (see Chapter V for discussion on the use of reverse osmosis and its cost).

IV.2.4 Hard Infrastructure

Elimination of Curbs and Gutters LID and Cities of the Future drainage minimizes or outright eliminates the use of underground sewers for stormwater conveyance. As Figures 4.4 and 4.23 show, pollutants accumulate near the curb and are then washed into the underground drainage without attenuation.

Storm Solids and Oil Separators If storm sewers cannot be replaced in high-density high-imperviousness urban zones, there are a number of pollution-removing units that can be used to remove a portion of the pollution load both in combined and separate storm sewers and, in most cases, send it to a treatment plant for removal (Novotny, 2003).

Regulators, Concentrators, and Separators These devices are capable of separating solids from the flow of stormwater. The dual-functioning swirl-flow regulator/solids concentrator has shown a potential for simultaneous quality and quantity control (Field, 1986, 1990). A helical-type regulator/separator has also been developed. These devices were primarily applied to CSOs; however, they can also be installed as storm runoff pollution control devices. The concentrated flows, which may amount only to a few percent of the total runoff flow, can be stored and subsequently directed towards sanitary sewers for treatment during low-flow periods

The vortex solids separator is a compact solids separation device. As early as 1932, an idea of separating solids from CSO in a vortex chamber was conceived in England (Brombach, 1987; Pisano, 1989). In the 1970s the idea was pursued in the U.S. and resulted in a device known today as a "swirl concentrator." Similar devices known as "fluidsep^{TM}" developed in Germany, and "Storm King®," developed in the United Kingdom, have also been implemented throughout the U.S. and Europe. Vortex separation devices have no moving parts (Figure 4.24). During wet weather

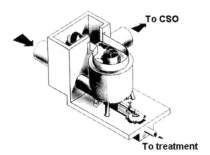

Figure 4.24 Swirl concentrator (Courtesy of W. Pisano).

the unit's outflow is throttled (typically only 3 to 10% of flow passes through the foul sewer outlet towards the treatment plant), causing the unit to fill and to self-induce a "swirling" vortexlike operation. Settleable grit and floatable matter are rapidly removed. Concentrated foul matter is sent to the treatment plant (or sent to temporary storage), while the cleaner, treated flow discharges to the receiving waters.

The design overflow rate for vortex separators installed in the U.S. ranged from 10 to 30 L/s-m^2 (swirl concentrator) and 18 to 140 L/s-m^2 (fluidsep), respectively. Up to 60% of suspended solids removal can be achieved, but generally, the performance of such units is less than that of primary treatment.

A **helical bend regulator/concentrator** induces helical motion in a curved separator with a bend angle of about 60° and the radius of the curvature equal to 16 times the inlet pipe diameter. Dry-weather flow passes through the lower portion of the device to the intercepting sewer. As the flow increases during a wet-weather period, the helical motion begins, and the particles are drawn to the inner wall and drop to the lower channel leading to the treatment plant. The excess cleaner flow overflows over a weir into a CSO. The removal efficiency of helical bend separators is about the same as that of swirl concentrators (Field, 1990).

Sand Filters   *Sand filters* represent a new application of an old technology for treatment of urban stormwater (Schueler et al., 1991). These units can be effectively used for control of runoff from small sites, or treatment of runoff prior to entering another stormwater management structure (such as a vegetative biofilter), or as a retrofit strategy in an urban stormwater system. The design includes both sedimentation and filtration components. The drainage area should be less than 2 ha of impervious surface, and the sedimentation chamber volume should be about 50 m^3 per hectare of connected impervious area. The sedimentation chamber is followed by a filtration chamber with the same volume.

Since sand is inert with little or no adsorbing capacity, dissolved fraction of priority pollutants is not removed (unless an organic-microbiological population develops in the top layer of filter containing organic matter which is then periodically removed and replaced by clean sand).

Slow-rate filters in which the development of a mostly anaerobic microbiological layer occurs, remove priority pollutants by microbiological action and by adsorption of organic particulates. Carlo et al. (1992) showed that a sand filter did not remove selenium when operated at higher loading rates, but when it was operated as an anaerobic slow-rate filter, removal efficiencies for selenium were 74 to 97%. Selenium was retained in the organic layer of the filter.

Enhanced (peat-sand) filters utilize layers of peat, limestone, and/or topsoil, and may also have a grass cover top. Peat-sand filters provide high phosphorus, BOD, and nitrogen removals, in addition to the removal of solids. For sizing the filters, Galli (1990) recommends that the filter area should be about 0.5% of the contributing watershed area, and the annual hydraulic loading should equal 75 m/year or less. As with most biological filtration systems, the peat-sand filter works best during the growing season of the year, since a part of the nutrient load is taken up by grass. Also the filter should remain aerobic.

Sizing Storage Basins   The objective of storage in the sustainable Cities of the Future is not just the attenuation of the peak flows and volume to reduce flooding, which is the main and sometimes only goal of urban watershed planning and abatement. COF goals also include water conservation, which is done by rainwater (urban runoff) harvesting and storage. Hence, the design parameters of the storage basins must be changed from the current designs, which require stored water above the permanent pool to be released in a period of 24 to 48 hours after the precipitation event, to much longer retention periods that should cover the entire dry period between precipitation events. Hence, the storage capacity should be determined by long-term simulations and not just by sizing storage for a certain design storm. As a matter of fact, the concept of single-design storms may not be appropriate in LID or COTF designs.

Ponds and wetlands may provide ample surface storage in some cases to accomplish the goals of the development. In some other cases, it may not be possible to find enough storage, especially when reuse of or rainwater is contemplated. Underground storage basins are needed and have been used in densely populated urban zones. Underground storage alternatives are:

- Hard infrastructure concrete or modular shallow-depth basins can be located under plazas or parking lots, or even in the basement of larger buildings. Examples include Battery Park in New York, Potzdamer Platz (Plaza) in Berlin, and Star City in Seoul, Korea (see Chapter V). These underground storage basins capture precipitation for reuse in flushing toilets and outdoor waterscapes, as in Potzdamer Platz (Dreiseitl and Grau, 2009). In Germany and elsewhere in Europe, more than 10,000 small storage basins were built to capture 1 cm or more of the rainfall to control the first flush effects of COSs.
- Shallow aquifer can be used where geological stratum allows storage and reuse without significant losses (see Chapter V).

IV.2.5   LID Urban Drainage—A Step to the Cities of the Future

LID drainage cannot rely on only one or two BMPs to achieve the sustainability goal of low (zero) water impact on the hydrology of the development or retrofitted urban area. It will take a portfolio and a train of BMPs and other modifications of the watershed to accomplish the goal. First, the goals of watershed abatement and improvement have to be formulated and extensively discussed with the stakeholders. LID and the more extensive and demanding goals of the Cities of the Future also require a change in the attitudes, behavior, and commitment of all involved—that is, regulators, city and county (district) development boards, elected government officials, citizens, and NGOS. Often existing building, zoning, and public health regulations will have to be changed. For example, it is very difficult under present regulations in the state of Massachusetts to develop sustainable urban drainage without incorporating massive granite curbs and gutters, which many city officials and developers consider as a necessary aesthetical feature (never mind that rain gardens look much better than granite curbs).

An example of the layout of a LID suburban development is shown in Figure 4.25. This figure shows a low-density LID development with created surface drainage paths that are not in front of the houses; they are abutting homeowners' yards. The ephemeral and perennial drainage channels are followed by a wetland and pond (not shown on the figure). In Atelier Dreiseitl's development in the Tianjin province in China (Figure 4.26) a shallow perennial drainage canals with stepping stones abut the buildings, as was also done in Sweden (Figures 3.13 and 3.17). Creating the drainage–water body system will also be shown in Chapter X. All street and house drainage is connected to the drainage corridors by bioswales.

The last figure in this chapter, Figure 4.27, is a conceptual rendering of the LID environmental corridor planned to protect Lake Ontario from runoff pollution from

Figure 4.25 Layout of the Willow Brook subdivision near Victoria, British Columbia. Courtesy Aqua-Tex Scientific Consulting, Victoria, BC.

Figure 4.26 Zhangjiawo New Town development in Tianjin province. Courtesy Herbert Dreiseitl, Atelier Dreiseitl, Uberlingen, Germany.

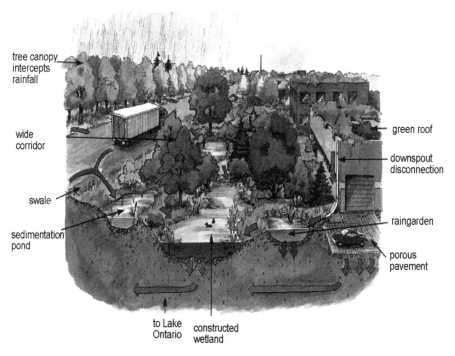

Figure 4.27 Regeneration of Toronto's Lake Ontario lakefront. Illustration by Heather Collins in "Greening the Toronto Portlands," 1997 by Michael Hough, Waterfront Regeneration Trust.

the lakefront development and road artery. This demonstrates the portfolio approach towards selection of LID practices. The general portfolio and the components of LID and COTF developments include:

Building BMPs
    Densely populated areas
        Green roofs
        Underground storage and reuse in the buildings
        Green areas with rain gardens
    Low-density areas
        Disconnecting impervious areas (roofs, driveways) from drainage
        Xeriscape to minimize or eliminate irrigation
        Rain harvesting for irrigation and swimming pools
Local drainage
    Densely populated areas
        Pervious pavements on parking lots, side streets, and courtyards
        Rain gardens for retention, treatment, and infiltration
        Surface conveyance wherever possible in bioswales
        Stream corridors with buffers, reclaiming floodplain
    Low-density development
        Surface drainage in bioswales with retention
Storage and treatment
    Densely populated areas
        Surface storage in waterscape basins
        Underground storage in infrastructure basins or shallow aquifer
        Treatment by wetlands or water reclamation plants
    Low-density developments
        Storage and treatment by ponds (lakes) and wetlands
        Stream corridors with buffers, reclaiming floodplain
Public access
    All LID best management practices should be pleasing to the public and accessible.

Achieving LID status is a necessary prerequisite, but not the only or final step. COTF criteria as presented in Chapter II have other aspects, such as carbon (GHG) neutrality, good transportation, recreation, walking and biking paths, leisure, green space, local agriculture, and so forth.

REFERENCES

American Association of State Highway and Transportation Officials (1980) *Design of Sedimentation Basins*, National Cooperative Highway Research Program Synthesis of Highway Practice, issue 70, Transportation Research Board, Washington, DC

Backstrom, M., U. Nilsson, K. Hakansson, B. Allard, and S. Karlsson (2003) "Speciation of heavy metals in road runoff and roadside total deposition," *Water, Air, and Soil Pollution* 147 (1–4):343–366

Barnes, K., J. Morgan, and M. Roberge (2001) *Impervious surfaces and the quality of natural built environments*, Department of Geography and Environmental Planning, Towson University, Baltimore, MD

Bass, B., E. S. Krayenhoff, A. Martilli, R. B. Stull, and H. Auld (2003) "The impact of green roofs on Toronto's urban heat island," in *Proceedings of the First North American Green Roof Conference: Greening Rooftops for Sustainable Communities*, Chicago, May 20–30, The Cardinal Group, Toronto, Canada.

Breen, P. F. (1990) "A mass balance method for assessing the potential of artificial wetlands for wastewater treatment," *Water Research* 24, pp. 689–697

Brezonik, P. L. and T. H. Stadelmann (2002) "Analysis and predictive models of stormwater runoff volumes, loads, and pollutant concentrations from watersheds in the twin cities metropolitan area, Minnesota, USA," Water Research 36 (7):1743–1757

Brombach, H. (1987) "Liquid-solid separation on vortex-storm-overflows," in *Topics in Urban Storm Water Quality, Planning and Management* (W. Gujer and V. Krejci, eds.), Proceedings of the Fourth International Conference on Urban Storm Drainage (IAWPRC-IAHR), Êĉole Polytechnique Fêdêrale, Lausanne, Switzerland, pp. 103–108

Carlo, P. L. et al. (1992) "The removal of selenium from water by slow sand filtration," *Water Science and Technology* 26 (9–11):2137–2140

Clark, M. L., B. J. Barfield, and T. P. O'Connor (2004) *Stormwater Best Management Practices Design Guide, Volume 2: Vegetative Biofilters*, #1C-R059-NYSX, U. S. Environmental Protection Agency National Risk Management Research Laboratory, Cincinnati, OH

Cunningham, S. D. and D. W. Ow (1996) "Promises and prospects of phytoremediation," Plant Physiology 110, pp. 715–719

Davis, A. P., M. Shokouhian, H. Sharma, and C. Minami (2001a) "Laboratory study of biological retention for urban stormwater management," *Water Environment Research* 73, pp. 5–14

Davis, A. P., M. Shokouhian, and S. Ni (2001b) "Loading estimates of lead, copper, cadmium, and zinc in urban runoff from specific sources," *Chemosphere* 44 (5):997–1009

Del Barrio, E. P. (1998) "Analysis of the green roofs cooling potential in buildings," *Energy and Buildings* 27, pp. 179–193.

Deutsch, B., H. Whitlow, M. Sullivan, and A. Savineau (2005) "Re-greening Washington, DC: A green roof vision based on environmental benefits for air quality and storm water management," in *Proceedings of the Third North American Green Roof Conference: Greening rooftops for sustainable communities*, Washington, DC, May 4–6, The Cardinal Group, Toronto, Canada, pp. 379–384

Dietz, M. E. (2007) "Low Impact Development Practices: A Review of Current Research and Recommendation for Future Directions," *Water Soil and Air Pollution* 186, pp. 351–363

Dreiseitl, H. and D. Grau (2009) *Recent Waterscapes: Planning, Building and Designing with Water*, Birkhäuser, Basel, Boston, Berlin

Driscoll, E. D. (1988) "Long term performance of water quality ponds," In *Design of Urban Runoff Quality Controls*, Proceedings of Engineering Foundation Conference (L. A. Roesner et al., eds.), American Society of Civil Engineers, New York, NY, pp. 145–162

Environment Canada Health Canada, Environment Protection Act (1999) *Priority Substances List Assessment Report — Road Salt*, Environment Canada Health Canada (ECHC), Ottawa, Canada

Field, R. (1990) "Combined sewer overflows: control and treatment," in *Control and Treatment of Combined-Sewer Overflows* (P. E. Moffa, ed.), Van Nostrand Reinhold, New York, NY, pp. 119–191

Field, R. (1986) "Urban stormwater runoff quality management: Low-structurally intensive measures and treatment," in *Urban Runoff Pollution* (H. C. Torno, J. Marsalek, and M. Desbordes, eds.), Springer Verlag, Heidelberg, Germany and New York, pp. 677–699

Galli, J. (1990) *Peat-sand filters: A proposed stormwater management practice for urban areas*, Department of Environmental Programs, Metropolitan Washington Council of Governments, Washington, DC

Galli, J. (1992) *Preliminary analysis of the performance and longevity of urban BMPs installed in Prince George's County, Maryland*, prepared for the Department of Environmental Resources, Prince George's County, MD

Getter, K. L and B. Rowe (2006) "The Role of Extensive Green Roofs in Sustainable Development," *Horticultural Science* 41 (5):1276–1285

Gobel, P., C. Dierkes, and W. G. Coldewey (2007) S"torm water runoff concentration matrix for urban areas," *Journal of Contaminant Hydrology* 91 (1–2):26–42

Godwin, K. S., S. D. Hafner, and M. F. Buff (2003) "Long-term trends in sodium and chloride in the Mohawk River, New York: The effect of fifty years of road-salt application," *Environmental Pollution* 124, pp. 273–281

Grolimund, D. and M. Borkovec (2005) "Colloid-facilitated transport of strongly sorbing contaminants in natural porous media: Mathematical modeling and laboratory column experiments," *Environmental Science and Technology* 39 (17):6378–6386

Hammer, D. A. (1989) *Constructed Wetlands for Wastewater Treatment: Municipal, Industrial, and Agricultural*, Proceedings from the First International Conference on Constructed Wetlands for Wastewater Treatment, held in Chattanooga, Tennessee, on June 13–17, 1988, Lewis Publishers, Chelsea, MI

Hammer, T. R. (1972) "Stream and channel enlargement due to urbanization," *Water Resources Research* 8, pp. 1530–1540

Hartigan, J. P. (1989) "Basis for design of wet detention basin BMP's," in *Design of Urban Runoff Quality Controls*, Proceedings of Engineering Foundation Conference (L. A. Roesner et al., eds.), American Society of Civil Engineers, New York, NY, pp. 122–143

Hatt, B. E., A. Deletic, and T. D. Fletcher (2007) "Stormwater reuse: designing biofiltration systems for reliable treatment," *Water Science and Technology* 55, pp. 201–209

Heinze, W. (1985) "Results of an experiment on extensive growth of vegetation on roofs," *Rasen Grünflachen Begrünungen* 16 (3):80–88

Henderson, C., M. Greenway, and I. Phillips (2007) "Removal of dissolved nitrogen, phosphorus and carbon from stormwater by biofiltration mesocosms," *Water Science and Technology* 55, pp. 183–191

Hill, K. (2007) "Urban ecological design and urban ecology: An assessment of the state of current knowledge and suggested research agenda," in Cities of the Future: Towards Integrated Sustainable Water and Landscape Management (V. Novotny and P. Brown, eds.), IWA Publishing, London, UK, pp. 251–266

Jackson T. J. and R. M. Ragan (1974) "Hydrology of porous pavement parking lots," *Journal of the Hydraulics Division* (Proceedings of the American Society of Civil Engineers) 12, pp. 1739–1752

Jones, P. H. and B. A. Jeffrey (1992) "Environmental impact of road salting," in *Chemical Deicers and the Environment* (F. M. D'Itrie, ed.) Lewis Publishing, Boca Raton, FL, p. 1

Kadlec, R. H. and R. L. Knight (1996) *Treatment Wetlands*, CRC/Lewis Press, Boca Raton, FL

Kadlec, R. H. and S. Wallace (2008) *Constructed Wetlands and Aquatic Plant Systems for Municipal Wastewater Treatment*, CRC Press, Boca Raton, FL

Kaushal, S. S., P. M. Groffman, G. E. Likens, K. T. Belt, W. P. Stack, V. R. Kelly, L. E. Band, and G. T. Fisher. (2005) "Increased salinization of fresh water in the northeastern United States," *Proceedings of the National Academy of Sciences* 102 (38):13517–13520

Lee, D., T. A. Dillaha, and J. H. Sherrard (1989) "Modeling phosphorus transport in grass buffer strips," *Journal of Environmental Engineering* 115 (2):409–427

Linker, L. C. (1989) "Creation of wetlands for the improvement of water quality: A proposal for the joint use of highway right-of-way," in *Constructed Wetlands for Wastewater Treatment: Municipal, Industrial and Agricultural* (D. A. Hammer, ed.), Lewis Publishers, Chelsea, MI, pp. 695–701

Liu, K. and B. Baskaran (2003) "Thermal performance of green roofs through field evaluation," in *Proceedings of the First North American Green Roof Conference: Greening rooftops for sustainable communities*, Chicago, May 29–30, The Cardinal Group, Toronto, Canada, p. 273–282

Lofgren, S. (2001) "The chemical effects of deicing salt on soil and stream water of five catchments in southeast Sweden," *Water, Air, and Soil Pollution* 130 (14):863–868

Makepeace, D. K., D. W. Smith, and S. J. Stanley (1995) "Urban stormwater quality: summary of contaminant data," *Environmental Science and Technology* 25 (2):93–139

Marsalek, J. and B. Chocat (2003) "International Report: Stormwater management," *Water Science and Technology* 46, pp. 1–17

Marsalek, J., Q. Rochfort, B. Brownlee, T. Mayer, and M. Servos (1999) "Exploratory study of urban runoff toxicity," *Water Science and Technology* 39 (12):33–39

Marsalek, J., G. Oberts, K. Exall, and M. Viklander (2003) "Review of operation of urban drainage systems in cold weather: Water quality considerations," *Water Science and Technology* 48 (9):11–20

Mentens, J., D. Raes, and M. Hermy (2003) "Greenroofs as a part of urban water management," in *Water Resources Management II* (C. A. Brebbia, ed.), WIT Press, Southampton, UK, pp. 35–44

Mentens, J., D. Raes, and M. Hermy M (2005) "Green roofs as a tool for solving the rainwater runoff problem in the urbanized 21st century?" *Landscape and Urban Planning* 77, pp. 21–226

Minnesota Pollution Control Agency (1989) Protecting Water Quality in Urban Areas, Division of Water Quality, St. Paul, MN

Mitsch, W. J. (1990) *Wetlands for the Control of Nonpoint Source Pollution: Preliminary Feasibility Study for Swan Creek Watershed of Northwestern Ohio*, Ohio Environmental Protection Agency, Columbus, OH

Mitsch, W. J. and J. G. Gosselink (2000) *Wetlands*, 3rd ed., John Wiley & Sons, Hoboken, NJ

Mitton, G. B. and G. A. Payne (1996) *Quantity and quality of runoff from selected guttered and unguttered roadways in northeastern Ramsey county, Minnesota*, Technical Report U.S. Geological Survey Water-Resources Investigations Report 96-4284, USGS Branch of Information Services (distributor), Mounds View, MN

Nieswand, G. H. et al. (1990) "Buffer strips to protect water supply reservoirs: A model and recommendations," *Water Resources Bulletin* 26 (6):959–966

Norrstrom, A. C. and G. Jacks (1998) "Concentration and fractionation of heavy metals in roadside soils receiving de-icing salts," *Science of the Total Environment* 218 (2–3):161–174

Novotny, E. V., D. Murphy, and H. G. Stefan (2008) "Increase of urban lake salinity by road deicing salt," *Science of the Total Environment* 406, pp. 131–144

Novotny, E. V. and H. G. Stefan (2008) "Chloride ion transport and mass balance in a metropolitan area using road salt," *Water Resources Research* (in press)

Novotny, V. (2003) *WATER QUALITY: Diffuse Pollution and Watershed Management*, John Wiley & Sons, Hoboken, NJ

Novotny, V., D. W. Smith, D. Kuemmel, J. Mastriano, and A. Bartosova (1999) *Urban and Highway Snowmelt: Minimizing the Impact on Receiving Water*, Project 94-IRM-2, Water Environment Research Foundation, Alexandria, VA

Novotny, V., D. Muehring, D. H. Zitomer, D. W. Smith, and R. Facey (1998) "Cyanide and metal pollution by urban snowmelt: impact of deicing compounds," *Water Science and Technology* 38 (10):223–230

Novotny, V. and H. Olem (1994) *WATER QUALITY: Prevention, Identification and Management of Diffuse Pollution*, Van Nostrand Reinhold, New York, NY (distributed by John Wiley & Sons, Hoboken, NJ)

Oberndorfer, E., J. Lundholm, B. Bass, R. R. Coffman, H. Doshi, N. Dunnett, S. Gaffin, M. Kohler, K. K. Y. Liu, and B. Rowe (2007) "Green Roofs as Urban Ecosystems: Ecological Structures, Functions, and Services," *Bioscience* 57 (10):823–833

Oberts, G. L. (2003) "Cold climate BMPs: Solving the management puzzle," *Water Science and Technology* 48 (9):21–32

Peck, S. W. (2005) "Toronto: A model for North American infrastructure development," in *EarthPledge. Green roofs: Ecological Design and Construction*, Schiffer Books, Atglen, PA, pp. 127–129

Pisano, W. C. (1989) "Swirl concentrators revisited: The American experience and new German technology," in *Design of Urban Runoff Quality Controls*, Proceedings of Engineering Foundation Conference (L. A. Roesner et al, eds.), American Society of Civil Engineers, New York, NY, pp. 390–402

Potts, R. R. and J. L. Bai (1989) "Establishing variable width buffer zones based upon site characteristics and development type, *Proceedings of the Symposium on Water Laws and Management*, American Water Resources Association, Bethesda, MD

Prince George's County (1999) *Low-Impact Development Design Strategies: An Integrated Design Approach*, http://www.epa.gov/owow/nps/lid/lidnatl.pdf (accessed December 2009)

Ramakrishna, D. M., and T. Viraraghocan (2005), "Environmental impact of chemical deicers - a review," *Water, Air and Soil Pollution* 166, pp. 49–63

Read, J., T. Wevill, T. Fletcher, and A. Deletic (2008) "Variations among plant species in pollutant removal from stormwater in biofiltration systems," *Water Research* 42, pp. 893–902

Salt, D. E., R. D. Smith, and I. Raskin (1998) "Phytoremediation," *Annual Review of Plant Physiology and Plant Molecular Biology* 49, pp. 643–668

Sansalone, J. J., S. G. Buchberger, and S. R. Al-Abed (1996) "Fractionation of heavy metals in pavement runoff," *Science of the Total Environment* 189–190, pp. 371–378

Schnoor, J. L., L. A. Licht, S. C. McCutcheon, N. L. Woolfe, and L. H. Carreira (1995) "Phytoremediation of organic and nutrient contaminants," *Environmental Science and Technology* 29 (7):318A–323A

Schueler, T. and M. Helfrich (1988) "Design of extended detention wet pond systems," in *Design of Urban Runoff Quality Controls*, Proceedings of Engineering Foundation Conference (L. A. Roesner et al., eds.), American Society of Civil Engineers, New York, NY, pp. 180–200

Schueler, T. R., P. A. Kumble, and M. A. Heraty (1991) *A Current Assessment of Urban Best Management Practices. Techniques for Reducing Non-point Source Pollution in the Coastal Zone*, Technical Guidance Manual Prepared by the Metropolitan Washington Council of Governments, Office of Wetlands, Oceans, and Watersheds, U. S. Environmental Protection Agency, Washington, DC

Song, Y., M. Fitch, J. Burken, L. Nass, S. Chilukiri, N. Gale, and C. Ross (2001) "Lead and zinc removal by laboratory-scale constructed wetlands," *Water Environment Research* 73, pp. 37–44

Stahre, P. (2008) *Bluegreen Fingerprints in the City of Malmö, Sweden – Malmö's Way towards Sustainable Urban Drainage*, VASYO, Malmö, Sweden, http://www.vasyd.se/SiteCollectionDocuments/Broschyrer/Publikationer/BlueGreenFingerprints_Peter.Stahre_webb.pdf

Stahre, P. and B. Urbonas (1990) *Stormwater Detention For Drainage, Water Quality, and CSO Management*, Prentice Hall, Englewood Cliffs, NJ

Steiner, G. R. and R. J. Freeman, Jr. (1989) "Configuration and substrate design considerations for constructed wetlands for wastewater treatment," in *Constructed Wetlands for Wastewater Treatment* (D. A. Hammer, ed.), Lewis Publishing, Chelsea, MI, pp. 363–378

Stockdale, E. C. and R. R. Horner (1987) "Prospects for wetlands use in stormwater management," *Proceedings of the Fifth Symposium on Coastal and Ocean Management*, vol. 4, American Society of Civil Engineers, New York, NY, pp 3701–3714

Sullivan, R. H. *et al.* (1982) *Design Manual: Swirl and Helical Bend Pollution Control Devices*, EPA-600/8-82-013, U. S. Environmental Protection Agency, Storm and Combined Sewer Section, Edison, NJ

Theodosiou, T. G. (2003) "Summer period analysis of the performance of a planted roof as a passive cooling technique," *Energy and Buildings* 35, pp. 909–917

Thunqvist, E. L. (2004) "Regional increase of mean annual chloride concentration in water due to the application of deicing salt," *Science of the Total Environment* 325, pp. 29–37

University of New Hampshire Stormwater Center (2009) *UNHSC Design specifications for asphalt pavement and infiltration beds*, UNHSC, Durham, New Hampshire

Urbonas, B. and W. P. Ruzzo (1986) "Standardization of detention pond design for phosphorus removal," in *Urban Runoff Pollution* (H. C. Torno, J. Marsalek, and M. Desbordes, eds.), Springer Verlag, Heidelberg, Germany, and New York, NY, pp. 739–760

U.S. EPA (1988) *Constructed Wetlands and Aquatic Plant Systems for Municipal Wastewater Treatment*, Washington, DC

U.S. EPA (2000) *Vegetated Roof Cover: Philadelphia, Pennsylvania*, United States Environmental Protection Agency Report no. 841-B-00-005D, USEPA, Washington, DC

U.S. EPA (2001) "Street Edge Alternative (SEA Streets) Project," http://www.epa.gov/greenkit/stormwater_studies/SEA_Streets_WA.pdf (accessed November 2009)

U.S. EPA (2003) *Cooling Summertime Temperatures: Strategies to Reduce Urban Heat Islands*, EPA 430-F-03-014, United States Environmental Protection Agency, Washington, DC

U.S. EPA (2006) *International Brownfield Study: Emscher Park*, Germany, http://www.epa.gov/brownfieds/partners/emscher.html (accessed May 2006)

U.S. EPA (2007) *Reducing Stormwater Costs through Low Impact Development (LID) Strategies and Practices*, EPA 841-F-07-006, United States Environmental Protection Agency, Washington, DC

U.S. Geological Survey (2007) *Salt Statistics and Information* (USGS, ed.), United States Geological Survey, Reston, Virginia, http://minerals.usgs.gov/minerals/pubs/commodity/salt/

VanWoert, N. D, D. B. Rowe, J. A. Andresen, C. L. Rugh, and L. Xiao (2005) "Watering regime and green roof substrate design impact Sedum plant growth," *Horticultural Science* 40, pp. 659–664

Vymazal, J. and I. Kröpfelová (2008) *Wastewater Treatment in Constructed Wetlands with Horizontal Sub-Surface Flow*, Springer Verlag, Dortrecht, Netherlands

Warren, L. A. and A. P. Zimmerman (1994) "Influence of temperature and NaCl on cadmium, copper and zinc partitioning among suspended particulate and dissolved phases in an urban river," *Water Research* 28 (9):1921–1931

Water Environment Federation (1990) *Natural Systems for Wastewater Treatment*, 2nd ed., Manual of Practice FD-16, Water Environment Federation, Alexandria, VA

Water Pollution Control Federation (2000) *Manual of Practice for the Design and Construction of Urban Stormwater Management Systems*, Water Environment Federation, Alexandria, VA

Westerlund, C., M. Viklander, and M. Backstrom. (2003) "Seasonal variations in road runoff quality in Lulea, Sweden," *Water Science and Technology* 48 (9):93–101

Wiesner, P. E., A. M. Kassem, and P. W. Cheung (1982) "Parks against storms," in *Urban Stormwater Quality, Management and Planning* (B. C. Yen, ed.), Water Resources Publications, Littleton, CO, pp. 322–330

Wilson, L.G. (1967) "Sediment removal from flood water," *Transactions of the American Society of Agricultural Engineers* 10 (1):35–37

Woodard, S. E. (1989) The Effectiveness of Buffer Strips to Protect Water Quality, Master of Science Thesis, Department of Civil Engineering, University of Maine, Orono, ME

V

WATER DEMAND AND CONSERVATION

V.1   WATER USE

V.1.1   Water on Earth

Life on earth cannot exist without water. While water covers almost two-thirds of the earth's surface, only a small portion of it is the usable freshwater needed for terrestrial life (Table 5.1), and the rest is salt or inaccessible water. However, in the last 15 years, conversion of salt water into usable freshwater by desalination has become an affordable reality in high-income countries (the U.S., the Middle East, Singapore, and others).

Water is a resource that is partially renewable, such as freshwater in lakes, rivers, precipitation, snowmelt, and the groundwater withdrawn in amounts less than or equal to the aquifer recharge. However most of groundwater is fossil water that entered groundwater zones thousands of years ago—for example, from melted glaciers during the last ice age—and the current recharge is very small. Today, unbalanced overdrafts of fossil water from groundwater aquifers, or water mining, are common throughout the world, causing many aquifers to be economically exhausted during this century. Many parts of the world with large urban areas are already suffering water shortages, which are especially serious in Southwest US, northeast China, African countries, Middle East, parts of India, and Peru. By the year 2025, almost 50 countries with 2.8 billion people are expected to have serious water shortages, and this number is expected to increase to 4 billion by 2050, caused primarily by population increases and pollution, mainly in urban areas, and by the increasing demand for water (see Chapter I).

In the first decade of the new millennium, the news media informed the public about extreme water shortages for both cities and agriculture, which obviously

Table 5.1   Water distribution on earth (from Gleick et al., 2008)

Total volume of water on earth	1.3×10^9 km^3
Total freshwater	35.0×10^6 km^3
Fresh groundwater	10.5×10^6 km^3
All glaciers	2.4×10^6 km^3
Freshwater lakes	91,000 km^3
Wetlands	11,500 km^3
River flow	2120 km^3
Atmospheric	12,900 km^3

are competing for scarce and diminishing water resources. Severe and lasting water shortages were reported in 2007–2009 from Atlanta, Georgia (Box 5.1) and in 2009 from Mumbai (India), China, and Australia, on top of the chronic water shortages in developing countries. Water shortages have become a major problem facing China's rapid urbanization.

Box 5.1 highlights the severe water situation in a growing urban area of Atlanta, Georgia, that has relatively ample rainfall but does not have a large water source such as a major river. Atlanta's drought situation improved in 2009 after a wet winter and spring, but the federal court ruling still makes future water shortages a real possibility because it established a precedent that cities cannot use all water from common federal resources and leave other downstream water users, including ecology and aquatic life, without water. Under the U.S. judicial doctrine covering the Eastern U.S., during a time of drought, shortages are shared by all users.

BOX 5.1   DROUGHT IN ATLANTA, GEORGIA

Most streams in the Atlanta area are small, and many are severely affected by prolonged droughts. The Chattahoochee River, the only sizable stream in the metro area, has insufficient size to supply a metropolitan area of approximately 5.4 million with water. The metropolis receives most of its water from Lake Lanier (Figure 5.1); additional sources are Lake Allatoona and the Flint and Altamaha Rivers. Although average annual rainfall in the Atlanta region is 1280 mm, Atlanta has suffered extreme water shortages during the 2007–2009 drought that almost completely exhausted the water supply from all sources. The future outlook is still very bleak because of (1) the growing population in the metropolitan area, and (2) a 2009 judicial ruling by a federal judge that forbids the city to take excessive volumes of water from Lake Lanier, which is a federal reservoir providing water to ecological, agricultural, and industrial interests in a large portion of the states of Georgia, Alabama, and Florida. By 2012, the water withdrawals from Lake Lanier by the metropolitan area will have to be dramatically reduced.

(continued)

Figure 5.1 Federal Lake (reservoir) Lanier near Atlanta is the major source of water for the Atlanta, Georgia, metropolitan area. It also provides water to large downstream areas of Georgia and Florida. The picture shows the situation in January 2008, when the water from the lake was almost exhausted (Photo courtesy of Bill Kinsland, National Oceanic and Atmospheric Administration).

Before the drought of 2007–2009, the municipal average per capita water use in metropolitan Atlanta was ranging between 415 (winter) to 680 (summer) liters/capita-day. Water conservation during nondrought years is voluntary, but mandatory restrictions on lawn sprinkling were implemented during the drought. Without significant cutbacks in water use, Atlanta will soon run out of water.

There are many other cases throughout the world where the situation in urban water supply is critical, and cities are looking for increasingly expensive ways to provide water to citizens. Singapore is a small city-state located on a 700-km^2 equatorial island between Malaysia and Indonesia. The population of Singapore reached 4.6 million in July 2008. The country has one of the highest standards of living in the world and also ample water from the average 2370-mm rainfall it receives. However, because most of the territory of Singapore is urbanized, its natural water resources in small streams and reservoirs (catchment water) have been insufficient, and a lot of rainfall used to leave the city as unusable runoff. Singapore has had two treaties for importing water from Malaysia, but one treaty expires in 2011 (see also Chapter VI). Consequently, to avoid water shortages, Singapore is focusing on developing water infrastructure and expensive additional sources in desalination and indirect potable reuse (NEWater) from reclaimed used water (wastewater). The city has also built dams (barrages) across two estuaries, changing them from tidal brackish water bodies into freshwater reservoirs collecting primarily treated urban runoff. The freshwater is then pumped into water supply reservoirs, some located in protected areas to supplement the available catchment water. More on the Singapore water management and reclamation effort will be presented in Chapter VI.

Australia has been facing severe drought for more than a decade, and millions of urban dwellers are coping with severe shortages. In 2008 the two largest cities, Brisbane and Adelaide, with a total population of three million, were running out of water. The government of Australia made a declaration that farmers would receive no irrigation water in the most fertile region of the country. The impact on cities was severe, but it was devastating on farming communities as the drought continued. The drought is being attributed to climate change. As in Atlanta, rainfall in the wetter period at the end of the first decade of this millennium has partially lessened the problem.

In China, the China Daily news agency reported (November 24, 2007) serious water shortages in two-thirds of the 641 largest cities in the country, based on the information provided by the Ministry of Water Resources. Of the about 400 cities with water shortages, about 100 face very serious problems. The shortages are due to (1) increasing population, (2) increasing living standards, which heighten water demand, (3) inadequate water resources (e.g., Beijing), and (4) water pollution. Regarding the last reason, it was stated that more than 70 billion m^3 of wastewater were released in 2006 into receiving waters, with about 45 billion pumped into lakes and rivers without treatment. This created secondary shortages from pollution calamities in many cities, the most highly publicized being cyanobacteria bloom outbreaks in Lake Tai (Taihu), which is the largest freshwater lake in China, providing water to four million people.

Since ancient times, urban planners and engineers have tried to resolve issues of water shortages and security by means of water transfers (see Chapter I), by building large infrastructure projects such as aqueducts, pipelines, and canals. Water losses by leaking and evaporation were large in these historic water transfer projects, but water was plentiful. More recent examples of water transfers to bring water to watershort areas include California and Arizona canals that bring water to farmers and communities in Southern California and Arizona, transcountry pipelines bringing fossil groundwater from southern to northern Libya in Africa, and the South to North Canal from the Yangtze River to northern populated areas in China.

Currently, about 7% of the energy used in the U.S. and, consequently, the same proportion of greenhouse gas (GHG) emissions are for providing water and used water disposal for the urban population. Many urban areas of the world do not have sufficient water supply to provide water to their inhabitants based on the current water demand for domestic indoor and outdoor, municipal, and commercial water uses. Population growth in many countries, including the U.S., is putting additional stresses on municipal water supply systems, which are often more than one hundred years old and lose water from leaks and inappropriate use by citizens. The Johns

Hopkins University Public Health Study, at the end of the last millennium (Anon., 1998), leaned towards the conclusion that it may be too late to avert serious water shortage problems and possible catastrophes. With global warming, increasing population, and rising living standards, this warning is becoming more serious today.

This chapter describes water demand, uses of water, and ways to reduce usage by water conservation and alternative land management. Water reclamation from used water and reuse is covered in Chapters VI and VII.

V.1.2   Water Use Fundamentals

Water uses are divided as follows:

- In-stream uses
  - Navigation, hydropower production, recreation
  - Ecological uses for propagation and support of aquatic life in water and surrounding zones
- Withdrawals
  - Uses whereby water withdrawn is returned to the water body after the use, near the point of withdrawal, such as municipal and industrial water supply without interbasin transfers, once through cooling water
  - Consumptive uses whereby water is not returned to the water body, such as evapotranspiration in irrigation systems, evaporation in cooling and water treatment and transport systems, water exported in products, and wastewater transported out of the watershed for treatment and discharge

Water should be used for beneficial purposes, and the use should not impair the ecological uses of the water body and downstream uses. Using surface and groundwater systems for accepting and transporting used (waste) water is a societal choice, but it is not a beneficial use. Gleick et al. (2008) separated the human (anthropogenic) and ecological (natural) uses and benefits. Figure 5.2 shows the concept. Economic benefits of water increase with the amount of water withdrawn. If the water is not returned back with a quality that would support the aquatic ecology or not returned at all, ecological benefits are reduced and at some point of human overuse the ecological benefits disappear. Most water resources providing water for anthropogenic uses have a limited resilient capacity to support human use without adverse effects on ecology known as assimilative capacity (see Novotny, 2003 for discussion). For example, as long as groundwater withdrawals are below the recharge safe yield, or concentrations of pollutants in the water body resulting from discharges of used water are safely below the toxic levels, the ecological benefits of the water body are retained. The goal of water use and withdrawals management is to keep the ecological impact below the threshold at which a healthy ecological function of the surface water body can continue or groundwater is not excessively mined.

Very often, human uses of water are competing and may interfere with each other. For example, streams and reservoirs used for water supply and their watersheds should not be extensively used for recreation or any other polluting activity, of which

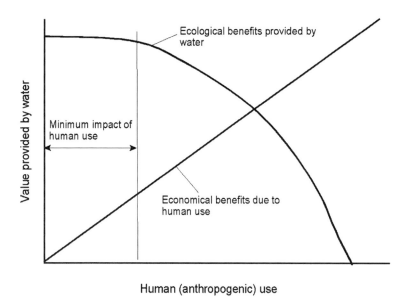

Figure 5.2 Balancing anthropogenic use of water with the ecological value of the water body (modified from Gleick et al., 2008).

intensive agriculture is the most damaging. If polluting activities cannot be avoided or removed, barriers protecting the source water bodies must be implemented, including buffer strips and best management practices to prevent and control diffuse pollution and eutrophication, and provide adequate water and used water treatment. Treated (or sometimes untreated, in developing countries) used water discharged into receiving waters creates so-called water-sewage-water cycle (see Chapters VI and IX), leading to uncontrolled indirect potable water reuse.

The solution to water shortages, including bringing new water to cities for the increasing population, was in the past generally achieved by hard infrastructure such as the aforementioned long-distance water transfers by canals, aqueducts, and groundwater mining from increasingly deep aquifer depths. The concepts of the Cities of the Future described in this book advocate giving increased attention to the "soft path," which is the path towards sustainability. Gleick et al. (2008) defined the soft path as a "comprehensive approach to water management, planning and use that uses water infrastructure, but combines it with improvement in overall productivity of water use, the smart application of economics to encourage efficiency and equitable use, innovative new technologies, and the strong participation of communities." As pointed out by Gleick et al., rather than looking for an endless enlargement of the water supply, which may not exist or be very expensive, the soft path matches water services to the scale of users' needs and takes environmental and social needs into account, to ensure that basic human and environmental needs would be met.

The soft path also includes matching the water quality of the water needed with the water use. The most nonsensical feature of current (third and fourth paradigm) water supply systems is using potable water for irrigation of lawns, which themselves are landscape features that demand large volumes of water. The soft path also encourages water reclamation and reuse. Water conservation seems to be a logical first step in water-short urban areas, followed, if needed, by water reclamation from used water and reuse. While technologies exist to treat reclaimed water to a high quality comparable to bottled water, direct potable reuse is not recommended and may be socially unacceptable (see Chapter VI).

Water overuse, overdrafts and groundwater mining, excessive losses, and waste are an unfortunate legacy of the third and fourth paradigm in the urban water supply field, in both arid and humid areas. For example, humid Massachusetts has an average rainfall of 1118 mm/year, yet water shortages and restricted water uses in some communities are common. The scenic Ipswich River in the northern part of the state, originally a river with a thriving trout population, runs dry every other year due to excessive withdrawals for water supply, and the quality fish have disappeared. The water withdrawn is conveyed to regional wastewater treatment facilities and not returned. Many urban rivers have either disappeared from the surface or been turned into effluent dominated streams (see Chapters I, VI, and IX). Consequently, managing water withdrawals and water use, reclamation, and reuse is a part of the overall comprehensive water management and revitalization effort focusing today on integrated resources management (IRM) rather than on providing water supply and waste disposal only.

Minimum water use criteria. In suggesting water conservation and sometimes outright drastic restrictions, in water shortage emergencies, it is necessary to know what the minimum water use criteria are. Globally, most water withdrawn from surface and groundwater sources is used in agriculture for irrigation, resulting in large consumptive losses. Agriculture worldwide accounts for 69% of annual water withdrawals (Anon., 1998). However, this varies among countries and even continents. In Europe, most water is used for industries.

A range of 20 to 40 liters/capita-day is generally considered as the lowest limit to meet basic water supply and sanitation needs in developing countries (Gleick, 1996). Gleick further proposed "an overall basic water requirement of 50 liters/person-day" as a minimum standard to meet four basic needs: drinking, sanitation, bathing, and cooking. Falkermark and Widstrand (1992) used 100 liters/person-day as a rough estimate of a minimally acceptable water availability standard for people living in developing countries, excluding uses for agriculture and industry. These estimates have been widely accepted in many hydrological studies (Anon, 1998). These low values do not consider water reclamation and reuse.

Drinking water quality standards. Each country regulates potable water quality. All sources of potable water must be protected from contamination by providing barriers against contamination, such as protection areas and several barriers in the treatment process. A difference between standards and criteria or guidelines should be explained. Standards are legally enforceable limits that are issued by national or state (in the U.S.) pollution control authorities. Criteria and guidelines are based on scientific knowledge and may be recommended by the authorities but not directly enforced. However, in the U.S., water quality criteria issued by the U.S. EPA are the upper limits, on which states must base and develop their enforceable standards.

Water quality of potable water in the U.S. is protected by binding water quality standards and international water quality guidelines. In the U.S., the Safe Drinking Water Act of 1974 (amended in 1986 and 1996) led to the publishing of the National Primary Drinking Water Standards issued by the U.S. Environmental Protection Agency, which apply to the water quality at the point of use (at the tap) (http://www.epa.gov/safewater/contaminants/index.html). The National Primary Drinking Water Standards apply to public water systems and are enforceable. They protect public health by limiting the levels of contaminants in drinking water. The primary standards are numerous and, in addition to the U.S. EPA web site given above, they have been included in water engineering texts (e.g., Nemerow et al., 2009). The recommended Maximum Contaminant Level Goals (MCLG) and enforceable Maximum Contaminant Levels (MCL) have been issued for the following categories of contaminants:

- Microorganisms (*Cryptosporidium*, total coliforms, *Giardia lamblia*, viruses)
- Disinfectants (chloramines, chlorine, chlorine dioxide)
- Disinfection byproducts (chlorite, bromite, trihalomethanes, haloacetic acids)
- Inorganic chemicals (toxic metals, arsenic, asbestos, cyanide, nitrate, fluoride)
- Organic chemicals (more than 50 chemicals such as pesticides, PAHs, solvents, etc.)
- Radionuclide alpha and beta particles, radium 226 and 228, and uranium)

MCLs are set as close to MCLGs as feasible using the best available treatment technology and taking cost into consideration.

The U.S. EPA (2008) has also issued Secondary Drinking Water Regulations, which include pH, additional less toxic metals, taste, odor, sulfates, chlorides, total dissolved solids (TDS), color, iron, manganese, and corrosivity.

The quality of raw source water is regulated in the U.S. by states. The U.S. Environmental Protection Agency has issued the general National Recommended Water Quality Criteria (http://www.epa.gov/waterscience/criteria/wqctable/), which serve the states as guidelines for developing their own standards. The criteria are available for 190 pollutants, many coinciding with the drinking water quality standards. International guidelines for drinking water quality were issued by the World Health Organization (1984). WHO water quality guidelines contain limits for treated and untreated water.

Many other countries, for example China and countries in the European Community, have issued water quality critera (EU) and country-specific standards based on the water quality classes. The best water quality is typically designated as Class I (suitable for water supply with minimal treatment), and the worst as Class V (quality unsuitable for most uses).

V.1.3 Municipal Water Use in the U.S. and Worldwide

The Commonwealth of Massachusetts is an example of a relatively water-rich state that has escaped the serious water problems in the arid and semi-arid areas of the

U.S. The large Quabbin and Wachusett Reservoirs in central Massachusetts have provided in most years ample water supply for domestic, municipal, and commercial uses in eastern and central Massachusetts, including lawn watering—and the utility operating the vast water supply system, the Massachusetts Water Resources Authority (MWRA), wants to sell more. The average water demand reported by MWRA in 2003 was 380 liters/capita-day in the 43 communities served in the Boston metropolitan area (served population two million). However, suburban residents use and waste most water. The affluent Boston suburban community of Weston draws treated water from the Quabbin-Wachusett Reservoirs system. Its residents use on average 415 liters/capita-day, obviously much more in summer. It should be pointed out that the large lots on which houses in Weston are built are mostly forested, with smaller or no lawns. In Plymouth, Massachusetts, a well-known historic community south of Boston, outside of the MWRA system, the use is 630 liters/capita-day. The lots are smaller, with lawns and flower beds.

The situation in water use or overuse is much worse in the Southwest U.S. communities located essentially in desert or semi-desert environments. In spite of arid conditions and a lack of local water resources, average water use in Scottsdale, Arizona, a suburb of Phoenix, is a whopping 890 liters/capita-day provided mostly by imported water, mainly from the Colorado River. Table 5.2 shows municipal water use in several U.S. cities. The U.S. has the highest worldwide per capita water use, followed by Canada and the United Arab Emirates. It is interesting to note the water use is highest in arid areas (Southwest U.S., UAE) that have very little rainfall but enough financial resources to obtain water from alternate often unsustainable (e.g., groundwater mining) sources, some benefiting from government-subsidized long-distance transfers of water from the Colorado River or other sources.

Table 5.2 Average per capita water use in selected North American urban areas

City/Metropolitan Area	Liters/Capita-Day
Atlanta (GA) (2004 before drought—no water conservation)	415 to 680
Boston (MA) (2003)	380
Boulder (CO)	540
Chicago (IL) (2009)	680
Denver (CO)	680
El Paso (TX) (2009)	502
Eugene (OR)	582
Phoenix (AZ)	669
Scottsdale/Tempe (AZ)	892
Seattle (WA)	336
Tampa (FL) (2009)	390
Waterloo (ONT)	284

Source: Heaney, Wright, and Sample (2000); later data obtained from the Internet and personal communication

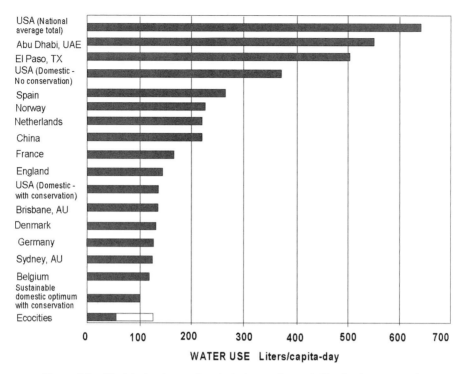

Figure 5.3 Municipal water use in selected countries and cities (various sources).

Water use in most European and Asian cities is much less than that in the U.S. For example, average water use in England is about 150 liters/capita-day and that in Germany is 120 liters/capita-day. Figure 5.3 shows the ranges of municipal domestic water use in several cities and countries.

According to the American Water Works Association study reported by Heaney, Wright, and Sample (2000), the average water use at the end of the last century, in nonconserving single-family homes in the U.S., was about 380 liters/capita-day, but it reached up to 1000 liters/capita-day in hot arid and semi-arid regions of the U.S. These are the areas with the greatest shortages of available local water resources. Water to Phoenix, Scottsdale, Los Angeles, Orange County (California) and San Diego (California) is brought from large distances by aqueducts. The last-century goal in the U.S. was to bring the per capita water use to 185 liters/capita-day at some future time by implementing water conservation and plugging the leaks. In water centric ecocities, by conservation and water reuse, use of water can be reduced to less than 100 liters/capita-day (see Chapters VII and XI). Hence, U.S. communities have a long—but achievable—way to go to reach sustainable water use, but the trend in domestic water use has been downward in the first decade of 2000.

The overuse of treated municipal water supplemented by long-distance transfers and pumping has global warming consequences as well. For example, to bring water to Phoenix, Scottsdale, or Tucson in Arizona, water is pumped from Lake Havasu on the Colorado River from an elevation of 128 meters above sea level (masl) 200 meters up to Phoenix (elevation 337 masl) and 640 meters to Tucson (elevation 777 masl). Losses by friction in the aqueduct must be added. The pumping energy use is proportional to flow times pumping head. Calculation of energy use and carbon emissions from pumping water is illustrated in Box 5.2.

BOX 5.2   ENERGY USE AND GHG EMISSIONS FROM DELIVERING WATER BY PUMPING AND LONG-DISTANCE TRANSFER

This approximate calculation illustrates the amount of energy needed to transfer the daily water supply volume of water from Lake Havasu on the Colorado River to Tucson by the Central Arizona Project (CAP) canal (Figure 5.4). The length of the CAP canal is 541 km. The water is pumped from Lake Havasu at an elevation of 128 masl to Tucson, to a terminal point southwest of the city, at an elevation of

Figure 5.4   Central Arizona Project canal delivering water from the Colorado River to Arizona cities, farmers, industries, and Native American nations. The canal is 541 km long, and it is the largest water project in the U.S. (Credit U.S. Bureau of Reclamation).

about 762 meters. Hence the elevation difference between the Colorado River and Tucson, which must be overcome by large pumping stations, is $\Delta H_e = 762 - 128 = 634$ meters. To carry the flow of approximately 85 m³/s and overcome the friction losses, the canal slope is 0.00008 m/m; hence friction head loss to be added to the elevation difference is $\Delta H_f = 0.00008 \times 541{,}000$ (meters) $= 42.3$ meters, rounded to 50 meters. The calculation will be carried out for delivering the daily water demand per person of $V = 650$ liters (0.65 m³) from the Colorado River to Tucson.

The work, W, done by the pumps to deliver the daily per capita water supply amount is

$$W = \frac{\gamma V H}{\varepsilon} = \frac{9810 \,[\text{N/m}^3] \times 0.65 \,[\text{m}^3] \times (634 + 50) \,[\text{m}]}{0.8}$$
$$= 5.45 \text{ MJ} = 1.51 \text{ kW-hour}$$

where $\gamma =$ specific weight of water in N/m³, and $\varepsilon =$ estimated efficiency of the pump system.

The CAP is powered by a coal-fired Navajo power plant. One kW-hour represents 0.960 kg of CO_2 emissions (see Section I.3.3 of this book); hence, delivering water from the Colorado River to Phoenix or Tucson will cause annual emissions of 529 kg of CO_2/person-year just for delivering water to the city. GHG emissions due to treatment and city delivery will have to be added to get the total carbon imprint.

The losses of water by evaporation in the arid Arizona semi-desert are about 4.4% of the water carried by the canal. The communities, including Phoenix, Mesa, Scottsdale, and Tucson, use about 30% of the flow from the CAP canal. The rest is for agricultural interests and replacing water taken from Native American nations by Arizona cities for their water supply.

Data source: Central Arizona Project, http://www.cap-az.com/about-cap/faq/

V.1.4 Components of Municipal Water Use

Water in cities is used for:

- Domestic water supply
  - In-house water use
    - Drinking, cooking, washing dishes
    - Shower and bath
    - Laundry
    - House cooling (by evaporation systems)
    - Toilet flushing

- Outside water demand
  - Lawn and garden irrigation
  - Swimming pools
  - Car washing
- Commercial water use in restaurants, hotel/motel establishments, car washes, commercial laundries, stores, small businesses, offices, etc.
- Public use for irrigation of parks and public golf courses, firefighting, street cleaning, schools, hospitals and other health-care facilities, government offices, sanitary facilities in train and bus stations, public restrooms, public drinking water fountains, etc.

Water use is also divided into potable and nonpotable water uses. Chapter VI, dealing with water reclamation and reuse, also categorizes use into direct and indirect potable uses and reuse. It is clear that the highest quality is required for potable use, which should comply with public health water quality standards. Water use for irrigation and swimming pools is almost all consumptive use because of evaporation and evapotranspiration losses.

V.1.5 Virtual Water

Virtual water, transfers, and trading refer to water use outside of the city to produce food and other goods to satisfy the needs of the people living in the city. Such water-demanding activities outside of the city include agriculture and production of electricity, construction materials, paper, and, today, biofuel and oil derived from tar sands. For example, the water use of an average U.S. citizen for direct household use is 380 liters/capita-day, but producing food for the same citizen, including irrigation and livestock, will require 1928 liters/capita-day, of which 61% is consumptive use—that is, water lost by evaporation and transpiration. Producing electricity requires 1780 liters/capita-day, mostly for cooling. The consumptive loss from cooling water is about 3 to 4% (see Chapter VIII); hence, the virtual water demand for producing electricity is about 53 to 73 liters/capita-day (McMahon, 2008; Gleick et al., 2008). Box 5.3 is an example of virtual water requirements: consuming bottled water.

The concept of virtual water has been extensively studied by the Pacific Institute (Gleick et al., 2008) in the US and in the Netherlands by Hoekstra et al. (2009) and Hoekstra and Chapagain (2007), who point out that the overall water footprint and, by the same reasoning, the energy footprint are related to the virtual water. The concept was introduced by Allan in the early 1990s (Allan, 1993, 1994) when studying the option of importing virtual water (as opposed to real water) to solve the problems of water scarcity in the Middle East. As pointed out by Hoekstra and Hung (2007), when assessing the water footprint of a city (or a country), it is essential to quantify the flows of virtual water leaving and entering the city (country). Similar estimates can be made for building materials, fruit, vegetables, and other products delivered from distant areas, some of them on the other side of the globe. While international trade is not discouraged in this book, the philosophy of ecocities emphasizes local (organic) fruit and vegetables, and recycled and local building materials.

> **BOX 5.3 EXAMPLE OF VIRTUAL WATER AND CARBON EMISSIONS: THE BOTTLED WATER DILEMMA**
>
> An interesting interlink between water and energy that should not be overlooked when considering water energy use is sales of *bottled water*. The growing consumption of bottled water has economical, environmental, and global warming consequences that are not widely publicized. From time to time, TV media broadcast the results of "tasting contests" that rate tap water from the grid as having better quality and taste than bottled water. As a matter of fact, many bottled water brands are more purified tap water, not spring water. The Pacific Institute (Gleick and Colley, 2009) estimated that in 2006:
>
> - Americans bought a total of 31.2 billion liters of bottled water, sold in mostly polyethylene terephthalate bottles of various sizes
> - Producing bottles just for American consumption required the equivalent of more than 17 million barrels (2.5 million of tons) of oil, not including the energy for transportation
> - Bottling water emitted more than 2.5 million tons of carbon dioxide
> - Producing 1 liter of bottled water requires 3 liters of water
>
> Plastic bottles are not degradable and will stay in landfills for thousands of years. Discarded plastic bottles also have an adverse impact on the water quality of receiving waters and biota.

V.2 WATER CONSERVATION

V.2.1 Definition of Water Conservation

Gleick et al. (2003) defined the term "water conservation" as reducing water use by improving the efficiency of various uses of water, which also includes the socially beneficial reduction of water use or water loss. They emphasized that conservation should not lead to unnecessary cutbacks causing deprivation. In the most general context, conservation includes both improving water use efficiency and, to a lesser degree, substituting reclaimed water for some end uses. This chapter deals with improving water use efficiency by eliminating or reducing inefficient uses leading to waste. Chapters VI and VII will then deal with water reclamation and reuse.

V.2.2 Residential Water Use

Indoor Urban Residential Use Heaney, Wright, and Sample (2000) report that the AWWA Research Foundation–sponsored nationwide project referred to as the North American Residential End Use Study (NAREUS) documented that indoor water use does not vary greatly among the 12 studied cities. The

Table 5.3 Indoor and outdoor water use in a single-family home in 12 monitored cities in North America

	Without Water Conservation*		With Water Conservation	
Water Use	Liters/Capita-Day	Percent	Liters/Capita-Day	Percent
Faucets	35	14.7	35	25.8
Drinking water and cooling	3.6	1.2	2.0	1.5
Showers	42	17.8	21	15.4
Bathtubs and hot tubs	6.8	2.0	6.0	4.4
Laundry	54	22.6	40	29.4
Dishwashers	3.0	1.4	3.0	2.2
Toilets	63	26.4	14	10.3
Leaks	30	12.6	15	11.0
Total Indoor	238	100	136	100
Outdoor	313	132	60**	44
Total	551	232	196	144

*Adapted from AWWA RF, 1999; Heaney, Wright, and Sample, 2000; and Metcalf & Eddy, 2007
**Reflects switch from lawn to xeriscape using native plants and ground covers with no irrigation. Water use is for swimming pools and watering flowers and vegetable gardens.

study monitored in detail 1200 households. The indoor water use ranged between 162 liters/capita-day (in Seattle, Washington) and 276 liters/capita-day (in Scottsdale/Tempe, Arizona). The standard deviation of the indoor water use was 38 liters/capita-day. This relatively stable indoor water use reflected the fact that indoor water use is for essential purposes that are culturally the same among the U.S. population. This range of water use does not reflect any water conservation—for example, toilet tank content generally was 15 to 19 liters in the 1990s (Heaney, Wright, and Sample, 2000), and no water-saving shower heads were used in the tested households. Table 5.3 shows the per capita volumes and proportions of the daily water use in a typical U.S. single-family home. The left part of the table is based on the AWWA RF (1999) study as reported by Heaney, Wright, and Sample (2000). The potential effects of future conservation of domestic water by users in the U.S. are presented on the right side of the table. The conservation options do not include water reclamation and reuse, which will be presented and discussed in Chapter VI.

Toilet flushing, which generates yellow and black water, has been the largest indoor water use. Only about 25% of flushing is related to disposal of fecal matter (Friedler et al., 1996). In the U.S., new plumbing codes mandate toilet flushing to be limited to 6 liters per flush; however, worldwide development is far beyond this very modest U.S. goal. The U.S. EPA promotes Water Sense toilets, now produced by hundreds of manufacturers, that reduce the water use to about 4 liters per flush. Dual-flushing vacuum-assisted toilets common in Japan and Singapore, and also slowly appearing in the U.S., can reduce the average water use per flush to 2 liters or less. Some models of dual-flush toilets manufactured and used in Japan, Sweden, and

Australia also enable urine (yellow water) and feces (brown water) separation, which is important as a resource recovery option. Urine contains most of the nutrients generated by domestic wastewater (75% of N, 50% of P and 70% of K), and phosphate recovery (see also Chapter VII) is one of the major goals of the integrated resources management and recovery to be implemented in the Cities of the Future (Mitchell, Abeysuriya, and Fam, 2008). If water-saving and dual-flush toilets are installed (50% each), water use can be cut by almost 80% over the next 10–20 years, as compared to the pre-2000 no-conservation situation.

Some proposals for future urban developments suggest changing to composting toilets. For example, companies from developed countries tried to introduce this technology to China, but this change was met with resentment from the new urban dwellers moving into experimental new developments. Composting toilets may be appropriate for rural villages, where villagers can use the compost on their fields, and for urban high-density shantytown-type developments in developing countries without sewers. Flushing toilets were invented more than two thousand years ago and will likely stay for the foreseeable future. Water for flushing is considered nonpotable use; hence, use of potable water is inefficient, wasteful, and unsustainable.

Showers. Showers in the U.S. are more popular and convenient than bathtubs. A standard shower head uses about 20 liters of water in a minute. A water-efficient shower head (Figure 5.5), now becoming standard (especially in hotels/motels and in water-short areas), uses about 10 liters of water a minute (Gleick et al., 2003). It is possible but very unlikely (and was not measured in the water use studies) that a person taking a shower would compensate for the reduced flow by taking alonger time to shower (Heaney, Wright, and Sample, 2000). A potential reduction of 50% in water use due to water conservation was entered in Table 5.3. Water from showers is designated as gray water, but it is polluted by soap, shampoos, and hair care

Figure 5.5   Water-saving shower head (Credit U.S. Environmental Protection Agency).

products; potentially also pathogens; and certainly by *Escherichia coli* bacteria washed from the body. Changing the shower heads can bring about significant savings on the electric energy required to heat water (Box 5.4).

BOX 5.4   ILLUSTRATION CALCULATION OF ENERGY SAVINGS AND GHG EMISSION REDUCTIONS DUE TO INSTALLATION OF WATER-SAVING SHOWER HEADS (BASED ON GLEICK ET AL., 2003)

Water savings 21 liters/capita-day = 7.665 m^3/year.
Water temperature differential in water heater: cold water 15.5°C, hot water 40.5°C.
Heater efficiency of converting electricity to heat = 80%
Energy saved

E = [7.665 (m^3/year) $*$ 1000 (kg/ m^3) $*$ 1000 (g/kg) $*$ (40.5 – 15.5)(°C) $*$ 4.2 (J/g$*$ °C)$*$1/(3.6$*$10^6) (kw-hr/J)]/0.8 = 348 kw-hr/year

If the electricity is provided by a coal-fired power plant, then the plant CO_2 emissions are 0.96 kg of CO_2/kw-hr (see Section I.3.3 in Chapter I). Therefore, the equivalent GHG emissions reduction attributed to switching from 19 liters/min (5 gpm) to 9.5 liters/min (2.5 gpm) shower heads would become

GHG reduction = 348 (kw-hr/year) $*$ 0.96 (kg CO_2 /kw-hr) = 334 kg CO_2/year

---

Laundry. Almost all commercial laundromats use front-loading washing machines, in contrast to many domestic models that are top-loading. Front-loading washers (Figure 5.6) use 35 to 40% less water. Water for laundry should not contain pathogens, and iron and manganese concentrations should be low because they could stain the laundry. Used water from laundry is categorized generally as gray water.

Bathtubs and hot tubs. Although most bathrooms are equipped with bathtubs, their use is infrequent. Large bathtubs require more than 200 liters of water. Hot tubs are luxury items that, in spite of their large size, actually may not require very large volumes of water because they are not used for washing with soap, and water in the tub stays hot for a long time. Without an insulating cover they may be significant users of energy and should be covered during off-use times to save water (minimize evaporation) and energy (heat lost by evaporation). Bathtub and hot tub outflows are categorized as gray water and may contain pathogens that also may multiply in warm/hot water. Hot tub water should be periodically disinfected like swimming pools.

Faucets. Water from faucets is used for drinking, cooking, washing food, and rinsing dishes, and for disposing of food waste by in-sink comminutors (garbage disposals). Use of a comminutor is not common outside of the U.S. In Table 5.3, water for drinking is listed separately. Faucet use requires the highest quality, which has to meet drinking water quality standards. Very often the taste of water is important, but generally in many news media contests, water from public water supplies has outperformed many commercial bottled water products. Most commercial bottled and softened water is produced from public water supplies by taking minerals such

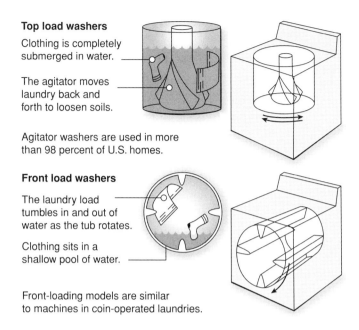

Top load washers

Clothing is completely submerged in water.

The agitator moves laundry back and forth to loosen soils.

Agitator washers are used in more than 98 percent of U.S. homes.

Front load washers

The laundry load tumbles in and out of water as the tub rotates.

Clothing sits in a shallow pool of water.

Front-loading models are similar to machines in coin-operated laundries.

Figure 5.6 Comparison of front-loading and top-loading washing machines (Credit U.S. Department of Energy).

as calcium from water, which diminishes taste. Kitchen sink water (and other liquids, e.g., from food), including water from comminutors, may be classified between gray and black waste. It may contain pathogens (e.g., from washing poultry), and significant quantities of organic solids.

Dishwasher water use is relatively small; generally it could be considered as gray water.

Leaks. Most leaks in domestic water use are indoor and occur as a result of running toilets or dripping faucets. Outdoor leaks generally occur as a result of pipes freezing, mainly in the municipal pipe distribution grid, and such leaks can be substantial. Household leaks are quite variable, and a small portion of houses accounts for a large portion of leaked volume of water (Gleick et al., 2003). Gleick et al. (2003) after summmarizing literature data estimate that approximately 30 to 47 liters per household per day can be saved by plugging domestic leaks and making repairs. The target rate of leaks of 15 liters/capita-day in Table 5.3, after reduction measures are implemented, is very similar to that used by Gleick et al. (2003) for California.

With water conservation, the data in Table 5.3 indicate domestic in-house water savings of about 43%, about the same saving as that estimated by the Pacific Institute for California (Gleick et al., 2003), and domestic overall savings of 65%.

Outdoor Water Use *Irrigation* is the largest domestic water use, which causes U.S. water use to go off the charts when compared with outdoor use in other countries. Table 5.2 indicates that the largest outdoor use occurs in warm, arid climatic conditions. Based on the AWWA RF–sponsored study of 12 U.S. cities (see Heaney, Wright, and Sample, 2000), outdoor use in Waterloo, Ontario, was 95 liters/capita-day during summer and was negligible in winter. The indoor use was 235 liters/capita-day. Hence, outdoor use was less than half of the indoor use during summer and negligible during the non-growing period. In contrast, in arid parts of California, while indoor use in households was about the same as that in the Ontario city, outdoor use was almost five times larger than indoor use.

The high water use for growing and maintaining urban lawns is due to the high evapotranspiration requirements of commercial urban grasses (e.g., Kentucky blue grass). A typical lot size in suburban communities located in the North-Central and Central U.S. is 0.4 ha (1 acre). A lawn requires about 25 mm of water each week during the growing season in the Central U.S., more in the arid Southwest. For example, potential evapotranspiration in semi-arid South Central California is between 5 to 8 mm/day (Hidalgo, Canyan, and Dettinger, 2005). Assuming that half of the lot is not irrigated (house, driveway, patios), the daily water use for irrigation still could be as high as 7 m^3/day, or roughly 2500 liters/capita-day, if it does not rain for more than one week. Before 1980, 10% of households with lawns had installed a sprinkler system. Today it is more than 80%. Since sprinkler system operation is automated and programmed, sprinklers may be activated even shortly before rainfall.

The reason for this dichotomy of growing water-demanding lawns in water-short areas of the U.S. is rooted in the population migration during the second half of the last century (and partially still continuing today). A large part of the population in many Southwestern U.S. cities relocated there from North Central cities (Chicago, Milwaukee, Minneapolis, Detroit, Pittsburgh) and brought with them a desire to grow lawns on their new (dry) lands. Originally, water was cheap (subsidized) and seemed to be plentiful. The situation changed at the end of the last century, and cities in Southern California, Arizona, and Texas (e.g., El Paso) were running out of water. The first response of urban planners, developers, and residents was to look for and bring more water from larger distances and depths. Today the marginal cost of new water in El Paso or Southern California is about the same as the cost of energy-demanding desalination (about $1.5 to $2 per cubic meter of new water). Is the very energy-demanding use of desalinated seawater or deep underground water the answer to current and increasing future shortages? Or would it be much cheaper to practice sustainable water conservation or even ecocity-type reuse that would bring water uses down to sustainable levels?

As for more humid areas, the disastrous water shortages in Atlanta and Florida, with their abundant average rainfalls, can be mostly attributed to water overuse for lawn sprinkling. As pointed out previously, almost all areas of the U.S. (with the exception of the Pacific coastal Northwest) will periodically have summer water emergencies caused by water overuse for lawn sprinkling, and water restrictions are common.

The outdoor water demand and water shortages in the U.S. are very different from those in almost all other developed countries. First, a majority of urban populations outside of the U.S. live in higher-density communities and city centers, where small gardens with flowers, fruit trees, and/or some vegetables are common, and large man-icured lawns are unknown or grow native grasses not requiring irrigation. Second, they see no reason to grow grass which, after cutting, may end up in landfills.

Gleick et al. (2003) investigated in California the effect of various practices that would improve the efficiency of irrigation of typical lawns, including turf mainte-nance, irrigation scheduling, adding compost to soils, and adding decorative plants with mulching. These practices reduced irrigation water use by 10 to 20%. Allowing lawns to go dormant obviously reduces the water use by 90% or more.

Current sprinkler systems are programmed and automated and have rainfall sen-sors, and sometimes soil moisture sensors. This may reduce irrigation water use by 10–30%. Drip/bubbler irrigation reduced water use by up to 50%. Gleick et al. (2003) also considered use of gray water and rainwater harvesting. However, gray water use may not be feasible in typical homes because of its possible contamination by pathogens and the inability of an average homeowner to reliably treat and disinfect gray water. Under normal circumstances, the public health department would not al-low use of gray water on private lawns. Rain harvesting to provide water is feasible, but the amount of water needed to water a lawn is very large, and a typical rainwater collection barrel would only provide a fraction of the water needed.

Energy use and pollution from lawn care. Overuse of water is not the only problem. The largest and mostly overlooked pollution—and, to a lesser degree, global warming—effects are due to (1) use of gasoline-powered lawn mowers, and (2) use of chemicals. Lawn "care" is a $30 billion industry in the U.S., and homeowners use excessive amounts of fertilizers and pesticides to maintain their green grass. The U.S. EPA estimates that the cost of chemicals and gasoline to care for the 0.4-ha (1-acre) typical lawn in the U.S. is about $400–700 each year (http://www.epa.gov/NE/lab/pdfs/lablandscaping-factsheet.pdf). An estimated 31,780 tons of pesticides are applied in the U.S. on lawns in a year, 10 times more per hectare than are applied to agricultural crops. *In 1997, sales of lawn-care pesti-cides in the U.S. accounted for one-third of total world expenditures on pesticides.* Forty to sixty percent of nitrogen fertilizer applied to lawns ends up in surface water and groundwater.

Gas-powered lawn mowing is far more polluting than driving an automobile. This is due to the fact that lawn mowers use inefficient, polluting two-cycle engines that also burn oil. An average gas-powered lawn mower uses about 12 liters of gasoline per 1 hectare cut (1.5 gallons/ha). To illustrate the energy use and GHG emissions, a home with a 0.2-ha lawn that is cut every 10 days for 5 months will use 37 liters of gasoline and at least 4 liters of oil. Burning 1 liter of gasoline (or oil) produces 2.33 kg of CO_2 (see Chapter I). Hence, the total CO_2 emissions from lawn mowing would be 85 kg/year. This does not include the energy and GHG emissions from the production and transport of lawn-care chemicals and irrigation water. The total lawn area in the U.S. is about 12 million hectares or 120,000 km^2, which is the area of the state of Pennsylvania, with 75% of the area being home lawns, and the rest golf courses and other landscaping.

Xeriscape. The final answer to reducing outdoor water use, pollution, and GHG emissions in the U.S. and elsewhere is switching from growing imported

Figure 5.7 Simple xeriscape landscaping of a home in Flagstaff, Arizona. Courtesy City of Flagstaff.

water-demanding grasses to *xeriscape*, which uses native plants and decorative ground cover not requiring irrigation in the more arid areas, and to grasses and wildflowers not requiring irrigation. Xeriscape landscaping essentially refers to creating a landscape design that has been carefully tailored to withstand drought conditions, and which could also be more aesthetically pleasing than a lawn. The term means "dry scape," but xeriscape can be used anywhere. For example, in more humid areas of the U.S., native prairie flowers and grasses can be grown, while in the Southwest U.S., desert plants and natural mineral ground cover result in far more pleasing urban landscapes with almost no irrigation water requirement (Figure 5.7). The "with conservation" outdoor water requirements in Table 5.3 reflect a switch from imported grass to xeriscape. Many communities provide subsidies and other incentives to homeowners and commercial establishments to install or convert lawns to xeriscape. Xeriscape also provides a habitat to small animals, insects, and the like. In El Paso, Texas, water use in new homes with xeriscape was reduced by 200 liters/capita-day. Gleick et al. (2003) estimate 75 to 85% water use reduction with xeriscape.

Swimming pools. Swimming pools are also a common feature of the U.S. suburban landscape. There may be water use differences between in-ground and aboveground pools. The major water loss is from evaporation, which can be minimized by covering the pool with a plastic cover between uses. Both types of pools are prone to leaks. Aboveground pools are typically assembled before the swimming season and refilled with water each season, while most of the water in in-ground pools is left under a cover over the winter and is replenished by snowmelt and rain. Hence, the overall water use (loss) from in-ground pools is less than water loss from grass turf of the same size.

Implementing outdoor water conservation as outlined herein, along with the indoor water-saving measures, can reduce the potable domestic water demand, as shown in Table 5.3, by almost 55% (from 550 liters/capita-day to less than 200 liters/capita per day). This would be a realistic goal and a remarkable achievement for the Phase I water reduction leading towards ecocities in water-stressed areas; it can be accomplished by retrofits without large investments in new "ecocity" infrastructure. The resource recovery from used water would be relatively minimal, the water supply and used water disposal would remain linear, and additional new water could be obtained by regional system-wide water reclamation and reuse, supplementing groundwater and surface water sources, which would then lead to indirect potable water reuse. Such systems are now under development in, for example, Singapore, Florida, and Orange County, California, and would be suitable for other stressed cities such as El Paso, Texas, San Diego, California, or Beijing, China.

While water use reductions by the measures and practices presented in this chapter would be significant and would bring large economical, ecological, and carbon footprint benefits, water use would still be at the higher end of the range of current water use, compared to that in other developed countries such as Australia and countries of Europe and Asia. That is why these water conservation measures are only a very important beginning of the process towards the sustainable water and energy use of the Cities of the Future.

V.2.3   Commercial and Public Water Use and Conservation

Standard environmental engineering textbooks (e.g., Nemerow et al., 2009; Metcalf & Eddy, 2003) contain ample information on per person water uses in commercial, institutional, recreational, and industrial facilities and establishments. The water use data are generally based on older information. Compared to the total amount of water used in the United States, the amount going to commercial purposes is not large. According to the U.S. Geological Survey (http://ga.water.usgs.gov/edu/wuco.html) in 1995, the last year when commercial water-use data were compiled, less than 1% of about 1285×10^6 m^3 per day of freshwater used was withdrawn by the commercial sector. Another 25×10^6 m^3 per day (2% of the total freshwater withdrawal) was delivered to commercial facilities by public suppliers, such as the local water department.

In the United States, almost all water delivered to a home is potable—treated to the highest drinking water standards in order to maintain human health free of water-related diseases. Exceptions can be found in some arid states like Utah, where many homeowners in Salt Lake City and other urban areas have a dual water supply, one for home use and the other for irrigation. As shown in the previous section, only a small portion of delivered water is used for drinking in households. Since the

1990s drinking use was further reduced by the sales of bottled water at exorbitant prices, compared to the cost of municipal water of the same or even better quality. The same high-quality potable water is delivered to urban commercial, industrial, and institutional water users for toilet flushing, watering landscapes, washing cars, landscape irrigation, and even large-scale industrial cooling (Gleick et al., 2003).

Metcalf & Eddy (2003) categorized nondomestic water uses as commercial, institutional, recreational, public service and system maintenance, and industrial, when water is supplied by the municipal water distribution grid. Many larger water-demanding industries such as chemical plants, heavy industries, refineries, paper manufacturing, and sugar refineries have their own water supply and water conservation and reuse systems. A permit is required for such uses. Table 5.4 presents some typical data for commercial water use as compiled by Gleick et al. (2003) for California, complemented by data from Metcalf & Eddy (2003) and Nemerow et al. (2009).

The water savings potential is again based on the Gleick et al. (2003) report, but the estimates are obviously approximate. The same assumptions as those used in estimating household water conservation effects were made. Many hotels already use water-saving shower heads and toilets, and are trying to reduce laundry water use by asking guests to use towels more than once.

Public services and maintenance. Cities also use water for firefighting, irrigating public parks and golf courses, street flushing, line and hydrant flushing, and other purposes.

Table 5.4   Commercial water use (year 2000)

Commercial Activity	Water Use Liters/Person-Day	
	Without Conservation	With Conservation
Office employees (toilets, faucets)	40–50	20–25
other uses (landscaping, cooling, other)	170	90
Hotels guests (including water used for food)**	190–280	115
Hospitals (per patient)	380	230
Laundry industry (per employee)	3700	1850
Restaurants (per employee)	1000	700
Restaurants (per customer)	50	35
Retail stores – groceries (per employee)	640	450
Retail stores – other (per employee)	575	345
Schools (K–12) (per employee)*	1160	580
Schools (K–12) (per student)**	95	
Airports (per passenger)**	10–20	
Automobile service stations (per vehicle served)**	40	
Public lavatory (per user)**	20	6
Theater (per seat)	10	4

*Based on 255 days
**From Metcalf & Eddy, 2003, and Nemerow et al., 2009

V.2.4 Leaks and Other Losses

Leaks in the water supply distribution system may be significant; they depend on system age, geographical location (pipe freezing), and lack of system maintenance. Metcalf & Eddy (2003) estimated that from 10 to 12% of the total water production in newer distribution systems (less than 25 years old) is unaccounted for (leaks and measuring errors), and that the figure may be 15 to 30% for older systems.

The City of Chicago and its Metropolitan Water Reclamation District (MWRD) have been conducting an extensive water conservation and leak-plugging program since 1987. The water distribution system in Chicago is old and had problems with winter freezing breaks. Figure 5.8 shows the results of a decade of water conservation efforts in the metropolitan Chicago area. From Figure 5.8 one can also calculate the overall water use converted to per capita, as follows

Water use in liters/capita-day	
1991	1155
2001	818
2009*	680

* Personal communication

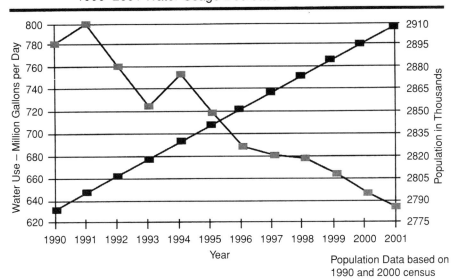

Figure 5.8 Population increase and water conservation impacts in the Chicago metropolitan area (to convert from million gallons per day to m³/day, multiply by 0.00378). From Lanyon (2007).

During the early 1990s, MWRD's leak detection and repair program established routine surveys performed by MWRD personnel. All MWRD distribution pipes (460 kilometers) are checked for leaks on a regular maintenance schedule, with repairs made promptly. Between 1991 and 2003, leaks amounting to 330,900 m³/day (87.54 mgd) were detected and fixed in Chicago and other MWRD contracting communities. Additional water conservation reduced the demand even further.

V.3 SUBSTITUTE AND SUPPLEMENTAL WATER SOURCES

The urban water supply grid is and will remain the major water source in cities. It is quite possible that in more humid and colder geographical areas, after implementing water conservation measures, this will be the only source needed until a comprehensive integrated resource recovery plan is implemented, which would require water reuse. Furthermore, citizens should be encouraged (by education and using smart pricing of delivered potable water) not only to conserve but also to implement some simple capture of water from other sources. For example, condensate from air conditioners and dehumidifiers has a composition of distilled water; it is not potable, but could be used for ironing, battery refilling, or plant irrigation in the house. Captured rainwater has been used as a source of potable and nonpotable water for millennia (see Figures 1.5 and 1.7 in Chapter I). The following section outlines possibilities of using other sources of water to supplement or replace potable water in the municipal grid. Chapter VI outlines and describes water reclamation and reuse, which is also extensively covered in Metcalf & Eddy (2007).

V.3.1 Rainwater Harvesting (RWH)

As stated, rainwater harvesting has been practiced by human beings for millennia and is now gaining popularity among the conservation-conscious public generally, not only in water-short areas. It is simple and inexpensive. In general, all a homeowner needs is one or two barrels to collect rainwater from roof downspouts and a small pump, which may not even be needed if the collection tank is positioned at some height above the ground. Rainwater harvesting is now becoming a legitimate and widespread topic of urban water management. Ecocities (see Chapter XI) located in water-short areas (for example, Tianjin in China or Masdar in the UAE) rely on captured rainwater, after treatment, as an important supplement of water for all urban uses, including drinking, laundry, and irrigation.

Rainwater outside of urban centers with air pollution is relatively clean and soft. Without air pollution it has a pH of about 5.6, and is free from disinfection by-products, salts, minerals, and other natural and anthropogenic contaminants found in surface water. Users with potable systems prefer the superior cleansing properties of rainwater (Texas Water Development Board, 2005). On the other hand, in areas with air pollution from traffic and power plants, rainwater can be acidic, with pH as low as 3.0, which makes it corrosive and damaging to plants, soils, and infrastructure. For example, many ancient marble statues and monuments in Rome and Athens have been severely damaged and eroded by precipitation made acidic by heavy traffic emissions and other urban air pollution. Urban rainfall can be made acidic by nitric acid formed by traffic emissions containing nitrogen oxides (NO_x). As a matter of fact, traffic emissions from freeways and other heavily traveled areas are a major source of acidity of precipitation (Novotny, 2003). On the other hand, nitrogen in precipitation can be counted as a source of fertilizing N. It should also be pointed out that if the rainfall has high acidity, roof covers such as asphalt shingles and zinc or copper gutters and downspouts become sources of pollution, namely of polyaromatic hydrocarbons (PAHs) from shingles and zinc or copper toxic metals (Novotny, 2003). Hence, for rain-harvesting programs to be effective, they must incorporate air pollution controls.

The Texas Water Development Board (TWDB) manual, quoting Krishna (2003), lists the following advantages of using harvested rainwater:

- The water is free; the only cost is for collection and use.
- The end use of harvested water is located close to the source, eliminating the need for complex and costly distribution systems.
- Rainwater provides a water source when groundwater is unacceptable or unavailable, or it can augment limited groundwater supplies.
- The hardness of rainwater is very small, which helps prevent scale on appliances, extending their use. Using rainwater eliminates the need for a water softener and the salts added during the softening process.
- Rainwater is sodium-free, important for persons on low-sodium diets.
- Rainwater is superior for landscape irrigation.
- Rainwater harvesting reduces flow to stormwater drains and also reduces diffuse (nonpoint) source pollution.
- Rainwater harvesting helps utilities reduce the summer peak demand and delay expansion of existing water treatment plants.
- Rainwater harvesting reduces consumers' utility bills.

Rain harvesting is most popular in single-family applications. However, with the emergence of ecocities and integrated resource management schemes, there is a progression towards cluster (ecoblock) management schemes rather than the more traditional practice of storage on individual properties (McCann, 2008a). Integrated water management in urban clusters can develop more surface and subsurface storage, can provide more reliable blending with water from other sources (e.g., reclaimed used water), and can treat collected rainwater all the way to potable reuse, which is not possible in small domestic applications where reuse is limited to laundry and irrigation. Storage, blending, and treatment of captured rainwater for cluster-wide (ecoblock) reuse is described in Chapter VII. Although storing captured roof rainwater in barrels is the most frequently used technique, aboveground and underground storage reservoirs and tanks implemented in the U.S. and elsewhere could be quite large. Rainwater harvesting also includes land-based systems with man-made

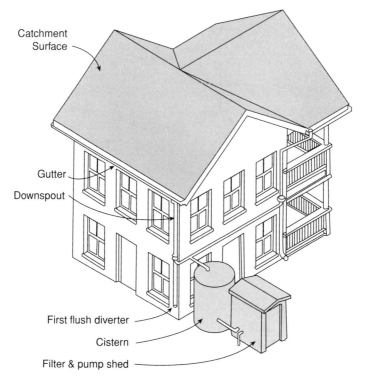

Catchment Surface

Gutter

Downspout

First flush diverter

Cistern

Filter & pump shed

Figure 5.9 A schematic of a simple rainwater-harvesting installation in a single-family home.

landscape features to channel and concentrate rainwater in either storage basins or planted areas. Rainwater capture and storage has also been used to provide water to urban "waterscapes," that is, pools, fountains, and cascades (Dreiseitl and Grau, 2009). A manual by König (2001) presents technologies and examples of rainwater harvesting.

Domestic Applications Figure 5.9 shows a simple rainwater-harvesting system for a single-family home, and Figure 5.10 shows a larger (10 m^3) rainwater-harvesting tank installation in Australia. Arizona taxpayers who install a "water conservation system" (defined as a system to harvest rainwater) after January 1, 2007, and before January 1, 2012, may take a one-time tax credit of 25% of the cost of the system (up to a maximum of $1000). Builders are eligible for an income tax credit of up to $200 per residence unit constructed with a water conservation system installed.

Australian Rules

1. The tank must not collect water from a source other than gutters or downpipes on a building or a water supply service pipe.

Figure 5.10   A 10,000-liter rainwater collection tank installed in Orange City, New South Wales, Australia. The city is located 270 km west of Sydney in a semi-arid area. Photo credit Orange County.

2. The tank must be fitted with a first-flush device (a device that causes the initial runoff of any rain to bypass the tank to reduce pollutants entering the tank).

3. Any overflow from the tank must be directed into an existing stormwater system.

4. The tank must be enclosed, and any inlet to the tank must be screened or filtered to prevent the entry of foreign matter or animals.

5. The tank must be maintained at all times so as not to cause a nuisance with respect to mosquito breeding or overland flow of water.

6. The tank and other rain-harvesting components of the system must be properly designed and installed by a licensed installer (plumber).

Cluster (Ecoblock) Applications   Two articles by McCann (2008a and b) present examples of larger-scale rainwater harvesting in Germany and Korea.

An apartment block in Muhlheim (Germany) installled18 in-ground cisterns ranging in capacity from 7 to 11 m^3. The volume capacity of the cisterns was related to the roof size (König, 2001). All roof water is directed into the cisterns, which are connected to washing machines in the common laundry room. The room is also equipped with a connection to the city water supply, but a majority of tenants prefer rainwater over tap water for washing. As pointed out, rainwater is soft and is free of charge, while using tap water requires a payment for used water.

Commercial development in Freiburg (Germany) has 34,000 m^2 of roof area. Rainwater from the roof is collected and stored in two underground storage basins, each holding up to 300 m^3 of harvested water. The industry uses harvested water blended with groundwater for cooling both offices and machines on the production floor. All used water is percolated into the ground. Because this system eliminates the need for air conditioning and is used also for industrial cooling, it saves energy and ultimately eliminates 870 tons of CO_2 emissions equivalent per year (McCann, 2008a).

Seoul's Star City (Korea) is a 6.25-ha mixed-use commercial/residential development, 35 to 57 stories high, located near Keokuk University in Korea's capital city. McCann (2008b) describes a project conceived by Professor Mooyoung Han of the Seoul National University Rainwater Research Center. In this project rainwater from all roofs is piped into three underground RWH storage cisterns, each having 1000 m^3 capacity, located under the garages of the 35-story-high development. A fourth basin of the same capacity is used to intercept stormwater runoff from the area; hence, this tank is kept empty between storms. Runoff from the drained green area is also routed into the stormwater tank. Water from the stormwater-collecting tank, which is polluted, is used for irrigation. Much cleaner RWH water from roofs is reused for toilet flushing, irrigation of green areas, cleaning of paved surfaces, and firefighting. Irrigation is the main demand. In an average year, the RWH and storm runoff attenuation system captures 67% of rainfall, which otherwise would be discharged. To ensure good quality of RWH water, the inlet line into the RWH tanks is equipped with a filter; however, it was found that the turbidity of settled RWH water in the tanks is very low—0 to 1.5 nephelometric turbidity units (NTU)—even without influent filtering. The water- and energy-saving features of the project are very impressive. In Professor Mooyoung Han's calculations, energy required to treat and deliver one cubic meter of water elsewhere in Seoul requires 1.82 kWh, while the same volume of water from the RWH system requires only 0.012 kWh of pumping energy to pump the water from the storage.

The Battery Park development in New York City has also implemented RWH as part of a large-scale residential development. This will be described in Chapter VI.

RWH practices not only provide water for reuse, they also attenuate urban flooding and have beneficial impacts by reducing CSOs. Therefore, rainwater harvesting is a part of the overall integrated water management and resource recovery plan.

V.3.2 Gray Water Reclamation and Reuse as a Source of New Water

Gray water represents about 50% of the total indoor used water discharge, based in the information given in Table 5.3. Consequently, counting it as a resource to replace potable water in a reuse cycle at first appears worthwhile, but it is not as straightforward as rainwater harvesting. Several ecocities now in planning design or already being built consider gray water reclamation and reuse mostly for nonpotable purposes, although two ecocity designs (Qingdao in China and Masdar in the UAE;

see Chapter XI) included highly treated reclaimed gray water in the potable water supply loop.

Gray water originates from showers, bathtubs, kitchen dishwashers, and from laundry washing machines. Untreated gray water contains soaps, shampoos, and detergents in addition to dirt washed from laundry and dishes, and pathogens and organic matter from human bodies. Soaps and detergents are present in gray water as emulsions that are difficult to remove by means of simple sand filters. In most cases soap and some detergent emulsion can be precipitated by adding hardness to the water (calcium-containing salts, including backwash water from water softeners), and a clear water effluent will result after sand filtration. COD and BOD are relatively high, between treated and untreated wastewater effluent quality, and the water goes foul and generates foul black water (sewage) odors during storage or ponding in puddles. Today soaps and detergents as well as food solids in gray water are biodegradable, and gray water will turn anaerobic after a couple of days of standing. A significant part of COD is soluble; furthermore, shower, laundry, and bath water may contain pathogens that can multiply in warm water. Note that "sewage"-smelling flows in squatter settlements (shantytowns) or poor neighborhoods in developing countries are mostly gray water flows, since feces and urine are disposed of mostly in latrines (Figure 5.11). The irrigation value of gray water is questionable because in many

Figure 5.11 Gray water discharged into street drainage in Tijuana, Mexico. The dark color indicates foul anaerobic conditions, in contrast with still oxygenated fresh gray water input from the house on the left. Photo and credit Art Ludwig, 2009.

developed countries soaps and detergents do not contain phosphates and, to treat gray water biologically, nitrogen and phosphorus would have to be added. Gray water containing soaps may also have higher pH; hence, plants preferring neutral or slightly acidic soil water will not grow when irrigated by gray water. Observations in squatter towns of developing countries have shown that not many plants will grow in gray water channels or puddles (Ludwig, 2009), as shown on Figure 5.11. However, it should be pointed out that the concentrations of BOD, COD, pathogens, and other pollutants in gray water drainage in shantytowns of the developing world may be much larger than those in gray water from U.S. households. Water use in the shanty towns of the developing world is around 20–40 liters/capita-day, depending on drinking water availability and the distance of the standpipes from the settlement.

In addition to soaps, detergents, and skin-care lotions, gray water may also contain household toxins such as bleaches, harsh cleansing products, and cleaning solvents. Information on concentrations of key constituents will be provided in the next chapter, "Water Reclamation and Reuse."

For these reasons public health departments are very reluctant to give a permit to individual homeowners for gray water reuse. In 2008, of an estimated 1.8 million households in the state of California reusing gray water, only about 200 have a permitted gray water reclamation system. Many more of the systems in use are illegal. To obtain a permit in California, gray water pipes must be buried 45 cm below the surface, and no effluent surfacing is allowed. Essentially, the disposal of untreated gray water is similar to that in a septic tank leaching field. Some states in the Southwest U.S. (Arizona, New Mexico) have relaxed or are considering relaxing the rules, but surfacing and ponding are still not allowed, and distribution pipes must be covered by soil or mulch. Use of gray water for irrigation is allowed in Australia as long as the sprinkler spray is less than 2 meters in diameter, and people, especially children, are prevented from coming into direct contact with the sprayed gray water.

Many "house"-scale gray water treatment schemes have been developed and are commercially promoted on the Internet. A typical system has a storage tank (barrel) followed by some kind of filter. Almost all small-scale systems are costly, and would not be approved by public health agencies. They could remove most turbidity, but the effluent would still be hazily turbid, have relatively high COD, and, without disinfection, could contain pathogens and viruses. Typically, claims for the efficiencies of small-scale packaged systems are overinflated.

Gray water reclamation (treatment) is feasible on a larger scale, as opposed to a single home. It could be done on a subdivision or cluster (ecoblock) or resort hotel scale. Regional applications would be uneconomical because they would require dual sewer and water distribution systems. Reasonably successful treatment schemes allowing reuse of reclaimed gray water are:

1. Physical treatment by sand or microfiltration, followed by reverse osmosis and ultraviolet disinfection (see Qingdao and other designs in Chapter VII and in the next section, on desalination) and possibly pH control. This scheme has been successfully used for larger-scale total used water reclamation and reuse

in many localities, most notably in Singapore. Such systems could work in a controlled environment in an urban cluster (ecoblock) but would be difficult to implement and very expensive in an individual home. Furthermore, crud in untreated gray water clogs the filters and pipes (Ludwig, 2009).

2. Simpler biological treatment by subsurface flow wetlands followed by filtration and, if needed, by disinfection. The needed area of wetland would be about 2–3 m^2/person. Caution should be taken with plant selection because of the possible pH and salt problem with gray water.

The first scheme above (#1) results in about 70% efficiency of water reclamation; that is, 20 to 30% of the water is lost as filter backwash and concentrate from the reverse osmosis, which have to be discharged into sewers. The effluent from such systems would have high, bottled-water quality, but such units would require a lot of energy for pumping, filtration, and—especially—reverse osmosis, to overcome the osmotic pressure and push water through the membrane (see the next section). The cost of the unit, energy, and maintenance, and the effect on global warming, in most cases would outweigh the benefits derived from saving water. Gray water reclamation and reuse also require dual indoor and outside sewers and, if the reclaimed water is to be used indoors (e.g., for laundry and bathtubs), dual water distribution systems. Technically stated, the cost of installing a house-scale gray water capture, treatment, and reuse system is far greater that the marginal cost of the utilities to obtain new water and/or implement large-scale (citywide) water conservation measures. This is a setback and, in addition to other difficulties encountered with individual home-scale gray water reuse, it may be more efficient and economical to consider implementing it on a cluster or regional scale, in a total used water reclamation scheme, and using it for nonpotable use in industrial and commercial developments (Singapore model) or in a controlled indirect potable reuse (see Chapter VII).

The second scheme (#2, above) is relatively inexpensive but has to be properly designed as a subsurface flow constructed wetland (for system design and design parameters, see Chapter VII). If designed properly, it would require minimum energy and could work in individual situations where water is very scarce and the reclaimed water would be used mostly outdoor for irrigation. However, as has already been pointed out, the nutrient value of gray water is relatively low, plants may not optimally develop, and fertilizing by commercial fertilizers would defy the purpose of gray water treatment. The only proven safe and reliable way of irrigating lawns with gray water is through underground drip tubing supplied by a backwashing sand filter type system—far beyond what most residences are likely to install. Again, direct human (children) contact with turf with surficial irrigation must be avoided. Unfortunately, turf accounts for the bulk of the irrigation needs in the typical landscape, and gray water irrigation of lawns is by far the most prevalent violation of commonsense gray water safety rules (Ludwig, 2009). Gray water irrigation of vegetables and other edible crops must be avoided. These rules are clearly not followed in developing countries. Wetlands also emit methane and nitrous oxides, which are higher-potency greenhouse gases contributing to global warming.

V.3.3 Desalination of Seawater and Brackish Water

Water obtained from seawater or from deep groundwater zones with a high salt content is becoming more and more common in water-short areas. As a matter of fact, building oasis-type ecocities in the wealthy Middle East countries is only possible with desalinated ocean water. Obviously (see Table 5.1), there is a great, almost an infinite, volume of salt water, compared to earth's freshwater resources, but until the end of the last century the technologies for desalting seawater were prohibitively expensive and were implemented only in water-short areas of countries with a high gross domestic product (e.g., Florida, Saudi Arabia, Dubai in the UAE).

In contrast, water from deep subsurface geological zones is mostly fossil water left by geological seas millions of years ago. This also includes shallow groundwater in the Northern U.S. left by melting ice age glaciers kilometers thick (similar to present Greenland), approximately 15,000 years ago. Originally the glacier-melt water was without salts (soft), but in most North Central U.S. areas it is located mostly in a dolomite (a calcium-magnesium carbonate rock) aquifer, which makes water extracted from the aquifer hard but still very good. However, in the same location below an impermeable shale layer are very thick sandstone deposits, containing remnants of a geological sea mixed with later freshwater recharge over millions of years. Because seawater is heavier than freshwater, the deeper the wells penetrate into the sandstone, the more salty the water is. Also the longer groundwater resides in the underground zones, the higher the measured salt content. This is a rule for both confined and unconfined aquifers. Also the basic chemical composition of groundwater changes with depth, from one similar to rainwater near the surface to a sodium- and chloride-dominated composition in deep zones (Freeze and Cherry, 1979). Mineral water has been generally defined as water with a total mineral content of more than 1000 mg/L. In the North Central U.S., recharge of the sandstone aquifer today is very small because the aquifer is confined—i.e., covered by the impermeable shale layer—and its recharge area (area where the confined aquifer outcrops) is very small (Novotny, 2003; Novotny and Olem, 1993).

Both coastal and some inland communities that rely on mining groundwater resources are facing the problem of salt water intrusion; they try to solve the problem of generating more water by:

- Pumping brackish water from deeper ground water zones and either mixing it with available freshwater from other sources or installing desalination
- Desalinating seawater
- Injecting reclaimed wastewater into the aquifer, either to form a barrier to salt-water intrusion or to supplement water withdrawn from the freshwater zones (see Chapter VI)

Before the introduction of membrane filters with molecule-sized pores (about 0.001–0.0001 μm), the only feasible way to desalinate seawater was distillation, which was very expensive, and only a few desalination plants were built. Today, the cost of the salt removal process has dropped, and desalination is becoming more common. At the end of the first decade of 2000, Florida led the U.S. with more than 130 desalination plants, followed by Texas (38) and California (33). Unlike Texas or Southern California, Florida is not an arid state; it has plenty of rainfall, but the upper aquifer containing freshwater is relatively thin and, due to the overdraft (mining), salt water intrusion is widespread. In the South Florida Water Management District (SFWMD) alone, the total desalination capacity was 530,000 m^3/day (140 mgd), with another 454,000 m^3/day under construction (SFWMD, 2009). Worldwide, the number of desalination plants may be reaching 10,000 in the near future, with about two-thirds located in the Middle East, where other freshwater sources are minuscule when compared to the water demands of growing affluent urban communities and expanding irrigated agriculture. The largest desalination plant in the world, with a capacity of 800,000 m^3/day (3024 mgd), opened for production in 2009 in Saudi Arabia. Water from this plant, located on the shores of the Persian Gulf, is delivered and distributed to the inland communities by 4000-km pipeline systems. The plant also generates 2750 MW of electric power from energy recovery. Part of the energy used to produce desalinated water can and should be recovered. This power (energy) is in the pressure of the concentrate that is released and should be safely discharged.

Desalination Process Before using reverse osmosis (RO), desalination was done by evaporation which required a lot of energy (approximately 25 kW-hrs/m^3). This technology s still used in the Middle East and even it Masdar (CH2M-Hill –personal communication). The desalination processes today use RO. RO treatment has been used in many applications in (bottled) water, food, pharmaceutical, and chemical industries to provide distilled or bottle-quality water free of salts and microorganisms. Obviously, RO units are on board of the space station and vehicles and in submarines. RO uses membrane technology whereby membranes with extremely small pores (Table 5.5) allow passage only of water molecules, and large molecules, ions, solids, and microorganisms are rejected into the "brine/concentrate" (Baker, 2004). Hence, in RO, the membrane has a high permeability for water but extremely low permeability for salts and other impurities; under sufficient pressure differential between the pressure on the raw water side of the membrane and the osmotic

Table 5.5   Operating parameters for reverse osmosis (from Metcalf & Eddy, 2007, and Rizutti et al., 2007, based on information from Crittenden et al., 2005, Taylor and Weisner, 1999, and Metcalf & Eddy, 2003)

	Reverse osmosis
Pore size (μm)	0.0001–0.01
Operating pressure differential (bar[1])	
TSS between 1000–2500 mg/L)	12–18
Sea and brackish water desalination	55–85
Energy consumption (kWh/m^3)	
Brackish water (TDS 1000–2500 mg/l)	1.5–2.5
Seawater desalination	5–25

[1] 1 bar = 100 kPascals = 100,000 N/m^2 = 0.97 atmospheres = 10 meters of hydraulic head

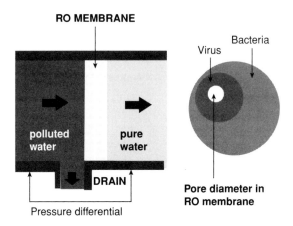

Figure 5.12   Reverse osmosis process schematics.

pressure, the membrane (Figure 5.12) allows water molecules to pass through but rejects and concentrates dissolved and suspended solids and ions. The purified water is called *permeate*. The rejection efficiency of mineral salts and ions in the RO process is around 99% with total dissolved solids (TDS) content in produced water ranging between 10 and 500 mg TDS/L. The quality of produced water depends on the pressure, the salinity of the input water, and the quality and characteristics (pore diameter) of the membrane (Pantell, 2003; Baker, 2004; Committee to Review the Desalination Process, 2004).

To reduce the salinity of seawater with salt concentration of 30 to 35 grams/L down to less than 500 mg/L, more than one pass through RO unit may be required, although today high-rejection membranes are commercially available. A desalination process generally has five major components, shown in Figure 5.13:

1. *Pretreatment*, in which suspended solids, down to very fine particles, are removed. This step may include coagulation and flocculation (for surface water sources such as seawater) and filtration by sand filters and microfilters (diatomaceous earth or membrane filters).

2. *High-pressure pump* generates pressure within the units to push water molecules through the membrane. The pressure must be significantly greater than the osmotic pressure, which is approximately proportional to the molar concentration of salt to be removed. For seawater the osmotic pressure is 27 bar (2700 kP). Typical operating pressures are given in Table 5.5. This process requires a lot of energy.

3. *Reverse osmosis*. In an RO unit, high pressure forces the pretreated water through a semipermeable membrane that lets only water molecules pass through and rejects salts, larger molecules, and other dissolved solids into the concentrate.

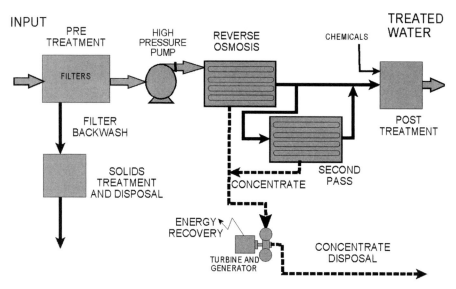

Figure 5.13 Treatment process by reverse osmosis for brackish water and seawater desalination.

4. *Energy recovery.* The energy contained in the concentrate can be recovered, for example, by a turbine with a large hydraulic head. The recovered energy is proportional to the specific density and flow rate of the concentrate and the difference in the pressure between the operating pressure within the RO unit and the pressure in the outlet pipe, either atmospheric (if the outlet is into a channel) or hydrostatic.

5. *Post-treatment* includes disinfection by ultraviolet light and adding hydrogen peroxide, pH adjustment, and adjusting taste of water by removing gasses and adding minerals, because produced water has lost most of its minerals in the RO process.

Figure 5.14 shows RO units installed in the Kay Bailey Hutchison Desalination Plant in El Paso, Texas, that removes excess salt from brackish groundwater.

Seawater and brackish (estuary or groundwater) salinity consists mainly of mono-valent sodium (Na^+) and chloride (Cl^-) ions. Brackish groundwater may also contain other anions such as sulphates (SO_4^{2-}) or bicarbonates (HCO_3^-) and cations such as hardness ($Ca^{++} + Mg^{++}$). Seawater salinity is in concentrations exceeding 25,000 mg/L as TDS; hence, for desalination of seawater, two RO units in a series may be needed. Brackish water has salinity between seawater and freshwater. Typically salt (total dissolved solids) content above 500 mg TDS/L in water is not acceptable for public water supply, and the water should be desalinated. Mineral water in spas, often tasty and perfectly acceptable for drinking, has salinity exceeding 1000 mg/L and would cause serious problems with scaling in pipes and appliances.

Figure 5.14   Reverse osmosis in the El Paso desalination plant (courtesy CDM).

It is evident that desalination of seawater requires a lot of energy to generate the necessary pressure differential and will have a large carbon footprint if it runs on conventional energy produced by coal-fired power plants. Consequently, energy could be recovered from the concentrate that stays under high pressure by a high-pressure turbine. Treated water leaves the RO unit at a low pressure; the energy is lost from overcoming the resistance of the membrane and osmotic pressure, so energy recovery is not possible.

Due to RO's high energy use and carbon imprint, the trend in arid regions is to use solar power (Rizutti et al., 2007). Reverse osmosis in some ecocities located in water-short areas is either run solely or supplemented by solar power. In Masdar and Qingdao the entire energy need for desalination is provided by a concentrated solar panel array, solar photovoltaics, and wind (see Chapter VIII).

Examples of Desalination *The Kay Bailey Hutchison Desalination Plant* (named after a U.S. senator from Texas), in El Paso, Texas, is the largest in the U.S., providing 103,950 m³/day (27.5 mgd) of freshwater, or 25% of the regional demand. The plant desalinates brackish water from a deeper aquifer that otherwise would be intruding on freshwater wells, the main source of freshwater in this large city on the Rio Grande ("Great River") (Hutchison, 2004; see also http://www.epwu.org/water/desal_info.html). The city also uses water from the river.

In addition to providing a supply of freshwater, the CDM designed and built facility provides other important benefits:

- The facility serves as a model and center of learning for other inland cities facing diminishing supplies of freshwater.
- The water pumped to the desalination plant protects the El Paso region's fresh groundwater supplies from brackish water intrusion by capturing the flow of brackish water toward freshwater wells.
- The desalination process not only removes salts but removes other potential pollutants from the water.
- The facility augments existing supplies to make sure the El Paso region has sufficient water for growth and development for 50 years and beyond.

With the freshwater sources in El Paso maxed out and actually diminishing because of freshwater aquifer mining and decreasing flow and quality in the river, the city also tries to implement water conservation because water use is still very high (Table 5.2). The Rio Grande is one of the two most stressed large river systems (the other is the Colorado River) in the U.S., with increasing municipal, agricultural, and industrial demands for water from both the U.S. and Mexico. Desalination was an attractive alternative because the marginal cost of new freshwater was high without expanding conservation efforts. This is true for many other cities that have built or are planning desalination plants.

The Tampa Bay Seawater Desalination Plant (Florida) is the second-largest desalination plant in the U.S. (http://www.tampabaywater.org). Building was initiated in 1996, but the plant was plagued for years with problems, needed repairs and redesigns, and was not operational until 2009. Nevertheless, today, operated by American Water, it provides 95,000 m³/day (25 mgd) of potable water, distributed to the communities by a 24-km pipeline (http://www.water-technology.net/projects/tampa). The intake to the desalination plant is located next to the cooling water discharge channel from a local power plant, from which a total 166,000 m³/day is taken into the desalination plant. Hence, the condensate flow is $166,000 - 95,000 = 71,000$ m³/day. This high-salinity (see Table 5.5) concentrate is then mixed with approximately 5,000,000 m³/day of cooling water discharge flow from the power plant, providing a dilution ratio of 70:1 to prevent any adverse ecological impact of salinity on the Tampa Bay aquatic community. The pressure power in the concentrate is recovered by a turbine as energy and returned to the power grid.

A comparison of the performance parameters of the El Paso (brackish groundwater) and Tampa Bay (seawater) desalination plants is presented in Table 5.6.

According to the South Florida Water Management District online press release, at the end of the last century the desalinated water cost was about $2.4/m³. Ten years later (2009), the cost has dropped to less than half of that amount ($0.5 to $1.3 per m³), which is comparable to, or even less than, the marginal cost for new freshwater in many growing communities, obtained from more distant, deeper, and

Table 5.6   Comparison of El Paso and Tampa Bay desalination plants

Parameter	El Paso	Tampa Bay
Design capacity-produced water m^3/day (mgd)	103,950 (27.5)	95,000 (25)
Water source	Brackish groundwater	Sea bay
Salinity of raw water, g/L	1.5	18–35
Concentrate salinity, g/L	6.0	58–88
Concentrate reject flow, m^3/day (% of input)	34,000 (27)	71,000 (43)*
Concentrate disposal	Deep well injection	Tampa Bay after 70:1 dilution
Operating pressure, bar	~15	43–73
Energy consumption kW-hr/m^3 **	1.5–2.5	5–15***
Per capita water use, liters/cap-day	500	390
Carbon footprint kg CO_2/capita-year	166–278	433–1,302***
Energy recovery	NO	YES

*Average, depending on the salinity of the raw water
**Per Metcalf & Eddy, 2007, and Rizutti et al., 2007, Table 5.4
***The upper limit may not include the effect of energy recovery. Carbon emissions calculated as 0.61 kg or CO_2 emitted per 1 kW-hour electricity produced in a coal-fired power plant (see Section I.3.3 in Chapter I).

polluted sources (SFWMD, 2009). While the cost of providing desalinated water is now affordable, there are two problems with desalination:

1. The process is energy intensive and has a high carbon footprint if electric energy is provided by fossil-fuel plants.
2. The by-product of the process is brine/concentrate containing salts and pollutants that have not passed the membrane. The brine/concentrate represents between 20% (low-salt source such as reclaimed used water) and 45% (high-salinity source such as seawater) of the total intake flow. This concentrate cannot be directly discharged into receiving water, including seas, without significant dilution and eventually treatment.

Brackish water RO is the most cost effective of the desalination technologies, followed by seawater, and finally used water. Used water salinity is the lowest of the three sources but requires as much energy as brackish water to produce freshwater because it necessitates additional treatments to remove other dissolved solids and micropollutants.

V.3.4   Urban Stormwater and Other Freshwater Flows as Sources of Water

Best management practices (BMPs) for capture, storage, infiltration, and controlled release of stored water into receiving water have been developed over the last

40 years (Novotny, 2003; Field, Heaney, and Pitt, 2000), and thousands of them have been implemented. BMPs are described in detail in Chapter IV. Originally, the purpose of the BMPs was to control urban flooding and remove pollution from urban runoff and snowmelt. The difference between rainwater harvesting and stormwater capture is the quality. Rainwater collected from roofs or other relatively clean surfaces is much cleaner than urban runoff from street surfaces.

The capture of stormwater and snowmelt has typically been done by ponds and dual-purpose storage basins designed to capture large storms and allow the small rainfalls to pass thorough with or without treatment. In the original designs, accumulated water following a large storm was generally released relatively soon, in less than 24 to 48 hours. Wet ponds with permanent pools function in a similar way; the storage volume is above the permanent water level in the pond, and the stored excess volume is released in a relatively short time. It has been rare for either type of pond to be used as a source of water supply, because water was available only for a short time after the storm. Furthermore, urban runoff is not clean and contains toxics from traffic (metals, PAHs, oil and grease, de-icing salts, and also pathogens), which makes the stored water less attractive as a source or raw water for water supply (Field, Heaney, and Pitt, 2000; Novotny, 2003). It could be a perfect source of water for irrigation, but in the absence of a special water distribution system for reclaimed stormwater, most urban irrigators use potable tap water, or, in some cities, reclaimed used water. The size of the ponds and the bleeding rate of the outlet led to installation of hundreds of small ponds scattered throughout some communities (e.g., Chicago) that interfered with each other and often diminished their effect on the reduction of flooding.

Special cases are storage basins for combined and sanitary sewer overflows (CSOs and SSOs) that range again from small subsurface basins to large deep tunnels (see Figure 1.18). Unlike the stormwater ponds designed to capture large storms, many CSO storage basins in Europe were built to capture the first 5 mm to 12 mm of rainfall from the impervious area of the watershed, called the "first flush," which is highly polluted. This stored mixture of sewage, scoured solids from sewers, and rainwater must then be diverted to the treatment plant; hence, it is of little use as a local source of water. Large rainwater-harvesting storage basins work on the same principle; however, they store cleaner water for a long time. As pointed out several times in this book, rainwater storage cisterns in sizes ranging from a household small (200 liters) barrel to 80,000 m^3 underground cisterns (Figure 1.7) have been built for more than two thousand years, mainly in the Middle East, southern Europe and China, and today in Australia.

BMPs employing infiltration recharge groundwater, and therefore contribute to the available groundwater in the aquifer. Today urban stormwater is considered as a resource to be used on-site instead of discharged into sewers or into the receiving water body without reuse. As will be shown in Chapters VI and VII, surface or, more likely, underground basins hold rainwater and stormwater for a long time, and these raw or reclaimed water supplements are the basic components of water management in ecocities. In ecocities, urban runoff in not collected by underground conduits; it is collected by surface drainage, treated by biofilters, and stored. Rainfall is also infiltrated by pervious pavements and infiltration raingardens and ponds into the underlying aquifer for further potable and nonpotable use. Captured stormwater and rainwater can satisfy a significant portion of the ecocity water demand.

Reasonable use of urban snowmelt is tied to de-icing practices and chemical use on streets, roads, and highways during winter snow and ice driving conditions (see Chapter IV). At the end of the last millennium, the prevailing practices for de-icing and snow control used heavy applications of de-icing salt (sodium chloride, sometimes mixed with calcium chloride) to melt snow and ice covering streets, and homeowners used them on their driveways and sidewalks. Such practices render the urban snowmelt useless for reuse due to high salinity and pollution. Other de-icing chemicals are also polluting. See Chapter IV for a more detailed discussion on pollution by de-icing chemicals and its abatement. The future ecocities must develop and use better snow- and ice-handling BMPs. This will be easier because the car and other vehicular traffic will be restricted or even eliminated within the city proper, being replaced by public transportation, convenient location of stops and stations, and walking.

The concept of surface capture, treatment, and storage of stormwater will have to be revised. First, stormwater must either be kept unpolluted or be treated. Storage should be long-term so that stored water is available and can be withdrawn for water supply over extended periods of drought. A pond designed to capture and store 2-, 10-, 25-, 50-, or 100-year floods does not have to be drained over a period of 48 hours. As a matter of fact, the entire methodology of stormwater capture for reuse must be revisited, and dimensioning of storage volumes should be done by long-term rainfall/runoff models with storage capture. Such models are available.

Singapore is expanding the use of (treated) urban runoff by converting brackish estuaries into freshwater bodies storing primarily urban runoff, by building barrages across the mouths of the estuaries (see Marina Barrage in Chapter VI). The stored urban runoff is pumped as another supplement into the freshwater storage reservoirs to provide water to the city. Such drastic changes of the ecological system may not be feasible in the U.S., but flood control barrages have been built to protect people and communities from tidal surges in The Netherlands and across the River Thames in England.

The decorative and conjunctive use of surface drainage channels, storage ponds, and basins, and the diverting of all clean water inputs into the ecocity surface drainage systems (see Chapters IV, IX, and X)—not into sewers—make the opportunities for finding additional freshwater more apparent: drainage of basements; condensate water from large air-conditioning units, dehumidifiers, and other cooling systems; urban springs and buried streams; and others.

REFERENCES

Allan, J.A. (1993) Fortunately there are substitutes for water otherwise our hydro-political futures would be impossible, In: *Priorities for Water Resources Allocation and Management.* London, ODA, 13–26

Allan, J.A. (1994) Overall perspectives on countries and regions. In: *Water in the Arab Worlds: Perspectives and Prognoses,* (P.A. Rogers and P. Lyndon, eds.), Harvard University Press, Cambridge, MA, pp 65–100

Anon. (1998) "Solutions for a Water-Short World," *Population Reports* 26 (1, September 1998), The Johns Hopkins University School of Public Health, Baltimore, MD, http://www.infoforhealth.org/pr/m14/m14chap7.shmtl

AWWA RF (1999) *Residential End Use of Water*, American Water Works Association Research Foundation, Denver, CO

Baker, R. W. (2004) *Membrane Technology and Applications*, John Wiley & Sons, Hoboken, NJ

Committee to Review the Desalination Process (2004) *Review of the Desalination and Water Purification Technology Road Map*, National Research Council, National Academies Press, Washington, DC

Crittenden, J. C., R. R. Russel, D. W. Hand, K. J. Howe, and G. Tchobanoglous (2005) *Water Treatment, Principles and Designs*, 2nd. ed., J. Wiley & Sons, Hoboken, NJ

Dreiseitl, H. and D. Grau (2009) *Recent Waterscapes: Planning, Building and Designing with Water*, Birkhäuser, Basel, Boston, Berlin

Falkermark, M. and C. Widstrand (1992) "Population and water resources – A delicate balance," *Population Bulletin* 47(3): 1–36

Field, R., J. P. Heaney, and R. Pitt (2000) *Innovative Urban Wet-weather Flow Management Systems*, TECHNOMIC Publishing Co., Lancaster, PA

Freeze, R. A. and A. Cherry (1979) *Groundwater*, Prentice-Hall, Englewood Cliffs, NJ

Friedler, E., D. M. Brown, and D. Buttler (1996) "A study of DC Derived Sewer Solids," *Water Science and Technology* 33 (9): 17–24

Gleick, P. (1996) "Basic water requirements for human activities," *International Water* 21(2): 83–92

Gleick, P. H., D. Haasz, C. Henges-Jeck, V. Srinivasan, G. Wolff, K. Kaocushing, and A. Mann (2003) *Waste Not, Want Not: The Potential for Urban Water Conservation in California*, Pacific Institute for Studies in Development, Environment and Security, Oakland, CA, http://www.pacinst.org/reports/urban_usage/waste_not_want_not_full_report.pdf

Gleick, P.H. and H.S. Cooley (2009) Energy implications of bottled water, *Environmental Research Letters* 4 014009 9 (6 pp)

Gleick, P., M. Palaniappan, M. Morikavwa, H. Cooley, and J. Morrison (2008) *The World's Water*, Island Press, Washington, DC, http://books.google.com/books/islandpress?id=_wd-s1FB7VEC&pg=PT13&dq=isbn:9781597265058&source=gbs_toc_r&cad=4

Heaney, J. P, L. Wright, and D. Sample (2000) "Sustainable urban water management," Chapter 3 in *Innovative Urban Wet-Weather Flow Management Systems* (R. Field, J. P. Heaney, and R. Pitt, eds.) TECHNOMIC Publishing Co., Lancaster, PA

Hidalgo, H. G., D. R. Canyan, and M. D. Dettinger (2005) "Sources of variability of evaporation in southern California," *Journal of Hydrometeorology* 6, pp. 1–19

Hoekstra, A.Y., Hung, P.Q., 2002. Virtual water trade: a quantification of virtual water flows between nations in relation to international crop trade. Value of Water Research Report Series No. 11. UNESCO-IHE, Delft, The Netherlands. Available from: www.waterfootprint.org/Reports/Report11.pdf [Accessed January 2008].

Hoekstra, A.Y., and A.K. Chapagain (2007) Water footprints of nations: water use by people as a function of their consumption pattern. *Water Resources Management* 21(1), 35–48

Hoekstra, A.Y., A.K. Chapagain, M.M. Aldaya, and M.M Mekonen (2009) Water footprint manual: State of the art 2009, Water Footprint Network, Enschede, the Netherlands. http://www.waterfootprint.org/downloads/WaterFootprintManual2009.pdf (accessed May 2009)

Hutchison, B. (2004) "El Paso's desalination efforts," in Water Desalination and Reuse Strategies for New Mexico, Proceedings *of the 49th Annual New Mexico Water Conference* (September 21, 2004), New Mexico Water Resources Research Institute,, New Mexico State University, Las Cruces, NM, http://wrri.nmsu.edu/publish/watcon/proc49/hutchison.pdf

König, K. (2001) *The Rainwater Technology Handbook – Rain harvesting in Buildings*, WILO Brain, Dortmund, Germany

Krishna, H. (2003) "An overview of rainwater harvesting systems and guidelines in the United States," Proceedings of the First American Rainwater Harvesting Conference (August 21–23), Texas Water Development Board, Austin, TX

Lanyon, R. (2007) "Developments towards urban water sustainability in the Chicago metropolitan area," Chapter 2 in *Cities of the Future – Towards Integrated Sustainable Water and Landscape Management* (V. Novotny and P. Brown, eds.), IWA Publishing, London, pp. 8–17

Ludwig, A. (2009) *Common Greywater Mistakes and Preferred Practices*, Oasis Design, http://www.oasisdesign.net/greywater/misinfo/index.htm (accessed June 2009)

McCann, W. (2008a) "Global prospects for rainwater harvesting," *Water21*, December 2008, pp. 12–14

McCann, W. (2008b) "Seoul Star City: a rainwater harvesting benchmark for Korea," *Water21*, December 2008, pp. 17–18

McMahon, J. E. (2008) "The energy-water nexus," PowerPoint presentation at University of California Berkeley, October 1, 2008

Metcalf & Eddy (2003) *Wastewater Engineering: Treatment and Reuse*, 4th ed. (revised by G. Tchobanoglous, F. L. Burton, and H. D. Stensel), McGraw-Hill, New York

Metcalf & Eddy (2007) *Water Reuse: Issues, Technologies, and Applications*, McGraw-Hill, New York

Mitchell, C., K. Abeysuriya, and D. Fam (2008) *Development of Qualitative Decentralized Systems Concepts for the 2009 Metropolitan Sewerage Strategy – Volume 1: Synthesis Report for Melbourne Water*, Institute for Sustainable Futures, University of Technology, Sydney, Australia

Nemerow, N., F. J. Agardy, P. Sullivan, and J. A. Salvado (2009) *Environmental Engineering: Water, Wastewater, Soil and Groundwater Treatment and Remediation*, John Wiley & Sons, Hoboken, NJ

Novotny, V. (2003) *WATER QUALITY: Diffuse Pollution and Watershed Management*, 2nd ed., John Wiley & Sons, Hoboken, NJ

Novotny, V. and H. Olem (1993) *WATER QUALITY: Prevention, Identification and Management of Diffuse Pollution*, John Wiley & Sons, Hoboken, NJ

Pantell, S. E. (2003) *Seawater Desalination in California*, California Coastal Commission, San Francisco, CA, http://www.coastal.ca.gov/desalrpt/dtitle.html

Rizutti L. *et al.* (2007) *Solar Desalination for the 21st Century*, Springer Verlag, Heidelberg and New York

SFWMD (2009) "Desalination in South Florida – Frequently asked questions," South Florida Water Management District, Tampa, FL, published online, http://www.sfwmd.gov/pls/portal/docs/PAGE/PG_GRP_SFWMD_WATERSUPPLY/SUBTABS%20-%20WATER%20CONSERVATIO%20%20-%20BRACKISH/TAB1610173/SFWMD

%20DESALINATION%20FREQUENTLY%20ASKED%20QUESTIONS.PDF (accessed 2009)

Taylor, J. S. and M. Wiesner (1999) "Membranes," Chapter 2 in *Water Quality and Treatment* (R. D. Letterman, ed.), McGraw-Hill, New York

Texas Water Development Board (2005) *The Texas Manual on Rainwater Harvesting*, 3rd ed., Texas Water Development Board, Austin, TX

U.S. Environmental Protection Agency (2008) *Fact Sheet, Office of Ground Water and Drinking Water*, U.S. EPA, Washington, DC

World Health Organization (1984) *Guidelines for Drinking Water Quality, Volume 1: Recommendations*, World Health Organization, Geneva, Switzerland

VI

WATER RECLAMATION AND REUSE

VI.1 INTRODUCTION

For thousands of years water was brought to cities first from local wells and streams and later by qanads, aqueducts, or pipes from distant water sources. Water then was used, converted to wastewater, and discharged with or without treatment into the receiving water bodies. This is the principle of the linear water/wastewater management system covered from the historical perspective in Chapter I and elsewhere throughout this book. Water has been required by human beings for drinking, washing, irrigating, producing food and other products (e.g., leather, paper, textiles), and sanitation (flushing toilets). An upstream community often used the majority of the water from the water body and/or polluted it to a point that downstream users were unable to fully use the surface water source, and aquatic life, including fish, was damaged or changed to an undesirable polluted state. This situation still impacts many countries, including the U.S. and China.

Concurrently, trees are harvested for wood or even clear-cut by people who also extract fossil fuel from the earth for their energy needs and produce organic and inorganic wastes to be discarded or disposed of in unsanitary and sanitary landfills. In these processes, energy, water food, and raw materials are used—unsustainably, in most cases—and the environment is polluted beyond its assimilative capacity. The process in which a city or urban area absorbs these inputs and transforms them into outputs as shown on Figure 1 is called urban metabolism which can be linear, cyclic or a hybrid (Kennedy, Cuddihi, and Engel-Yan, 2007; Daigger, 2009). The previous paradigms of urbanism were almost strictly linear. The demands of a city on the available resources and the impact of pollution on the environment are not limited to the metropolitan area, typically defined by the commuting distance of people who work daily in the city. The repercussions extend far beyond to rural areas producing basic food and to watersheds providing water. The total area of productive land needed to support an urban community, including waste assimilation, then represents an ecological footprint (Rees, 1997; Wackernagel and Rees, 1996). See Chapter X for a more detailed discussion on urban footprints.

The goal of sustainable development, in the context of the future ecocity, as presented in Chapter II, is to meet the water and energy needs of the current and future generations in way that would (1) be equitable, but still result in economic development, (2) protect the environment, even under the scenario of diminishing nonrenewable resources (e.g., fossil fuel), and (3) benefit society. Because resources are finite, sustainability under the (third and fourth) paradigms of the last century cannot be achieved without severe curtailment of society's demands on these resources, which would result in inconveniences and even hardship. Unlike people living in rural communities, urban people purchase food, other merchandise, and energy, all of which produce trash, waste, polluted water, and emissions of air pollution, including carbon at elevated levels. This is a linear process of transport of mass and energy from the resource to pollution and other waste emissions and discharges (Figure 6.1). The objective of the pollution controls of the last century was to minimize the effects of pollution and the throwaway practices of society in both developing and developed countries. It is significant to note that, under the hardship of poverty, the poorest segments of the population in the megalopolises of developing countries have become a caste of "recyclers" who depend for their subsistence on unsanitary landfills, where entire familes spend their days searching for throwaway food and trash.

Societies have realized that the throwaway polluting practices of the last century cannot continue at the current increasing pace. Water resources have been severely polluted or even lost, there is no more landfill space available, and global warming is

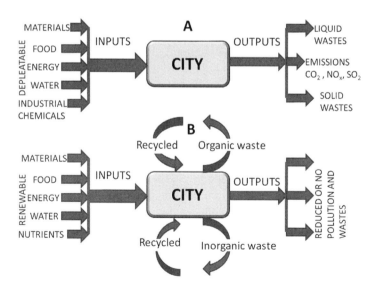

Figure 6.1 Mass and energy transformation in an urban area – the urban metabolis A) linear and B) cyclic.

accelerating. Ideas of recycling and reuse are again attracting attention as the main methods for achieving sustainability.

The bottom part of Figure 6.1 is the cyclic urban metabolisms processes that would reduce water pollution, landfill space, and greenhouse gas (GHG) emissions. The fundamental premise of effective reclamation and reuse is a change of philosophy concerning urban material and energy use and its by-products. Under the new philosophy of sustainability, there is no such thing as "waste"; a reasonably large portion of the materials output from urban living processes can be recovered and reused, dramatically reducing not only pollution from urban areas but also the demand for raw materials for the city—above all, energy. In the ecocities, in combination with resource conservation, the use of renewable sources of energy can reduce carbon and pollution emissions to a net zero carbon footprint, and the residual emissions into air, water, and soil will be within the assimilative capacity of the entire environment. Water use can be reduced and demand on landfill space minimized or even eliminated. This can be accomplished in harmony with improved health, leisure, recreation, and other benefits of better urban living. All of this is a premise of the ten One Planet Living (World Wildlife Fund, 2008) goals for sustainable ecocities (Chapter II).

Under this new paradigm of cyclic or hybrid urban metabolisms, the notion of "waste," "throwaway," and "pollution" will become obsolete and be replaced by "renewable sources," "resource recovery," "reclamation," "reuse," and "low impact," "pollution free," and the like.

The four processes of resource recovery and conservation of diminishing resources are:

1. Water conservation and reclamation and reuse of used water
2. Energy use savings and reclamation from various sources, such as heat, electricity, methane recovery from wastewater and organic wastes, and renewable wind, solar, and geothermal power sources
3. Recycling of organic solid waste for power generation by incineration or methane biogas production, or cardboard or paper production
4. Recycling of inorganic waste from metal, asphalt, glass, insulation, construction materials, and other products

VI.2   WATER RECLAMATION AND REUSE

VI.2.1   The Concept

Many people—especially in developing countries, developed parts of China, and even the U.S. Southwest—live in areas that anticipate severe water shortages, which will be exacerbated by population increases and migration (see Chapter I), and also by global warming. The World Bank (2001) has estimated that, during this century, available water must increase by 25–60% to meet the basic needs of the population, but most of it will have to come from water conservation and reuse. In many communities that have already reached the limit of the availability of freshwater, water reclamation, in addition to desalination (in coastal communities), has already become a viable option for providing additional water for their increasing population. Water reclamation and reuse also create opportunities to build new sustainable cities in areas that would be unsuitable for urban development under the current linear paradigm, such as the desert coastal areas in the Middle East (see Chapter XI).

In addition to clean potable water brought from areas outside of the urban area, which includes water delivered by streams to the urban area and withdrawn in the city, the sources of reclaimed water are:

1. Used (waste)water from buildings, sports venues, and public facilities, potentially separated into black and gray or even yellow water streams
2. Harvested rainwater
3. Urban stormwater stored and treated by best management practices in the urban landscape and underground, including infiltration into groundwater zones underlying the urban area
4. Groundwater pumped from basements, underground commercial areas, and parking garages
5. Condensation from air conditioners, (which, however, has the chemistry of distilled water lacking minerals and cannot be used directly as potable water, in spite of its purity)
6. Desalination of sea water
7. Irrigation return flow after infiltration into groundwater zones

Generally, reclaimed water can be used (Metcalf & Eddy, 2007) for the following purposes:

1. As a substitution in applications that do not require high-quality water supplies, such as toilet flushing, irrigation, street and car washing, and pavement cooling to reduce the urban heat island effect
2. For augmenting water sources and providing alternative sources of supply to assist in meeting present and future needs
3. For protecting aquatic ecosystems by providing ecological flows and avoiding ephemeral flows created by excessive withdrawals (see Chapter IX)
4. For recharging groundwater to enhance available water and/or prevent subsidence of historic buildings built on wood piles
5. As cooling water
6. As a source of heat

During the water reclamation (treatment) process, other valuable products besides water can be produced, recovered and reclaimed, such as:

1. Nutrients in the form of nitrogen, phosphorus (struvite), and organic solids for crop fertilization, soil conditioning, and compost production

2. Heat from warmer used water and groundwater

3. Organic solids for production of methane

4. Electricity and/or hydrogen gas produced from methane by hydrogen fuel cells, and directly from used water or anaerobic digestion by microbial fuel cells (MFCs)

5. Waste carbon dioxide and residual nutrients can be sequestered in algal reactors and farms which produce new biomass that can be converted to biogas and energy (see Chapter 8).

Figure 6.2 shows the difference between conventional wastewater treatment and disposal, which is the endpoint of the linear "potable water to wastewater" transformation, and the use of urban water systems and concepts of water reclamation proposed for the new ecocities. The former treatment system presents a two-step (primary and secondary) biological wastewater treatment using a conventional activated sludge biological unit process (Figure 6.2A). This process requires a significant amount of energy and produces treated effluent that, in most cases, can be discharged into the receiving water body, but its reuse is limited. Energy is required for turning mechanical scrapers in the primary and secondary sedimentation basins and sludge thickener, for compressors or aerators in the aeration basin, and for providing heat and mixing in the sludge digester. The process emits GHG carbon dioxide from the aeration tank and digester, as well as methane from the digester. The global warming potential (GWP) of methane emitted to the atmosphere is more than 25 when averaged over one hundred years. The global warming potential relates the global warming effect of the compound to that of carbon dioxide, which has GWP = 1.

The simple water reclamation and reuse system (B) proposed for ecocities (see Chapter XI) produces reclaimed water that is fit for discharging as ecological flow into a receiving water body and, after some additional treatment, can be reused for other nonpotable or indirect potable water uses. Process B is simple, based on current commonly known and widely used technologies, and does not use separation. More efficient and revolutionary reclamation and reuse processes will be presented in Chapters VII and VIII. In the advanced systems, water/wastewater streams are separated into black (toilets and kitchen used water) and gray (from laundries and washing) flow. Process B requires only a minimum amount of energy, far less than the system produces in various forms, and recovers nutrients and organic solids for further reuse. In the process, energy and nutrients are recovered not only from the wastewater but also from the organic solids in the kitchen food by-products and from the collected organic residues harvested from constructed wetlands with subsurface flow and vegetation residues such as fallen leaves and clippings in parks, gardens, and open spaces. In general, process B and the more advanced systems are clean energy net producers rather than users. Table 6.1 contains definitions and terminology describing reclamation and reuse.

The anaerobic digester—an energy, nutrient, and solids recovery unit—is at the center of the water–energy reclamation system. A similar process unit is the upflow anaerobic sludge blanket (UASB) reactor or even an anaerobic mixed liquor

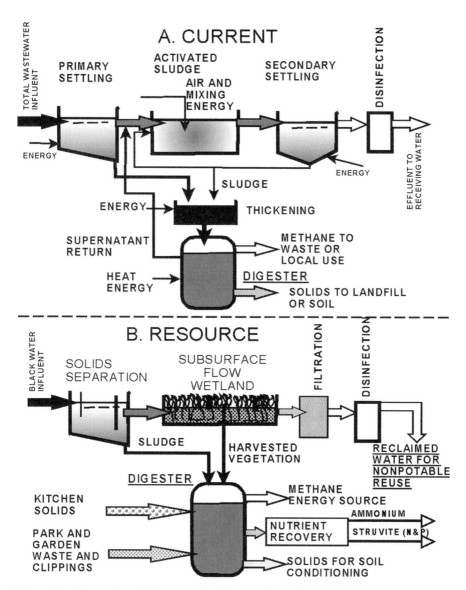

Figure 6.2 Comparison of the conventional energy-demanding linear wastewater treatment process (A) and a basic process (B) to reclaim water, energy, and nutrients from black water, similar to that proposed for the Qingdao ecocity, that is GHG-emissions neutral.

reactor similar to the classic activated sludge process; however, this one requires more energy for mixing. The unit's processes for water, nutrient, and energy reclamation will be discussed in Chapters VII and VIII. Reclamation means that used water and organic vegetation residues in most cases are not waste, they are resources. Therefore, throughout the book we refrain from using the terms "wastewater" and

Table 6.1   Definitions of water reuse (adapted from Levine and Asano, 2004, and other sources)

Term	Definition
Reused/reclaimed water classification	
Blue water	Water quality suitable for potable use from springs and wells, from high-quality surface water bodies, unpolluted rainwater, and water treated to potable water quality
White water	Mildly polluted surface runoff
Gray water	Medium polluted water from laundries, bathrooms (shower and bathtub), washing, emulsified with soaps and detergent; may have significant concentrations of biodegradable and nonbiodegradable organics, and may also contain pathogens
Black water	Highly polluted wastewater from toilets and urinals, containing urine and excreta and drain water from kitchen sinks with disposal units
Yellow water	Separated urine (about 1% of the total flow)
Brown water	Black water after yellow water is separated
Green water	Water use in irrigation
Processing	
Water reuse	Use of the used water effluent for the same or another beneficial use, with treatment before the reuse
Water reclamation	Capture, treatment, or processing of used water or stormwater to make it reusable
Water recycling	Repeated recovery of used water from a specific use, and redirection of the water either back to the original or to another use
Product	
Reclaimed water	The end product of used water reclamation that meets water quality requirements for biodegradable materials, suspended matter, and pathogens for the intended reuse
Direct reuse	The direct use of reclaimed water for applications such as agricultural and urban landscape irrigation, cooling water and other industrial uses, urban applications for washing streets or firefighting, and in dual water systems for flushing toilets (direct potable reuse is still rare)
Indirect reuse	Mixing, dilution, and dispersion of treated wastewater by discharge into an impoundment, receiving water, or groundwater aquifer prior to reuse, such as groundwater recharge with a significant lag time between the use and reuse
Potable water reuse	Use of highly treated used water to augment drinking water supplies
Nonpotable reuse	Includes all water reuse applications other than direct or indirect use for drinking water supply

"solid waste" and replace them with the terms "used water" and "resources." Sustainable cities of the future are being or will be designed to approach a closed-loop system (see Figure 2.10 in Chapter II). Produced methane will be fully used as an energy source, not emitted into the atmosphere or flamed out by methane burning torches in landfills or anaerobic treatment plants.

The focus of this chapter is to describe the concepts and barriers to reclamation and reuse, ending with examples of systems. Chapters VII and VIII will present processes for reclamation and reuse technologies that will have a small or no carbon footprint, or will be net producers of clean energy. These processes are resources not only of potable and nonpotable water but also of energy, electricity, nutrients, compost, and other useful products and by-products. In addition to reducing energy consumption, these have other environmental benefits, including providing the ecologic flow that urban streams need to maintain their high-quality ecological status. Recycled water can also be used to create or enhance wetlands and riparian habitats. Water recycling can decrease the diversion of water from sensitive ecosystems and prevent or reduce discharging of partially contaminated freshwater sewage into sensitive brackish or saltwater marine areas. For a comprehensive design manual on water reuse technologies, the reader is referred to the Metcalf & Eddy publication *Water Reuse: Issues, Technologies, and Applications* (Metcalf & Eddy, 2007). For design of individual used water treatments, consult monographs by Metcalf & Eddy (2003); Imhoff and Imhoff (2007); and Novotny et al. (1989).

VI.2.2   Reclaiming Rainwater and Stormwater

Stormwater in ecocities is also a very valuable commodity—a resource that can be inexpensively reclaimed for nonpotable uses and is a valuable supplement to the sources of potable water. The key is to prevent excessive pollution, which occurs when rain falls on contaminated impervious surfaces. There are numerous stormwater best management practices that can be used to capture, store, and eventually treat stormwater for use (See Chapters IV and V). Large-scale capture and reuse of urban stormwater from a densely populated area is practiced in Singapore (see Section VI.3.6 in this chapter). The difference between reclaimed rainwater and urban stormwater is that the former is captured immediately after rain falls on a relatively clean surface, such as roofs or special impervious capture areas. Stormwater runoff often moves through contaminated urban surfaces, including street gutters, ditches, and storm sewers. Reclaimed rainwater/stormwater is the third-best source of water for urban water supply, after groundwater aquifers and good-quality surface sources. Rainwater/stormwater reclamations processes are:

- Rainwater harvesting, storage, and direct use for potable (after disinfection) and nonpotable uses (see Chapter V)
- Natural or engineered infiltration and groundwater recharge of rainwater and stormwater, followed by indirect potable (after disinfection) or nonpotable use
- Capture, storage, and natural treatment of runoff, followed by treatment and direct nonpotable use (Chapter IV)

As discussed in Chapter I, rainwater harvesting has been practiced for millennia and is becoming popular again in many water-short areas of the world. In the ecocities of Tianjin and Qingdao, in China, and Masdar in Abu Dhabi (see Chapter XI), located in arid or semi-arid regions, captured and treated stormwater represents a significant portion of the potable and nonpotable water supply. All ecocities infiltrate either all or a major portion of rainfall into groundwater aquifers, which are then tapped into for water supply. Surface storage in ponds and underground reservoirs is also common (see Chapter IV).

VI.2.3 Water-Sewage-Water Cycle—Unintended Reuse

Karl Imhoff, a pioneer of European sanitary (environmental) engineering (see Novotny, 2003) almost one hundred years ago realized that once an effluent is discharged into a receiving water body it will be reused, in a more or less diluted form, as a source of water somewhere downstream. This uncontrolled indirect reuse was called the "water-sewage-water" (WSW) cycle (Imhoff, 1931; Imhoff and Imhoff, 2007) and it occurs in almost every municipal water management system, with an exception of wastewater disposal into the ocean. Environmental protection agencies and municipalities relying on water supply from watercourses must always consider the fact that raw water contains residuals of upstream wastewater discharges and nonpoint pollution. Even groundwater cannot be safe from wastewater contamination; in general, shallow groundwater was once surface water subjected to pollution. Almost every wastewater discharge into continental waters may later be reused by the population downstream, with no controls on its use (Imhoff, 1931). Imhoff also stated that the only thing engineers can do is maximize the lag time between the discharge and reuse. The WSW cycle is essentially an indirect reclamation and reuse of effluents diluted by river flows.

The problems with the water-sewage-water cycle came to light many times during the last century. In the Ruhr area, during the dry periods of 1929 and 1959, back-pumping of river water was activated. Water shortages in 1959 lasted four months, and 109 million m^3 of the Rhine-Ruhr Rivers mixture were pumped upstream towards the water intakes of Essen and other Ruhr cities (Koenig et al., 1971). As a result, the quality of potable water greatly worsened, concentrations of nitrate (NO_3^-) reached 35 mg/L, detergents 1.5 mg/L, and chloride 490 mg/L. At the end of 1959, the water demand could only be met by back-pumping the Rhine and Ruhr water. This resulted in a sewage reuse factor of 0.86 (Koenig et al., 1971), and about 10% of the people in Essen suffered gastrointestinal problems. In the succeeding years, the Ruhr River Management Association (Ruhrverband) increased the capacities of storage reservoirs, and an occurrence of a similar back-pumping and effluent reuse has not happened and is unlikely to happen. Nevertheless, water-sewage-water cycles occur everywhere. Downstream users of the Mississippi River water in New Orleans, Louisiana, reuse wastewater effluents from about one-half of the U.S. population mixed with the flow of the river.

Another example of a large water-sewage-water cycle is the Trinity River in Texas (Figure 6.3), which in the upper reaches receives treated effluents from the Dallas–Ft. Worth metropolitan area (population 6 million). As a result of these discharges, the river becomes effluent dominated, and during dry periods more than 90% of flows are treated effluents from the metropolitan area and other cities. Downstream, the river

Figure 6.3   Effluent dominated Trinity River downstream of Dallas–Ft. Worth. Picture by the U.S. Army Corps of Engineers.

is impounded and becomes Lake Livingston, which is a source of potable water for the City of Houston, the fourth-largest city in the U.S., with 5.8 million inhabitants.

The lag time between the discharges of Dallas–Ft. Worth effluents and the use of water in Houston, due to a large volume in Lake Livingston, is more than six months. The reuse lag time in ecocities may be very short, likely days or a week for direct reuse and up to a month for indirect reuse by reusing infiltrated reclaimed used water. The state of the art of treatment technology is today already at the level that direct reuse after a short lag time is possible, but it is not recommended, and extreme precautions and monitoring will be necessary when direct reuse is contemplated.

Today, many large cities such as Cincinnati, St. Louis, and New Orleans, drawing water from the Ohio, Missouri, and Mississippi Rivers respectively, with several multimillion megalopoli located on them; several cities on the Rhine River in Germany; and cities located on the Danube River and its tributaries (e.g., the Morava River in Czech Republic), the Elbe River in Germany and Czech Republic, the Yangtze and Pearl Rivers in China, are practicing indirect potable water reuse.

VI.2.4   Centralized versus Decentralized Reclamation

A linear system does not allow much reclamation and reuse except for downstream flow enhancement that could be used for agricultural irrigation and reuse by downstream communities in the WSW cycle. Typically, after water is withdrawn from a river, it is redirected to controlled recharge of alluvial sediments from which the water can be pumped and treated for potable use. Most regional "water reclamation" effluent treating facilities, such as the Stickney (Illinois) Water Reclamation Plant or the Metropolitan Water Reclamation District plant in Denver, Colorado, are far downstream from the city, and to reuse the reclaimed water back in the city would be very costly. Levine and Asano (2004) state the logical truth that cost-effective use of reclaimed water for municipal, irrigation and industrial uses requires it to be produced relatively close to the potential user. In the future sustainable cities, the users of reclaimed water, especially in arid regions, will be individual homeowners and tenants in apartments and condominiums.

A centrally located water reclamation plant is not such an anachronism as it may seem. The treatment plant for 33,000 people living in the Principality of Monaco on the Mediterranean coast is located in the high-density affluent city-state, which has an area of 2 km^2. People in the LEED platinum Dockside Green development in Victoria (BC-Canada) will also have the water and energy resource reclamation plant in the community (see Chapter X). Most of the new ecocities have their resource recovery plants either centrally located (Tianjin, China) or partially distributed in each ecoblock (Qingdao, China). The concepts of cluster distributed water/stormwater/used water management of water centric ecocities are presented in Chapters II and X, and also described by Mitchell, Abeysuriya, and Fam (2008).

VI.2.5   Cluster Water Reclamation Units

Satellite Withdrawal and Treatment. Metcalf & Eddy (2007) suggest alternatives for retrofitting decentralized water management into existing urban environments. All urban sewerage systems are continuously upgraded and rebuilt over a period of 25–40 years. By starting with building satellite treatment in upstream portions of the urban drainage area, used water (wastewater from the local collection system) can be intercepted and treated to the high degree required for various reuse alternatives such as nonpotable in-house uses, landscape irrigation, groundwater recharge, and ecological flow enhancement. Diverting a portion of the reclaimed flow into the existing deficient flow or daylighted streams provides opportunities for reuse downstream, supplied by gravity flow, and eliminates the necessity of pumping water upstream from a distant regional water reclamation facility. A typical regional wastewater treatment plant is almost always located at the lowest point of the urban watershed and often in an uninhabited area far from the center of the city.

In the satellite type of treatment plant, the used water withdrawn from the sewer is treated to the quality required for reuse, and the residuals produced in the treatment process can either be converted to useful products (e.g., energy) or redirected back to the sewer and conveyed to a central/regional resource recovery facility for energy, methane, and nutrient recovery. Satellite (cluster) treatment water reclamation plants would typically be compact packaged treatment units placed underground in garages, under parking lots, or in basements of commercial buildings, or they can even be disguised in a city building, as was done in Monaco or Dockside Green.

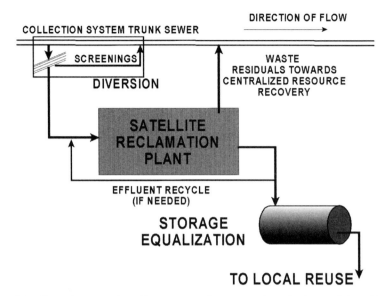

Figure 6.4   Extraction type of satellite treatment for water reclamation in retrofitting older cities (from Metcalf & Eddy, 2007)

Metcalf & Eddy (2007) categorize satellite treatment plants into three types:

1. Interception
2. Extraction (Figure 6.4)
3. Upstream of the use

For example, the North Shore Water Reclamation Plant in Chicago is located on the upper part of the North Chicago River. The reclaimed water from the plant will be discharged into the river to augment the flow and also for reuse. In Beijing, where there are 16 treatment plants with more being planned, effluents are heavily reused for park irrigation and other uses. It is clear that the satellite facility pictured in Figure 6.4 reclaims water from the combined or sanitary sewer flow that may contain black and gray sewage, white infiltration, and surface runoff flows. Hence, the utility of sewer flow extraction is limited and less efficient.

Cluster (Ecoblock) Management. In the most common ecocity application, the flows are separated and urban runoff is conveyed almost exclusively on the surface. For black water, the decentralized (cluster) treatment plants in Qingdao ecoblock use septic tanks, intermittent (batch) and recirculating packed bed reactors (filters), or subsurface flow wetlands (see Chapter VII) for treatment. Compact treatment technologies for traditional or innovative secondary treatment are also available (Metcalf & Eddy, 2007), typically followed by additional treatment (e.g., filtration by either granular media or membrane filters and/or reverse osmosis followed by disinfection) to bring the flow into compliance with the standards and criteria for reuse.

Satellite or central facilities also include flow diversion from sewers with screens, storage, and equalization basins. Flow diversion with self-cleaning screens for combined sewer overflows was developed in Switzerland (Krejci and Baer, 1990; Novotny, 2003), but the same principle can also be used for tapping sewage flow from underground conduits. Storage basins are designed to store excess peak flow and provide flow during dry periods. Storage is accomplished either in surface or underground basins and ponds, or in the groundwater aquifer. Equalization basins facilitate water supply during peak hours and store produced reclaimed water during off-peak hours or when the use is intermittent, such as irrigation. In most applications in retrofitting older cities, designers will opt for underground storage with pumping, which requires energy. Reclaimed water can also be stored under controlled and managed conditions in groundwater zones. However, because the reclaimed water should be of high quality, surface storage architecturally blending with the landscape may be a far more ecological, economical, and sustainable alternative.

Figure 6.5 shows a variant of the separation of used water proposed for Qingdao (Chapter XI) in each cluster (ecoblock) into dual interconnected reuse systems, one for potable blue water, another one for black water (sewage with excreta and kitchen waste), another for gray water (laundry, bath and shower washing water), and finally reclaimed white water. The ecocity has two treatment systems: a local treatment system connected to a central water reclamation and resource recovery (methane) facility (Fraker, 2008). Storage reservoirs and UV disinfection would be

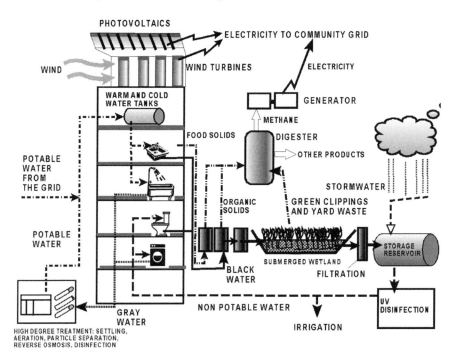

Figure 6.5 A variant of black and gray water separation loops proposed for water reclamation and reuse in the Qingdao ecoblock. Replotted and modified from Fraker (2008) and Arup. The ecoblock is represented on the figure by one house.

centralized in the ecocity; the rest of the multiple-cycle reclamation and reuse system is decentralized in each ecoblock. The potable blue water input from the city grid represents only 25% of the total water demand (see Chapters VII and XI). In the Qingdao reuse system, drinking water would contain about 50% reclaimed water, part of which would be direct reuse of treated black water. In Section VI.3 we will propose and document that direct reuse water for drinking and food preparation can be avoided.

The Qingdao ecoblock proposed by the University of California, Berkeley, team from the School of Environmental Design, led by Dean Fraker and shown on the figure (see also Chapters VII and XI), has 600 apartment units (along with local commerce) in buildings ranging in height from seven to twenty-four floors. These ecoblocks, represented in the figure by a one-building schematic, also produce enough electricity from renewable sources (solar photovoltaic, wind, and methane) to satisfy all its energy needs and, on average, will have net zero GHG emission footprint. In the near future, it is expected that electricity and hydrogen gas can also be reclaimed from wastewater during treatment by microbial fuel cells (Liu, Ramnarayanan, and Logan, 2004; see also Chapter VIII).

A more centralized water reclamation system similar (but not identical) to that proposed for the Tianjin ecocity is presented in Figure 6.6. The difference between

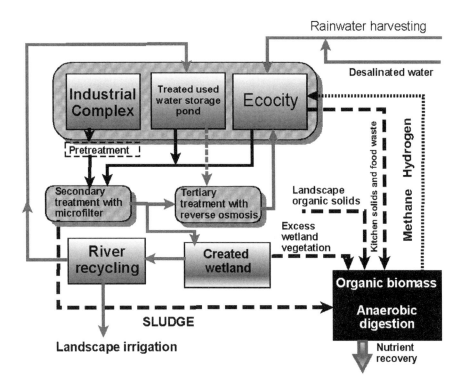

Figure 6.6 More centralized water and energy reclamation system with reuse, based on the Tianjin system (*Source:* www.tianjinecocity.gov.sg).

the Tianjin and Qingdao systems is use of surface water bodies, streams, canals, and ponds in Tianjin, which is not as evident in the Qingdao ecoblock and ecocity.

VI.3   WATER QUALITY GOALS AND LIMITS FOR SELECTING TECHNOLOGIES

VI.3.1   Concepts

Figures 6.5 and 6.6 provide an indication of the relative complexity and logic incorporated into a "near perfect" but still not fully tested complete reclamation and recycling system. Similar high-efficiency systems are used or proposed for the other ecocities, especially for those located in water-short or previously contaminated brownfield areas (see Chapters X and XI). For example, the only consistently available water source for Tianjin (China) and Masdar (UAE) is desalinated seawater, which is still relatively expensive to produce and energy demanding. It is, however, questionable whether or not the treatment level of used water all the way to potable water quality is needed in all ecocities, when some cities with better availability of inexpensive fresh water sources can reach their sustainability goals by water conservation and plugging leaks in the water supply system (Chapter VIII). As a matter of fact, mandatory requirements of potable quality and reuse for all reclaimed water could be a deterrent to the ecocity idea in more humid areas. Hydrologically, municipal water supply withdrawals must be balanced with other uses, but water is a renewable source, and as long as the water withdrawals are below the safe yield and not infringing on other important uses—for example, sustaining healthy aquatic life, recreation, and agriculture—natural surface and groundwater sources are far more preferable and economical than a complete closure of the urban hydrologic cycle by treating the reclaimed water to potable use quality. On the other hand, the present practice of wasting potable water and applying it for nonpotable usage (irrigation, toilet flushing, cooling, street and car washing, firefighting) is wasteful and unsustainable in any urban area. In a regional linear system, such wasteful uses cannot be avoided, but in a distributed cluster (ecoblock) water management, water separation based on use and reuse is highly efficient under any circumstances.

The hydrologic cycle can never be 100% closed because of other factors described later in this and subsequent chapters. The most important factor is the accumulation of conservative or difficult to remove harmful pollutants in the cycle. That may include endocrine disruptors and pharmaceutical compounds. For example, the Qingdao reclamation/reuse system proposes that only 25% of the water demand be brought from the municipal potable water grid. This would imply that some water in the cycle may be reused three to four times with a very short lag time as potable water. The amount of importing of clean potable water is a key decision parameter.

A better alternative is to include natural (or recreated) urban surface water bodies in the reuse cycle that will provide storage, extend the lag time, and also provide natural physical and biochemical attenuation processes for further purification. This alternative has been proposed for the ecocities Tianjin (Figure 6.6) and Dongtan in

China (see Chapter XI). As discussed in the preceding section on the water-sewage-water cycle in natural water bodies, reuse of treated effluents has been practiced for more than a century, and in a great majority of cases it has been safe. Furthermore, one of the fundamental characteristics of an ecocity should be to mimic natural systems to the greatest extent possible.

Reclaimed water has been used in many countries and states. In the arid and semi-arid states such as Arizona, Utah, California, Nevada, and Texas, but also in China, the Middle East (including Israel), Japan, Singapore, Australia, and South Africa, the demand for reclaimed water is driven by water shortages in burgeoning cities and suburbs trying to provide the same, often wasteful, water services as in the more humid states, such as those growing and irrigating "blue grass" lawns and maintaining swimming pools. In the arid states of the U.S., reclaimed water is used mainly for watering golf courses and suburban landscapes.

The biggest difference between reclamation and reuse in the ecocities and in traditional urban areas in the arid and semi-arid U.S. is the magnitude of water demand. A typical design of the total water demand for an ecocity is between 100 and 200 liter/capita-day, which by reclamation and reuse is reduced to 100 liter/capita-day, or even less. Water demand and use in the arid and semiarid U.S. are at least five to ten times higher. The water use in many communities in water-short areas of the U.S. and oil rich Middle East countries is still characterized by high water demand (and a lot of waste) which often cause severe water shortages.

It is important that the designer of the reclamation system have the water demand and quality goals in mind when selecting the sequence of unit processes leading to the goals, and that these considerations be included in the design. For example, it may be wasteful and unsustainable if reclaimed water to be reused for irrigation has nutrients (nitrogen and phosphorus) removed by treatment, and then industrial fertilizers are applied on the land or crops. By similar reasoning, nitrate levels in drinking water are perfectly safe up to the relatively high concentrations of 10 mg NO_{3-} - N/L, which is the worldwide standard for health safety of potable water, established by the World Health Organization (WHO) for preventing methemoglobinemia (blue baby disease). On the other hand, high nitrate levels could be devastating to the ecology of surface water bodies, including especially coastal zones, where they accelerate the eutrophication process of the receiving water bodies. Today severe algal (eutrophic water bodies) and cyanobacteria (hypertrophic water bodies) blooms have caused anoxia and other damages contributing to biota and fish and shellfish die-off (see Chapter II). A "bloom" is a massive accumulation (10^4–10^6 cells/L) of a single or coexisting nuisance species that is manifested by scum; failure of filtration systems in water treatment plants from excessive clogging; smell of water due to anoxic conditions, resulting in emanation of hydrogen sulfide; bad aesthetics; and toxicity to people (swimmers developing rashes and gastrointestinal sickness) and animals. Systems with a high degree of recycle may be susceptible to the occurrence of algal or cyanobacteria blooms. Irrigation water can tolerate low levels of contamination by biodegradable organics, but a high dissolved solids content (>500 mg/L) may be an impediment to irrigation (see Krenkel and Novotny, 1980; Novotny and Olem, 1994).

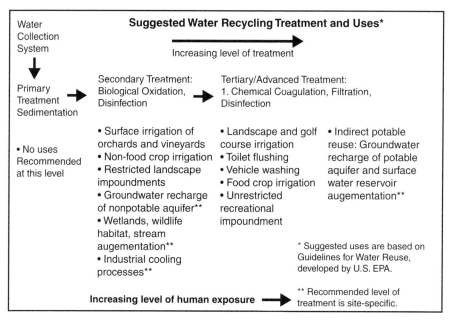

Figure 6.7   Standard secondary and tertiary treatment recommended by the U.S. Environmental Protection Agency for nonpotable uses of water reclaimed from municipal sewage.

All reclaimed water should be free of pathogenic microorganisms, yet the least expensive disinfection by chlorine is not acceptable, and disinfection processes under the current state of the art are limited to ultraviolet light and/or ozonization units. Water quality criteria for various uses are contained in most countries' national water quality standard publications and manuals (e.g., U.S. EPA, 1994). The U.S. EPA guidelines for treatment of typical municipal sewage are presented in Figure 6.7.

There are two major references on water reclamation and reuse: the U.S. EPA *Guidelines for Water Reuse*, produced by CDM (U.S. EPA, 2004), and the extensive manual *Water Reuse: Issues, Technologies, and Application* by Metcalf & Eddy (2007). A committee of the National Research Council (1998) extensively evaluated the problems with and technologies for the use of reclaimed water for augmenting potable water supplies, and numerous additional references will also be mentioned in the text. Because of rapid urbanization and population increases, many cities and urban areas—not only those located in arid or semi-arid regions of the world—are running out of water and are implementing effluent reclamation and reuse. Water reclamation from used freshwater and also reclamation of potable water by desalinating seawater have become a fast-growing trend in the water industries. This chapter focuses on urban uses in the future water centric ecocities. However, it should be realized that an "ecocity" is not only a medium-density built-in community, it also extends to suburban and near-urban agriculture that may be using the reclaimed water, fertilizers, and soil-conditioning solids for growing crops and vegetables for the city.

VI.3.2 Landscape and Agricultural Irrigation

In addition to providing for the water needs of urban grasses, gardens, and plants, landscape irrigation in ecocities may have a dual purpose. First, water satisfies the evapotranspiration requirements of the landscape flora, and, second, irrigation excess recharges groundwater zones from which water can be subsequently withdrawn for reuse. Storage in groundwater zones also provides beneficial lag time between use and reuse.

In traditional urban areas located in water-short areas, reclaimed water is used for golf course and public park irrigation. In Beijing, Singapore, and many other cities, reclaimed water is trucked to irrigate green areas and trees in parks and along the streets that are not reached by a reclaimed-water pipeline. An example is the system in the Beijing Olympic Park, where the use of potable water for irrigation is prohibited. Irrigation can be done by surface spraying, drip irrigation, or subsurface. Direct irrigation of parks and public areas with reclaimed water typically requires tertiary treatment with odor removal and disinfection by UV or ozone; however, this additional tertiary treatment can be avoided if high-quality disinfected (by UV light) secondary effluents and/or outflows from created wetlands are discharged into receiving surface waters such as streams and ponds from which water can be withdrawn for subsequent irrigation. The subsurface or drip irrigation is less water quality–demanding.

It is important that the landscape of ecocities contain mostly native plants and be created as xeriscape (see Chapters III and V). Local native plants are accustomed to the precipitation conditions of the area, minimizing the need for irrigation. Plants, trees, and vegetative land covers with low water demand and higher resistance to salinity are obviously recommended in areas irrigated with reclaimed water. As stated before, the nutrient levels in the reclaimed water are not an issue; as a matter of fact they are acceptable and may reduce or eliminate the need for fertilizer applications. It should be noted that commercial irrigation companies rarely factor the nutrient content of irrigation water into their estimates of fertilizer requirements, which leads to overfertilization. Use of industrial fertilizers should be avoided. If fertilizers are needed, ecocity water and resource recovery systems are capable of producing fertilizers for landscape irrigation, citizens' gardens, and suburban (organic) agriculture (Chapter VII). If nutrients are removed by treatment in the water reclamation system, then nutrients recovered on-site, in the integrated resource recovery unit (Chapter VIII), should be used instead of commercial industrial fertilizers.

Growing lawns in water-short areas, where lawn grass is not native, is highly unsustainable and water demanding. During warm/hot dry weather, a typical lawn requires about 2.5 cm ($250 \ m^3$/ha) of irrigation water per week. Assuming per capita water demand in an ecocity to be about 150 liters/capita-day, the lawn water demand is about the same as the total water demand of 240 people/ha living in the same area. Spray irrigation also uses a lot of energy because of the high pressure requirement in the system. Less energy and water are needed for subsurface and drip irrigation. The amount of pumping energy is proportional to the product of irrigation flow multiplied by the pressure in the irrigation nozzle divided by the pump efficiency, or

$$\text{Pumping energy used } E = \frac{0.273\ p\ Q}{\sigma}\ t$$

where E = energy used in kW-hours; p = pressure in N/m^2 (Pascal); Q = pumped flow in m^3/s; σ = pump and motor efficiency in %; and t = pumping time in hours. To convert energy into equivalent carbon emission in grams from a coal-fired fossil fuel power plant, multiply E by 0.96 kg of CO_2/kW-hour (Keller and Hartley, 2003; U.S. Department of Energy, 2000). For example, if the pumped flow is 0.1 m^3/s and sprinkler nozzle pressure is 1,000,000 N/m^2 = 10 bar (1 bar = 1.03 atmosphere), pump efficiency is 0.7 (70%) and daily sprinkling lasts for 5 hours, then, as a result of pumping, 195 kg of CO_2 would be emitted from the power plant during the 5 hours of irrigation. Friction losses in the pipeline would have to be added to the pressure.

Irrigation demand for an ecocity is not limited to landscape irrigation within the city proper. An ecocity is also connected to the suburban agriculture which, logically, could also be a recipient of the reclaimed water. For example, Dongtan (China) and Masdar (UAE) plan to use reclaimed water to grow crops, vegetables, and flowers in the suburban agricultural areas. All ecocities are planning to incorporate urban organic agriculture, which is also one of the criteria to receive One Planet Living certification (see Chapter II).

Golf course irrigation systems typically operate at higher pressures (7.5 to 15 bar) and, if directly connected to the reclaimed water system, a booster pump station will be required.

Agricultural irrigation. One of the Leadership in Energy and Environmental Design (LEED) and One Planet Living criteria (See Chapter II) is local food production, preferably produced by organic agriculture and, if needed, irrigated by the reclaimed water. Irrigation of agricultural land (land application) has also been used for wastewater treatment (see Imhoff and Imhoff, 2007; Novotny, 2003). The chemical constituents of concern in reclaimed water for agricultural irrigation are salinity, sodium, trace elements, excessive chlorine residual, and nutrients. Plant sensitivity to such constituents in irrigation water is generally a function of a given plant's tolerance to the constituents encountered in the root zone or deposited on the foliage. Reclaimed water (irrigation return flow) tends to have higher concentrations of harmful constituents than the groundwater or surface water sources from which the water supply is drawn.

Table 6.2 lists the chemicals of concern for agricultural and urban landscape irrigation. The types and concentrations of constituents in reclaimed wastewater depend upon the municipal water supply, the influent waste streams (i.e., domestic and industrial contributions), the amount and composition of infiltration in the wastewater collection system, the wastewater treatment processes, and the type of storage

Table 6.2   Recommended Limits for Constituents in Reclaimed Water for Irrigation

Constituent	Long-Term Use (mg/l)	Short-Term Use (mg/l)	Remarks
Aluminum	5.0	20	Can cause nonproductiveness in acid soils, but soils at pH 5.5 to 8.0 will precipitate the ion and eliminate toxicity.
Arsenic	0.10	2.0	Toxicity to plants varies widely, ranging from 12 mg/L for Sudan grass to less than 0.05 mg/L for rice.
Beryllium	0.10	0.5	Toxicity to plants varies widely, ranging from 5 mg/L for kale to 0.5 mg/L for bush beans.
Boron	0.75	2.0	Essential to plant growth, with optimum yields for many obtained at a few tenths mg/L in nutrient solutions. Toxic to many sensitive plants (e.g., citrus) at 1 mg/L. Usually sufficient quantities in reclaimed water to correct soil deficiencies. Most grasses are relatively tolerant at 2.0 to 10 mg/L.
Cadmium	0.01	0.05	Toxic to beans, beets, and turnips at concentrations as low as 0.1 mg/L in nutrient solution. Conservative limits recommended.
Chromium	0.1	1.0	Not generally recognized as an essential growth element. Conservative limits recommended due to lack of knowledge on toxicity to plants.
Cobalt	0.05	5.0	Toxic to tomato plants at 0.1 mg/L in nutrient solution. Tends to be inactivated by neutral and alkaline soils.
Copper	0.2	5.0	Toxic to a number of plants at 0.1 to 1.0 mg/L in nutrient solution.
Fluoride	1.0	15.0	Inactivated by neutral and alkaline soils.
Iron	5.0	20.0	Not toxic to plants in aerated soils, but can contribute to soil acidification and loss of essential phosphorus and molybdenum.
Lead	5.0	10.0	Can inhibit plant cell growth at very high concentrations.
Lithium	2.5	2.5	Tolerated by most crops at concentrations up to 5 mg/L; mobile in soil. Toxic to citrus at low doses. Recommended limit is 0.075 mg/L.
Manganese	0.2	10.0	Toxic to a number of crops at a few tenths to a few mg/L in acidic soils.
Molybdenum	0.01	0.05	Nontoxic to plants at normal concentrations in soil and water. Can be toxic to livestock if forage is grown in soils with high levels of available molybdenum.
Nickel	0.2	2.0	Toxic to a number of plants at 0.5 to 1.0 mg/L; reduced toxicity at neutral or alkaline pH.

(Continued)

Table 6.2 (*Continued*)

Constituent	Long-Term Use (mg/l)	Short-Term Use (mg/l)	Remarks
Selenium	0.02	0.02	Toxic to plants at low concentrations and to livestock if forage is grown in soils with low levels of selenium.
Tin, tungsten, & titanium	–	–	Effectively excluded by plants; specific tolerance levels unknown.
Vanadium	0.1	1.0	Toxic to many plants at relatively low concentrations.
Zinc	2.0	10.0	Toxic to many plants at widely varying concentrations; reduced toxicity at increased pH (6 or above) and in fine-textured or organic soils.

Constituent	Recommended Limit	Remarks
pH	6.0	Most effects of pH on plant growth are indirect (e.g., pH effects on heavy metals' toxicity described above).
TDS	500–2000 mg/l	Below 500 mg/L, no detrimental effects are usually noticed. Between 500 and 1000 mg/L, TDS in irrigation water can affect sensitive plants. At 1000 to 2000 mg/L, TDS levels can affect many crops, and careful management practices should be followed. Above 2000 mg/L, water can be used regularly only for tolerant plants on permeable soils.
Free Chlorine Residual	<1 mg/l	Concentrations greater than 5 mg/L causes severe damage to most plants. Some sensitive plants may be damaged at levels as low as 0.05 mg/L.

[1]*Source:* Adapted from Rowe and Abdel-Magid, 1995

facilities. Conditions that can have an adverse impact on reclaimed water quality may include:

- Total dissolved solids levels that are elevated by upstream irrigation
- Industrial discharges of potentially toxic compounds into the municipal sewer system
- Saltwater (sodium) infiltration into the sewer system in coastal areas

Water applied on land in the form of treated effluent, with or without being mixed with water from nearby surface water bodies, always contains dissolved salts and dissociated ions. After application on an irrigated area, sometimes substantial portions of the applied water are lost by evapotranspiration in the form of pure H_2O, leaving the salts and ions in the soil. The portion of water lost by evapotranspiration may range from 20% in humid climatic conditions, to almost 100% in arid or semi-arid climatic zones. This leads to salt buildup in soils, which, if unchecked, could lead to loss of soil fertility.

To maintain an acceptable salt content of soils to sustain crop growth and fertility of the soil, excess irrigation must be applied if natural precipitation is not sufficient to control the salt buildup. The excess water (called *irrigation return flow*), containing increased salinity and also nutrient content, is collected by drainage if returned to the water reuse system, or infiltrates into soil. Nitrate levels of irrigation return flow may be quite high. Irrigation return flow is polluted and should be treated before reuse unless return flow quality is troublesome in arid zones of the world, where many ecocities relying on extensive water reclamation and reuse may be located.

VI.3.3 Urban Uses Other Than Irrigation and Potable Water Supply

Some cities and commercial/sports/recreation establishments are reclaiming water and reusing it. The most widespread uses of reclaimed used water are:

- Toilet and urinal flushing in commercial high-rise buildings and several sports stadiums throughout the world
- Street cleaning and surface spraying (e.g., at construction sites to reduce dust, or onto city pavements to reduce the heat island effect)
- Fire protection
- Air-conditioning cooling towers
- Commercial car washing and laundries
- Providing flow to augment flow in urban streams and lakes

Metcalf & Eddy (2007) found consistency in the water quality requirements among several states in the U.S. for all of the above uses, namely that they all require, at a minimum, secondary treatment followed by filtration and disinfection (Figure 6.7). It is implicitly assumed that secondary treatment in ecocities will not consist of energy-requiring aerated activated sludge processes combined with aerobic digestion of the sludge; the processes of choice should be those that rely on mixed aerobic/anaerobic treatment and digestion, the best of which are anaerobic processes with biogas production. Constructed subsurface flow wetlands are also attractive due to their zero energy demand and carbon sequestering; however, the problems with methane and nitrous oxide emissions from wetlands must first be resolved.

There are also specific requirements for some uses; for example, commercial car washes and laundries use reclaimed water for the first wash cycle, but the final rinse is done with water that has potable water quality. Dual water supply and plumbing is one of the implicit features of the future ecocities that maximize water conservation and reuse of reclaimed water.

If reclaimed water is used for ecologic flow, control of eutrophication and public health concerns are the major issues. Unsightly eutrophication and hypertrophy exhibited by the dense algal (cyanobacteria) growths of effluent dominated urban

Table 6.3 **Quality requirements for some urban uses (adapted from Metcalf & Eddy, 2007)**

Parameter	Unit	Intended Reuse of Reclaimed Water		
		Toilet/Urinal Flushing	Street Washing or Spraying on Ground	Recreation Uses, Ecologic Flow
Escherichia coli	No/100 ml	Nondetectable (?)	Nondetectable (?)	126* (536)
Turbidity	NTU	2	2	2
pH		5.8–8.6	5.5–8.6	5.6–8.6
Appearance		Not unpleasant	Not unpleasant	Not unpleasant
Color	Color units			<10
Odor		Not unpleasant	Not unpleasant	Not unpleasant
Total phosphorus	mg/L			<0.1
Treatment		See Figures 6.7 and Chapter VII		

*See water quality criteria (U.S. EPA 1986, 2004) and standards. Numbers are expressed as geometric means. The number in parentheses is an absolute maximum.

waters are common; hence, phosphorus should be removed from the reclaimed water. Swimming safety is a concern even if the water body is not specifically designated for primary recreation. Accidental or infrequent primary recreation by children is unavoidable, unless the water body is fenced off or is designated as not swimmable. Water quality requirements for some urban uses are presented in Table 6.3.

The specific treatment processes will be presented and discussed in the following sections. *Escherichia coli* bacterial criteria for recreation are based on U.S. surface water quality standards for infrequent/incidental swimming. The approximate equivalent of total fecal coliform bacteria density, which is used as a test for pathogens in many countries outside of the U.S., to the *Escherichia coli* would be a geometric mean of about 200 TFC/100 mL, which is the old U.S. standard. More stringent standards for fecal coliform, not used anymore nationwide, are still in force in California and some other states.

By presenting the required treatment processes in Figure 6.7, it is implied that the reclaimed water will be of a high quality. The selection of treatment will also depend on the source of reclaimed water, whether it is the total wastewater flow from the city, or separate flows in the ecocity. Specific unit processes will be introduced and discussed in Chapter VII. Gray water is treated mostly by physical processes: particle separation by microfiltration, which removes fine particles and colloids; nanofiltration, removing multivalent ions and large molecules; or reverse osmosis, removing dissolved compounds and ions. Due to its high concentrations of biodegradable organic matter (BOD, COD) black water is typically treated by anaerobic treatment and total wastewater (smaller BOD and COD concentrations) is treated by anoxic/aerobic treatment, followed by physical treatment (the term "aerobic" does not necessarily imply "aeration"). None of the required treatment unit processes listed in Figure 6.7 removes nitrogen significantly; however, reverse osmosis can remove nitrate ions. The treatment of used water in Battery Park (Box 6.1 and Figures 6.8 and 6.9) removes both nitrogen and phosphorus in an anoxic/aerobic

BOX 6.1   SOLAIRE APARTMENTS, BATTERY PARK, NEW YORK – NONPOTABLE WATER RECLAMATION AND REUSE

Building Type: Multiunit residential

Building Size: 33,160 m^2

Location: New York City, New York, United States

Occupancy Date: Completed August 2003

LEED Rating: U.S. Green Building Council, Platinum

This building has its own central water reclamation advanced wastewater treatment system with a capacity to recycle 95 m^3/day of reclaimed water for nonpotable uses of toilet and urinal flushing and air-conditioning cooling systems. In addition, a separate stormwater collection system captures and stores up to 45 m^3 of runoff from the building roof to irrigate the building's rooftop gardens. These water-saving measures, along with energy and water-efficient appliances, allow for 55% less potable water in the building than a residential development of similar size. The 33-story, 251-unit tower was in early 2000s considered to be the greenest high-rise residential condominium in the United States.

The water reclamation system collects, treats, and reuses all water used within the building—including "black water" from toilets and organic food waste from kitchen sink disposals. The used water treatment plant is located in the basement of the building (demonstrating that modern cluster water reclamation does not have to be remote). Figure 6.9 shows the schematic of the treatment plant. The black water first enters the collection tank, which also separates plastic and other inorganic solids. The sewage then undergoes an anoxic/aerobic biological treatment that removes biodegradable organics and nutrients. ZeeWeed ultrafiltration membranes are immersed directly into the bioreactor, which eliminates the need to settle solids, and significantly decreases the necessary size of the treatment tanks. Permeate pumps are used to pull the wastewater through membrane fibers. Each fiber is filled with billions of microscopic pores that physically block suspended solids, bacteria, and viruses from passing through—guaranteeing an exceptional water quality and clarity on a continuous basis. The treated water is then further disinfected by ultraviolet lights. Any remaining color and odor are removed using an ozone generator that also provides residual disinfection during water storage. The storage tanks serve as reservoirs for the treated water, which is used as flush water, as make-up water for the cooling towers, and for irrigation.

Typical water quality

BOD (mg/L)	raw water	230	treated	< 2
TP (mg/L)	raw water	10	treated	< 1
TN (mg/L)	raw water	45	Treated	< 3

In another reuse system, rainwater is collected from the green roof and used in the cooling system.

Source: General Electric, 2006; Engle, 2006

Figure 6.8 Solaire Apartments, Battery Park, in New York City. A LEED Platinum certified "green" building with nonpotable water reclamation and reuse.

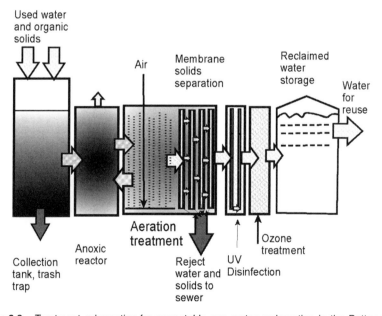

Figure 6.9 Treatment schematics for nonpotable use water reclamation in the Battery Park (New York City) development by Alliance Environmental LLC (based on Engle, 2006; General Electric, 2006, and other sources).

treatment process that produces methane and converts nitrogen to nitrogen gas. In other applications still employing the standard activated sludge or extended aeration, it is expected that most of the ammonium will be converted to nitrate, but without denitrification. Nitrate content may not be an issue unless nitrogen levels are controlled by the discharge limitations to control eutrophication if the reclaimed water becomes a significant effluent flow into the receiving waters. The European Community's Nitrate Directive limits nitrogen discharges to control eutrophication of the North, Baltic, Mediterranean, and Black Seas.

Box 6.1 presents an example of used water reclamation and reuse in the Solaire Apartments in the Battery Park neighborhood of New York City (Figure 6.8). This system could be considered a good example of sustainable cluster water management; however, the resource recovery is still incomplete in comparison to the current ecocity developments outside of the U.S. The treatment units of the water reclamation plant for the building are shown on Figure 6.9.

VI.3.4 Potable Reuse

At the onset of presenting the concepts of reclamation and reuse for potable use, one has to admit that there is still a lot of public resistance to the idea of potable reuse of used water that contained excreta before treatment. Even in Singapore, the NEWater, which is the reclaimed water suitable for direct potable reuse, is not directly reused for drinking. There had been a lot of hesitation among Singapore officials about the direct and indirect reuse, and until 2009 NEWater was primarily used for nonpotable purposes (Lee, 2005; personal observation by V. Novotny in 2009 and 2010). On the other hand, the indirect WSW reuse discussed previously has been accepted and is very common. Direct water reuse has been defined by the National Research Council (1998) as follows:

> *Direct potable reuse refers to the introduction of treated wastewater (after extensive processing beyond usual wastewater treatment) directly into a water distribution system without intervening storage.*

Hence, indirect potable reuse refers to projects that discharge recycled water to a surface or groundwater body before reuse and mix it with a cleaner freshwater source before additional treatment and reuse. Almost all potable water reuse systems have been based on indirect reuse. Even the first used water reclamation system for potable reuse mixes the reclaimed water with water from other sources, both surface and groundwater, and implements intercepting storage of the treated mixture before it is reused (see Box 6.2). The National Research Council (NRC) guidelines (1998) were even more direct and specified that "direct use of reclaimed wastewater for human consumption, without the added protection provided by storage in the environment, is not currently a viable option for public water supplies," and the committee writing the report focused only on indirect use, which refers to the intentional augmentation of a community's raw water supply with treated municipal wastewater. However,

BOX 6.2   DIRECT/INDIRECT POTABLE WATER REUSE IN NAMIBIA, AFRICA

The first indirect potable reuse in the world is currently practiced in Windhoek, the capital of Namibia. This city originally used direct potable reuse on an intermittent basis only. The city of 200,000 is located in a very arid environment, surrounded by the Namibia and the Kalahari Deserts. The population has been increasing at a rate of about 5% over the last 30 years. For its water supply, the city uses both groundwater resources and impounded rivers. However, the surface sources are ephemeral and cannot provide enough water. Furthermore, these streams are polluted, and, because of the dryness of the area, water stored in reservoirs is subjected to very high evaporation losses. Consequently, the city has focused on development of reuse of their used water as potable water.

With the help of the European Union, the U.S. Environmental Protection Agency, and the World Health Organization, the city developed and implemented in 1968 a system that took effluent water from the city's wastewater treatment facility and augmented it with flow from one of the city's reservoirs. The wastewater from residential and commercial settings is treated in the city's treatment plants by trickling filters and activated sludge with enhanced phosphorus removal. The effluents from each of these processes go to two separate maturation ponds for 4 to 12 days of polishing. Only the polished effluent from the activated sludge system is directed to the Windhoek reclamation facility, where it is mixed with water from the Goreangab Dam (blending ratio 1:3.5), to be treated to drinking water standards. After tertiary treatment, reclaimed water is blended again with bulk water from different sources.

Advanced treatment processes (including ozonizationand activated carbon) have been added to the initial separation processes of dissolved air flotation, sedimentation, and rapid sand filtration. A chlorine residual of 2 mg/L is provided in distribution systems. Membrane treatment has been considered, as well as an additional 140 days of storage of the secondary effluent from the maturation ponds in the Goreangab Dam.

The water reclamation plant originally produced 4800 m³/day of reclaimed water. This plant was recently replaced by the New Goreangab Water Reclamation Plant (NGWRP) that produces more than 40,000 m³/d. The plant practices a "multiple barrier" approach. Each barrier addresses more than one approach to each critical element of water treatment. At least three barriers are used for microbial pollutants, and at least two barriers are used for aesthetic conditions. In some extreme instances, several barriers may be used; for example, when treating for *Crytosporidium* removal, ozonation, enhanced coagulation, dissolved air flotation (DAF), dual media filtration, ultrafiltration and chlorination are used.

Sources: Lahnsteiner and Lempert, 2007; U.S. Environmental Protection Agency, 2004

at least two ecocity projects, Qingdao in China (see Figure 6.5) and Masdar in the United Arab Emirates (see Chapter XI) plan to reuse a portion of the reclaimed used water directly for drinking after advanced treatment.

The resistance to direct reuse may be slightly diminishing, and a more recent guideline report by the U.S. EPA (2004) concluded that direct potable reuse will seldom be necessary in the U.S., which could mean that it could be possible in very rare instances. It should be noted that only a small portion of the water used in a community needs to be of potable quality. While high-quality sources will often be inadequate to serve all urban needs in the future, the use of reclaimed water to replace potable-quality water for nonpotable purposes will release more high-quality potable water for future use. Typically, the reclaimed water might be added to a water course, lake, water supply reservoir, or underground aquifer and then withdrawn after mixing with the ambient water and undergoing modification by natural processes in the environment. The mix of reclaimed and ambient water would then be subjected to conventional water treatment before entering the community's distribution system (National Research Council, 1998).

In the potable reuse projects, multiple independent barriers that address a broad spectrum of microbiological and organic chemical contaminants must be implemented, and sufficient lag time between the reclamation and reuse must be provided to avoid contamination of tap water. These systems must be continuously chemically, bacteriologically, and toxicologically monitored in real time, and the lag time must be sufficient to avoid contamination. The NRC (1998) states that indirect potable reuse is an option of last resort. It should be adopted only if other measures— including other water sources, nonpotable reuse, and water conservation—have been evaluated and rejected as technically or economically infeasible.

Contaminants of Concern in Potable Water Reuse   Municipal wastewater could contain the following chemical and bacteriological contaminants of concern:

Chemical

1. Inorganic chemicals and natural organic matter that are naturally present in the potable water supply
2. Chemicals created by industrial, commercial, and other human activities in the wastewater service area
3. Chemicals that are added or generated during water and wastewater treatment and distribution processes, such as halogenated carcinogenic compounds

Pathogenic Microorganisms

1. Coliform group (fecal coliforms, *Escherichia coli*)
2. Salmonella
3. *Giardia*
4. *Cryptosporidium*
5. Viruses

According to the National Research Council (1998), organic chemicals in wastewater present one of the most difficult challenges a public health engineer or scientist faces in considering potable reuse. The challenges arise from the large number of compounds that may be present, the inability to analyze all of them, and the lack of toxicity information for many of the compounds. Efforts to account for the total mass of organic carbon in water are further frustrated by the fact that the bulk of this material is aquatic humus, which varies slightly in structure and composition from one molecule to the next and cannot be identified like conventional organic compounds. These challenges are not unique to potable reuse systems. In fact, the most protected water supplies are those for which the smallest fraction of the organic material can be identified. For potable reuse systems, however, anthropogenic organic compounds pose the greatest concern and should be the major focus of monitoring and control efforts.

VI.3.5   Groundwater Recharge

This section addresses planned groundwater recharge using reclaimed water with the specific intent of replenishing groundwater that, in the ecocities, could then be reused for potable and nonpotable uses. Although practices such as irrigation may contribute to groundwater augmentation by excess irrigation return flow, replenishment is a byproduct of the primary activity of irrigating crops, and irrigation return flow recharging the aquifer contains more dissolved pollutants (for example, nitrates) than the irrigation water. In cases where water is injected into groundwater zones for future reuse, evapotranspiration effects and losses must be minimized. Figure 6.10 shows three infiltration methods for injecting reclaimed water into groundwater aquifers. Recharge from infiltration basins in arid areas is not recommended because of the losses by evaporation.

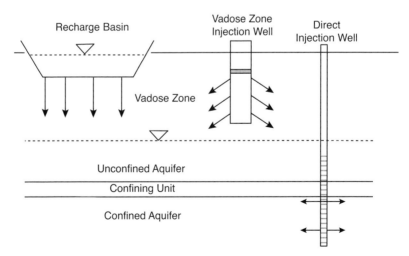

Figure 6.10   Three methods for recharging reclaimed water into aquifers (U.S. EPA, 2004).

The purposes of groundwater recharge using reclaimed water may be (U.S. EPA, 2004):

1. To establish saltwater intrusion barriers in coastal aquifers
2. To provide further treatment for future reuse
3. To augment potable or nonpotable aquifers
4. To provide storage of reclaimed water for subsequent retrieval and reuse
5. To control or prevent ground subsidence

All five reasons are applicable in the context of sustainable ecocity development.

Infiltration and percolation of reclaimed water take advantage of the natural removal mechanisms within soils, including biodegradation and filtration, thus providing the overall wastewater management system with additional in situ treatment of reclaimed water and also additional barriers and reliability before the reuse. The treatment achieved in the subsurface environment may reduce the need for costly advanced wastewater treatment processes.

Aquifers provide a natural mechanism for the storage and subsurface transmission of reclaimed water. Irrigation demands for reclaimed water are often seasonal, requiring either large storage facilities or alternative means of disposal when demands are low. Aquifer storage and recovery (ASR) systems are being used in a number of states to overcome seasonal imbalances in both potable and reclaimed water projects. The large volumes of storage potentially available in ASR systems mean that a greater percentage of the resource, be it raw water or reclaimed water, can be captured for beneficial use. Aquifer storage may also provide a long lag time between the time of reclaiming the water and the time of reuse, adding additional safety for the reuse.

Groundwater recharge with reclaimed used water is carried out in Southern California (Box 6.3). The largest project has been implemented in Orange County, south of Los Angeles, in San Diego, and in other communities.

BOX 6.3 INDIRECT NONPOTABLE VS. POTABLE REUSE STUDY IN SAN DIEGO, CALIFORNIA

The City of San Diego has conducted a study on indirect reuse spurred by the growing population of the metropolis and a lack of natural water sources. Currently, most water is imported by an aqueduct from the Colorado River and by the California State Water Project from Northern California. Court decisions have reduced California's allocation of water from the Colorado River. At the beginning of this millennium, effluent from the secondary wastewater treatment plant was discharged into the ocean by a deep outfall.

(*continued*)

In 1997 the city built a water reclamation plant for nonpotable use, providing the northern city service area with 114,000 m³/day of reclaimed water for unrestricted urban use, followed in 2002 by a 57,000 m³/day capacity southern service area water reclamation plant. However, the city soon realized that building a separate distribution system for reclaimed water, connected to a centralized water reclamation plant, would be prohibitively expensive; hence, the city initiated a study to decentralize and diversify its sources of reclaimed water and increase the use of locally reclaimed water. The city focused on two alternatives (Metcalf & Eddy, 2007):

1. Diversified nonpotable use
2. Indirect potable use

The reclamation treatment sequence operated in San Diego, after secondary treatment, includes the following processes:

- Coagulation with ferric chloride
- Multimedia filtration
- Ultraviolet disinfection
- pH adjustment with sulfuric acid
- Cartridge filter
- Reverse osmosis

Microbial analysis, performed over a 2.5-year period, showed that the water quality of the advanced wastewater treatment effluent was low in infectious agents. Specifically, research showed (City of San Diego, 2006; U.S. EPA, 2004):

- An overall virus removal rate through the primary, secondary, and advanced wastewater treatment plants of between 99.9999% and 99.99999%. Levels of removal were influenced by the number of viruses introduced.
- Enteric bacterial pathogens (that is, *Salmonella*, *Shigella*, and *Campylobacter*) were not detected in 51 samples of advanced wastewater treatment effluent.
- Protozoa and metazoa of various types were absent in the advanced treatment effluent.
- *Giardia lamblia* were not recovered, and based on recovery rates of cysts from raw wastewater, removal rates were estimated to be 99.9%.
- All measured constituents listed in the drinking water quality standards were below the limit.

In spite of these high removals, the produced reclaimed water may need additional barriers and increased lag time between reclamation and reuse.

Source: City of San Diego, 2006

Orange County, California. Orange County imports water from the Colorado River, Northern California, and the Santa Ana River, which contains highly treated sewage from the communities located on the river. Orange County could not exist without these imported water sources containing treated sewage effluents, such as that from Las Vegas, Nevada, on the Colorado River. The Santa Ana River, in several sections, is essentially either a perennial effluent dominated stream or an ephemeral water body. Generally, these upstream communities use tertiary treatment for their sewage purification before discharging the effluent into the receiving water bodies. When this water reaches Orange County, it is again purified before being distributed to residents. But these sources are not sufficient to provide enough water to the growing population.

CDM designed and Orange County built the new Groundwater Replenishment (GWR) System that implements a high-efficiency purification system that receives on average 325,000 m^3/day of secondary treated sewage effluent from the Orange County Water District. The effluent in the GWR unit process is filtered by microfiltration, which uses 60,000 m^3/day (18.5%) for backwashing. The remaining 265,000 m^3/day is treated by reverse osmosis and disinfected by ultraviolet light. Hydrogen peroxide is added to oxidize the remaining trace organics and improve taste and odor before the highly purified effluent is injected into the groundwater aquifer for further potable and nonpotable reuse (Figure 6.11). The water before injection has a quality similar to or better than bottled water and meets all drinking

Figure 6.11   Reverse osmosis water reclamation plant, treating used water of the Orange County, California, GWR system (photo courtesy CDM).

water quality standards. Further treatment is provided by percolation through the aquifer. Aquifer percolation improves the taste of water by adding "good" minerals that were lost in the reverse osmosis process.

VI.3.6   Integrated Reclamation and Reuse—Singapore

Singapore, officially the Republic of Singapore, is a small city-state located on an island at the tip of the Malay (Malaysia) Peninsula. With an area of 719 km^2 and a population of about 4.8 million, the country is one of the few small independent city-states in the world, the others being Vatican City, Monaco, and San Marino. In the early 19th century, due to its strategic location, Singapore became one of the most important commercial and military centers of the British Empire and the hub of British power in southeast Asia. In 1965, Singapore became an independent city-state. Since gaining independence, Singapore's standard of living has risen dramatically. Foreign direct investment and a state-led drive to industrialization have created a modern economy focused on industry, education, and urban planning. Singapore is the fifth-wealthiest country in the world in terms of GDP (purchasing power parity) per capita (World Bank, 2008). The island is located about 137 km north of the equator between Malaysia and Indonesia.

In spite of its relatively large rainfall (2400 mm/year), Singapore has insufficient water resources to satisfy the daily water demand of 1.36 million m^3, which translates to about 180 liters/capita-day. Water scarcity is not due to lack of rainfall, but to limited area where water can be captured and stored (Tortajada, 2006). Before gaining independence, Singapore negotiated two treaties with Malaysia to import water from the Malaysian state of Johor at a relatively low price of 0.25 cent per m^3. The water from Malaysia is brought by three pipelines over the Straits of Johor. The treaties will expire in 2011 and 2061. However, in 2009–2010, negotiations with Malaysia for extending the treaties under the same favorable conditions were at a stalemate.

For this reason and for the sake of strategic water self-sufficiency, Singapore has developed a highly sophisticated plan and system of water and used water capture, reclamation, storage, treatment, and reuse. The water management system is comprehensive and relies heavily on the most advanced infrastructure and the latest developments in water/used water treatment. Water management is conducted by the Public Utilities Board (PUB), which manages the entire water cycle of Singapore. This enabled the PUB to develop a holistic policy, which includes protection and expansion of water resources, stormwater management, desalination, demand management, catchment management, and community-driven education about water, desalination, and used water collection and treatment (Tortajada, 2006). The country is fully sewered and collects all used water and practices water reuse on a large scale.

Figure 6.12 is a schematic of the water management system in Singapore. The sources of water for Singapore are divided into:

1. *Catchment water.* Since 1976 Singapore has promulgated a system of protected catchments that provide water supply to the island. Fully protected catchments are demarcated, and no polluting activity is allowed in the

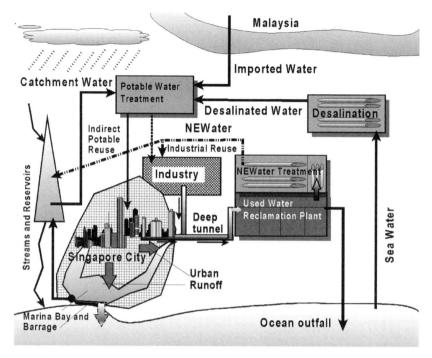

Figure 6.12 Schematic of water sources and comprehensive water/stormwater/used water management in Singapore.

protected areas. These fully protected catchments cover about 5% of the island area. Partially protected catchments now cover almost two-thirds of the island area. The latest addition to source water was the Marina Barrage (Figure 6.13) and pumping station, which converted a tidal estuary of the Singapore and Kallang Rivers into a freshwater body from which water is delivered to the water supply reservoir to augment the available catchment water.

2. *Imported water* is the fresh water delivered from Malaysia, in accordance with the treaties.

3. *Desalinated water.* The first large desalination plant, with a capacity of 113,400 m³ of water per day (30 mgd), was put in operation in 2005. The current technology, developed and installed locally, enables it to produce freshwater at an affordable price of U.S. $0.55/m³ (Lee, 2005).

4. *NEWater.* All used water from the island is collected by sewers and brought by a tunnel or interceptors to regional water reclamation plants. After further advanced treatment reclaimed water is purified to potable water quality.

Used water reclamation (Bufe, 2009) is done by full sewering of the island, which includes 600 km of link sewers ranging in diameter from 0.3 to 3 meters. The sewers convey collected used water into a 48-km-long deep tunnel with a diameter

Figure 6.13 Marina Barrage and pumping station at the mouth of the Singapore River changed the tidal estuary into a freshwater body collecting treated urban runoff (to the left of the barrage). Water is pumped into upstream water supply reservoirs to augment the available fresh catchment water. The green roof of the pumping station is used as a vista and for picnicking (Photo courtesy CDM.).

ranging from 3 to 6 meters. The new water reclamation plant in Changi has been designed with capacity to treat 800,000 m³ of used water daily. The Changi plant is compact and replaced several previous reclamation plants, which freed large areas for other development. The treatment plant designed by CH2M HILL is a state-of-the-art plant consisting of (1) an influent pumping station that pumps influent from the 60-meter-deep tunnel, (2) headworks (fine screens, grit removal, and oil and grease removal), (3) a liquid treatment module (primary sedimentation, anoxic step-feed bioreactors, secondary sedimentation tanks), (4) a solids treatment facility (thickening centrifuges, sludge-blending tanks, anaerobic digesters, dewatering centrifuges, sludge-drying system), and (5) two 5-km-long sea outfalls. The water reclamation plant produces high-quality effluent that can be safely discharged into the sea. The tunnel and the new Changi water reclamation plant, designed and built by CH2M HILL, was completed in 2008 and inaugurated in June 2009.

NEWater is an ultra-clean water produced from reclaimed water by an advanced physical multibarrier treatment scheme consisting of *membrane microfiltration*, which removes all suspended particles. The effluent from microfiltration contains only dissolved salts and organic molecules. Microfiltration is followed by *reverse osmosis*, which uses a semipermeable membrane allowing only water molecules to pass through. Undesirable substances such as bacteria (mostly removed by microfiltration), viruses, and large molecules and ions (e.g., nitrates) cannot penetrate the membrane. Disinfection is accomplished by ultraviolet light, which acts as a safety barrier, because it is assumed that all pathogenic microorganisms and viruses are removed by reverse osmosis. The final step is adjustment of the chemistry of the NEWater product to restore the pH balance. The NEWater treatment was also designed by CH2M HILL, with technology by Siemens.

About 30% of the influent flow into the NEWater production goes to filter backwash and brine wastewater from the reverse osmosis. These flows are returned to the influent of the water reclamation facility. In Singapore, the newest NEWater production plant will be located above the Changi water reclamation plant, which saves a lot of space. The overall toxicological status of reclaimed water before reuse is being detected by flow-through aquaria with sensitive fish (e.g., trout). The fish should show no distress, and after one year of living in an aquarium their fatty tissues can be analyzed for excessive accumulation of organic compounds.

Since 2003, NEWater has been used for production of nonpotable water in manufacturing and for other nonpotable reuse purposes such as cooling and toilet flushing in commercial buildings. This has dramatically reduced the potable water demand. A small amount of NEWater is blended with raw water from the reservoirs and then sent for further treatment at the waterworks before being supplied to consumers as tap water in the indirect potable reuse schemes. NEWater is meeting 30% of the daily Singapore water needs.

Marina Barrage. This project completed in 2008 was designed and built by the CDM. The project had three major objectives:

1. Water supply. The barrage is built across the 350-meter wide Marina Channel at the mouth of the Singapore River. It converted a previously tidal brackish estuary into a freshwater pool receiving river flow and urban runoff. Freshwater will be pumped into a reservoir and, after blending, treated with advanced membrane technology. This has increased the available catchment water. Marina water will provide more than 10% of the Singapore water demand.
2. Flood control. The Marina barrage provides storage of flood water and alleviates flooding by high tides and floods in the city's low-lying areas.
3. Recreation and lifestyle attraction. The cleanup of the Singapore River flow, urban runoff, and the fact that the water level in the bay and former estuary will by unaffected by tides make the bay and the upstream estuarine part of the river entering the bay an ideal place for active (water skiing, tourist boat traffic) and passive (stream bank restaurants and entertainment) activities (Figure 6.14).

Singapore today has the most advanced system of water management, reclamation, and reuse in the world on a large-city scale. Although NEWater is not applied to direct reuse, it is perfectly safe for drinking and other domestic uses. In spite

Figure 6.14 Marina Bay in Singapore with the barrage in the background. The area between the bridge and the barrage is a fill being converted to a park. Photo by V. Novotny.

of severe natural water shortages, the city-state has secured adequate water supply for its current and future needs. The system is heavy on infrastructure and energy-demanding (use of energy recovery and renewable energy sources were not fully considered), but, judging from the enormous technological capability of this small nation, one should not be surprised if the Singapore engineers address the global warming problem with the same vitality as they did the water issue. Singapore has become a worldwide hub of the water industry.

REFERENCES

Bufe, M. (2009) "In Singapore, used water is NEW," *Water Environment & Technology,* 21(8):16–20

Burton, F. L. (1996) *Water and WasteWater Industries: Characterization and Energy Management Opportunities*, CR-10691, Electric Power Institute, St. Louis, MO

City of San Diego (2006) *City of San Diego Water Reuse Study 2005: Final Draft Report*, Water Department, City of San Diego, CA

Daigger, G. (2009) Evolving urban water and residuals management paradigms: Water reclamation and reuse, decentralization, resource recovery, *Water Environment Research* 81(8):809–823

Engle, D. (2006) "Onsite treatment, City-style," *Onsite Water Treatment*, May-June, Forester Media, http://www.onsitewater.com/ow_0605_treatment.html (accessed December 2009)

Fraker, H., Jr. (2008) "The Ecoblock-China Sustainable Neighborhood Project," PowerPoint presentation, the second Connected Urban Development Conference, September 24, 2008, Amsterdam, http://bie.berkeley.edu/ecoblocks

General Electric (2006) *Solaire Apartments Battery Park – Case Study*, GE Water & Process Technologies, Trevose, PA

Imhoff, K. (1931) "Possibilities and limits of the water-sewage-water-cycle," *Engineering News Report*, May 28

Imhoff, K. and K. R. Imhoff (2007) *Taschenbuch der Stadtentwässerung* (*Pocket Book of Urban Drainage*), 30th ed., R. Oldenburg Verlag, Munich, Germany

Keller, J. and K. Hartley (2003) "Greenhouse gas production in wastewater treatment: Process selection is the major factor," *Water Science and Technology* 47(12):43–48

Kennedy, C., J. Cuddihy, and J. Engel – Yan (2007) The changing metabolism of cities, *Journal of Industrial Ecology*, 11(2):43–59

Koenig, H. W., G. Rincke, and K. R. Imhoff (1971) "Water re-use in the Ruhr Valley with particular reference to 1959 drought period," Pap I-4, *Proceedings of the IAWQ Water Pollution Research Conference*, San Francisco, CA, in *Advances In Water Pollution Research*, Pergamon Press, Oxford, UK

Krejci V. and E. Baer (1990) "Screening structures for combined sewer overflow treatment in Switzerland," *Proceedings of the Fifth International Conference on Urban Storm Drainage*, IAWPRC-IAHR, Suita-Osaka, Japan, pp. 927–932

Krenkel, P. and V. Novotny (1980) *Water Quality Management*, Academic Press, Orlando, FL

Lahnsteiner, J. and G. Lempert (2007) "Water management in Windhoek, Namibia," *Water Science and Technology* 55, pp. 441–448

Lee, P. O. (2005) "Water management issues in Singapore," Paper presented in Mainland Southeast Asia, November 29–December 2, 2005, Siem Reap, Cambodia. International Institute for Asian Studies and the Center for Khmer Studies, Cambodia

Levine, A. D., and T. Asano (2004) "Recovering sustainable water from wastewater," *Environmental Science and Technology* 38(11):201A–208A

Liu, H., R. Rammarayanan, and B. Logan (2004) Prodiction of electricity during wastewater treatment using a single chamber microbial fuel cell, *Environ. Sci. Technol.* 38:2181–2285

Metcalf & Eddy (2003) *Wastewater Engineering: Treatment and Reuse*, 4th ed. (revised by G. Tchobanoglous, F. L. Burton, and H. D. Stensel), McGraw-Hill, New York

Metcalf & Eddy (2007) *Water Reuse: Issues, Technologies, and Applications*, McGraw-Hill, New York

Mitchell, C., K. Abeysuriya, and D. Fam (2008) *Development of Qualitative Decentralized System Concepts – For the 2009 Metropolitan Sewerage Agency*, Institute for Sustainable Futures, University of Technology, Sydney, Australia

National Research Council (1998) Issues in Potable Reuse: The Viability of Augmenting Drinking Water Supplies with Reclaimed Water, National Academies Press, Washington, DC

Novotny, V. (2003) *WATER QUALITY: Diffuse Pollution and Watershed Management*, 2nd ed., John Wiley & Sons, Hoboken, NJ

Novotny, V., K. R. Imhoff, M. Olthof, and P. A. Krenkel (1989) *Handbook of Urban Drainage and Wastewater Disposal*, John Wiley & Sons, Hoboken, NJ

Novotny, V. and H. Olem (1994) *WATER QUALITY: Prevention, Identification and Management of Diffuse Pollution*, Van Nostrand Reinhold, New York, NY, 1994,reprinted 1997 by John Wiley & Sons, Hoboken, NJ

Rees, W. E. (1997) Urban ecosystems: the human dimension, *Urban Ecosystems* 1:63–75

Rowe, D.R. and I.M. Abdel-Magid (1995) Handbook of Wastewater Reclamation and Reuse. CRC Press, Inc., Boca Raton, FL, 550pp.

Tortajada, C. (2006) "Water management in Singapore," *Water Resources Development* 22(2):227–240

U.S. Department of Energy (2000) *Carbon Dioxide Emissions from the Generation of Electric Power in the United States*, U.S. DOE, Washington, DC

U.S. Environmental Protection Agency (1986) *Ambient Water Quality Criteria for Bacteria*, EPA 440/5-89-002, Office of Regulations and Standards, Washington, DC

U.S. Environmental Protection Agency (1994) *Water Quality Standards Handbook*, 2nd ed., EPA-823-b-94-005A, Office of Water, Washington, DC

U.S. Environmental Protection Agency (2000) *Improved Enumeration Methods for the Recreational Water Quality Indicators: Enterococci and Escherichia Coli*, EPA/821/R-97/004, Office of Science and Technology, Washington, DC

U.S. Environmental Protection Agency (2004) *Guidelines for Water Reuse*, EPA/625/R-04/108, Office of Wastewater Management, Office of Water, Washington, DC

Wackernagel, M. and W. Rees (1996) *Our Ecological Footprint: Reducing Human Impact on the Earth*, New Society Publishers, Gabriela Island, BC, Canada

WEF (1996) *Wastewater Disinfection, Manual of Practice FD-10*, Water Environment Federation, Alexandria, VA

World Bank (2001) *Water and Sanitation Programs, 99-00 Report*, Water and Sanitation Program, Washington, DC

World Bank (2008) World Development Data Base – PPP and Population, http://siteresources.worldbank.org/DATASTATISTICS/Resources/GDP_PPP.pdf, accessed May, 2010

World Wildlife Fund (2008) "One Planet Living," http://www.oneplanetliving.org/index.html

VII

TREATMENT AND RESOURCE RECOVERY UNIT PROCESSES

VII.1 BRIEF DESCRIPTION OF TRADITIONAL WATER AND RESOURCE RECLAMATION TECHNOLOGIES

VII.1.1 Basic Requirements

Based on the discussion in the preceding chapter on the water reclamation– reuse sequence, water is treated by two trains of treatment unit processes separated by storage that provides lag time. Traditionally, the *first water reclamation treatment sequence* involves (1) primary biological treatment, removing large and smaller settleable solids such as grit (if present) and organic and inorganic particulate matter, and (2) secondary biological treatment, removing biodegradable organics and nutrients (if needed). At a minimum, high-efficiency secondary biological treatment should be provided before storage for reuse. Additional tertiary steps such as high-efficiency filtration or coagulation/precipitation may be required to remove toxic compounds, residual particulate matter, and pathogens to meet the surface water quality and groundwater disposal standards and criteria. Effluent quality from this first treatment sequence must comply with effluent standards, and if the effluent is discharged into water-quality-limited water bodies, more stringent effluent limitations can be imposed by the Total Maximum Daily Load planning process and limitations (U.S. EPA, 1991; Novotny, 2003). Water-quality-limited classification arises in the U.S. when the effluent limitations, derived from the effluent criteria, are not sufficient to achieve the designated status of the receiving water body, as expressed by the ambient water quality standards for that use. Furthermore, in many situations in which water reclamation and reuse are considered, excess reclaimed used water may be discharged into water bodies that could be naturally or anthropogenically ephemeral (see Chapters V and IX). In such cases, reclaimed water should be at the level of surface water quality and should comply with the surface water criteria and standards. After the first sequence of treatment unit processes, the water is (1) stored in natural surface water bodies or artificial surface ponds or reservoirs, and/or (2) injected into alluvial deposits along the river or other shallow aquifer from which water is withdrawn for potable use after treatment, and/or (3) injected from wells into deep underground aquifers. Water from this sequence of treatment unit processes can also be reused for irrigation and toilet flushing.

In the *second indirect potable water reuse treatment sequence*, water from the storage is withdrawn and treated again, by physical/chemical treatment processes, to potable water quality. Reclaimed used water should only be a supplement to the flow of raw water from a better and cleaner source, which can be either a natural surface water or groundwater resource or desalinated seawater. The level of treatment depends on the type of reuse. The second sequence of treatment and preparation of potable water provides treatment efficiencies that are commensurate with conventional potable water treatment or better. The water quality of the water produced for various uses, including indirect potable reuse, has been discussed in Chapter VI (Section VI.3). Potable reuse must comply with drinking water quality standards. Because the reclaimed used water contained microorganisms and potentially pathogens extra barriers and precautions must be added if used water is considered for potable reuse.

Injection of reclaimed water into deeper underground aquifers (in contrast to shallow alluvial aquifers) is probably the safest method, from the public health protection standpoint, providing needed lag time between the time of reclamation and potable reuse. While discharging/injecting reclaimed water into deeper aquifers may provide water of superior quality, pumping the water back to the surface for reuse may be energy demanding. The same formula given in Section VI.3.2 for sprinkler design can be used to calculate energy use, except that the pressure is replaced by the total pumping head, H_t (elevation difference between point of use and the aquifer groundwater table plus all friction losses in the pipeline in meters) times the specific weight of water, γ, or

$$p = \gamma H_t$$

where

$\gamma = 9810 \ \text{N/m}^3$ for clean 4°C cold freshwater.

VII.1.2   Considering Source Separation

Typical urban sewage contains the following quantities of key quality constituents expressed in mass (weight) per capita (Imhoff and Imhoff, 2007)

	P	N	TOC	TSS
	grams/capita-day			
Raw sewage	1.8	11	40	90
After biological treatment	1.1	7.6	6	10
Incorporated into sludge	0.7	1.7	6.8	variable

This indicates that about 70% of the incoming organic carbon in raw sewage is converted to carbon dioxide in a typical aerobic treatment process. Also, most phosphorus and nitrogen removals are only due to sludge uptake. Biochemical oxygen demand (total BOD) to total organic carbon (TOC) ratios are highly variable; theoretically BOD/TOC = 1.85. While concentrations of the above compounds will vary among nations, according to water use, the per capita loads are more stable and vary less.

In Section II.3.3 of Chapter II we defined several used water flows from urban areas, namely (1) black water, containing urine and feces from toilets, lavatories, and shredded solids from kitchen sinks, (2) gray water, containing used water from baths, showers, kitchen dishwashers, and washing machines, (3) white water (runoff), and (4) blue water, which is rainwater and groundwater. It was also emphasized that mixing black and gray water (sewage) with white and blue water should be avoided and remedied when it happens in separate sewer systems.

Recently, further separation of black water into urine (yellow) and feces (brown) used water streams has been introduced for consideration for the following reasons:

- Urine contains 80% of the nitrogen, 70% of the potassium, and about 50% of the phosphorus in less than 1% of the used water flow and has no pathogens; it is sterile (Mitchell, Abeysuriya, and Fam 2008). Nitrogen and phosphorus recovery is relatively simple by converting ammonium and phosphate into struvite mineral solids, which are easy to handle (Ganrot, 2005; Ganrot, Slivka, and Göran, 2008; Tettey-Lowor, 2009). Struvite can also be recovered from the supernatant of the sludge digestion process (see more detailed discussion in "Nutrient Recovery," Section VII.3 in this chapter, and in Chapter VIII).

- Conventional nitrification, followed by denitrification that removes nitrogen but does not recover it, requires 60% to 80% more energy than is needed for a process without nitrogen removal (Mauer, Schweigler, and Larsen (2003). These authors also provide overall energy requirements for nitrogen and phosphorus removal and recovery processes and struvite production and suggest treatment schemes. Their detailed analysis shows that in many cases recovery of N and P from used water (with or without triplicate collection systems) is more energy efficient than simple removal and new N and P fertilizer production from natural sources.

Zheng, Tu, and Fei (2010) provided the following information of concentrations in undiluted urine:

pH	9.1
Total nitrogen, mg/L	9200
Ammonium N, mg/L	8100
Total phosphorus, mg/L	500–800
COD, mg/L	1000

Urine concentrations from household and public toilets with flushing would be diluted by a factor 3 to 4. Urine separation would require triplicate used water separation (brown, yellow, and gray) collection and conveyance systems. Tettey-Lowor (2009) describes an experimental reuse of urine from public urinals in Accra, Ghana, on suburban farms. Considering the population of Accra that uses public toilets, the amount of nutrients (nitrogen, phosphorus, and potassium) obtainable from these toilets would suffice to provide all the fertilizing needs of the suburban agriculture of the Ghana capital city. The problem indicated in the report was the diluted nature of nutrients in urine, which would make the operation of collecting, transporting, and spraying urine uneconomical. Although urine is sterile, spraying it would be objectionable and odorous. Economical methods of concentration and nutrient recovery from urine separation must be developed. An alternative would be not to separate black used water into yellow and brown, but to maximize incorporation of N and P into sludge—for example, by adding inexpensive phosphate-precipitating chemicals. Subsequently most of the N and P could be recovered as a fertilizer either (1) by sludge drying and pasteurization by heating (energy demanding) conversion into fertilizer solids (e.g., Milorganite in Milwaukee, Wisconsin), or (2) by a recovery of N and P from the supernatant of the sludge digestion or upflow anaerobic sludge blanket (UASB) processes by struvite crystallization.

Information on concentration strength of gray water published by Li, Wichmann, and Otterpohl (2009) indicates that gray water cannot be reused without treatment. Typical concentrations for mixed gray water (kitchen dishwasher, bathroom, laundry) are:

pH	6.3–8.1
Total suspended solids (mg/L)	25–183
COD (mg/L)	100–700
BOD (mg/L)	47–466
Total nitrogen (mg/L)	1.7–34.3
Total phosphorus (mg/L)	0.11–22.8
Fecal coliform bacteria (CFU/100 ml)	$0.1–1.5 \times 10^8$

Henze and Comeau (2008) reported that raw wastewater from settlements with a high level of water conservation or simply water shortage at a level of used water discharge of 80 L/cap-day would have chemical oxygen demand (COD) of untreated discharge of 2750 mg/L and biochemical oxygen demand (BOD) of 1125 mg/L. This would essentially be a concentration of black water.

Malmqvist and Heinicke (2007) reported on black water separation in Göteborg, which obviously means that a second sewer pipe is installed in each household, collecting only the water from the toilets and later kitchen sinks because household in sink garbage disposals are just being introduced in the city. The black water flows are transported in pipes to four local plants in the outskirts of Göteborg, in which the water is concentrated about 10 times and sanitized before being transported to farms for direct use in agriculture. This system is introduced in the "New Göteborg," that is, the parts of Göteborg where new buildings are expected to be built until 2050, and in districts where far-reaching renovations to old buildings will be carried out. Altogether, about 70% of the inhabitants in Göteborg by 2050 will live in such buildings.

VII.1.3 Low-Energy Secondary Treatment

In this section we will focus on reducing energy use and GHG emissions.

Secondary treatment of used water can be done by numerous treatment processes that are generally divided into the following categories:

Primary Treatment Removing Inorganic and Organic Settleable Particulate Matter

- Screening and grit removal
- Primary settling in ponds and clarifiers
- Fine-screen solids separation
- Air flotation
- Septic tanks

Secondary (Biological) Treatment   In the biological processes, biodegradable organic matter expressed as BOD (or TOC) and removed by the biochemical action of microorganisms is converted to either carbon dioxide and water in aerobic processes or carbon dioxide, methane, and water in anaerobic processes. In these treatment processes, new bacteria cells (biomass) are produced. Secondary biological treatment units are aerobic, anaerobic, or anoxic, and can be categorized as:

- Aerobic processes (for BOD and suspended solids removals, conversion of ammonium to nitrate, small total nitrogen and phosphorus removals)
  - Suspended growth flow-through mixed reactors such as various modifications of the activated sludge processes, including membrane bioreactors (MBR)
  - Suspended growths in upflow clarifiers
  - Sequencing batch reactors
  - Attached growth trickling filters, packed media units, and submerged attached growth units
  - Combined activated sludge and attached growth process units
- Anoxic/aerobic processes (for BOD, TSS, and nitrogen removal; small phosphorus removal without an anaerobic step or chemical precipitation)
  - Suspended growth flow-through mixed reactors with a high recycle
  - Sequencing batch reactors with modified operation
  - Upflow or downflow packed bed reactors
  - Fluidized bed reactors
  - Subsurface flow wetlands

- Anaerobic/anoxic/aerobic processes (for BOD, TSS, nitrogen and phosphorus removals)
  - Suspended growth such as upflow anaerobic sludge blanket reactor
  - Sequencing batch reactors
  - Membrane systems (in addition to BOD, TSS, TN, and TP removals these reactors also remove colloidal solids)

Constructed wetlands are special biological treatment units that involve several physical and biological processes. For more detailed information and design of treatment processes, the reader is referred to standard wastewater treatment texts such as those published by Metcalf & Eddy (2003, 2007) or Imhoff and Imhoff (2007).

Figure 6.2A in the preceding chapter shows the traditional secondary activated sludge wastewater treatment plant with conventional sludge handling. In the treatment process, organic and inorganic suspended solids are removed first by settling in the primary sedimentation unit process, and in the secondary aerobic step the suspended and soluble organics are decomposed, oxidized by bacteria into carbon dioxide (CO_2) and water, producing new bacterial cell biomass. In a typical municipal sewage treatment, nitrogen and phosphorus serve as the necessary nutrients for bacterial growth and are incorporated into new growth cells, the sludge yield. The organic content of the incoming wastewater (sewage) is measured by the chemical oxygen demand and/or biochemical oxygen demand and/or total or dissolved organic carbon (TOC and DOC). Conventional single-step aerobic and anaerobic biological treatment processes are not efficient for removing nitrogen and phosphorus, which are only removed by incorporation into the excess sludge produced in the process. Nitrogen is present in sewage in a form of organic particulate nitrogen and ammonium ion measured as total Kjeldahl nitrogen (TKN), which can be incorporated into biological organic matter or adsorbed on particulate matter (e.g., ammonium) or dissolved. Phosphorus is also incorporated into biological organic solids, adsorbed, or dissolved.

In an anoxic/anaerobic treatment process, oxygen is not provided, and the bacteria decompose the incoming organics into carbon dioxide (using the oxygen present in the organic molecules or from nitrates and sulfates) and methane (CH_4). Primary and secondary sludge containing the settled solids and excess sludge growth is further thickened, and then anaerobically or aerobically digested to reduce the organic mass of the solids. The residual solids are dewatered and disposed of on land or incinerated. Aerobic digestion requires oxygen, and by-products of the process are again carbon dioxide, water, and residual solids. In an anaerobic digestion process, the by-products are methane, carbon dioxide, residual solids, and nutrient, and dissolved organic-matter-rich supernatant liquid that is generally recycled back to the biological treatment step.

Various combinations of the secondary unit processes have been used for treatment of municipal and industrial wastes, and the reader is referred to the appropriate technical literature such as Metcalf & Eddy (2003, 2007) or Imhoff and Imhoff (2007).

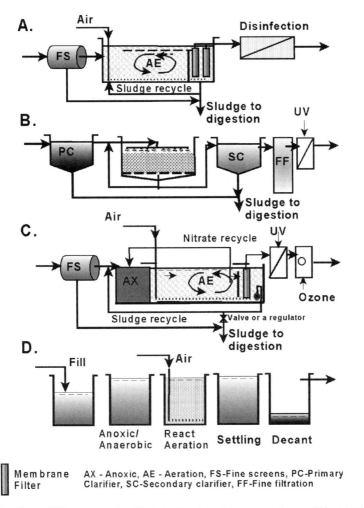

Figure 7.1   Four additional alternative biological treatment processes that could be considered in used water treatment and reuse for treating mixed sewage or separated black water: (A) membrane bioreactor, a modification of the conventional activated sludge process, (B) trickling filter, (C) Bardenpho process for BOD, TSS, and nitrogen and phosphorus removal, and (D) sequencing batch reactor. Adapted from Metcalf & Eddy (2007).

Four examples of biological treatment unit processes are depicted in Figure 7.1. System A is known as a *membrane bioreactor (MBR)*, which is a combination of the conventional activated sludge system (shown in Figure 6.2A) with a membrane liquid separation process. Membrane filters consist of a bundle of long hollow fibers with pore diameters ranging from 0.04 to 0.4 µm (Metcalf & Eddy, 2007), hence such filters can remove colloidal particles and microorganisms. The filter unit can be installed in the aeration tank or in a separate tank, and filtration is accomplished by creating a small negative pressure differential (Metcalf & Eddy, 2007).

Manufacturers and experience with this membrane technology indicate that, if properly designed and installed, MBR systems are robust to withstand variations in operating conditions due to fluctuating sludge concentrations. In this configuration, however, the system will require energy. Therefore, it is necessary to use methane produced by sludge digestion to reduce the energy demand.

The second (B) process is a *conventional trickling filter*. In the trickling filter, rock or lately plastic media provide conditions for the microorganisms to develop as a slime layer on the surface of the media, over which water is dispensed by a sprinkler system rotated by the hydraulic inertia force of the water exiting horizontally from the sprinkler nozzles. Trickling filters do not typically provide water quality commensurate with potable or high-quality-demanding nonpotable reuse. However, in combination with microfiltration or membrane filtration and disinfection, they can provide effluent of adequate quality for irrigation, street cleaning, or supplemental ecological flow. The trickling filter with a recycle can be designed to provide BOD removal and nitrification. Low or no energy use by trickling filters is the main advantage in the context of minimizing carbon footprint.

System (C) is the *activated sludge process removing BOD (COD), TSS, total nitrogen*, and some phosphorus. It consists of an anoxic tank followed by an aeration tank with membrane microfilters. Besides suspended solids and biodegradable organics, nitrogen and phosphorus are pollutants of concern when effluents are discharged into receiving waters that are vulnerable to or are already suffering from excessive eutrophication or even hypertrophic conditions, exhibited by dense and obnoxious algal and cyanobacteria blooms. Excess nitrogen can be removed from used water by nitrification that converts organic and ammonium N to nitrate (NO_3^-), followed by denitrification converting nitrate to inert nitrogen gas that escapes from the tank. Nitrification/denitrification is a three-step process and originally was designed with three tanks. Denitrification requires organic carbon as a source of energy of the reaction that originally was supplied by adding methanol or a similar inexpensive carbon source. Current designs have eliminated the need to add organic carbon by reversing the process and sequence of the unit processes (Figure 7.1 C) whereby the first anoxic step removes a part of biodegradable BOD (COD) and converts organic nitrogen to ammonium. In the second aeration step, remaining COD and BOD are further removed, and ammonium is oxidized by nitrification to nitrate. The nitrate from the aeration step is then returned by a high recycle back to the first anoxic step, where nitrate-nitrogen is denitrified to nitrogen gas. BOD is removed in both basins by anoxic and aerobic decomposition. Such treatment plants are very popular in smaller packaged operations as treatment plants for small communities (1000–20,000 people) or recreational facilities (e.g., the stadium of the New England Patriots in Foxboro, Massachusetts). Battery Park in New York used water reclamation, shown in Figure 6.8, removes BOD and nitrogen in the sequence of the anoxic and aerobic tanks.

To remove phosphorus biologically, an anaerobic contact basin is added ahead of the anoxic basin. Under anaerobic conditions, polyphosphate–accumulating microorganisms selectively accumulate large quantities of polyphosphate within their cells, which enhances the phosphorus removal. The phosphorus accumulated in the bacterial biomass is thus increased from about 1 to 2% without the anaerobic step to 5 to 7% with the enhanced phosphorus removal. Removed phosphate incorporated into the sludge can be recovered. However, this makes the operation relatively complex and less suitable for decentralized reclamation and reuse. Nutrient recovery from sludge could generally be done in a central integrated resource recovery facility (Chapter VIII) to which the sludge is conveyed by underground conduits (old sewers) or pumped for sludge handling and biogas (energy) and nutrient recovery. Decentralized biogas recovery has been done in rural communities in China and India (from manure), but can lead to safety problems such as explosions in small unsupervised cluster recovery units. Phosphorus can be removed more simply and efficiently by chemical precipitation with added iron salts. Phosphorus and nitrogen removals may be needed when reclaimed water is used for flow enhancement of streams entering impounded water bodies susceptible to excessive algal or cyanobacteria blooms and, in general, for phosphorus recovery.

The introduction of membrane filters and converting the reactors into membrane bioreactors was a breakthrough that brought efficient treatment of used water (both total mixed flow and black water) closer to a cluster distributed application. In these systems, effluents of high quality are reliably and consistently produced in compact units that can be installed close to the points of reuse. The filtration capability of membranes is very high due to small pore sizes (0.04 to 0.4 μm) in the membrane, capable of removing colloidal particles and microorganisms. After membrane filtration and disinfection (by ultraviolet light and/or ozonization), the reclaimed water is suitable for a number of nonpotable unrestricted urban uses. After storage and further physical treatment—for example, by nanofiltration and/or reverse osmosis—and pH and taste adjustment, water can be made suitable and safe for indirect potable reuse, typically after mixing with water from a quality freshwater source.

System (D) in Figure 7.1 is *a sequencing batch reactor* (SBR). The picture shows the phases of the operation in a single tank. For continuous flow at least two rectors will be needed. For BOD removal only, aeration begins once the reactor is filled, followed by settling and decanting the clean supernatant. Sludge wasting occurs during the aeration period. For nitrogen removal, the filling period is followed by several periods of aeration and no aeration, so that nitrification and denitrification can occur. SBR treatment can be automated.

The subsurface flow wetland system shown in Figure 6.2B is a suitable water reclamation unit process for cluster decentralized applications because of its reliability, little need for constant qualified supervision and control, and zero energy use. Wetlands are not just soil filters with uptake of pollutants by vegetation; removal and attenuation of pollutants is accomplished by a multiplicity of processes such as (Kadlec and Knight, 1966; Vymazal, 2005; Mitsch and Gosselink, 2007; Novotny, 2003; Šálek, Žáková, and Hrnčíř, 2008):

- *Physical processes* include sedimentation, filtration in the porous substrate, adsorption onto soil organic matter and clay, volatilization, and evaporation and evapotranspiration.

- *Chemical processes* include precipitation, decomposition, oxidation, reduction, and chemical adsorption.

- *Bacteriological processes taking place on the active biological layer surrounding roots (rhizomes)*: Wetland plants have the capability of transferring oxygen to their roots and creating oxygen-containing layers in the otherwise anaerobic environment of the substrate. This promotes partial nitrification/denitrification but, in general, removal efficiency for ammonium is relatively low (<30%). On the other hand, removal of incoming nitrate is very high (>95%). Hence, to achieve high nitrogen removals, the TKN in the incoming stream must be nitrified first, or a recycle from a follow-up aerobic oxidization treatment process (e.g., a pond) should be considered. In addition to nitrifying and denitrifying microorganisms, the rich microbial population in wetlands can decompose cellulose, lipids and fats, starch, and sugars. Microorganisms in the anaerobic environment of the substrate decompose organic biodegradable carbon compounds into carbon dioxide and methane and produce peat.

- *Biological processes* include plant uptake of carbon dioxide, nutrients, and dissolved compounds such as toxic metals that are incorporated into the tissue of the plants, photosynthesis, and transpiration.

Constructed wetlands are divided into those with surface or subsurface flow. In some ecocities it was suggested that the treatment wetlands be close to the housing development, or even a part of the ecoblock. However, surface flow wetlands with standing surface pools of treated sewage are not hygienically acceptable and represent a public health risk and nuisance. Surface flow wetlands work in combination with a pond (settling and oxidation) for controlling pollution by urban runoff and also provide aesthetical amenities to the area. The use of surface flow wetlands was covered in Chapter IV.

Subsurface flow wetlands, highly suitable for used water treatment, can have either horizontal or vertical flow (Figures 7.2 and 7.3). Wetlands with vertical flow may

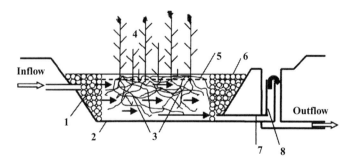

Figure 7.2 Schematics of a constructed wetland with subsurface rizontal flow (Source Vymazal, 2005, 2009). 1. distribution zone, 2. impermeable liner, 3. filtration medium (gravel, crushed rock), 4. vegetation, 5. water level in the bed, 6. collection zone filled with large stones, 7. collection drainage pipe, 8. outlet structure for maintaining water level in the bed.

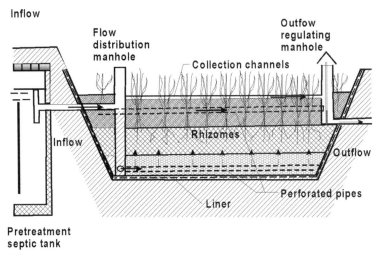

Figure 7.3   Schematics of a constructed wetland with vertical subsurface upflow (Source Šálek, Žáková, and Hrnčíř, 2008).

have better winter operating conditions and require less area than those with horizontal flow. A horizontal flow–constructed wetland for a population of 800 equivalent load from a combined sewer is shown in Figure 7.4. This wetland consists of four compartments working in two parallel series, each having two compartments. It was constructed in 1994, and after 15 years of operation the treatment efficiency for BOD (COD) and TSS removal has remained high (Vymazal, 2009).

Constructed wetlands require primary treatment or pretreatment, which may be accomplished by anaerobic septic settling tank systems (small operations), or screens. Biological pretreatment is not necessary and will not improve the treatment efficiency. For treatment of black water or raw mixed sewage, ponding must be avoided and the flow should be all subsurface. This is accomplished by perforated collection pipes below the surface of the wetland.

Design parameters of subsurface wetlands are given in Table 7.1, and typical treatment efficiencies and effluent concentrations are listed in Table 7.2. Based on this information and typical loading, Vymazal (2005, 2009) and Šálek et al. (2008) reported that about 4 to 5 m^2 of wetland area is required per one population equivalent. A synthesis of the wetland efficiencies by Kadlec and Reddy (2001) found that wetlands, when properly designed, work year-round, and the temperature has a minimal effect on BOD and P removals. Because of the very high particulate organic content in the substrate, wetlands are also very efficient for removing and detoxifying absorbable toxic compounds such as divalent ions of heavy metals and organic chemicals with a high octanol partitioning number (pesticides, PCBs) (Novotny, 2003).

To gain knowledge about removal of nitrogen, Vymazal (2005) evaluated hybrid systems consisting of two or more subsurface wetlands in series. The first, larger, horizontal subsurface flow (SF) wetland removes BOD and TSS and converts organic nitrogen to ammonium. The second, vertical SF wetland polishes the effluent

Figure 7.4 Constructed wetland with subsurface horizontal flow consisting of two parallel series, each with two compartments, built in Rakovnik, Czech Republic. The load coming from a combined sewer has a population equivalent of 800, and the area of the wetland is 3080 m². Courtesy Jan Vymazal.

and—because the BOD in the wetland is relatively low—heterotrophic bacteria do not outcompete the nitrifying microorganisms, and the oxygen supply delivered by rhizomes of plants nitrifies ammonium. Part of the flow is then recycled back to the first wetland; hence, the system works on the same principle as the two-step anoxic-oxic system presented in Figure 7.1C. The hybrid system is a new development

Table 7.1   Design parameters of subsurface flow wetlands (after Water Environment Federation, 1990; Šálek et al., 2008; Vymazal, 2005)

Hydraulic loading	0.04 to 0.08 m/day
Hydraulic retention time (HRT)	5–7
Aspect ratio (Length/Width)	2:1–10:1
Configuration	Multiple beds in parallel series
Substrate gravel size	8–16 mm
BOD$_5$ load	80–120 kg/ha-day
Suspended solids loads	up to 150 kg/ha-day
Total nitrogen load	10–60 kg/ha-day
Total phosphorus load	up to 15 kg/ha-day
Typical wetland area	4–5 m²/population equivalent

Table 7.2   Typical treatment efficiencies and effluent concentrations of single-step subsurface flow systems

Parameter Treatment	Effluent Efficiency %	Concentration mg/L	Comment
BOD_5	85–97	<15	Removal efficiencies are not related to HRT, effluent BOD is from production in the wetland and not from influent
TSS	80–97	< 20	Same as above
Nitrogen Ammonium N Nitrate	40–50	>95	Variable, wetland can produce ammonium
Phosphorus	30–40		
Fecal coliforms	90–99		Removals insufficient to result in FC of <200/100 ml

Source: U.S. EPA, 1991; Vymazal, 2009; Vymazal and Kröpfelová, 2008

that can provide high-quality effluent with very low BOD_5 (<10 mg/L), Total N (<10 mg/L), and Total P (<5.0 mg/L). These systems are very promising for simple design and operation distributed treatment systems with energy saving (the only cost and energy used are for pumping the recycle).

The solids production and buildup in subsurface flow wetlands can be quite large. The primary productivity of freshwater marshes by photosynthesis without large BOD and TSS input, according to Mitsch and Gosselink (2007), is about 2 kg of biomass/m^2-year (20 tonnes/ha-year). On top of this, one should include organic solids from the influent and bacterial biomass growth produced by the BOD removal. Hence, organic solids buildup in wetlands will be substantially larger than that in natural freshwater wetlands. Natural wetlands can have an organic content of the substrate reaching 30–40% and, as a result, their capability of sequestering organic carbon is very large (see next section and Chapter VIII on GHG emissions and carbon sequestering). Because of the high organic carbon in the substrate (peat), wetlands also provide excellent removals of toxic metals and trace organics. These compounds are tightly held on the particulate organic matter in the substrate, and their uptake into above ground vegetation is minimal. Therefore, harvested vegetation is safe for reuse as a source of organic solids for agriculture. Wetlands essentially detoxify toxic metals by converting toxic divalent metallic cations into particulate metal-sulphide complexes (Vymazal, 2009; Vymazal and Kröpfelová, 2008; Novotny, 2003).

Carbon sequestering and GHG emissions. In general, wetlands contain reservoirs of carbon in above ground biomass, litter, peat, soils, and sediments. SF wetlands remove CO_2 from the atmosphere by sequestering organic carbon in water-saturated substrate where the reduced conditions in the zone outside of rhizome influence inhibit decomposition and denitrify nitrate. Net carbon sequestering occurs as long as rates of conversion of CO_2 into organic carbon exceed decomposition. However, anaerobic/anoxic decomposition also produces methane, which has a GWP (global warming potential) of 25, and nitrous oxide (N_2O), both being potent GHGs. Sovik et al. (2006) measured GHG emissions from several constructed wetlands in northern Europe (Scandinavia and Poland) and found these emissions quite high, the average nitrous oxide fluxes from vertical subsurface flow wetlands being 17 mg N_2O m^{-2} day^{-1}; methane 110 mg CH_4 m^{-2} day^{-1}; and CO_2 4500 mg m^{-2} day^{-1}. These are high fluxes, but they still were found to be less than emissions from conventional aerobic/anaerobic used water reclamation plants that do not capture methane. Hence, the goal of the wetland design and plant selection is to maximize carbon sequestering and minimize carbon dioxide, nitrous oxide, and methane production.

Harvesting the produced vegetation for energy production also enhances the nitrogen and phosphorus controls (Vymazal, 2005). Vegetation growth can also be removed for composting (see Qingdao ecoblock and Dongtan ecocity concepts, presented in Chapter XI) and or for biogas production (see Chapter VIII). The removed vegetation may contain up to 50% organic carbon (Lamlon and Savidge, 2003).

Subsurface flow wetlands may also reduce evaporation losses, in comparison to the aerated open air biological treatment systems where water losses are not only caused by evaporation from the water surface but also multiplied by the evaporation from splashing and escaping bubbles from aeration devices, which is much larger than surface evaporation from open water. Both types of water loss from open air aerated basins are increased by wind and dryness of the air above the basin (Novotny and Krenkel, 1973).

Creating an artificial wetland is an engineering and ecological endeavor that must be planned and executed properly to avoid development of unwanted species (for example, obnoxious algal mats in shallow standing waters, or nuisance growths of water hyacinths in subtropical and tropical areas). Other benefits of constructed wetlands are aesthetic value, attraction of water fowl and wildlife, growth of organic matter for bioenergy production, and carbon sequestering.

VII.1.4 New Developments in Biological Treatment

One hundred years ago, typical treatment units consisted of primary solids separation combined with anaerobic solids digestion—for example, the Imhoff tank (Figure 1.10). After 1920 secondary aerobic treatment was introduced that provided better effluent quality but required a lot of energy to blow air into the aeration tanks and, in the latest development, produce pure oxygen for high-efficiency biologic treatment. Almost 5% of all energy used in developed countries is spent on moving water and treating and disposing of used water and residual solids.

Anaerobic Sludge Blanket Reactor An *upflow anaerobic sludge blanket (UASB) reactor*, shown in Figure 7.5, was developed by Lettinga and his coworkers (Lettinga et al., 1980) more than 30 years ago. In appearance it is similar to the old Imhoff tank (Figure 1.10). The major difference between the two reactors is that the Imhoff tank is a primary settler where influent and effluent are on the top, and the primary solids settle into the sludge digestion compartment. In contrast, influent

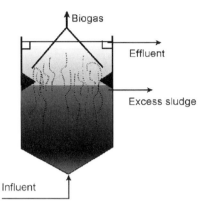

Figure 7.5   A variant of an upflow anaerobic sludge blanket (UASB) reactor.

into the UASB reactor enters the tank through the bottom inlet, creating upflow conditions that keep solids in the digestion compartment suspended, and the effluent is at the top. Hence, the upflow velocity of the influent flow should be about the same as the settling velocity of the granules formed in the reactor. This reactor requires supervised control and may not be suitable for small, unsupervised operations. While the Imhoff tank removes only organic particulate solids by settling, the UASB is a suspended growth reactor, and the primary COD removal mechanism in the UASB reactor acts by absorption, biological action of bacteria, and fermentation of organic suspended and dissolved solids in the active biological sludge blanket. In this process most of the nutrients are released into the supernatant (effluent), resulting in higher N and P concentrations. If nutrient removal is desired, the UASB process has to be followed by nutrient removal (see Section VII.2.1, describing struvite formation). The removal efficiencies for COD (BOD) are comparable to those of conventional activated sludge plants, and the advantage of the reactor is that it requires no energy and produces methane biogas that can be captured and converted to energy (see Chapter VIII). The anaerobic methanogenesis will be introduced in Section VII.2.

UASB rectors treat a variety of domestic and industrial wastes of different strengths, from typical sewage to highly concentrated industrial organic wastewater (municipal black water, brewery, sugar refineries) with high input COD values. A typical UASB reactor consists of three parts: (1) sludge bed, (2) sludge blanket, and (3) solid-liquid-gas separator, which is installed on top of the reactor. The influent is pumped into the bottom and passes upward through the granular sludge bed, in which the organic compounds in the influent are biologically converted to biogas. Mixing with the sludge bed is due to influent flow and biogas production. While anaerobic digesters work best in the mesophilic (35–45°C) range, UASB reactors work well at temperatures of 20°C and up. Heating may be necessary if the temperature drops below 20°C.

Sludge is withdrawn from the blanket at the rate it is produced, either continuously or daily. Daily sludge production depends on the characteristics of the raw influent and equals the sum of (1) sludge produced from COD removal (sludge yield) and

Table 7.3   **Basic design parameters for UASB reactors**

Input COD concentration	0.5–15 g/L
Max Input TSS	6–8 g/L*
Temperature in the reactor	> 20°C
Concentration of solids in the sludge blanket	30–50 g VSS/L
Hydraulic residence time (HRT) at average flow	
Temperature range 20–26°C	7–10 hours
Temperature > 26°C	~6 hours
Upflow velocity	0.6–0.9 m/hour
Solids residence time (SRT)	30–50 days
Organic loading	
Influent COD 0–2 g/L	8–14 kg COD/m^{3-}day
Infl. COD 2–6 g/L	12–24 kg COD/m^{3-}day
COD removal	75–95%
Sludge yield	0.08 kg VSS/kg of COD removed
Sludge blanket depth	2–3 meters
Methane production	0.2–0.4 m^3/kg of COD removed
Energy content of methane at 25°C	9.2 kW-hr/m^3 of CH$_4$

Source: From Lettinga and Hulshoff Pol, 1991; Metcalf & Eddy, 2003; Agrawal et al., 1997
* Preclarification is needed for influent TSS greater than 8 g/L.

(2) nondegradable residuals. Table 7.3 contains basic information on the design of UASB reactors.

The two main design parameters are hydraulic residence time (HSR) and solids residence time (SRT), which are different. *HRT* is defined as

$$HRT = V/Q$$

where V is the digester volume (m^3), and Q is the daily influent flow (m^3/hour).

The solids residence time on the other hand is

$$SRT = MSS/SY$$

where M_{SS} is the mass of sludge solids in the reactor (kg) and SY is the mass of solids withdrawn per day. $M_{SS} = V_{SS} \times C_{SS}$ where C_{SS} is the average concentration of solids in the sludge blanket.

The influent COD is converted to volatile suspended solids (VSS), methane, and residue solids. Methane production and COD removals will be covered in Section VII.2 and in Chapter VIII (Section VIII.4). The importance of the UASB process will be emphasized in Chapter VIII in connection with developing and proposing an integrated resource recovery facility (IRRF). In this context, UASB (or a similar anaerobic process) is the first important unit process at the beginning of treatment and resource recovery, which is a revolutionary departure from the standard sequence of biological wastewater treatment. In most Cities of the Future applications, the UASB reactor will treat concentrated (black) used water, often mixed with other organic solids such as liquid sludge or predigested vegetation and food residues. Primary settling may only be necessary if the influent contains larger amounts of inorganic suspended solids. This will bring about the best biogas production, which can then be converted either directly to energy by combustion or, more efficiently, to hydrogen and electricity by hydrogen fuel cells.

Direct Electricity and Hydrogen Production from Used Water   Electricity and hydrogen can be directly reclaimed in treatment tanks by specific actions of bacteria in the *microbial fuel cells* (see Chapter VIII). Such cells can produce far more of the electric and/or hydrogen energy that is used by the treatment process itself. This groundbreaking technology, emerging and being investigated in several laboratories throughout the world in this century, would remove organic content expressed as BOD without converting it into GHG methane. In the new process, bacteria working in the microbial fuel cells can convert biodegradable organics either into hydrogen, or directly into electricity. In this context the notion of biological oxygen demand (BOD) becomes an anachronism, and the strength of the organic content of wastewater could instead be expressed in joules/liter or, even better, in watt-hours/m^3 rather than (or in addition to) the archaic BOD_5.

Algae to Biogas Processes   During his early career in the 1960s, the primary author witnessed pioneering research on the growing of algae using wastewater, by a research team at the Czechoslovak Academy of Science in Třeboň (today in the Czech Republic). The team, led by Dr. Petr Marvan, successfully cultivated algae using the effluent from the wastewater treatment plant (WWTP) in Brno. The use of algae was for animal feed and other purposes. Later the interest in algae utilization was driven by problems of algal blooms associated with eutrophication of lakes, reservoirs, and seas. Theoretically, uncontrolled massive algal blooms and subsequent sedimentation of algal biomass could be a (dubious) solution to global warming. As a matter of fact, this is what occurred approximately three billion years ago when oxygen was produced by the first photosynthetic organisms (most likely cyanobacteria) from the carbon dioxide–rich atmosphere. Before these geologic times, the atmosphere had no oxygen. The geologic remnants of this period are carbonate rocks, oil, and natural gas billions years old. Figure 7.6 shows a massive algal bloom of green algae in an urban pond in China. After die-off, the dead algae settle to the bottom of the pond and slowly anaerobically decompose, producing methane and carbon dioxide.

   Today, there is an enormous interest among scientists, large oil companies (e.g., ExxonMobil), and venture investors, who want to repeat the process of massive algal production under controlled conditions in reactors and ponds, producing biogas (methane) to be converted into biofuel. Use of algal ponds for wastewater treatment (called originally facultative ponds) is not new and is still popular in many developing countries. In algal ponds, BOD (COD) is removed by facultative bacteria that use oxygen provided by algae thriving on nutrients in the water. Such algal ponds also have an anaerobic zone that may provide denitrification. Alternatively algal ponds and reactors could be used as tertiary treatment to remove nutrients. In most cases,

Figure 7.6 Massive algal bloom in a polluted nutrient laden urban pond in China. Photo Charles Melching.

concentrations of nutrients in traditional algal ponds are far in excess of algae needs; hence, their growth is limited by available CO_2 (alkalinity), and the algae are not specifically selected to maximize production of biomass and biofuel.

The focus of producing algae under the new paradigm (Oilgae, 2009) is to:

- Remove nutrients
- Sequester carbon
- Grow biomass for energy

The product from algal biomass can be methane first, then converted by a hydrogen fuel cell (see Chapter VIII) to hydrogen gas or biofuel. Research in the 1990s by the National Renewable Energy Laboratory (Sheehan et al., 1998) documented under controlled conditions that algae are capable of producing 20 times the amount of biofuel (biodiesel) per unit area of land produced by agricultural oilseed crops such as soy and rape (colza). Older research indicated that some algae could switch from producing oxygen to the production of hydrogen. The best combination of algal reactors is in a sequence following anaerobic digesters or UASB reactors that yield supernatants rich with nutrients. The best algal fast-growing species are small-sized

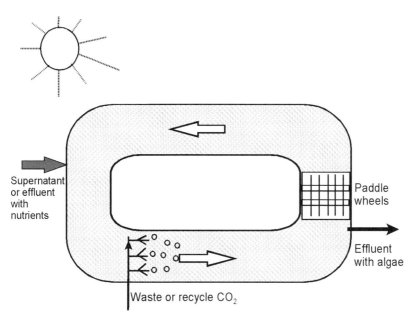

Figure 7.7   A standard circular ditch can be used for algal biomass production (Sheehan et al., 1998).

algal microorganisms that in nature are capable of overgrowing in a short time into massive algal blooms (Figure 7.6). CO_2 needs to be supplied to the reactor. Deublern and Steinhamer (2008) suggested *Chlorella vulgaris* as one of the species. Others have suggested *Cyclotella cryptica* as the best diatom out of about 100,000 to produce biofuel from supernatant of digesters. Blue-green algae (cyanobacteria) produce harmful toxins (Carmichael, 1997). A standard recirculation (oxidation) ditch type reactor, shown in Figure 7.7, or a series of reactors is one of the possible configurations of the algal production unit process. The difference between the circular algal production reactor and an oxidation ditch is that oxidation (commonly provided by the Kessener brush wheel) is not desirable in algal reactors, because algae produce oxygen and prefer low turbulence.

If algae are digested to produce biogas or biodiesel, the process is carbon neutral. In Chapter VIII, carbon sequestering (carbon positive) will be suggested.

VII.2 SLUDGE HANDLING AND RESOURCE RECOVERY

Almost all treatment processes listed in the preceding sections of this chapter produce organic and inorganic sludge residues that must be treated; their products must be reclaimed and reused, and residual solids disposed of. An average person produces about 90 grams of solids per day (Metcalf and Eddy, 2003). In some countries, including the U.S., residual solids volume is increased by food residues discharged by garbage disposals/comminutors into used water kitchen flow, which increases sludge solids mass by about 30%. In commercial operations, food waste is still disposed of into garbage containers but it could and should be recycled—for example, as animal (hog) food. Another alternative is to process food waste with the sludge and other organic solids and recover energy. The traditional view of sludge as waste that, at best, may have some soil-conditioning benefit can now be replaced by considering sludge as a valuable resource, not only for its fertilizing and soil-conditioning values but also for its energy reclamation potential, which could be carbon neutral or even reduce carbon dioxide emissions. The potential for reclaiming energy from organic sludge will be discussed in Chapter VIII. The residue solids from the treatment processes are generated by solids separation in the pretreatment units, including screens; in the primary solids separation units; and as excess sludge from the secondary biological treatment units. Wetland treatment does not generate excess sludge; all solids are incorporated into substrate soil, where they decompose on-site, but wetland vegetation can be harvested and used for biogas production.

The traditional secondary activated sludge treatment process (Figure 7.1A) produces the largest mass of excess sludge. In the traditional biological treatment process, about 40 to 70% of BOD is converted to activated sludge solids. Modifications such as *extended aeration* reduce the sludge mass but emit more CO_2 and require more energy and larger volume of treatment tanks. Sludge volumes from nitrification/denitrification processes are also smaller, but adding iron salts to precipitate phosphorus increases sludge production. *High-loading activated sludge treatment* systems require less aeration energy and produce larger sludge mass and volumes, but their BOD removal efficiency is less than 75% (Imhoff and Imhoff, 2007).

Sludge handling in distributed systems with a large number of local water and energy reclamation units could be costly because of the complexity of the sludge-handling system, potential for offensive odor, and problems with combustible and explosive methane. For this reason, solids separated as sludge in the distributed or satellite water and energy reclamation units could be returned into existing sewers and delivered by gravity or pumped to a central sludge-handling and resource recovery facility. In the "ecovillages" of the developing world, solids from composting toilets can be mixed with manure and vegetation residues and safely digested in small digesters to produce methane for nearby houses (see Chapter VIII). It is quite likely that in the Cities of the Future, residual sludge disposal from the water reclamation facility will be combined with disposal of other organic solids, including vegetation residues, food waste from households and commercial establishments, excess manure from suburban agricultural animal operations, and the like. In this sense, organic solids management will be a resource recovery operation potentially quite different from traditional sludge handling in the fourth paradigm treatment plants and water reclamation facilities. The end product may also change. Although today most designs end with production of methane biogas (a product of anaerobic digestion), which would be a carbon neutral process, the future lies with hydrogen gas production and carbon sequestering (Schnoor, 2004; Halmann and Steinberg, 1999).

VII.2.1 Types of Solids Produced in the Water Reclamation Process

Several unit operations separate solids from liquid in the treatment process (Metcalf & Eddy, 2003), including:

- *Screenings* contain large particles delivered to combined sewers, such as wood, dead animals, sanitary waste, plastics, etc., trapped by bar racks. Screenings are usually shredded by comminutors and sent into the treatment process, where the shreddings are mostly removed with the primary sludge.

- *Grit* is large, heavy, mostly inert particles that are removed in grit chambers by settling. Grit may also contain fats and grease.

- *Scum/grease* is a general category of floatable materials removed in the treatment process and sent to sludge handling.

- *Primary sludge* is removed in the settling tanks before biological treatment. Settling tanks can be anaerobic or aerobic. Anaerobic tanks combine particle separation with digestion of organic solids (Imhoff tanks, septic tanks); in aerobic tanks, particles settle by gravity and are removed as primary sludge to sludge handling. Removal of solids from aerobic settling tanks (clarifiers) is continuous, while sludge is removed from anaerobic tanks periodically, once or several times in a year, and hauled for sludge handling or discharged on land. Primary sludge from anaerobic tanks has an objectionable odor.

- *Secondary activated sludge* is separated from the aeration tank or trickling filter effluents in a secondary clarifier, then partially recycled back into the aeration or anoxic tanks and partially separated into sludge handling. More modern and efficient membrane filters leave the separated solids in the aeration tank, and they are removed with the sludge excess.

- *Sludge from chemical precipitation and coagulation* originates from precipitation of toxic chemicals and phosphorus by iron or aluminum salts and/or polyelectrolyte coagulant aids. Sometimes it has a slimy or gelatinous texture. In the distributed systems of the future cities, coagulant chemicals for precipitation of phosphate can be added directly to the aeration tank, and the precipitate will be removed with the secondary sludge.

The chemical and physical characteristics of sludge are described in the key reference manuals such as Metcalf & Eddy (2003), Imhoff and Imhoff (2007), and Novotny et al. (1989). The characteristics, composition, and quantity of the sludge depend obviously on the treatment process. Treatment wetlands do not generate sludge under the common definition, although they generate solids buildup that eventually may have to be removed. Treatment processes that generate less sludge volume include *extended aeration* and *anoxic-oxic* processes; however, these processes require larger volumes of treatment tanks and, in the case of extended aeration, more aeration energy. They also result in more GHG emissions.

Metcalf & Eddy (2003) and Imhoff and Imhoff (2007) divide conventional sludge handling into the following sequential processes:

Sequence without Sludge Digestion

Thickening and dewatering	Typical processes are gravity, flotation, centrifuging, rotary drum, and belt thickening
Stabilization	Lime stabilization, heat treatment
Disinfection	Pasteurization by heat
Disposal	Farms, commercial fertilizer (e.g., Milorganite), mixing with compost, incineration

The above sequence has a high energy demand. It retains the nutrients and organic carbon in the final products; however, the high energy demand increases GHG emissions.

Sequence with Anaerobic Digestion   Both aerobic and anaerobic digestions have been proposed for sludge stabilization. However, aerobic digestion has a high energy demand and no energy recovery. Consequently aerobic stabilization will not be proposed.

The design of anaerobic digestion needs a new look, as indicated in Figures 7.1B and 7.5. Typically, the anaerobic digestion process is a three-stage process in which organic matter is first broken down by hydrolysis and fermentation into organic acids, then further decomposed to acetic acid and hydrogen gas. In the last fermentation step, methane is formed by acetogenesis decomposing acetic acid to methane, carbon dioxide, water, and hydrogen gas, which combines with carbon dioxide to form additional methane. Hence, methane is theoretically the only final gas product of the process (Figure 7.8). However, the process is not 100% perfect, and the biogas produced by anaerobic digestion still contains about 35% carbon dioxide. Metcalf &

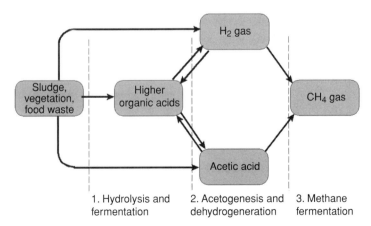

Figure 7.8   Anaerobic digestion process schematics (adapted from Metcalf & Eddy, 2003).

Eddy (2003) estimated the ratio of the methane produced to the total BOD removed as 0.25 kg of CH_4/kg of BOD_t. Theoretical biogas production is 0.4 kg of CH_4 per kg of BOD_t or COD. The heating value of the produced biogas is about 22,000 kJ/m^3.

The process is sensitive to pH, and sufficient alkalinity should be present to maintain pH in the range of 6.6 to 7.6 (Metcalf & Eddy, 2003). The bacteria in the process are also sensitive to the presence of toxic compounds (e.g., toxic metals) or low pH caused by sulfate conversion to sulfide.

Anaerobic digestion requires heating to provide the best results. For mesophilic digestion the reactor should be kept in the temperature range of 30 to 38°C. For thermophilic decomposition the reactor is kept in the temperature range of 49 to 57°C, which obviously requires more heating but less reactor volume. Most current treatment (water reclamation) plants reclaim methane and use it for heating of the anaerobic digesters and generating electricity for the plant.

The traditional sludge-handling unit operations with anaerobic digestion in a traditional water-from-sewage reclamation plant include in a sequence:

Thickening: Gravity thickener similar to a settling tank or a centrifuge

One- or two-tank anaerobic digestion: With a supernatant returned back to treatment and methane recovered for heating of digesters and for local electricity use or biofuel (Figure 7.9)

Chemical conditioning

Drying: On sand beds or mechanical belt filters or centrifuges with the filtrate returned back to treatment

Disposal: According to US EPA (1999) land application (41%), incineration (22%), landfill (17%), advanced treatment (12%), and other (8%).

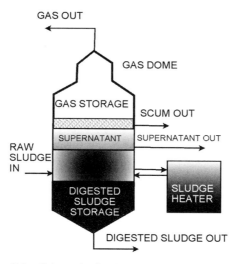

Figure 7.9   Schematic of a simple one-tank sludge digester.

Table 7.4 **Concentrations of COD and nutrients in the supernatant of the BNR* and conventional activated sludge systems (from Parsons et al., 2001)**

Parameter	Units	BNR Sludge	Activated Sludge
Total COD	mg/L	15,320	13,640
Total solids	mg/L	12,620	11,080
Total nitrogen	mg/L	486	399
Total phosphorus	mg/L	335	143
Soluble phosphorus	mg/L	55	34
Total calcium	mg/L	686	247
Total magnesium	mg/L	108	39
pH	Units	7.2	7.4

* BNR – Biological nutrient removal

None of the processes in the above sequences is overly energy efficient, although methane energy recovered from the digesters can make up for some energy lost in the process and provide heating and energy for the plant. In these processes, sludge is considered as a waste and the primary goal is to disposeof it in a sanitary way.

In addition to gases (methane and carbon dioxide) and digested solids, the sludge digestion process produces high-strength liquid—the supernatant (Table 7.4), which has very high contents of BOD, nitrogen, and phosphorus. In the conventional treatment process, supernatant is returned to the biological treatment units for treatment, but it is far better to treat it separately because nitrogen can be removed or recovered with phosphorus in the form of struvite (magnesium ammonium phosphate, MAP), which is a valuable fertilizer. Uncontrolled, struvite growth can clog pipes and other components of the treatment plant. In the Cities of the Future, both methane and nutrients will be recovered, as is already being done in the Hammarby Sjöstad ecocity in Sweden (see Chapter XI).

The mass load of biodegradable organics in the returned supernatant is significant and should be accounted for in planning the dimensions of the biological treatment units and estimating the aeration capacity and GHG emissions.

VII.2.2 A New Look at Residual Solids (Sludge) as a Resource

The first postulate in this deliberation is that residual solids containing organic carbon are a valuable source of energy. If one uses simple economics, at the end of the last millennium, the cost of energy per kw-hour recovered with conventional wastewater treatment technologies was more than the cost of energy derived from fossil fuels (coal, oil, or natural gas). However, burning fossil fuels contributes to global warming (see Chapter II), while deriving energy from sludge (or algae) is carbon neutral and could be carbon positive (sequestering carbon). With the technologies available at the beginning of the 21st century, the cost of energy recovery from biomass would be about the same as or less than that of producing ethanol for fuel, which is subsidized but does not have many environmental benefits. If the social and environmental costs are included in the triple bottom line assessment (Chapters II and X), the total cost and benefits will include not only the cost of treatment and disposal of sludge and organic solid waste but also the environmental cost of carbon emissions from the greater energy use in the treatment and disposal processes—and the potential for cost reductions once the new GHG-neutral or GHG-reducing processes become more common and accepted. When all three dimensions of the triple bottom line assessment are used, the change will become attractive. Even today, many European countries have for years been steadily building anaerobic digestion facilities for generating electricity from methane produced from manure, sewage, and garbage. Villagers in many undeveloped countries use very simple technology to convert animal and human wastes to biogas for cooking and heating (see Chapter VIII). Also hundreds of farms in China, Mexico, and South America have installed anaerobic digesters to collect and use methane from manure to provide energy for farm use. Many of these digesters have been paid for by a company that aggregates carbon credits and sells them to factories and utility companies in countries that signed agreements under the Kyoto Protocol to reduce greenhouse emissions. Carbon credits are earned by reducing greenhouse gas emissions and these credits have considerable value. In the U.S., which has not signed the Kyoto Protocol (which in 2009 was to be replaced by an agreement reached among the world nations in Copenhagen), the Obama administration has expressed commitment to the control of global warming. Nevertheless, most of the methane escaping from liquid or solid waste treatment and disposal processes in the U.S. is allowed to be wasted or is flamed out. Few landfills have installed methane recovery systems to produce electricity (Figure 7.10).

Figure 7.10 A power plant producing electricity from biogas (methane) collected from the Omega Hill landfill operated by Waste Management in Germantown (Wisconsin) in 1990s. Photo V. Novotny.

Three energy-yielding gases are produced by converting biomass to energy (Halmann and Steinberg, 1999):

1. The most common is methane produced by thermal gasification or anaerobic digestion.
2. The future is in hydrogen gas production.
3. Carbon monoxide is also produced by thermal gasification.

The three steps of anaerobic digestion of sludge and manure were presented in Figure 7.8. In typical anaerobic digestion, methane is the main product. In thermal gasification, in which coal, cellulose, and wood chips are the sources, organic carbon is converted into methane and carbon monoxide. Energy production from biomass, including sludge from water reclamation plants, will be presented in Chapter VIII. The possibility of converting methane as well as acidic acids (products of acidic fermentation—see Figure 7.8) into hydrogen gas has been intriguing scientists for some time.

Figure 7.8 shows hydrogen gas as an intermediate product of the sludge digestion process, but scientists can now recover hydrogen from methane and sequester carbon, or recover it directly from sludge and wastewater in microbial fuel cells (Chapter VIII). Producing hydrogen and electricity in hydrogen fuel cells is now a common technology, especially in manufacturing hybrid and plug-in cars. Furthermore, it makes sense to recover energy not only from sludge but also from other current organic waste residuals such as kitchen and food waste, vegetation growth excess and residuals from harvested crops, harvested vegetation from constructed wetlands and vegetated riparian buffers preventing diffuse pollution from reaching urban and rural water courses, and also from solid organic wastes and manure from suburban agriculture.

For the ecocities of the future, sludge produced by water reclamation plants represents only a part of the biomass processed as a resource, from which energy, nutrients, and biosolids can be recovered. Such resource recovery facilities have been or are being built not only in Hammarby Sjöstad in Sweden, Masdar (UAE), and other ecocities but also in developing countries (India and China) and in developed countries with large dairy production such as The Netherlands. Currently, discarded biomass from forestry, agriculture, and municipal sources is a potential source of 22% of total energy needs in Canada (Levin et al., 2006) and the proportion is similar elsewhere throughout the world. Biomass is a renewable source, and even if CO_2 is emitted, it is carbon neutral.

VII.3 NUTRIENT RECOVERY

Nitrogen and phosphorus stimulate the eutrophication process in the receiving waters. Eutrophication is characterized by excessive growths of algae and a diminishing value of the resource for the water supply. Recently, the quality status of many receiving waters in Europe, China, Latin America, and the U.S., as well, has reached an emergency state characterized by noxious algal blooms of cyanobacteria that render water useless for drinking, reuse, and recreation, and severely impair aquatic life. In the water centric ecocities, buildup of nutrients in the recycled water discharged into the water bodies from which water is then reused can have very serious adverse consequences. We have already mentioned that cyanobacteria would be a candidate species for algal biomass and biofuel production from used water and discarded organic solids.

The removal of nutrients is first accomplished in the treatment process by nitrification/denitrification processes for nitrogen removal without nitrogen recovery. Phosphorus is removed from water either by enhanced uptake by sludge bacteria in the anaerobic tank or by chemical precipitation by adding precipitating salts (e.g., iron sulfate) to the treatment tanks. In this case all of the removed phosphorus and part of the nitrogen end up in sludge and are transferred to sludge handling. In wetlands, nitrogen is removed by nitrification/denitrification in the substrate and, with phosphorus, by incorporation as organic nitrogen into vegetation and soil.

Nitrogen is in abundant supply in the atmosphere and biosphere, and recovering it for reuse makes sense only if water is reused for irrigation. In this case, the nutrient removal from reclaimed water is not an economic benefit. This is not the case for phosphorus. There is only a limited quantity of mineral phosphorus left in the world, and the sources of mineral phosphate (apatite, guano) will be gone before the end of this century. Loss of phosphorus availability would obviously have an adverse impact on agriculture. Therefore, recovering phosphorus is a necessity for sustainability.

As stated, phosphorus and some nitrogen and organic carbon in used water are removed into sludge or into the effluent. Consequently, removed sludge from biological nutrient removal (BNR) or conventional biological nutrient removal (by sludge uptake) has high concentrations of nitrogen in the form of total Kjeldahl nitrogen (organic + ammonium N) and phosphorus (Table 7.6). If magnesium is also present—for example, if the potable water has higher hardness content—this causes formation of a precipitate mineral called struvite, which is chemically magnesium ammonium phosphate ($Mg\,NH_4\,PO_4 \times 6H_2O$) (MAP). Struvite is chemically similar to kidney stones and normally causes problems with scaling development in pipes or pumps. The natural source of struvite is guano deposits, which already have been greatly depleted.

The same struvite production process has been proposed for recovering N and P from urine (Section VI.4.2), which would require triplicate used water separation, collection, and treatment systems. Struvite production could be, most likely, accomplished on a house or small cluster scale, but it could be more economically incorporated into the centralized integrated resource recovery facility (IRRF).

Struvite can be recovered from sludge digestion supernatant liquid, which is laden with both ammonium and phosphate. Struvite precipitation occurs in two stages (Parsons et al., 2001): (1) nucleation, and (2) growth of the struvite crystals. The precipitation/crystallization is controlled by pH, supersaturation, temperature, and presence of impurities such as calcium, and occurs when concentrations of ammonium,

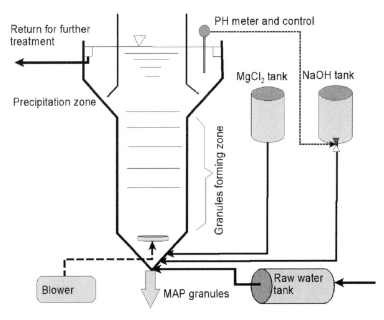

Figure 7.11   Upflow fluidized bed reactor for MAP production. *Source* Kurita Water Industries.

phosphate, and magnesium ions exceed the solubility product of struvite given by the formula

$$K_{sv} = [Mg^{2+}][NH_4^+][PO_4^{3-}]$$

The equilibrium constant is related to pH, and the relationship indicates that the solubility of struvite decreases with increased pH and reaches an optimum at about pH = 9.

Struvite is recovered in fluidized bed or pellet reactors (Figure 7.11). Figure 7.12 is a schematic of simple sludge handling with struvite recovery. Since the supernatant liquid does not have enough magnesium or an optimum pH, magnesium is added in a form of magnesium hydroxide, which also elevates pH, or magnesium chloride with pH adjustment with sodium hydroxide (Figure 7.11). Adding magnesium slightly above the 1:1 molar ratio and operating pH in the range of 8–8.5 will result in reliable 80 to >90% precipitation of phosphate in the form of struvite. Chemical additions can be reduced by air stripping CO_2 from the supernatant liquid. Up to 86% phosphorus removals were achieved from the belt press filtrate dewatering digested sludge in a full-scale wastewater treatment operation (Cecchi, Battistoni and Bocadoro, 2003). Struvite products from a full-scale operation in Treviso, Italy, have low organic content (<2%), nitrogen content >4%, and phosphate content >12%, which is comparable to commercial fertilizers. Increasing ammonium removal into struvite can be achieved by adding both magnesium hydroxide and some phosphate in the form of phosphoric acid or phosphate superfertilizer (note that the added phosphate will be mostly recovered as additional struvite).

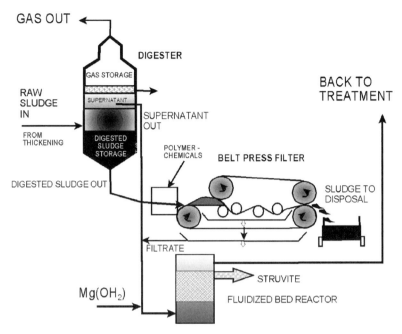

Figure 7.12   Simple sludge handling with struvite (ammonium and phosphate) recovery.

In an ecocity with a distributed system and a centralized integrated resource recovery facility, a small treatment plant may be needed to treat the polluted supernatant after struvite is extracted, or the liquid should be transferred to the nearest water reclamation facility designed to handle this load.

VII.4   MEMBRANE FILTRATION AND REVERSE OSMOSIS

In membrane filtration processes, water pretreated by biological or physical/chemical treatment passes through a thin membrane, which removes particulate materials, organic matter, nutrients, large and smaller molecules, and ions. Membrane processes include *microfiltration (MF), ultrafiltration (UF), nanofiltration (NF), and reverse osmosis (RO)* (Metcalf & Eddy, 2007). Reverse osmosis was described in Chapter V. This section will briefly describe the application of the membrane filtration technology to the treatment of reclaimed water that has much lower salinity than the brackish water and seawater used as a source of water. All processes are driven by the difference in pressure before and after the membrane. Hence, force or energy is needed to generate the pressure (before the membrane) or vacuum differential (after the membrane). The difference between these processes is in the size of millipores in the membrane. Table 7.5 presents the typical characteristics of the processes.

MF and UF membrane systems generally use hollow fibers that can be operated in the outside-to-in or inside-to-out direction of flow. Typical flux (rate of finished water permeate per unit membrane surface area) at 20°C for MF and UF ranges from

Table 7.5   Parameters of membrane treatment processes (Sources AMTA, 2007; Metcalf & Eddy, 2007)

Process	Pore Size (μm)	Operating Pressure (bar)	Energy Use (kW-hr/m^3)	Compounds Removed
Microfiltration	0.008–2.0	0.07–1	0.4–0.8	Fine particles down to clay size, bacterial, some colloids
Ultrafiltration	0.005–0.2	0.7–7	Wide range	All of the above plus viruses, large molecules
Nanofiltration	0.001–0.01	3.5–5.5	0.6–1.2	All of the above plus large molecules
Reverse osmosis of used water	0.0001–0.001	12–18	1.5–2.5	Ions, small molecules

1 bar = 100 kPascals = 100,000 N/m^2 = 1.03 atmospheres = 14.5 psi (at 0 elevation)

1.7 to 3.4 m/day (50 and 100 gallons/sq ft-day). Backwashing waste volumes can range from 4 to 15% of the permeate flow, depending upon the source water quality, membrane flux, frequency of backwashing, and type of potential fouling (AMTA, 2007). Microfilters can be submerged in the effluent liquid, as shown in Figures 7.1, and 7.13.

NF and RO systems are capable of removing molecules and ions. They should not be used for removal of particulates; hence, MF or UF can serve as pretreatment before RO. The difference between NF and RO processes is in removal of monovalent ions such as chlorides and sodium (seawater salt) by RO but not by NF. As the permeate (concentrate) water is produced, a concentrated stream containing all of the impurities rejected by the membranes is also produced. This concentrated stream, sometimes referred to as brine or reject, is a polluted flow that cannot be directly reused but could be treated, and some compounds (e.g., nitrate) removed by NF-RO units can be handled by the anoxic unit of the biological treatment (see Section VII.3). Reject water ranges from less than 5% of flow for MF up to 25% or more of water treated by RO. An RO system installed in the Singapore NEWater plant is shown on Figure 7.14.

Nanofiltration is used mainly for treatment of used water. The operational parameters are given in Table 7.5. Similar data for reverse osmosis were presented in Table 5.5. Reverse osmosis is used for both (1) final purification of the reclaimed water to remove monovalent ions, and (2) for desalination of seawater for input water (see Chapter V).

VII.5 DISINFECTION

Before reuse for almost all purposes, the reclaimed water must be disinfected. For many decades, chlorine gas and compounds (e.g., hypochlorous acid) were

Figure 7.13   Microfiltration. Siemens press picture, reproduced with permission.

disinfectants of choice in treatment plants because of their relatively low cost. Chlorine provided fast and reliable kill of pathogenic microorganisms and has been used to disinfect potable water and swimming pools. It has been used as bleach, as an oxidizer, and, during WWI, as a terrible weapon. When chlorine gas is used in treatment plants, special handling is required.

However, after a time, it was realized that the beneficial disinfecting effects of chlorine and chlorine compounds are outweighed by their adverse effects on human health and ecology. First, chlorine, when used in treatment plants, can combine with residual organic substances present in the treated water and form carcinogenic disinfection by-products known as Tri-halo methanes (TMH). Chemically, chlorine or bromine ions replace in THMs up to three hydrogens in the methane molecule. Second, chlorine was found ineffective as a disinfectant for protozoan pathogens such as *Cryptosporidium* or *Giardia* cysts. In 1993, a *Cryptosporidium* outbreak in one of the water supply treatment plants of Milwaukee, Wisconsin, made 400,000 people

Figure 7.14 Installed reverse osmosis units in NEWater plant in Singapore. Photo V. Novotny.

sick, and more than 100 died. As a result ozonation was implemented in an expensive overhaul of the water treatment system. Third, chlorine disinfection also adds a bad taste and odor to the produced water; consequently, ammonium is added to the produced water and effluent to convert the free chlorine into chloramines.

The required dosages of chlorine to conventionally treat effluents (activated sludge with secondary clarifier) are quite high, exceeding 20 mg/L (Metcalf & Eddy, 2007), and with these high dosages, aquatic life in receiving water has been adversely affected. For example, the Metropolitan Water Reclamation District of Greater Chicago (MWRDGC) found and argued that aquatic life in the effluent dominated Des Plaines River receiving the MWRDGC reclaimed water from the Stickney water reclamation plant has improved dramatically since effluent chlorination was discontinued. Subsequently, the Illinois Water Pollution Control Board granted a temporary waiver to discontinue chlorination.

For the above reasons, chlorination is not the disinfection methodology of choice for water reclamation plants of the Cities of the Future. Two acceptable and tested methods are presented below. However, it has to be emphasized that in the reuse system, disinfection is not a substitute for high-efficiency treatment.

Disinfection with ozone. Due to ozone's disinfection efficiency and the fact that chlorine generates carcinogenic compounds that are potential health risks, ozonization has been widely used in Europe for water treatment and is becoming the only method of water disinfection permitted in Sweden. The only reason that chlorination has remained the preferred methdod in the U.S. is the cost. Ozonization also uses more energy and, hence, there may be concerns with resulting increases in GHG emissions. These concerns are being addressed with new inventions and improvements that will make ozone generation more affordable, such as fitting the ozone generators with ceramic-coated metal conductors that increase energy efficiency while minimizing the cost. It is expected that more affordable ozonization processes will eliminate chlorination as the most commonly used disinfection method in water and used water treatment.

Ozone (O_3) is a gas with three atoms of oxygen in its molecule, instead of the two in the gas in the atmosphere. The oxygen molecule with two atoms is far more stable; hence, the third oxygen atom can detach from the ozone molecule and reattach to molecules of other substances. Consequently, ozone is a strong oxidizing and disinfecting compound. Recent advances in technologies for producing and using ozone and its superior disinfecting characteristics have made ozone the disinfectant of choice not only for potable water treatment but also for disinfecting the reclaimed water before reuse.

In nature, ozone is produced by lightning electricity during thunderstorms and by ultraviolet light. In the stratospheric layer, ozone shields the earth from harmful excessive ultraviolet rays, but also reacts with man-made fluorocarbons, which results in the loss of the protective ozone layer, especially in the far north and south latitudes. In the lower atmosphere, ozone produced by traffic air pollution during inversion is harmful to the respiratory systems of people, especially the elderly and those with asthma. During inversion (a stable atmospheric near-ground layer caused by overheating of the earth's surface, especially in cities), concentrations of other air pollutants are also high, and respiratory problems are exacerbated. Ozone has a distinct odor and can be detected by smell at very low concentrations.

In potable water treatment and reclamatiuon plants, ozone is produced from air by electrical disharge. It can also be produced by adding hydrogen peroxide (H_2O_2) in emergency situations. In large water treatment operations, ozone is produced by electricity from pure oxygen or oxygen-enriched air. Hence, producing ozone requires power. A typical ozone-generating system will require 22–26 kWh to produce 1 kg of O_3 from air and 8–12 kWh/kg of O_3 from pure oxygen. Table 7.6 provides information on dosages of ozone. Figure 7.15 shows a typical ozone disinfecting system. Ozone is introduced into water in ozone contact tanks by porous diffusers in multiple-chamber enclosed units. The off gas emitted from the system still contains residual ozone, which has to be removed.

The main advantage of ozone disinfection is that it can be produced on-site from air in a relatively small unit. Ozone is also readily soluble, reactive in water, and is a more effective disinfectant than chlorine. The disadvantage is that it is not stable, and its disinfecting time is relatively short.

Disinfection by ultraviolet radiation. The eradication properties of ultraviolet radiation have been known for more than one hundred years. UV radiation is a "clean" disinfectant—no chemicals are added. UV rays penetrate the cells of

Table 7.6 Ozone dosages and energy use for achieving required effluent quality from water reclamation plants (from Metcalf & Eddy, 2007, based on WEF, 1996, and White, 1999)

Type of effluent	Effluent Coliform Count (MPN/100 ml)	Ozone Dose (mg/L.)		Energy Use (kW-hr/m^3) for Human Use	
		For Ecological Flow and Urban Irrigation	For Direct Human Nonpotable and Indirect Potable Reuse	Total System O$_3$ Produced from Air	Total System O$_3$ Produced from O$_2$
Filtered activated sludge effluent	10^4–10^6	5–7	16–24	0.4–0.6	0.16–0.24
Filtered nitrified effluent	10^4–10^6	3–5	10–16	0.24–0.4	0.1–0.16
Microfiltration effluent	10^1–10^3	2–3	6–8	0.14–0.2	0.06–0.08
Reverse osmosis	0		1–2	0.02–0.05	0.01–0.02

microorganisms and damage DNA in the cells, which in essence sterilizes rather than kills them. Hence, pathogens cannot proliferate and cause sickness. However, some microorganisms can reactivate after exposure to the UV radiation and repair the damage. Ultraviolet light occupies the spectral range of 4000 to 1200 angstroms, which is from just below visible light to soft x-rays. Ultraviolet radiation at about the center of the range has been found to kill or deactivate many pathogens, including viruses, bacteria, and protozoa. UV light is produced by UV lamps, and new

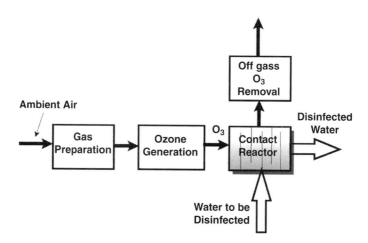

Figure 7.15 Ozone system schematic.

Figure 7.16 Ultraviolet disinfection installation. UV units are small and have very short radiation residence time. Courtesy Siemens press picture.

technologies and types of lamps are emerging. Also the residence time of water in the UV units is very short, in seconds. Figure 7.16 shows a UV disinfection installation.

The main categories of UV application systems are:

- *Unsubmerged systems,* where the process flow stream is routed through transparent tubes placed adjacent to externally mounted low-pressure mercury arc lamps.
- *In-channel submerged systems* consisting of lamps mounted on racks that can be lowered into the flow channel. Each lamp is inserted into a transparent tube and sealed to protect it from exposure to the water. For this type of system, water level variations should be kept to a minimum to avoid short-circuiting.
- *Enclosed submerged lamp systems, which are* much like the other submerged systems except that the lamps are fixed in place. This arrangement allows for the possible use of new high-intensity lamps for enhanced treatment capabilities.

Both dissolved and suspended solids can affect the efficiency of UV disinfection. For example, manganese and natural organic matter are strong absorbents of UV radiation, and other compounds can cause scaling on the tubes with UV lamps

(Metcalf & Eddy, 2007). Another problem with UV disinfection is that once water leaves the UV radiation chamber there is no residual effect; hence, it is recommended to install a follow-up disinfection such as ozonization or adding hydrogen peroxide. On the other hand, if RO precedes UV radiation, adding another disinfecting step might be overkill. All pathogens and viruses are removed by RO units.

Energy use for UV disinfection is comparable to that for ozonization or microfiltration.

VII.6   ENERGY AND GHG EMISSION ISSUES IN WATER RECLAMATION PLANTS

In water reclamation plants, energy is needed:

1. For transport and pumping (including overcoming energy losses in pipes by friction and local loses in valves, bends, and other pipe fittings) of treated water, liquid sludge, and supernatant from dewatering sludge through the various treatment units and storages to the point of reuse or disposal

2. For mixing treated water and liquid sludge in various treatment units, and supplying oxygen to aerobic biological treatment units (activated sludge, membrane bioreactors, for UV disinfection and ozonization)

3. For processing, drying, and dewatering solids and biosolids by thickening, centrifuging, heat drying, and aerobic digestion

Anaerobic digestion produces methane biogas, which is a GHG emission–neutral energy source for heating and a fuel for vehicles (buses), but a highly potent GHG with GWP of 25 if allowed to escape into the atmosphere.

Aerobic treatment processes require the oxygen supply to be provided by various types of aerators, of which surface aerators and diffused bubble aerators are the most common. Metcalf & Eddy (2007) based on Burton (1996) presented energy use estimates for the aerobic treatment process, given here in Table 7.7. For daily treated

Table 7.7   Energy use of treated volume of municipal used (waste) water (based on Metcalf & Eddy, 2007, and Burton, 1996)

Treatment Process	Energy Use in MJ/1000 m^3 (kW-hr/m^3) Daily Flow Volume of Treated Used Water (m^3/day)		
	10,000	25,000	>50,000
Activated sludge without nitrification	2000 (0.55)	1250 (0.38)	1000 (0.28)
Activated sludge with nitrification with filtration	2500 (0.7)	1800 (0.5)	1500 (0.42)
Membrane bioreactor with nitrification	3000 (0.83)	2600 (0.72)	2300 (0.64)

used (waste) water volume of less than 50,000 m3/day, the energy use per unit treated volume increases.

Energy use of aerobic biological treatment units depends on the BOD in the wastewater and on the aeration equipment used to supply oxygen, which can be either provided by air or pure oxygen. Obviously, pure oxygen is far more energy demanding than air aeration and is typically used for high-strength industrial wastewater that may not be amenable for municipal reuse. In general, air aeration systems can be divided into those in which air is supplied by surface aerators and those that use submerged diffusers. Imhoff and Imhoff (2007), Metcalf & Eddy (2003), and others recommend the following estimates of the energy use per total BOD removed (including both carbonaceous and nitrogenous oxygen demands):

For surface aeration, oxygen energy requirement is 1.2 to 1.4 kg O_2/kWh

For fine bubble diffused air aeration 2.0–2.5 kg O_2/kWh

Imhoff and Imhoff (2007) also provide a rule-of-thumb estimate of the oxygen requirement for oxidation of the BOD in the influent to the aeration system as:

For surface aerators, the oxygen requirement is 2.5 kg of O_2/kg of BOD_5 removed

For fine bubble aeration by diffusers 2.0 kg of O_2/kg of BOD_5 removed

According to the U.S. DOE (2000) report, the CO_2 emissions from power plants are (see also Chapter I):

Coal-fired	950 g of CO_2/kWh
Oil fuel power plants	890 g of CO_2/kWh
Natural gas	600 g of CO_2/kWh

However, in Chapter I, Section I.3.3, it was documented that in the U.S. about 30% of the electric energy is produced by nuclear power plants, hydropower, and renewable sources which do not emit GHGs. A weighted average of 620 g of CO_2/kWh was suggested for the calculation of the carbon imprint in the U.S. A much smaller weighted average is typical for France or Austria, which produce their energy from non-GHG-emitting power plants and renewable energy sources.

For each mole of oxygen consumed in the aeration process, one mole of carbon dioxide is emitted. Hence CO_2 emitted $= (12 + 2 \times 16)/(2 \times 16) = 1.37 \, O_2$ consumed. For example, if the BOD_5 concentration in used water is 300 mg/L $= 0.3 \, \text{kg/m}^3$ and 70% of the removed BOD is converted to CO_2, then the CO_2 emission in an aeration unit removing 95% of BOD_5 will be

$(CO_2 \text{ emitted } (\text{kg/m}^3)) = 1.4(BOD_{ultimate}/BOD_5) \times 0.95 \times 0.3(\text{kg } BOD_5/\text{m}^3 \times 1.37$

$(CO_2 \text{ emitted}/O_2 \text{ consumed}) \times 0.7 = 0.37 \, \text{kg of } CO_2 \text{ emitted per one m}^3.$

This value has to be added to the CO_2 due to the energy use listed in Table 7.7.

Table 7.8   Greenhouse gas production from selected treatment processes (adapted from Keller and Hartley, 2003) in a Queensland, Australia, treatment plant serving a community of 100,000 inhabitants

Treatment Process Sequence	t of CO_2/d from Aerobic COD Removal	t of CH_4/d from Anaerobic COD Removal	t of CO_2/d from External Power generation	Total t of CO_2/d Produced, Including CH_4 Burning	kg CO_2/ kg COD removed from COD Removal	kg CO_2/ kg COD Removed from Power Generation
AS and aerobic digestion	11.9	0.0	17.7	29.5	0.91	1.36
AS and anaerobic digestion	7.6	0.9	7.7	20.5	0.99	0.59
Extended aeration AS	13.7	0.0	17.8	31.4	1.05	1.37
High-rate aeration and anaerobic digestion	5.4	1.3	4.0	16.5	0.96	0.31
Anaerobic/aerobic process	2.4	2.1	0.3	13.5	1.02	0.02

AS = Activated sludge; BOD_5/COD ≈ 0.7; t=ton (SI)

Keller and Hartley (2003) investigated the energy requirement and carbon GHG emissions linked to CO_2 and methane production from several conventional treatment processes due to electric energy used for aeration and mixing in each process, and ranked several biological treatment process options according to the COD removed (Table 7.8). The data in the table include biological treatment and sludge digestion. Such analyses show the disadvantages of the traditional aerobic treatment processes requiring energy for both mixing and aeration. Trickling filters are less energy demanding but may not provide high removal efficiencies.

Wetlands are a special case in respect to GHG emissions. First and foremost, wetland operations use gravity and require no energy to move or mix water. Wetlands do sequester large amounts of carbon—but they also produce methane, which has a GWP of 25. Because they represent a natural system, it is unlikely that methane could be captured and converted to carbon dioxide or hydrogen. Research should focus on development of wetland systems that would maximize carbon sequestering while minimizing methane emissions by suppressing anaerobic methanogenesis decomposition in the substrate and converting most of the accumulating organic carbon to peat. In general, wetlands sequester more GHGs than they emit.

VII.7   EVALUATION AND SELECTION OF DECENTRALIZED WATER RECLAMATION TECHNOLOGIES

VII.7.1   Closed Cycle Water Reclamation

In conclusion we will summarize the knowledge presented in the chapter and apply it to a conceptual design of a water centric cluster water reclamation and recycle

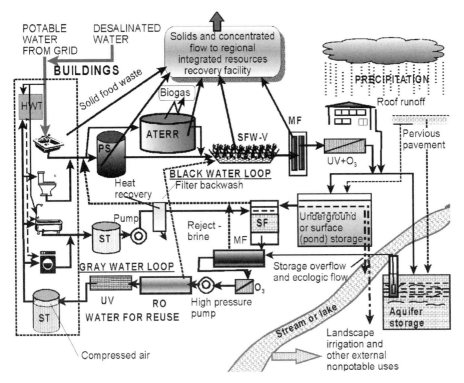

Figure 7.17 Cluster (ecoblock) closed cycle water reclamation and reuse. Legend: ATERR = anaerobic treatment and energy recovery reactor; SFW-H, horizontal subsurface flow wetland; SFW-V, subsurface flow wetland, vertical flow; HWT, reclaimed hot water tank; ST, storage tank; PS, primary settler with solid removal; MF, membrane filter; SF, sand filter; NF, nanofilter; RO, reverse osmosis; UV, ultraviolet disinfection; O₃, ozone addition.

system (Figure 7.17). Figure 7.18 is a water balance flow sheet for a dry day (no precipitation) with no pumping from groundwater storage. The values in the water mass flow chart are based on Table 5.3 in Chapter V, based on the Heaney et al. (2000) synthesis. It is assumed that some standard water conservation such as low-flush toilets and shower restrictors have been installed in the building. The following assumptions and hypotheses are considered in the design of the sustainable closed cycle water and solids management system in a generic cluster/ecoblock:

1. Water reclamation and reuse are conducted in a distributed fashion in clusters/ecoblocks. It is recommended that black and gray water flows be separated and that gray water and stormwater be managed locally in the cluster.

2. It is recommended that recovery of energy, nutrients, and biomass from solids and black water be done in a centralized (integrated) resource recovery facility to which solids in a liquid form (black water, sludge, shredded screening, and kitchen food waste) would be delivered by existing or installed sanitary sewers,

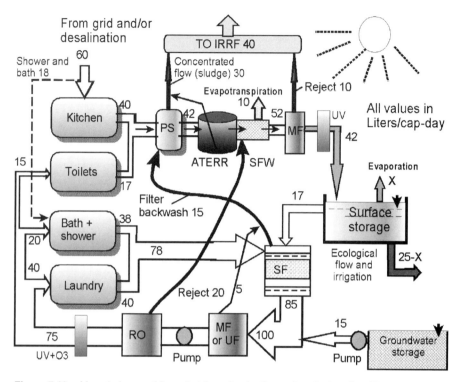

Figure 7.18 Mass balance of flows in L/cap-day for the system featured on Figure 7.17 on a dry day.

and organic solids for recycle such as vegetation residues or excess manure from suburban farming would be delivered by trucking. Heat is best recovered from gray water on-site or in the ecoblock.

3. Potable water from the municipal or regional grid and/or from desalination will be delivered only to kitchen and bathroom sinks and drinking water fountains in the buildings. Under normal circumstances reclaimed water will be perfectly safe for drinking, but direct potable use of reclaimed water is not recommended. Heaney (2007) and Heaney et al. (2000) in studying details of domestic potable water use estimated that kitchen water demand is about 40 L/cap-day, which represents about 17% of the total domestic in-house water use (Chapter V, Table 5.3). With sink disposals, used water volume from kitchen sinks is increased by 20–25%, and some water is used for drinking or included in food. We estimated that 60 L/cap-day will be delivered from the municipal water supply and/or desalination which is commensurate to Masdar ecocity (see Chapter XI) potable water delivery. The estimate of the total U.S. in-house water demand by Heaney (2007) was about 240 L/cap-day without water conservation, but he pointed out that with an aggressive water conservation program (shower flow restrictors, water-conserving toilets) the total use can be reduced to about 115–150 L/cap-day (see Chapter V).

4. It is assumed that water conservation devices (low-flush toilets plus shower flow restrictors) will be installed. This water demand does not include outdoor irrigation, which is much larger and varies greatly over the U.S. Smaller in-house water demand may be typical for other countries.

5. The total black water flow to be treated is kitchen sink flow (excluding dish-washer) and toilet flushes. Based on Heaney's (2007) data, toilet flushing is the largest in-house water use, amounting to about 80 L/cap- day. With standard water conservation (using toilets with 6 liters/flush or less), the black water toilet flow can be reduced to less than 30 L/cap-day. We estimated that 25 L/cap-day will be black water.

6. The primary settler for black water followed by a subsurface flow wetland, is the most simple alternative for black water treatment. It is a covered anaerobic tank promoting settling of solids from the kitchen garbage disposal and toilets into a separate sludge compartment (similar to Imhoff's tank, Figure 1.10), but without digestion of settled solids, to prevent or significantly reduce formation and emissions of methane. Hence solids should be continuously or very frequently removed from the settled solids compartment and conveyed to a central integrated resource recovery facility (IRRF). The primary settler will also receive sand filter backwash water from the graywater recycle loop that will have elevated TSS content.

7. An anaerobic treatment and energy recovery reactor (ATERR) is an alternative unit that could be used for biological treatment of the black water in the cluster and on-site biogas production. It is assumed that a portion of the black used water flow would be directed with the solids to the central IRRF. Verstraete, Van de Caveye, and Diamantis (2009) proposed for this treatment and energy recovery step an upflow anaerobic sludge blanket reactor with a septic tank (UASB-ST) which would produce biogas locally in distributed systems. ATEER type reactor does not need a primary settler.

8. Subsurface flow wetlands, preferably two in a series, with or without a recycle may provide needed post-treatment after primary settling or biogas producing ATERR. Effluent from the wetlands, after membrane filtration and disinfection, should have a quality allowing reuse for external nonpotable uses such as providing ecological flow to the preserved or restored urban water bodies in the green ecotone zones, landscape irrigation, street washing, and so forth. Because constructed wetlands emit GHG methane and nitrous oxides, the wetland should receive only water needed as make-up flow in the graywater recovery cycle (see Figure 7.18).

9. Water is lost from subsurface flow wetlands by evapotranspiration. It is a significantly smaller loss per unit area than that from open air aerated systems, from which water is lost because surface evaporation is stimulated by wind and multiplied by the effects of aeration splash, and because vapor is lost by saturated air from aerators leaving the basin (Novotny and Krenkel, 1973). Nevertheless, in dry summer months, evapotranspiration losses may be significant. For example, in central California (Davis, Fresno), summer evapotranspiration losses range from 5 to 8 mm/day (Hidalgo et al., 2005) and, assuming

3 m^2 of wetland per one connected person in the ecoblock (Vymazal, 2005, 2009), daily loss by evapotranspiration could amount to 15 to 25 L/cap-day.

10. Residual solids and colloidal particles are removed from the wetland effluent by membrane filters. Sludge, about 10% of the flow, is diverted to centralized sludge treatment and resource recovery.

11. Recommended disinfection of the locally treated used water from the black water stream is by ultraviolet rays, followed by ozonization. Adding ozone (mixed with air as it comes from ozone generation) will not only disinfect, it will also oxidize the wetland effluent. Outflows from subsurface flow wetlands almost always have low oxygen concentrations.

12. For nonpotable reuse of reclaimed water in buildings, further treatment is needed. Public health regulations generally require potable water quality for almost all nonpotable uses in the buildings, with the exception of toilet flushing, but even for this use, water should be free of pathogens and not be of objectionable quality. We have concluded that it is not economical to separate toilet reuse water from the other nonpotable building reuses, since it would essentially require a triplicate pipeline system and cause operational difficulties. Regular building plumbing has dual piping, one for hot water and the other one for cold water. The piping system would remain in place with the addition of a cold potable water pipe for kitchen sinks and public drinking fountains, with the alternative of a separate small flow through electric or gas water heaters for the kitchen and bathroom sinks.

13. Rainwater would be collected by roof water collection and either reused on-site or conveyed to the reclaimed water storage. Water from the subbase of pervious pavements would either recharge the groundwater aquifer or be collected and discharged into the cluster surface or underground storage basin. All stormwater conveyance would be on the surface. In the recycle more rainwater can be added to the gray water loop to reduce salinity and other residual conservative compounds accumulating over the cycle. In extended dry weather, water in the gray water recycle loop could be reused up to three times but without any human health problems. See Figure 7.19 for the flow mass balance during a typical average rainy day.

14. While storage of reclaimed used water in underground aquifers has been recommended by many as the safest and best way to enhance water quality for indirect potable reuse, pumping water back from deeper aquifers requires a lot of energy. Since in the system proposed on Figure 7.18 direct or indirect potable reuse is not contemplated, aquifer storage is needed as a supplemental source of reclaimed water during times of irrigation water needs.

15. To minimize losses by evaporation, underground storage basins are preferred. To save space, underground storage as well as treatment can be installed under the constructed wetlands, like the Hudson River WWTP in New York, which is under a created state park. Underground storage, in addition to the treated reclaimed black water, also receives reclaimed rainwater and surface and subsurface runoff. Ecological flow and other nonpotable external water needs in

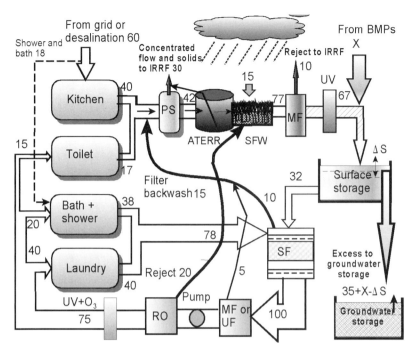

Figure 7.19 Mass balance of flows in L/cap-day during a rainy day with precipitation of 7 mm/day.

the ecoblock (irrigation, etc.) could be extracted from the storage basin or from the surface water body into which the excess effluent and rainwater flows are discharged. Reclaimed water from the storage basin is (1) combined with gray water effluent, (2) supplemented (during dry periods) by water pumped from surface stream and lake/pond storage, and (3) combined with groundwater. The mixture is then treated by the graywater recycle treatment units.

16. Figures 7.17 to 7.19 feature sand filter, macro- or nanofilter, and reverse osmosis sequence for treatment of the recycled gray water with the addition of water from the storage basin. In most cases only two solids removal units may be needed, and the sequence may have to be determined by a treatability study. The effluent should be clear and, if RO is included, it could be soft (Ca^{++} and Mg^{++} ions are easier to remove by RO process than Na^+ and Cl^-) and have dissolved solids content of less than 500 mg/L. This is beneficial to laundry and shower/bath waters because it reduces the soap and detergent requirement. But such water, though perfectly safe for drinking, may not have a good taste. Taste is not relevant because the water is not intended for drinking and should be marked as "not for drinking" at the points of use. Adding ozone improves taste and also odor. It should be noted that used water from bathing, showering, and laundry has medium-level BOD and pathogen contents; therefore, sand filtration or microfiltration is needed as a first step. For this reason,

backwash from the sand filter should be returned to the primary settler of the black water stream.

17. Reverse osmosis requires high pressure. The residual pressure after RO should be high enough to elevate the effluent into the top floors of a medium-height building. A standard pressure tank with compressed air is a flow equalizer and flow controller. Brine water from RO may contain nitrates, phosphates, pathogens, organic colloids,and the like; therefore, it should be returned to the wetland if nitrate content is high. About 15 to 30% of flow is lost as brine (reject) from nanofiltration and/or RO.

The water reclamation/reuse scheme presented herein will reduce water demand to about 60 L/cap-day. A near 100% recycle, though possible (e.g., the water recycle loop on the International Space Station), is not desirable because reclamation and reuse in closed-loop systems work best when there is a constant input of fresh make-up water and corresponding release of concentrated wastewater and sludge liquids with impurities that were not removed in the system. This system does not include direct or indirect use of reclaimed water for drinking. Considering the make-up potable water input is 60 L/cap-day and the average domestic indoor water demand in the U.S. is about 200 L/cap-day, implementing water conservation, reclamation, and reuse would reduce water demand by 70%. However, it should be realized that this type of reclamation and reuse/recycle requires significant energy input that, to meet net zero carbon footprint balance, would have to be offset by energy production in the integrated resource recovery facility (see Chapters VIII and X) and by renewable energy sources (solar, wind) throughout the development.

REFERENCES

Agrawal, L. K., Y. Ohashi, E. Mochida, H. Okui, Y. Ueki, H. Harada, and A. Ohashi (1997) "Treatment of raw sewage in a temperate climate using a UASB reactor and the hanging sponge cubes process," *Water Science and Technology* 36 (6–7):433–440

AMTA (2007) *Membrane Filtration*, American Membrane Technology Association, Stuart, FL, http://www.amtaorg.com/amta_media/pdfs/2_membranefiltration.pdf

Battistoni, P., R. Boccadoro, P. Ravan, and F. Cecchi (2001) "Struvite crystallisation in sludge dewatering supernatant using air stripping: the new full-scale plant at Treviso (Italy)," *Second International Conference on Recovery of Phosphorus from Sewage and Animal Wastes*, Noorwijkerhout, Holland, March 12–13

Burton, F. L. (1996) *Water and Waste Water Industries: Characterization and Energy Management Opportunities*, CR-10691, Electric Power Institute, St. Louis, MO

Carmichael, W. W. (1997) "The cyanotoxins," *Advances in Botanical Research* 27, pp. 211–256

Cecchi, F., P. Battistoni, and R. Boccadoro (2003) *Phosphate Crystallisation for P. Recovery Applied at Treviso Municipal Wastewater Treatment Plant* (Italy), Universita degli Studii di Verona (University of Verona, Italy) http://www.nhm.ac.uk/research-curation/research/projects/phosphate-recovery/Treviso2003.pdf, accessed June 2010

Deublern, D. and A. Steinhamer (2008) *Biogas from Waste and Renewable Resources*, Wiley-VC, Weinheim, Germany

Ganrot, Z. (2005) *Urine Processing for Efficient Nutrient Recovery and Reuse in Agriculture*, PhD Thesis, Department of Environmental Sciences and Conservation, Göteborg University, Sweden

Ganrot, Z., A. Slivka, and D. Göran (2008) "Nutrient recovery from human urine using pre-treated zeolite and struvite precipitation in combination with freezing-thawing and plant availability tests on common wheat," *Clean Soil and Water* 36 (1):45–52

Halmann, M. M. and M. Steinberg (1999) *Greenhouse Gas Carbon Dioxide Mitigation: Science and Technology*, CRC Press, Boca Raton, FL

Heaney, J. P. (2007) "Centralized and decentralized urban water, wastewater systems & storm water systems," in *Cities of the Future: Towards Integrated Sustainable Water and Landscape Management* (V. Novotny and P. Brown, eds.), IWA Publishing, London, UK

Heaney, J. P., L. Wright, and D. Sample (2000) "Sustainable urban water management," Chapter 3 in *Innovative Urban Wet-Weather Flow Management Systems* (R. Field, J. P. Heaney, and R. Pitt, eds.), TECHNOMIC Publishing Co., Lancaster, PA, pp. 75–122

Henze, M., and Y. Comeau (2008) Wastewater concentrations, in *Biological Wastewater Treatment - Principles, Modeling, Design* (M. Henze, M. C,M, van Looschrecht, G.A. Ekama, and D. Brdjanovic (2008), IWA Publishing, London, UK

Hidalgo, H. G., D. R. Canyan, and M. D. Dettinger (2005) "Sources of variability of evapotranspiration in California," *Journal of Hydrometeorology* 6, pp. 1–10

Imhoff, K. and K. R. Imhoff (2007) *Taschenbuch der Stadtentwässerung* (*Pocket Book of Urban Drainage*), 30th ed., R. Oldenburg Verlag, Munich, Germany

Kadlec, R. H. and R. L. Knight (1996) *Treatment Wetlands*, CRC/Lewis Press, Boca Raton, FL

Kadlec, R. H. and K. R. Reddy (2001) "Temperature effects in treatment wetlands," *Water Environment Research* 73 (5):543–557

Keller, J. and K. Hartley (2003) "Greenhouse gas production in wastewater treatment: Process selection is the major factor," *Water Science and Technology* 47 (12):43–48

Krenkel, P. and V. Novotny (1980) *Water Quality Management*, Academic Press, Orlando

Lamlon, S. H., and R. A. Savidge (2003) "A reassessment of carbon content in wood: Variations within and between 41 North American species," *Biomass and Bioenergy* 25 (4):381–388

Levin, D. A., H. Zhu, M. Beland, N. Cicek, and B. Holbein (2006) Potential for hydrogen and methane production from biomass residues in Canada, *Bioresource Technology* 96 (3):654–660

Lettinga, G., A. F. M. Van Velsen, S. W. Hobma, W. de Zeeuw, and A. Klapwijk (1980) "Use of upflow sludge blanket (USB) reactor for biological wastewater treatment," *Biotechnology and Bioengineering* 22, pp. 699–734

Lettinga, G. and L. W. Hulshoff Pol (1991) "UASB – process design for various types of wastewater," *Water Science and Technology* 24 (8):97–109

Levine, A. D. and T. Asano (2004) "Recovering sustainable water from wastewater," *Environmental Science and Technology* 38 (11):201A–208A

Li, F. Y., K. Wichmann, and R. Otterpohl (2009) "Review of the technological approaches for grey water treatment and reuses," *Science of the Total Environment* 407, pp. 3439–3449

Malmqvist, P. A. and G. Heinicke (2007) "A strategic planning of the sustainable future wastewater and biowaste system in Göteborg, Sweden," Chapter 18 in *Cities of the Future: Towards Integrated Sustainable Water and Landscape Management* (V. Novotny and P. Brown, eds.), IWA Publishing, London, UK, pp. 284–299

Maucr, M., P. Schweigler, and T. A. Larsen (2003) "Nutrients in urine: energetic aspects of removal and recovery," *Water Science and Technology* 48 (1):37–46

Metcalf & Eddy (2003) *Wastewater Engineering: Treatment and Reuse*, 4th ed. (revised by G. Tchobanoglous, F. L. Burton, and H. D. Stensel), McGraw-Hill, New York

Metcalf & Eddy (2007) *Water Reuse: Issues, Technologies, and Applications*, McGraw-Hill, New York

Mitchell, C., K. Abeysuriya, and D. Fam (2008) *Development of Qualitative Decentralized System Concepts – For the 2009 Metropolitan Sewerage Agency*, Institute for Sustainable Futures, University of Technology, Sydney, Australia

Mitsch, W. J. and J. G. Gosselink (2007) *Wetlands*, 4th ed., John Wiley & Sons, Hoboken, NJ

Novotny, V. (2003) *WATER QUALITY: Diffuse Pollution and Watershed Management*, 2nd ed., John Wiley & Sons, Hoboken, NJ

Novotny, V. and P. A. Krenkel (1973) "Evaporation and Heat Balance in Aerated Basins," American Institute of Chemical Engineers, *Water* 73, and 74th National Meeting, March 11–15, 1973, New Orleans, LA

Novotny, V., K. R. Imhoff, M. Olthof, and P. A. Krenkel (1989) *Handbook of Urban Drainage and Wastewater Disposal*, John Wiley & Sons, Hoboken, NJ

Oilgae (2009) *Oilgae Guide to Algae-based Wastewater Treatment*, Oilgae, Tamilnadu, India, www.oilgae.com (accessed December 2009)

Parsons, S. A., F. Wall, J. Doyle, K. Oldring, and J. Churchley (2001) "Assessing the potential for struvite recovery at sewage treatment works," *Environmental Technology* 22, pp. 1279–1286

Šálek, J., Z. Žáková, and P. Hrnčíř (2008) *Přírodní čištní a využívání vody* (*Natural Treatment and Reuse of Water*), ERA publishing, Brno, Czech Republic

Schnoor, J. L. (2004) "A hydrogen-fueled economy," *Environmental Science & Technology* 38 (11):191A

Sheehan, J., T. Dunahy, J. Benemann, and P. Roesler (1998) "*A Look at the U.S. Department of Energy's Aquatic Species Program – Biodiesel from Algae*," National Renewable Energy Laboratory, NREL/TP-580-24190, Golden, CO

Sovik, A. K. *et al.* (2006) "Emissions of greenhouse gases nitrous oxide and methane from constructed wetlands in Europe," *Journal of Environmental Quality* 35, pp. 2360–2375

Tettey-Lowor, F. (2009) *Closing the Loop between sanitation and agriculture in Accra, Ghana*, SWITCH report, UNESCO, Delft, http://www.switchurbanwater.eu/outputs/results.php?pubtype_select=6

U.S. Department of Energy (2000) *Carbon Dioxide Emissions from the Generation of Electric Power in the United States*, U.S. DOE, Washington, DC

U.S. Environmental Protection Agency (1991) *Guidance for Water Quality-Based Decisions: The TMDL Process*, EPA 440/4-91-001, U.S. EPA, Washington, DC

U.S. Environmental Protection Agency (1999) *Biosolids Generation, Use and Disposal in the United States*, EPA 530-R-99-009, Office of Solid Waste, Washington, DC

U.S. Environmental Protection Agency (2000) *Improved Enumeration Methods for the Recreational Water Quality Indicators: Enterococci and Escherichia Coli*, EPA/821/R-97/004, Office of Science and Technology, Washington, DC

Verstraete, W., P. Van de Caveye, and V. Diamantis (2009) Maximum use of resources present in domestic "used water," *Bioresources Technology* 100:5537–5545

Vymazal, J. (2005) "Horizontal sub-surface flow and hybrid constructed wetland systems for wastewater treatment," *Ecological Engineering* 25, pp. 478–490

Vymazal, J. (2009) Kořenové čistírny odpadních vod: Dvacet let zkušeností v České Republice ("Wetland wastewater treatment systems: Twenty years of experience in the Czech Republic") *Vodní Hospodářství (Water Management)* May 2009, pp. 113–118

Vymazal, J. and I. Kröpfelová (2008) *Wastewater Treatment in Constructed Wetlands with Horizontal Sub-Surface Flow*, Springer Verlag, Dortrecht, The Netherlands

Water Environment Federation (1990) *Natural Systems for Wastewater Treatment, Manual of Practice FD-16*, Water Environment Federation, Alexandria, VA

Water Environment Federation (1996) *Wastewater Disinfection, Manual of Practice FD-10*, Water Environment Federation, Alexandria, VA

White, G. C. (1999) *Handbook of Chlorination, and Alternate Disinfectants*, 4th ed., John Wiley & Sons, Hoboken, NJ

Zheng, G. N., W. Tu, and P. Y. Fei (2010) "Development and practice of ecological sanitation system," in *Water Infrastructure for Sustainable Communities: China and the World* (X. Hao, V. Novotny, and V. Nelson, eds.), IWA Publishing, London, UK

VIII

ENERGY AND URBAN WATER SYSTEMS— TOWARDS NET ZERO CARBON FOOTPRINT

VIII.1 INTERCONNECTION OF WATER AND ENERGY

VIII.1.1 Use of Water and Disposal of Used Water Require Energy and Emit GHGs

The most obvious nexus between water and energy is the fact that delivering water to urban (and other) users, recovering the resources from used water, and safely disposing of the residuals into the environment require energy. Energy production, if done in conventional fossil fuel power plants, emits greenhouse gases (GHGs), leading to global warming. The consequences of GHGs on life on earth have been discussed in Chapter I. While GHG emissions from urban water systems are relatively small in comparison with other economic sectors, no sector of the economy can be excluded. The National Science and Technology Council (2008), established by the U.S. president, estimated that the U.S. building sector accounts for 40% of total primary energy use and GHG emissions, compared to 32% for the industrial sector, and 28% for transportation. To some degree, urban water systems and energy production compete for water and are, after agriculture, the most important users of water. However, power production and urban water use can and must work together symbiotically because of increasing tapping less conventional sources of water, such as seawater and brackish high-salinity water, which requires a lot of energy. It was shown in Chapter V that large desalination plants (e.g., in countries located in the Gulf of Arabia and Florida) require large quantities of electric energy to generate the pressure required for reverse osmosis, but a significant portion of the used energy can be returned to the grid as the reject water is depressurized and returned to the environment. However, many wealthy Middle Eastern countries use thermal evaporation for producing fresh water which requires five times more energy for desalting than

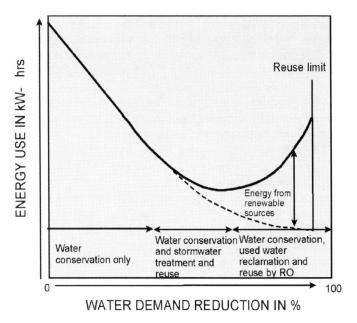

Figure 8.1   The relation of water demand reduction to energy use.

reverse osmosis. The used water can conceivably be diluted with the cooling water from the power plants (see Tampa Bay desalination in Chapter V). The increased use of energy-demanding water reclamation and reuse processes and pumping brackish water from deeper geological groundwater zones may offset the energy savings from water conservation, as is shown in Figure 8.1. This figure shows three possible stages of the water-energy nexus. From the discussion in Chapter V, it follows in most cases that water conservation—that is, reduction of demand by using more efficient appliances, xeriscape, and plugging the leaks and losses—does not require large amount of extra energy; hence, the energy use reduction is directly proportional to the reduction of the water demand. In the second phase of saving water, reclaimed stormwater will be used, which will require pumping and treatment. Both require energy (see Chapter V). The third phase leading to substantial reduction of water demand from the grid will require production of NEWater (using the Singapore acronym for highly treated reclaimed water) which, under current accepted practices, will be mixed with the surface water or groundwater, pumped, and retreated again—which requires a lot of energy. Reverse osmosis, microfiltration, and disinfection by UV and ozone are energy-demanding unit processes. This extra energy required for high-level reclamation and reuse should be derived from solar and wind renewable sources and by reclaiming energy from wastewater (Chapter VII). The discussion in Chapter VII indicated that there must be a limit on the percentage of water use reduction. If "closed" recycle systems were to be fully closed, there is nothing to prevent accumulation of the nondegradable potentially harmful compounds that may pass the reverse osmosis barrier. The reuse limit is not known and must be ascertained by research. Ozonization (another high user of energy) can to some degree break down nonbiodegradable organic molecules.

This chapter focuses on energy conservation and identification of alternate energy sources that would lead to meeting the One Planet Living (OPL) ecocity criteria and sustainable water and GHG emissions footprint goals (Chapter II), notably achieving net zero carbon emissions and providing adequate water for all, without irreversibly overmining the earth or overusing water resources. The discussions and presentations in this chapter follow those of Chapters V and VII, which reported energy use and potential savings in the processes of water treatment, domestic water use, water reclamation, and water reuse. This chapter will also outline new potential energy sources: recovering energy from used water, sludge, and algae feeding on used water; ground/geothermal sources; wind; photovoltaics; microbial fuel cells; biogas (methane); and hydrogen recovery from used water, sludge, and other organic solids. It will also briefly describe regional energy and resource recovery units that can only be installed on a cluster-wide basis (ecoblock) or on a small regional scale, but would not be feasible in individual homes, as also discussed in Chapters V and VII.

VIII.1.2 Greenhouse Gas Emissions from Urban Areas

CO_2 emissions vary widely among nations. Until recently the U.S. was the largest emitter of GHG gases, but recently it was overtaken by China. If statistics are presented in emissions per person, the Middle Eastern states are the largest emitters (Table 8.1), but the U.S., Australia, and Canada are also in the top ten. It should be pointed out that various statistics differ, and the emissions vary from year from year, but generally, they seem to be leveling off or even decreasing in this century, in some countries. It is expected that this decreasing trend will continue following the Kyoto Protocol and the 2009 Copenhagen summit (nonbinding) agreement.

The data in Table 8.1 should be complemented by information on undeveloped countries. The median value in the full Wikipedia table, which lists more than 200 countries, is 3.2 tons of CO_2 per capita in a year, and 30% of the world's poorest countries had per capita CO_2 annual emissions of less than one ton.

It would be natural to associate the highest per capita emissions with the cities; however, Dodman's report (2009) shows that population density and type of city affect the per capita emissions (see also Chapter X). For example, Washington, DC, has the highest emissions, but this may be a result of the city having far more office space, with more commuters per one real inhabitant than any other city in the world. Dodman also found that large cities emit per capita less GHG than the national average. For example, London's emissions (6.2 tons/capita-year) are 50% less than the national average (9.4 tons/capita-year). The same is true for U.S. data. The average for the one hundred largest cities analyzed by Gleaser and Kahn (2008) is 8.5 tons/capita-year, without considering industries, while the national average is 19 tons/capita-year (which includes industries).

Table 8.1   Per capita CO$_2$ emissions statistics

Top Ten Countries in the CO$_2$ Emissions in Tons/Person-Year in 2006[1]									
Qatar	UAE	Kuwait	Bahrain	Aruba	Luxembourg	U.S.	Australia	Canada	Saudi Arabia
56.2	32.8	31.8	28.8	23.3	22.4	19.1	18.8	17.4	15.8
Selected World Cities Total Emissions of CO$_2$ Equivalent in Tons/Person-Year[2]									
Washington, DC	Glasgow, UK	Toronto, CA	Shanghai, China	New York City	Beijing, China	London, UK	Tokyo, Japan	Seoul, Korea	Barcelona, Spain
19.7	8.4	8.2	8.1	7.1	6.9	6.2	4.8	3.8	3.4
Selected U.S. Cities Domestic Emissions of CO$_2$ Equivalent in Tons/Person-Year[3]									
San Diego, CA	San Francisco, CA	Boston, MA	Portland, OR	Chicago, IL	Tampa, FL	Atlanta, GA	Tulsa, OK	Austin, TX	Memphis, TN
7.2	4.5	8.7	8.9	9.3	9.3	10.4	9.9	12.6	11.06

[1]Wikipedia (2009)
[2]Dodman (2009)
[3]Gleaser and Kahn (2008)
[4]Values include transportation, heating, and electricity

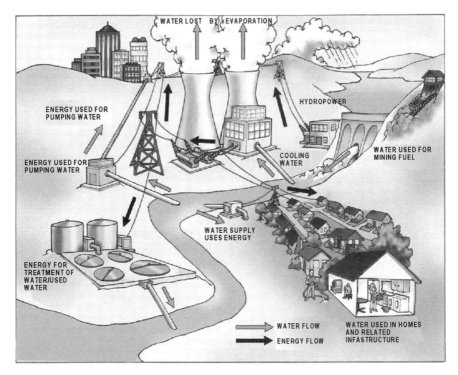

Figure 8.2 Water-Energy nexus under current standard conditions, showing also the virtual water demand for cooling and other use and consumptive losses (Credit U.S. Department of Energy).

VIII.1.3 The Water-Energy Nexus on the Regional and Cluster Scale

Figure 8.2 shows the water-energy nexus and competing uses of water under normal current conditions. The water-energy nexus cannot be evaluated on a single-house level. Energy in cities is used, as shown in Figure 8.2, to deliver and treat water, including used water, and water is lost in fossil fuel power plants by evaporation from cooling towers, by evapotranspiration of irrigated lands, and by leaks in delivery systems.

It has been pointed out that the current interlinking of water and energy is not sustainable; it emits significant quantities of carbon and also methane (e.g., from sludge digestion, landfills, and anoxic processes in wastewater treatment plants). The sustainability goal for urban areas—Cities of the Future—calls for implementation of water reclamation and reuse, water and energy conservation, and replacing power production from fossil fuels with renewable energy sources (wind, solar, energy recovery from used water and sludge, geothermal), in addition to reduction of energy use and GHG emissions from traffic and industrial and commercial operations.

Figure 8.3   A large 1.14-MW solar (photovoltaic) panel array in 2006 at Sonoma Mountain Valley (California) was installed as a start to achieve 100% of power with renewable energy and zero carbon emissions One Planet Living (OPL) goals. Courtesy Sonoma Mountain Village -SOMO.

On a cluster (ecoblock) scale, energy and other resources can be recovered from reclaimed used water and organic solids, from wood chips from tree pruning, and/or from trash in the integrated resource recovery plant (see section VIII.6). Also installation of solar panels and wind farms is less expensive and more efficient on a cluster scale than in individual homes, as proven in ecocities/villages Masdar (UAE) and Sonoma Mountain Village (Figure 8.3), described in Chapter XI. In both locales, large solar panel arrays were installed before construction began. However, many environmentally conscious "green" homeowners and businesses in developed countries are finding it very attractive to install photovoltaic solar panels and energy-saving appliances or to buy buildings that already come with net zero carbon footprint rating and low water usage. Such homes or installations are already on the market in Europe and are emerging in China, the U.S., Australia and elsewhere.

In 2007, 55 billion m^3 of water were used by the population of 301.3 million in the U.S. Using the U.S. EPA estimate of 3% energy use for water would result in a unit energy use of 2.26 kWh/m^3 attributed to water use. The corresponding carbon emission was 1.37 kg CO_2/m^3.

Box 8.1 contains units and parameter values for estimating and evaluating power and energy of the water-energy nexus.

BOX 8.1 UNITS OF POWER AND WORK/ENERGY (HEAT) USED IN ANALYSES OF ENERGY EFFICIENCY

Force

Newton N	$F = \text{mass} \times \text{acceleration}$	Newton $1\,N = 1\,kg \times m/sec^2$
Weight	$G = \text{mass} \times \text{gravity acceleration}$	$9.81\,m/sec^2 \times 1\,m = 9.81\,N$

Work/Energy

Joule J	Work done when a force of 1N moves the point of application a distance of 1 meter	$J = 1N \times m = 1\,kg \times m^2/sec^2$
Calorie	Energy (heat) required to increase temperature of 1 cm^3 of water with temperature of 4°C by 1°C	$Cal = 4.19\,J$
Specific heat of water	$C_p = 1\,Cal/(g \times °C)$	$1\,Cal/(g°C) = 4.19\,J/(g°C)$
British thermal unit BTU	$1\,BTU = 1.055 \times 10^3\,J$	
Watt-hour	Energy used when 1 watt of power is applied for one hour	$W\text{-}hr = 3600\,J$
Kilowatt-hour	$kWh = 3.6 \times 10^6\,J = 3.6\,MJ$	

Power

Watt W	Power that gives rise to 1 joule of energy in one second $W = J/sec$
Kilowatt	$1\,kW = 1000\,W$
Megawatt	$1\,MW = 10^6\,W$
1 ton of air conditioning (TAC) (1 ton of ice melted in 24 hrs)	$1\,TAC = 3594\,W$
1 horsepower HP in SI units (1 HP lifts 75 kg of weight up one meter in one second)	$HP = 75\,kg \times 9.81\,m\,sec^{-2}/sec = 736\,watts$
$1\,cm^3 = 1\,ml = 0.001\,L$	1 kWh delivered from U.S. grid in 2010 $= 0.62\,kg$ of CO^2 emitted
1 terawatt(TW) $= 10^{12}$ watts	1 megawatt $= 10^6$ watts $= 1000$ kilowatts

VIII.1.4 Net Zero Carbon Footprint Goal for High-Performance Buildings and Developments

The requirement for ecocities is not just to reduce GHG emissions, as expressed in the LEED energy efficiency standards (Section II.4.2 in Chapter II). The demand set forth in the One Planet Living (OPL) criteria by the World Wildlife Fund calls for ecocities to become carbon neutral. The same call has also been issued by the National Science and Technology Council (NSTC) (2008) of the U.S. president and by the British government, which called for development and implementation of both net zero carbon footprint and high-performance building technologies. **Zero energy building (ZEB)** or **net zero energy building** is a general term applied to a building with net zero energy consumption and zero carbon emissions on an annual basis. This implies that all energy used within these high-efficiency developments and ecocities must be produced by the renewable energy sources, and the demand for electricity must be much smaller than that in typical urban areas of the world. To meet this criterion the majority of the buildings within the ecocity development should have net zero carbon emissions; consequently, ecocities must be frugal with their energy use and, because energy use is closely tied to water use, must implement water conservation and reuse. Several ecocities being built are complying with the OPL criteria, including the net zero carbon and very small water footprints (see Chapter XI).

Water and energy. The water-energy nexus can also be looked at from the other end. All thermal and even some solar power plants need cooling water that takes the excess heat from the power production and releases it into the environment, directly by once-through cooling into water bodies or by cooling towers into the air and water. Thermal efficiency of thermal power plants, defined as a ratio of energy produced divided by the energy in fuel, is relatively low, about 45% for fossil fuel plants and 35% for nuclear. This means that 55 to 65% of energy in fuel from a thermal fossil fuel or nuclear power plant is released in the cooling system into either air (dry cooling) or water (wet cooling) or both. An example of the heat released into the environment is calculated in Box 8.3. Cooling water demand for energy production is the second-largest demand for water, after agriculture. The waste heat transfer to the cooling system occurs in the condenser, in which steam, after running the turbines, is converted back to liquid. A phase change (condensation) releases approximately 540 calories for each cm^3 (2257 kJ/kg) of liquid water equivalent of steam condensed into cooling water at the boiling point temperature. If the plant uses fossil fuel, about 10% of the waste heat will be lost through the smokestack, and the rest through the cooling system. Box 8.1 shows the calculation of power and energy use, and Box 8.2 presents a calculation of cooling water needed for a power plant.

Implicitly, a net zero carbon footprint building would be in the category of **high-performance green building**, which NSTC (2008) defines as one that, during its life cycle:

- Reduces energy, water, and material resource use
- Improves indoor environmental quality by reducing air pollution and improving thermal comfort, lighting, and the acoustic environment

---

BOX 8.2   HEAT EJECTED INTO THE ENVIRONMENT BY COOLING

A thermal power plant has electric energy power capacity, PC, of 1000 megawatts (MW). Calculate the amount of heat ejected into the environment. Efficiency of the plant is $e = 40\%$.
Amount of heat ejected into the environment is

$$C_E = \text{PC}/(1 - e/100) = 1000/(1 - 40/100) = 600 \text{ MW} = 1666 \times 10^6 \text{ watts}$$

In fossil fuel plants, 90% of the heat ejected is conveyed to cooling water and 10% directly to air.
The amount of cooling water needed can be calculated, assuming that the temperature differential through the condenser is 8°C. The specific heat of water is 4.19 J/(g°C), and 1 m^3 of water has a mass of 10^6 grams. Hence one cubic meter can absorb from the condenser the amount of heat that equals

$$C_H = 8[°C] \times 4.19 \text{ [J/(g°C)]} \times 10^6 \text{ [g/m}^3] = 33.52 \times 10^6 \text{ J/m}^3 \text{ of heat energy}$$

Since watt = joule/sec, the cooling water flow required to take the heat is

$$Q = C_E/C_H = 0.9 \times 1666 \times 10^6 \text{ [J/s]}/33.52 \times 10^6 \text{ [J/m}^3] = 44.73 \text{ m}^3/\text{sec}.$$

This would represent 0.045 m^3/sec of flow per 1 MW power capacity of the cooling water flow.
Cooling water is generally recycled by employing cooling towers that take the excess heat from the recycled water in the cooling system and release it in the form of vapor in the air, as shown in Figure 8.2. The water loss by evaporation represents about 2% of the cooling water recycle that must be replaced by make-up flow. Additional make-up flow is needed to prevent accumulation of salts in cooling water; hence, the total make-up water requirement is 4%, or $0.04 \times 0.045 = 0.0018$ m^3/(sec) per one MW power plant capacity, which is provided from a nearby water body or even from the public water supply.

- Reduces negative impacts on the environment, including air and water pollution and waste generation
- Increases the use of environmentally preferable products, including bio-based, recycled content, and nontoxic products
- Increases reuse and recycling opportunities
- Integrates systems
- Reduces environmental and energy impacts of transportation through building location and site design, by supporting a full range of transportation choices for occupants
- Considers its indoor and outdoor effects on human health and the environment, including: (1) improvements in workers' productivity, (2) the life cycle impact of materials and operations, and (3) other factors that are appropriate

The agenda of NSTC contains the following six major technology goals, which define the major transformational advances needed for energy, water, and material use for net zero energy and high-performance green buildings:

1. Develop the enabling measurement science to achieve the net zero energy, sustainable high-performance building technologies.
2. Develop net zero energy building technologies and strategies.
3. Develop the scientific and technological bases for significant reductions in water use and improved rainwater retention.
4. Develop processes to minimize waste generation from building construction, renovation, and demolition.
5. Develop the knowledge and associated energy efficiency technologies and practices needed to promote occupant health, comfort, and productivity.
6. Enable technology transfer for net zero energy, high-performance green buildings.

Under goal 3, NSTC envisions reductions of water use through more efficient water-saving appliances, fixtures, and water systems (covered in Chapter V); developing analyses and technologies to overcome environmental, health, and technological barriers to widespread water recycling and increased water harvesting (covered in Chapters V and VII); and development of Low Impact Development (LID) practices to significantly reduce stormwater runoff (covered in Chapter IV).

By comparing OPL criteria with the above NSTC goals, it can be seen that most of the NSTC goals, including the net zero carbon footprint, are also included in the One Planet Living criteria for ecocities presented in Chapter II. In the context of high-performance buildings and ecocity developments, *life cycle* represents all stages of the buildings development's useful life (including components, equipment, systems, and controls), beginning at the conception of the project and continuing through the site selection, design, construction, landscaping, commissioning, operation, maintenance, renovation, and demolition, removal, and recycling. *Life cycle assessment* is a comprehensive system approach for measuring the environmental performances, impacts, and consequences, beginning at raw materials acquisition and continuing through manufacturing, transportation, installation, use, reuse, and end-of-life waste management.

Sometimes, definitions of net zero carbon footprint imply that the buildings and other associated infrastructure must be independent of the municipal grid—that is, no energy would be supplied by the public energy-producing systems. While this would be theoretically possible, it is impractical for the time being. For example, net zero developments may produce excess energy on sunny and/or windy days, but they will not have enough energy during dark and/or windless hours. The technology for storing excess energy on-site has not yet been fully developed; converting energy into hydrogen fuel from water by excess energy and then using hydrogen as a source of energy during the times of low or no solar and wind energy production seems promising, but it could be costly (see section VIII.5). Smart network electricity grids are generally capable of regulating power production during the day, that is, reducing production when excess electricity would be provided by green developments and renewable sources, and activating more production (e.g., from hydropower storage power plants) during times when green energy would be insufficient.

Current research and studies sponsored by governments in several countries have shown (as also reported in this book) that using current already tested technologies can reduce a building's energy consumption and CO_2 emissions by 30–50% (NSTC, 2008; U.S. GBC, 2007). Furthermore, laboratory studies on new technologies integrated within the green building designs have shown that they can reduce energy consumption and GHG emissions by as much as 70% (NSTC, 2008). At this level of energy reduction, on-site renewable energy sources featured in this chapter, such as wind, solar, geothermal, and heat extraction from used water, can provide the remaining energy needs. Based on these estimates, NSTC (2008) has outlined the path towards the net zero carbon footprint of U.S. green buildings as shown in Figure 8.4.

The council's assessment has included the impact of the U.S. EPA's and Department of Energy's Energy Star program of licensing energy-using appliances, and also considered the potential for implementing nationwide solar energy generation. They estimated that current and future efficiency improvements could attain 60 to 70% reduction of the energy demand, with the remaining 30–40% to be provided by renewable (wind and solar) power.

On top of these estimates, energy can be provided by bioreactors and microbial fuel cells in cluster or regional water reclamation plants (see Section VIII.5), which

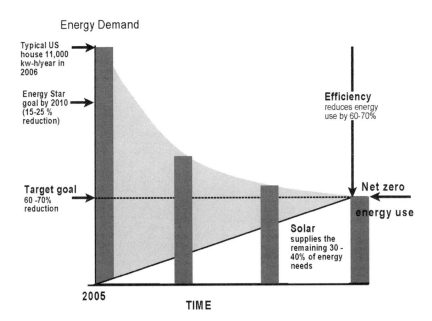

Figure 8.4 Achieving net zero energy performance of green buildings, as outlined by the National Science and Technology Council (2008) of the U.S. president.

can then run the water reclamation plants, returning the excess energy back to the electric grid, and by biogas conversion to hydrogen and heat. These improvements will reduce the overall energy use and GHG emissions in the urban area and could be counted towards the net zero GHG emission goals. Chapter XI describes successful net zero energy city development designs in China, the UAE, and the U.S. Other developments are being planned in the U.K., the Netherlands, and elsewhere. For example, the British government has a goal that by 2016 all buildings will have net zero carbon emissions. The U.K. government's definition of a zero-carbon home is net zero emissions from all energy used by the building over one year. This means that energy needed for heating, lighting, hot water, and all electrical appliances in the house, such as TVs and computers, must be obtained from renewable sources. Many architects have already designed and built net zero carbon footprint houses. The green net zero buildings and cluster (ecoblocks/subdivisions) developments will usually include:

- Passive architectural features for heating and cooling
  - Southern exposure with large windows regulated by shutters
  - Cross ventilation
  - Green roofs
  - A lot of insulation
  - Energy-efficient lighting
- Landscape features
  - Shading trees (planting trees also sequesters carbon), urban forests
  - Pavement cooling by reclaimed water to reduce urban heat island effects
- Renewable energy sources
  - Solar photovoltaic and concentrated solar thermal roof panels
  - Wind turbines
  - Heat in used water (especially in grey water and sludge) and stormwater
- Water conservation and reuse, addressing the entire water (hydrologic) cycle within the development, including rainwater harvesting and storage
- Distributed stormwater and used (waste) water management to enable efficient used water reuse and renewable energy production
- Xeriscape of the surroundings that reduces or eliminates irrigation and collects and stores runoff from precipitation
- Energy-efficient appliances (e.g., water heaters), treatment (e.g., reverse osmosis), and machinery (e.g., pumps, aerators)
- Connecting to off-site renewable energy sources such as solar power and wind farms
- Organic solids management for energy recovery
- Connection to low or no GHG net emissions heat/cooling sources such as heat recovered from used water or from the ground

- Smart metering of energy and water use and providing flexibility between the sources of water and energy
- Sensors and cyber infrastructure for smart real-time control

Kibert (2007) said that green buildings virtually always make economic sense on a life-cycle cost (LCC) basis, though they may be more expensive on a capital, or first-cost, basis. Sophisticated energy-conserving lighting and air-conditioning systems responding to interior and exterior climates will cost more than their conventional, code-compliant counterparts. Rainwater-harvesting systems collecting and storing rainwater for nonpotable uses will require additional piping, pumps, controls, storage tanks, and filtration components. Kibert stated many green building systems will recoup their original investment within a relatively short time. As energy and water prices rise due to increasing demand and diminishing supply, the payback period will decrease. LCC provides a consistent framework for determining the true economic advantage of these alternative systems by evaluating the performance over the course of a building's useful life.

Figure 8.5 shows the passive architectural components of green buildings, such as cross ventilation and southern exposure for heating. Additional examples of

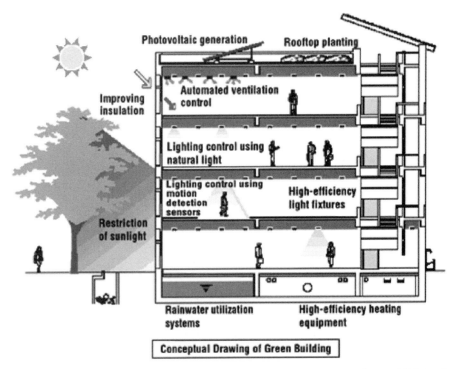

Figure 8.5 Passive energy-saving features in high-performance buildings. Source: Ministry of Land, Infrastructure, Transport, and Tourism, Tokyo (Japan), http://www.mlit.go.jp/english/2006/p_g_b_department/05_env-report/02_organization.html.

architectural passive features will be presented in Chapter XI. Implementation of green roofs that provide insulation and cooling is discussed in Chapter IV.

Many excellent texts are available on green building designs and achieving LEED (platinum) energy efficiency certifications, such as monographs by the American Institute of Architects (2004), Kibert (2007), Gevorkian (2007), and Yudelson (2009).

VIII.2   ENERGY CONSERVATION IN BUILDINGS AND ECOBLOCKS

VIII.2.1   Energy Considerations Related to Water

The Energy Information Center (2009) of the Department of Energy reported average monthly electricity use for an average U.S. household as 935 kW-hours, ranging from a low 530 kWh in Maine (and generally in the New England area) to 1344 kWh in Tennessee (and generally in Southern states). The Northern and Northeast U.S. use oil and natural gas for heating and less air conditioning. In Chapter I we estimated that the average GHG emission from electric power plants in the U.S. is 0.62 kg of CO_2/kWh of electric energy delivered to the grid. It was also pointed out that some countries, such as France, Austria, Japan, Brazil, and Argentina, emit far less CO_2 per kWh because of the higher proportion of energy being produced by nuclear and/or hydropower plants that do not emit GHGs. For example, just one very large hydropower plant on the Parana River in South America produces one-third of all power use in Brazil and almost 100% of that in Paraguay. France relies heavily on nuclear power plants, and Austria and Switzerland have plenty of hydropower in the Alps. Unfortunately for the U.S., most of the available hydropower sites have already been developed. China is working very hard to reduce their proportion of GHG-emitting power plants and is finishing the largest hydropower dam and plant in the world, the Three Gorges on the Yangtze River.

The average per capita use of water without water conservation in the U.S. was reported in Chapter V, Table 5.3, as 551 L/capita-day. Citizens in most European countries, China, and Singapore use about half of the U.S. demand, and undeveloped countries use far less. With 2.6 members in an average U.S. family, this translates to about 1432 liters of water used each day per household. Using the information in Table 5.3, just implementing water conservation, the energy use for water services can be reduced by 65% due to water conservation alone. The U.S. EPA estimates (conservatively, considering the fact that compliance will not be 100%) that practicing water-conserving techniques can be reduce domestic water use by 30%. The water conservation measures outlined in Chapter V included:

- Vacuum-assisted low-flush toilets
- Shower restrictions
- front-loading washing machines
- Xeriscape
- Plugging and eliminating leaks

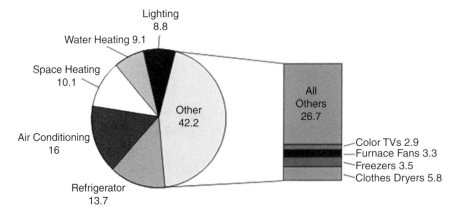

Figure 8.6 Proportions of energy use in an average U.S. household. Source: US Department of Energy.

The above practices reduce water use and, hence, the energy requirement to produce, treat, and transport water and used water. As shown in Box 5.3, replacing potable water from the tap by bottled water is a move in the wrong direction. Water reuse is not included because water reclamation and reuse (see Chapter VII) has energy demands that could be significant and could offset the energy use reduction from the lower water demand. Such energy demand considerations should be made in the overall energy balance of the development or block retrofit. The following energy-saving measures in the water system reduce domestic energy demand but not necessarily water use:

- Reduction of energy for heating water for showers, baths, kitchen, dishwashing, and laundry
- Installing more efficient food waste disposals
- Heat (energy) recovery from used water

Heating water is the third-largest energy user in domestic water use (Figure 8.6) and represents about 10% of the total domestic electricity use in households with electric water heaters. In most conventional applications, homeowners use a standard water heater, which is a 115- to 300-liter (30–80 gallon) tank heated by an electric coil or gas burner. Typically, electrically heated storage tanks have more volume than gas-heated tanks.

The U.S. EPA and U.S. DOE sponsor the Energy Star program (http://www.energystar.gov/), which explains the alternatives for selecting water-heating systems. The energy-efficient water-heating systems available on the market are:

- **Gas condensing heated storage tank.** Instead of venting the combustion gases directly outside, hot gases are captured and utilized to heat the water even more. A standard gas-heated water tank has its flue running straight up through the

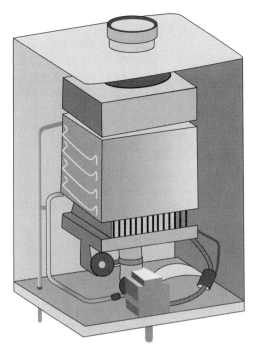

Figure 8.7   Tankless water heater. Courtesy U.S. EPA–DOE Energy Star.

middle, exiting at the top. A gas condensing water heater has its flue designed with greater surface area than a standard tank with a flue. The heat and combustion gases have much farther to travel before they exit the water tank, so more heat is transferred to the water in the tank. The gas condensing water tank uses 30% less energy that a standard tank.

- **Tankless water heaters.** Such heaters (Figure 8.7) have been widely used in Europe and elsewhere outside of the U.S. for more than 80 years. By heating water only when it is needed, gas tankless water heaters cut water-heating expenses by 30%, while also providing continuous hot water delivery. Gas tankless heaters are becoming popular as a replacement for gas storage tank water heaters. When a hot water tap is turned on in the home, cold water is drawn into the water heater and a flow sensor activates the gas burner, which warms the heat exchanger. Incoming cold water encircles the heat exchanger and leaves the heater at its set-point temperature. Combustion gases safely exit through a dedicated, sealed vent system.

- **Solar water heater.** Solar water heaters (Figures 8.8 and 8.9) come in a wide variety of designs, all including a collector and storage tank, ands all using the sun's thermal energy to heat water. The design principle is the same as that of the concentrated heat solar power panels discussed in Section VIII.3.1. Evacuated roof-mounted tube collectors are the most efficient collectors available. Each evacuated tube is similar to a thermos in principle. A glass or metal tube

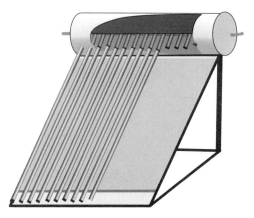

Figure 8.8 Energy Star evacuated tube solar hot water heat collector. Courtesy U.S. EPA and DOE Energy Star.

containing the water or heat transfer fluid is surrounded by a larger glass tube. The solar radiation is concentrated to heat the fluid in the tubes which, in a closed-loop system, is transferred into the water-heating tank, where it releases heat to the water in the tank. The space between the tubes in the solar panel is in a vacuum, so very little heat is lost from the fluid. These collectors can even work well in slightly overcast conditions and operate in temperatures as low as $-5°C$. Hence, the solar water heaters have the greatest efficiency and return on investment in arid regions such as the Southwest United States, the Middle

Figure 8.9 Installation of concentrated solar water heater on a rural home in Beijing province, China. The water heaters were provided by the government. Photo V. Novotny.

Figure 8.10   Solar panels in Hammarby Sjöstad in Stockholm. Picture credit Malena Karlsson, GlashusEtt, Stockholm

East, arid regions of China, and the like. *Energy Star*–certified systems should, at minimum, pay for themselves in less than five years and achieve 50% energy reduction on an annual basis when compared to standard water heaters, and should reduce GHG emissions by the same proportions. The arrangement of the building's hot water solar heating system is shown in Figure 8.10. In most cases, auxiliary electric (heat pump) may be necessary.

Consumer Reports (2009) magazine tested solar water-heating systems in Northeastern U.S. conditions (New York) and found that, during a relatively cool and rainy summer, two commercially distributed systems achieved 30% energy savings. There are a number of solar energy texts on the market (e.g., a book by Ramlow and Nusz, 2006).

- **Heat pump water and space heaters.** Energy Star–sanctioned heat pump water heaters (HPWH) are revolutionizing the way we heat and cool water and space. Heat pumps are used both for heating water air inside the building and for cooling. New models introduced on the market in 2009 utilize super-efficient technology to cut electric water heating costs by more than half. Tree types of heat pumps (Accent Energies, 2009) are available on the market

Air-to-Water. The typical heat pump design is air-to-water (Figure 8.12), or air-to-air for space heating. In this design, the unit is characterized by its large evaporator area. Heat pumps work on the same principle as refigerators, but in reverse. While a refrigerator removes heat from an enclosed box and expels that heat to the surrounding air, a HPWH takes the heat from surrounding air and transfers it to water in an enclosed tank. In a heat pump, a low-pressure liquid refrigerant is vaporized in the heat pump's evaporator and is passed into the compressor. As the pressure of the refrigerant increases, so does its

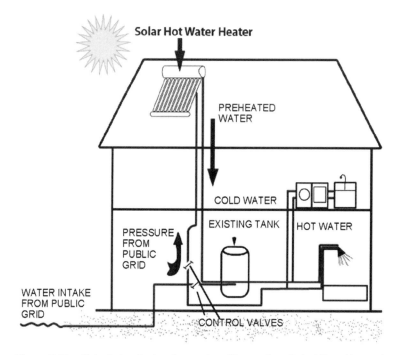

Figure 8.11 Solar hot water heating system (Source Best Solar Water Heaters)

temperature. The heated refrigerant runs through a condenser coil within the storage tank, transferring heat to the water stored there or to air blown through an heat exchanger unit. As the refrigerant delivers its heat to the water (or air), it cools and condenses, and then passes through an expansion valve, where the pressure is reduced and the cycle starts over. It takes heat from air at a high efficiency and transfers the heat to water or air in the internal heat exchanger. An air-to-water heat pump works very efficiently at temperatures as low as 5°C. Typically, at colder temperatures, supplemental heating must be provided by an electric heater, which greatly reduces the efficiency (see the discussion on efficiency of heating below).

Water-to-Water. In this design, the heat pump uses water as its heat source. The water may be subterranean or from an open source such as a river, lake, or used water, or it may be warm recycled gray water. Water provides a relatively constant heat source, and this approach may provide a more constant and higher average efficiency than the typical air-to-water heat pump. The unit itself has no evaporator and is more compact and quieter than the air-to-water heat pump. If the source of heat or cooling energy is a deeper well, the system is sometimes called *geothermal*, although this category is broader.

A simple modification of the water-to-water heat exchange is a *pipe-in-pipe* heat exchanger, in which a hot water pipe with high thermal conductivity walls (e.g., copper) is put inside a cold water pipe which is insulated. The heat from

Heat Pump Water Heater

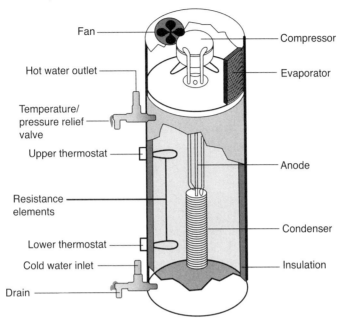

Figure 8.12   Air-to-water heat pump. Courtesy U.S. Environmental Protection Agency and Department of Energy.

the warm pipe is transferred to water in the cold pipe. No compression by electricity is involved; however, when the water temperatures in both pipes are about equal, the heat exchange stops, while far more heat (or cooling) can be extracted by the heat pump at a cost of some energy use.

Ground-sourced or ground-coupled. The ground absorbs nearly half of the thermal energy reaching the earth from the sun. In the ground-sourced design, the heat pump transfers this constant temperature source to water heating via a ground loop. Because the heat is extracted from ground and not from ambient air, the temperature does not vary much and is above freezing; hence, this type of heat pump works efficiently even in cold weather. The pipes containing heat transfer fluid in the primary loop are buried deeper below the freezing depth. This system provides much better efficiency for heating than the air-to-water systems that are rendered inefficient at low temperatures.

NSTC (2008) states that the use of ground energy sources and heat sinks (cooling) increases efficiency and offers significant potential for savings and peak load reductions for buildings. As a source of the heating/cooling energy, the report suggests the earth below the freezing depth, recycled gray water, effluents from used water reclamation plants, retention basins for stormwater, harvested rainwater, and water from a subsurface aquifer—either alone or in combination with outdoor air in a hybrid configuration. Such systems are far more efficient than outdoor air for heat exchange. The reason these are more efficient than air-to-air (space heaters) or air-to-water heat exchange is that the aforementioned heat/cooling energy sources are colder than the ambient air in summer and warmer than the ambient air in winter. The NSTC report also states that the heat pump hybrid systems can and should be implemented in individual buildings as well as on the cluster (neighborhood/ecoblock) scale. See Box 8.3 for efficiencies and energy considerations of heat pumps and other heaters.

BOX 8.3   ENERGY CONSIDERATIONS OF HEAT PUMPS AND OTHER HEATING DEVICES

The cooling/heating capacity of heat pumps (and also air conditioners or other heaters) in the U.S. is expressed by "tons of cooling" units (see Box 8.1), which is the cooling power than can melt 1 ton of ice in 24 hours. 1 TOC = 3594 watts. A medium window air conditioner requires a power of 1 kW. The heat pump unit for heating and cooling a typical U.S. home with 200 m^2 living area would typically require a 4-ton heat pump unit or a power of about 14 kW. Assume that on a warm or cold day the heat pump works 8 hours; hence, it gives up $E = 14 \times 8 = 112$ kWh of energy (E) in a day. The same calculation can be done for water heaters (see calculation below).

Efficiency of heating/cooling units is expressed by a coefficient of performance, COP, which is

$$COP = \frac{E_{ac}}{E_{ap}}$$

where E_{ac} is energy acquired, and E_{ap} is energy applied. For heating, COP for air-to-air or air-to-water heat pumps is related to the temperature of the ambient air. For very cold temperatures below 5°C, air-to-air or water heat pumps are inefficient, COP is 2 or less, and heating must be supplemented by electric heaters incorporated in the heat exchange units or separately in the tank. COP increases with the increased ambient air temperature. For water-to-water or ground-to-water heat pumps, the COP is in the higher range of 4 or more. Heat pumps with higher COP are being developed. Hence, COP for heat pumps ranges from 2 to 5, for gas, and for electric heaters COPs are less than 1.

What is the GHG equivalent of the energy used by appliances, assuming that power to electric appliances is provided by the U.S. electric grid?

Heat pump: Assuming that the COP for the heat pump is 4, the energy use to drive the heat pump will be $E_{ap} = E_{ac}/COP = 112/4 = 28$ kWh. Since the applied energy will be provided by the electricity, the GHG emissions will be

$$GHG = 28 \text{ [kWh]} \times 0.62 \text{ [kg } CO_2/kWh] = 17.4 \text{ kg } CO_2/day$$

Electric space heater or water tank: Assuming COP = 0.9 then

$$E_{ap} = E_{ac}/\text{COP} = 112/0.9 = 124.4 \text{ kWh}$$
$$\text{GHG} = 123.4 \text{ [kWh]} \times 0.62 \text{ [kg CO}_2\text{/kWh]} = 103.8 \text{ kg CO}_2$$

Gas heater or water tank: Because the electric grid in the U.S. includes energy produced by power plants that do not emit appreciable amounts of GHGs (hydro-, nuclear, and renewable power sources), the GHG equivalent of 1 kWh of a gas-condensing water or space heater could be about the same (see Section I.3.3) as that from the U.S. electricity grid.

Clearly, water-to-water or ground-to-water or air heat pumps are far more efficient than electric or gas heaters.

For air conditioning, the heat pump works in reverse, like a refrigerator—that is, it extracts heat from the air in the building space and releases it into the air (or water or ground) outside. Most heat pumps on the market work both ways.

Heat pumps for cooling? Commercial heat pumps are used for both heating and air conditioning (cooling). The air or water leaving the heat pumps used for heating (note that heat pumps are generally used for cooling and heating houses) is about 10°C colder or warmer, respectively, than that in the input.

VIII.2.2 Heat Recovery from Used Water

Water-to-water heat pumps can also recover heat from used water that would otherwise be wasted as it is done in several ecocities (for example, Hammarby Sjöstad or Dockside Greens – see Chapter XI). This could be done in a house (preferably with multiple apartment or condominium units) or on a cluster (ecoblock) scale in the resource recovery facility. Regional energy recovery from used water is not feasible due to losses in transmission lines of hot water. Used water is relatively warm even during winter conditions and has a relatively constant temperature, ranging from 10°C to 20°C, with an average of 15°C, if it is stored and equalized over a 24-hour period (Metcalf & Eddy, 2003). Gray water will be warmer, about 30°C; therefore, it may be more efficient to retrieve heat from it. Box 8.4 presents the calculation of the recoverable heat from used water or gray water.

BOX 8.4   HEAT RECOVERY FROM USED WATER

What is the heat or cooling energy equivalent of the energy extracted by the heat pump from a volume V = 1 m³ of used water, with the temperature differential between incoming and leaving water ΔT = 12°C?

From Box 8.1 the specific heat of water is $C_p = 4.2$ J/(g°C). Mass of 1 m³ of water is $\rho = 1000$ kg/m³ $= 10^6$ g/m³.

(*continued*)

Energy extracted from 1 m³ of water is then

$$E_{ac} = V\rho\Delta TC_p = 1[\text{m}^3] \times 10^6[\text{g/m}^3] \times 12[^\circ\text{C}] \times 4.2[\text{J/g}^\circ\text{C}]$$
$$= 50.4 \times 10^6 \text{ J} = 50.4 \text{ MJ} = 50.4 \text{ MJ}/(3.6 \text{ MJ/kWh}) = 13.95 \text{ kWh}$$

If COP = 4 then the energy applied to extract the heat (cooling) from 1 m³ corresponding to the 12°C temperature change will become

$$E_{ap} = E_{ac}/\text{COP} = 13.95 \text{ kWh}/4 = 3.48 \text{ kWh.}$$

And the net energy gain is $\Delta E = E_{ac} - E_{ap} = 13.95 - 3.48 = 10.47 \text{ kWh/m}^3$

VIII.3   ENERGY FROM RENEWABLE SOURCES

Worldwide, the total energy used in 2006 was 19,015 terawatt-hrs, which is the daily energy use of 52 TW-hours. The power of one TW is one trillion (10^{12}) watts. The sun energy potential is the largest of all renewable energy sources (Table 8.2), which include also wind (land-based and ocean-based), hydropower, and geothermal sources. In the previous section the possibility of extracting heat energy from the ground was mentioned. This combined energy potential far exceeds the world's energy use now and in the foreseeable future.

VIII.3.1   Solar Energy

Each day the energy from the sun reaching earth is 6000 times greater than the total world electricity use. The potential of capturing this energy obviously varies geographically, but it ranges from 2 to 6 kWh m⁻² day⁻¹ in temperate inhabited zones (Johnson, 2009). The high incoming solar radiation of 6 kWh m⁻² day⁻¹ is

Table 8.2   Potential renewable energy sources (from Johnson, 2009)

Source	Electric Energy Generation Potential Worldwide Terawatt-Hours in a Year
Solar photovoltaics	470,278
Concentrating solar	275,556
Wind (land-based)	105,556
Ocean (tidal and wave)	91,398
Hydropower	13,889
Geothermal	12,500
Wind (offshore)	6111

Sources: NASA; World Energy Statistics and Balances, OECD/IEA, 2008; National Renewable Energy Laboratory

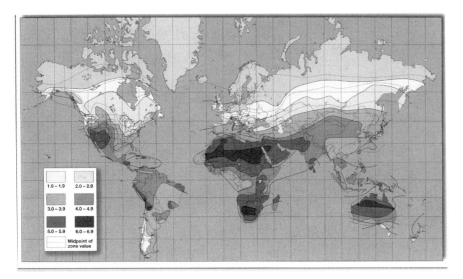

Figure 8.13   Average daily insolation of the world. The values are in kWh m^{-2} day^{-1}. Reprinted with the permission. All rights reserved © SunWize Technologies. This map shows solar energy input received each day on an optimally tilted surface during the worst month of the year (based on accumulated worldwide solar insolation data).

for deserts in Africa, Australia, and the Middle East; the middle range would be typical for the Southwest U.S.; and lower incoming radiation would be typical for the rest of the densely inhabited regions of the world outside of the Arctic (Figure 8.13).

Solar energy can be harvested by (Johnson, 2009; Stine and Geyer, 2001):

- Concentrating solar radiation by computer-guided mirrors; heliostats focus the sunlight on a receiver containing water or heat-transferring fluid. The solar collectors concentrate sunlight to heat the heat transfer fluid to a high temperature. The hot heat transfer fluid is then used to generate steam that drives the power conversion subsystem, producing electricity. Thermal energy storage provides heat for operation during periods without adequate sunshine (Figures 8.9 to 8.11). This works like a magnifying glass focusing the sun ray energy onto a smaller area and creating high temperatures capable of igniting paper or wood. Figure 8.14 shows a detail of a concentrated solar panel array with mirrors and pipes containing the heat transfer fluid.
- Converting sunlight directly into electricity with photovoltaic panels made of semiconductors (Figures 8.3 and 8.15). Photovoltaic panels are becoming very popular and can be seen providing wireless electricity for everything from pocket calculators to entire cities (see Figure 8.3 and the photovoltaic array in the ecocity Masdar in the United Arab Emirates, shown in Chapter XI). The advantage of photovoltaic energy is that it is produced by flat panels assembled to form a large power plant or installed on roofs of of buildings (Figure 8.15).

Figure 8.14   Trough mirror array of the solar panel in Kramer Junction, one of the largest concentrated solar energy farms in the U.S. Photo courtesy of Sandia National Laboratory.

Figure 8.15   Roof solar panels (concentrated heat for hot water and photovoltaic for electricity) on the roof of a county park visitor center in Rockport, Massachusetts. Photo V. Novotny

European countries such as Germanyand Spain, and South Africa as well, have built installations with large solar power capacities. Germany, in spite of its northerly location and often cloudy skies, produced 2.22 TW-hours from solar sources in 2006, followed by the U.S. (0.56 TW-h), South Africa (0.53 TW-h), Spain (0.12 TW-h), and China (0.1 TW-h). Most of the U.S. solar energy capacity is installed in California. China has invested heavily in small household water heaters (Figure 8.10), but in 2009 the country signed an agreement with a U.S. solar developer (First Solar) for a 2000 megawatt photovoltaic farm to be built in the province of Inner Mongolia. It is going to be a part of an 11,950-megawatt renewable energy park planned in the Inner Mongolian desert.

Concentrated heat plants include dish/engine systems, parabolic troughs, and central power towers. Trough systems (Figure 8.14) predominated among the commercial solar power plants in 2010. Because of their parabolic shape, trough mirrors can focus the sun at 30 to 60 times its normal intensity on a receiver pipe located along the focal line of the trough. Synthetic oil captures this heat as the oil circulates through the pipe, reaching temperatures as high as 390°C (735°F). The hot oil is pumped to a generating station and routed through a heat exchanger to produce steam. Finally, electricity is produced by a conventional steam turbine with a generator. Because these systems operate similarly to a fossil fuel power plant, they eject heat into the environment and require cooling by water (in about the same volume as thermal power plants—see Box 8.3).

Because of the location of the largest solar facilities in arid desert environments, plants compete with other users for scarce water, which leads to sociopolitical controversies. For example, requests for solar power plant installations asking for water in one county in Nevada (a desert state) far exceeded the amount of available water, as was noted in a *New York Times* article of September 29, 2009. The same article also pointed out that solar power installers have made numerous requests for drinking water for cooling, which in California requires a lot of energy for production and transfer. Water is also used for washing the mirrors. The alternative to wet cooling is dry cooling by forced air, requiring more energy at increased cost.

First-generation solar panels used in *photovoltaic plants* were made either of crystalline or amorphous silicone. The creation of the crystalline panel involves cutting crystalline silicon into thin disks (less than 1 cm thick). After repairing any damage from the slicing process and polishing, dopants (materials that alter the charge in the photovoltaic cell) and metal conductors are added across each disk. The conductors are spread across each disk and aligned in a thin, gridlike matrix. A thin layer of cover glass is then bonded to the top of the photovoltaic cell to protect the panel.

Amorphous silicon panels differ in structure and manufacture from the crystalline panels. Amorphous silicon solar cells are developed in a continuous roll-to-roll process by vapor-depositing silicon alloys in multiple layers, with each thin layer specializing in the absorption of different parts of the solar spectrum. The result is much higher efficiency and reduced material cost.

The most successful second-generation materials have been cadmium telluride (CdTe), copper indium gallium selenide, amorphous silicon, and micromorphous silicon. These materials are applied in a thin film to a supporting substrate such as glass or ceramics, reducing material mass and therefore costs. These technologies do hold promise of higher conversion efficiencies and offer significantly inexpensive production costs.

The photovoltaic panels can come in several shapes and forms. Of note are photovoltaic solar roof shingles. The photovoltaic shingle is textured to blend with and complement the granular texture of roof shingles. Each shingle produced by OK Solar, Inc. is 30 cm wide by 219 cm long and is nailed down, in place of roof decking. Electrical wires from the shingle photovoltaic panel pass through the roof deck, allowing interior roof or attic space connections.

Solar panels are usually connected in series in modules, creating additive voltage. Connecting cells in parallel will yield a higher current. Modules are then interconnected, in series or parallel, or both, to create an array with the desired peak DC voltage and current. Typical efficiencies of solar panels in 2010 were less than 20%—that is, max 20% of the intercepted solar energy is converted into electricity. However, third-generation solar panel materials are being developed in the world's laboratories that will double to quadruple the efficiency. Potentially, they could become inexpensive, imprinted on flexible materials, and draw energy after the sun has set from radiation heat. The new approach being developed by the Idaho National Laboratory and the University of Missouri uses a special manufacturing process to stamp tiny loops of conducting metal onto a sheet of plastic. Each "nanoantenna" is as wide as 1/25 the diameter of a human hair. Because of the size, the nanoantennas absorb energy in the infrared part of the spectrum, just outside the range of what is visible to the eye. The sun radiates a lot of infrared energy, some of which is absorbed by the earth and later released as radiation for hours after sunset. Nanoantennas (or nantennas) can take in energy from both sunlight and the earth's heat, with higher efficiency than conventional solar cells. The team has estimated that individual nanoantennas could absorb close to 80% of the available energy. The circuits themselves can be made of a number of different conducting metals, and the nanoantennas can be printed on thin, flexible materials like polyethylene, a plastic that's commonly used in bags and plastic wraps (Kotter et al., 2008).

Before China announced the Inner Mongolia solar power project, the largest single photovoltaic power plant was the 550-MW Topaz solar plant in California north of Los Angeles, built by OptiSolar, with operation beginning in 2011. It consists of thin-film silicon photovoltaic panels that use only about 1% of the silicon of competing crystalline panels, with nontoxic and recyclable equipment made largely of glass, metal, and concrete. The panels can be placed close to the ground. Large-scale solar power plants can be matched by dense small installations on roofs of individual houses connected to the grid (selling excess electricity to the public utility). As of 2010 this was happening in many European countries, in China and elsewhere in Asia, in California, and in several developing countries in Latin America. In poor rural and suburban communities in Peru and throughout Central America, small solar panels connected to a rechargeable battery provide electricity for powering TVs and light fixtures. In the Czech Republic developers provide space on the roofs and outsides of buildings to install solar panels. This is driven by financial stimuli provided by the government and the carbon trading credits under the Kyoto Protocol. The interest is so high that the utility grid cannot keep up with the connections. Box 8.5 contains an example of the calculation of the size of solar panels, assuming current electricity use. It is expected that future solar panel efficiency will double. According to a study by the U.S. president's National Science and Technology Council (2008), by implementing energy efficiency, electric power use by buildings can be reduced by 60–70%, and doubling of the efficiency (by 2030) would result in a panel of less than 3.5 m^2/person in the U.S. Southwest.

Photovoltaic plants do not require water for cooling because they produce electricity directly from sun energy. Water is used only for washing the panel.

Recommended references on photovoltaics systems are by Patel (1999), Messenger and Ventre (2004), and Green (2003). A good web site with a free book and educational materials on solar power systems is maintained by Power From The Sun (www.powerfromthesun.net).

BOX 8.5   SOLAR PANEL SIZE

Estimate the size of solar panels for a family of four in a house located in the Southwest U.S. that would provide 100% of electricity need. Average electric energy use by one person is 4370 kWh/year.

Assumptions: 320 sunny days

From Figure 8.13 the average insolation in the Southwest U.S. is 4 kWh m^{-2} day^{-1}. The average efficiency of solar panels in 2010 was about 20%.

$$\text{Solar panel area} = \frac{4,370[(\text{kw} - \text{hrs})/(\text{cap} - \text{year})]}{[320 \text{ days}] \times 4[(\text{kW} - \text{h})/(m^2 - \text{day})] \times 0.2} = 17.1 \text{ m}^2$$

or 68.3 m^2 for the house (a panel of 7 × 10 meters that would fit on one side of the roof).

It is expected that the efficiency of the photovoltaic solar panels will at least double in the next ten years, which would cut the size of the panes by half.

Most likely, such panels would produce significantly more energy during the daytime than the family would need, but far less energy during the dark hours. The most simple solution is to sell the excess electricity to the power company and buy it during the hours when it's needed. This requires a "smart" grid.

Solar energy conversion systems. A block diagram showing three of the most basic system types is shown in Figure 8.16. In the first diagram, the solar resource is captured and converted into heat, which then provides thermal energy (thermal load) for uses such as house heating, hot water heating, or heat for industrial processes. This type of system may or may not include thermal storage, and usually includes an auxiliary source of energy to meet demand during periods with no sunshine.

If the demand (load) to be met is electricity (an electrical load) rather than heat, there are two common methods of converting solar energy into electricity. As pointed out in the preceding paragraphs one method is by collecting solar energy as heat and

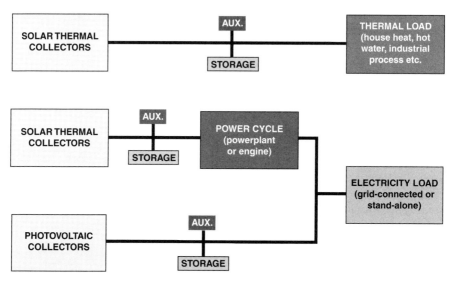

Figure 8.16 Solar to conventional (heat or electricity) energy conversion (courtesy Stine and Geyer, 2001).

converting it into electricity, using a typical natural gas burning power plant or engine; the other method is by using photovoltaic cells to convert solar energy directly into electricity. Both methods are shown schematically in Figure 8.16. In general, if solar energy conversion systems are connected to a large electrical transmission grid, no storage or auxiliary energy supply is needed. If the solar energy conversion system is to be the only source of electricity, storage and an auxiliary energy supply are usually both incorporated. If the thermal route is chosen, storage of heat rather than electricity may be used to extend the operating time of the system. Auxiliary energy may be supplied either as heat before the power conversion system, or as electricity after it. If the photovoltaic route is chosen, extra electricity may be stored, usually in storage batteries, thereby extending the operating time of the system. For auxiliary power, an external electricity source is the only choice for photovoltaic systems (Stine and Geyer, 2001).

As pointed out, in some countries a smart municipal grid can "buy" the excess solar electricity from individual small or large "producers" during excess power production, which typically coincides with the hours of peak electricity demand, and then provide needed power during the times when solar power is not produced (the same applies to wind power). If both wind and solar power are used on the premises, they complement each other, as will be documented in Section VIII.6.2 on a regional scale.

A novel system for power storage is being developed by the Massachusetts Institute of Technology (MIT). MIT researchers from the Department of Chemistry have developed a process by which sun energy is used to split water into hydrogen and oxygen gases (Trafton 2008; Kanan and Nocera, 2008). They have identified this process as "artificial photosynthesis." During times of a lack of solar power, the hydrogen and oxygen can be recombined inside a fuel cell, creating carbon-free electricity to power a house or automobile. This will be part of the hydrogen era of power and energy, which will be discussed in Section VIII.5. The key component in the new Nocera and Kanan process is a new catalyst that produces oxygen gas from water, and another catalyst that produces hydrogen. This new catalyst consists of cobalt metal, phosphate, and an electrode placed in water. A platinum catalyst produces hydrogen gas. After hydrogen and oxygen are recombined to produce energy and water, the water is recycled.

VIII.3.2 Wind Power

Wind energy has been used by human beings for centuries, and old wind mills still dot the countryside in many countries. Wind energy has been used for milling flour or pumping water from canals (e.g., in Holland). Wind energy has the same problem as solar, that is, it is not available all the time, and energy production is not stable. Hence the problem of energy storage is the same.

Today's wind turbines are a more efficient technology than the windmills of the past. They have fewer blades, usually two or three, that are aerodynamically designed to capture the most wind energy. When the wind blows, a pocket of low-pressure air forms on the downwind side of the blade, which pulls the blade toward it and causes the lift. This lift force causes the rotor to spin, which turns a generator that makes direct electric current. Sophisticated power electronics convert the direct current electricity into high-quality alternating current electricity transmitted through the power grid (U.S. DOE, 2006). As wind speeds increase, the amount of electricity generated increases exponentially. Because faster, less turbulent winds are found higher off the ground, new utility-scale wind turbines are normally placed on towers 30 meters tall or taller, or on roofs. At high wind speeds, a controller on the turbine shuts it down so turbine components will not be damaged by high winds.

The United States in 2010 generated more than 25,000 megawatts (MW) of electricity from the wind, which is enough to power about 7 million average American homes. Industry experts predict that, with proper development, wind energy could provide 20% of the nation's energy needs by means of larger, more efficient, utility-scale wind turbines in land-based and offshore installations, as well as more efficient, quieter small wind turbines for distributed applications (National Renewable Energy Laboratory, http://www.nrel.gov/wind/, accessed September 2009).

There are two basic types of wind turbines:

- Horizontal axis wind turbines, which are the type most commonly in use today and the focus of U.S. DOE research on wind power. They come in two varieties:
  - Two-blade horizontal axis turbines spin downwind.
  - Three-blade horizontal axis turbines spin upwind (Figure 8.17).
- Vertical axis wind turbines (Figure 8.18)

Figure 8.17 Horizontal axis wind turbine by Siemens. Large-megawatt turbines have rotor diameter up to 90 meters.

In most residential or small cluster (ecoblock) applications, wind turbines are used as a supplemental source of power in combination with local, on-the-grid, utility power. Below the wind speed of 11–16 km/hour, the wind turbine will cease to provide an output, and the utility grid takes over to supply energy. Above this cut-off speed, the wind turbine begins to operate, and the grid power supply is proportionately reduced. In hybrid systems featured in several ecocities (see Chapter XII), for example in Dongtan and Qingdao designs, a system consisting of vertical wind turbines, solar photovoltaics, and hot water solar panels would provide enough electricity and heat to make the ecocities net zero carbon footprint communities. However, even in these most progressive cases, the electric system of the ecocity is connected to the regional power grid.

In a typical U.S. residence that uses under 10,000 kilowatt hours per year of electric power, a 5–15 kilowatt wind turbine should more than suffice (see Box 8.6). The larger wind turbines have capacities from 50 kilowatts to severalmegawatts. A typical 250-kW horizontal turbine has a rotor with a diameter of 25 meters installed at a height of 25–30 meters. Megawatt-size turbines have rotor diameters of 60–70 meters (Pasqualetti et al., 2002). These windmills, known as utility-scale turbines, can be grouped together and connected to central lines for transmitting and distributing power in bulk to the local utility grids, which, in turn, sell power to homes

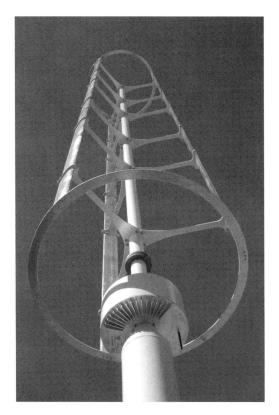

Figure 8.18   Vertical axis turbine by Mariah Power. The efficiency of the turbine is about the same as that of a turbine with a horizontal propeller. The turbine on the figure is 10 meters high with a diameter of 0.6 meters. Its power rating is 1.2 kW, and it can produce 2000 kilowatt-hours per year, Picture courtesy Mariah Power.

BOX 8.6   WIND ENERGY ESTIMATE

An average family in the top 100 urban areas of the U.S. uses 11,362 kWh in a year. With the average family size of 2.6, the per capita energy use is 4370 kWh/year. If the family is located in Boston, Massachusetts, how much wind turbine power would be needed to provide all needed electric power?

The average and maximum wind record for Boston is shown in Figure 8.20. Because the average wind is always in excess of the cut-off wind speed of 7–10 mph (11–16 km/hr), one can safely assume that a turbine open to wind would produce energy for more than 50% of time. Calculate the needed wind power. One year has 8760 hours.

The needed power $= (4370\ [\text{kWh}])/(0.5 \times 8760\ [\text{hrs}]) = 1\ \text{kW/person}$

Figure 8.19   Wind farm in northern Austria. Photo V. Novotny.

and businesses across the land. These are called wind power plants or wind farms (Figure 8.19).

Vertical wind turbines can be installed on roofs of buildings or elevated above streets in the "street canyons" between high-rise buildings that magnify wind power (e.g., Chicago, the windy city). The performance does not depend on wind direction.

Utility-scale wind generation is well suited to the Great Plains and Upper Midwest regions of the United States and countless wind-swept regions of the world. Both U.S. areas have good wind resources and wide-open spaces. There are large utility wind power plants in California, Colorado, Iowa, Kansas, Minnesota, Oregon, Texas, Washington, North and South Dakota, and Wyoming. Figure 8.20 shows the average and maximum wind in Boston, Massachusetts. Turbines used in these wind power plants range in size from 50 kW to 2 MW. A 3-MW wind turbine was unveiled in Spain in 2008, so the race is on. The advantage of wind over solar energy is that a wind turbine in wind farms occupies only a very small area of valuable land (a couple of square meters at most; see Figure 8.19), while solar power farms cover hundreds of hectares of land that cannot be used for anything else, unless the panels are put on the roofs of buildings.

European nations—including Austria, Denmark, Germany, the Netherlands, Spain, and Sweden—emphasize distributed wind generation. These heavily popu-lated nations have far less open land for wind power than the United States. For

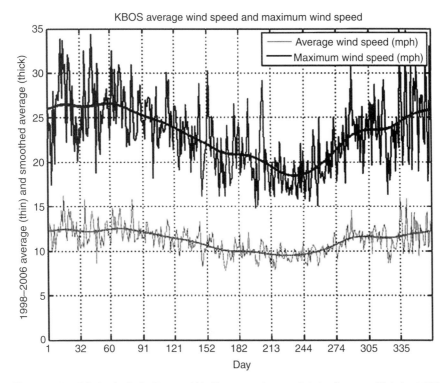

Figure 8.20 Wind velocity in Boston, MA. To convert from mph to km/hour, multiply by 1.608.

this reason, they are pioneering distributed generation with large, multimegawatt turbines, and have also begun experimenting with offshore wind turbines installed in shallow water near the coastline, to take advantage of strong offshore winds. The first large offshore wind power electricity generation was approved in the US near the Nantucket Island (Massachusetts) in 2010. It will contain 130 3.6 MW (Siemens) turbines which will provide 75% electricity for the Nantucket and Martha's Vineyard islands and Cape Cod.

Wind power is affordable. Electricity from wind typically costs between $0.04/kWh and $0.06 /kWh, a sevenfold decrease since 1980. This cost is highly attractive when compared to the average price from the grid, which in 2010 was about $0.1/kWh. New blade designs, which have increased wind turbine performance by 30% over the past decade, are helping make wind power one of the most cost-competitive renewable technologies. Researchers believe additional technological improvements could cut prices for wind-generated electricity by another 30% to 50% (NSTC, 2008). The leaders in the use of wind power are Germany, Spain, the U.S., Denmark, and India. Good references on design and applications of wind power are books by Burton et al. (2001), Pasqualetti et al. (2002), Hau (2006), and the web sites of the National Renewable Energy Laboratory of the U.S. DOE and the

American Wind Energy Association. Similar sources of information are available in many other countries.

VIII.4 ENERGY FROM USED WATER AND WASTE ORGANIC SOLIDS

VIII.4.1 Fundamentals

Used water and solids treated in the water reclamation process are a resource. Energy in the form of heat and electricity can be produced, and nutrients and soil-conditioning solids can be recovered as well. The technologies are known and have been practiced for decades. However, since the beginning of this century, a new look at resource recovery is emerging, driven by (1) the goal of reducing GHG emissions, (2) water shortages, and (3) discoveries (or rediscoveries) of new technologies that could revolutionize the reuse, recovery, and management of used water. These technologies will produce energy without adding GHGs to the atmosphere, recover phosphate, and heat (and cool) communities and industries, including those providing water and reclaiming used water for reuse. This is another component to urban sustainability and net zero carbon footprint efforts.

Chapter VII, Section VII.2, presented basic fundamentals of sludge residuals and energy recovery in water reclamation plants. In a traditional secondary treatment water reclamation plant, methane gas produced by the anaerobic digestion process is typically used on the premises to heat digesters and plant buildings. In some cases, methane gas is flamed out. Similarly, organic solid waste in landfills decomposes (slowly) to methane gas that must be flamed out to prevent explosions. Some landfills collect landfill gas and produce electricity that is then sold to the grid (Figure 7.10).

There are two processes by which organic solids can be decomposed to flammable gases:

- *Production of "wood gas"* from low-quality coal, wood chips, and garden residues (leaves) by gasification. After drying, organic sludge can also be used for the production of this gas. The gas is produced by heating the organic materials at higher temperatures between 650 and 1000°C in a low-oxygen atmosphere. The gas is also known as "syngas," and its production is now increasing rapidly. The produced gas is mainly carbon monoxide, hydrogen, and nitrogen, with traces of methane. It is interesting to note that during World War II in German-occupied territories, public buses were equipped and run with gas from wood chips produced by generators, and each station of the bus line had a container with woodchips. Oil and gasoline were used mostly for war machines. Syngas, because it is mostly highly poisonous, odorless carbon monoxide, is not safe for use in households.

- *Anaerobic digestion (fermentation)* produces biogas from biodegradable materials such as sewage sludge, manure, strong wastewater (leachate from landfills, wastewater from high-density operations), green waste from gardens, energy crops and agricultural residues (grass, algae, etc.). In addition, many organic chemicals, such as some pesticides, can be biodegraded by anaerobic processes. This type of biogas production occurs in digesters or landfills and in nature in wetlands (swamp gas) and anoxic sediments. Natural gas is mostly methane and is found underground, sometimes causing deadly explosions in mines and deep tunnels (e.g., Milwaukee's deep tunnel, Figure 1.18) during construction.

The final product of digestion is relatively inert organic humus-type residue and a liquid concentrate that contains high concentrations of dissolved organic solids (BOD/COD) and recoverable high concentrations of ammonium and phosphates that can be precipitated into struvite (see Chapter VII). Figure 8.21 shows digesters in the Changi wataer treatment plant in Singapore that produce biogas for heating the digesters and plant buildings.

In the subsequent discussion the term "biogas" will be used for methane-containing gases formed by anaerobic fermentation of organic matter. "Syngas" is used to refer to hydrogen- and carbon monoxide–containing gas produced by heat gasification (pyrolysis) of organic matter in a low-oxygen atmosphere. The chemical energy of oxidation of organic carbon (burning) in solids (dried sludge, combustible refuse, crop residues, wood chips) is traditionally converted to electricity in several steps that may involve gasification to wood gas and then combustion in an incinerator

Figure 8.21   Sludge digesters in the Changi Water Reclamation Plant in Singapore, designed by CH2M HILL (photo V. Novotny).

or internal combustion engine. The conversion includes the following steps (Barbir, 2005):

1. Gasification and combustion or direct combustion of organic dry solids convert the chemical energy into heat.
2. Heat boils water and produces high-pressure steam.
3. Steam runs a turbine, generating mechanical energy.
4. Mechanical energy is conveyed to a generator, producing electricity, which accounts for only a fraction (35 to 45%) of the energy in fuel.
5. Steam in the primary water-steam-water recirculation is condensed back into water, releasing heat (more than 50% of the energy in fuel) from the condenser to the cooling system. Released heat can be used to heat nearby buildings and industries, but in most cases this is not done.

In this process all carbon in the fuel in a well-functioning furnace is converted to carbon dioxide, which is a GHG. Natural gas and coal are fossil fuels, and the emitted CO_2 adds to the carbon footprint, causing global warming.

Gas produced by gasification or biogas or natural gas can be converted into mechanical energy using an internal combustion engine that runs a generator, producing electricity. The efficiency of the internal combustion engine is also relatively low, about 42 to 45%. Hence, most of the energy is converted into heat that is normally wasted but could be recovered. All organic carbon in the combustion energy-producing processes is converted into carbon dioxide, which is a GHG.

The fundamentals of digestion were presented in Chapter VII, which described biogas production from sludge produced in water reclamation plants. In such applications, the amount of produced biogas suffices to provide heat energy and/or electricity to the reclamation plant, but not much beyond. Regional or cluster (ecoblock) integrated resource reclamation systems, in which biogas production units accept both sludge and other organic residues, can produce significant amounts of energy for the community in the form of biogas and/or electricity. Biogas, after treatment and in-situ purification, can be added to the natural gas grid or provided as gas fuel to public transportation or converted into electricity and sold to the electric utility grid. Such a resource recovery system was implemented in the Swedish ecocity Hammarby Sjöstad; it recovers heat from used water, recovers energy from combustible solids in an incinerator, and uses produced methane as a fuel for public transportation (see Chapter XI). This system emits significant volumes of GHGs, but since they do not originate from fossil fuel the emissions are considered as carbon neutral.

VIII.4.2   Biogas Production, Composition, and Energy Content

Methane generation by digestion is a three-step process that is identical to sludge digestion, introduced in Chapter VII. In this process, the feed does contains not only primary and secondary (activated) sludge from the water reclamation (wastewater treatment) processes, but may also include manure, vegetation residues, or, in developing countries, fecal matter from latrines mixed with sludge. In addition, if algae are grown on wastewater to produce biogas, digestion is one step in the process. Anaerobic treatment processes can also be used to treat concentrated municipal and industrial used water. If the organic matter is dry, water or sewage has to be added for the digestion to start and progress. In fact, it takes several weeks to start the digestion process because digestion bacteria grow slowly and are sensitive to upsets. For these reasons, a "home" digester may not be a great idea, but it could be implemented on a farm or a village scale. Biogas production is best carried out under the supervision of an operator who is familiar with the process.

Typically, and especially in small operations, digestion operates in a thermophilic range of 35 to 45°C, which may need heating in colder climatic conditions. A part of the produced biogas can be used to heat the reactors.

The organic matter in the digester is in the form of slurry with organic solids content of about 10%. Methane formation progresses in two steps, shown below (see also Chapter VII):

Step 1: Acid formation, in which complex organic waste is converted to volatile acids by acid-fermenting bacteria Step 2: Methane forming, in which methane-forming bacteria convert volatile acids to methane and carbon dioxide

The composition and the energy content of biogas vary depending on the source of organic solids and the process. Landfill gas typically has a methane content of 50%, and advanced water reclamation (wastewater treatment) technologies can produce biogas with 55–75% methane gas. Syngas has very little methane. Letting landfill gas or any other methane (with the exception of methane production by nature) escape into the atmosphere contributes significantly to global warming because the GWP (global warming potential) of methane is 25, as compared to 1 for carbon dioxide.

According to information from Naskeo, a large French biogas plant manufacturer and installer, renewable energy from biogas worldwide could potentially be 50 to 100% of the total world natural gas production (Naskeo, 2009). Only a small fraction of this biogas potential has been developed. Unlike fossil natural gas burning, which has a GHG footprint, burning biogas produced from waste sludge and other biomass is carbon neutral. The typical composition of biogas produced from different sources is given in Table 8.3.

Biogas is a gas appreciably lighter than air, and it produces 40% less combustion energy when compared with an equal volume of natural gas, as shown in Table 8.4 below.

The presence of H_2S or CO_2 and water makes biogas corrosive, requiring the use of materials that resist corrosion. Because hydrogen sulphide is poisonous and odorous, biogas should not be used directly in households for cooking. Both carbon dioxide (a GHG) and sulfur can be removed biologically in large-scale treatment plants (Jensen and Webb, 1995). The best way of using biogas is to generate electricity or use it as fuel for buses, as has been done in Sweden and elsewhere. The produced gas is about $^2/_3$ methane, as shown in Table 8.3. Syngas is a low-energy gas (Friends of the Earth, 2009; Yun et al., 2003; Planet Power, 2009) mainly because of

Table 8.3   Composition of biogas from several sources (from Naskeo, 2009)

Components	Household Waste	Wastewater Treatment Plant Sludge	Agricultural Waste	Waste from Agrifood Industry
CH_4 % volume	50–60	60–75	60–75	68
CO_2 % volume	38–34	33–19	33–19	26
N_2 % volume	5–0	1–0	1–0	-
O_2 % volume	1–0	< 0.5	< 0.5	-
H_2O % volume	6 (@40°C)	6 (@ 40°C)	6 (@ 40°C)	6 (@ 40°C)
H_2S mg/m^3	100–900	1000–4000	3000–10,000	400
NH_3 mg/m^3	-	-	50–100	-

its high nitrogen content. There are several patented processes by which syngas can be cleaned up and converted into a higher-energy-content gas.

The use of biogas in developing countries is different, to some degree, from that in developed countries, the former still relying heavily on natural gas for cooking and heating. In developing countries, most biogas in rural areas is produced from cow and other animal manure, including vegetation residues. Production of biogas is now being widely introduced in developing countries as an alternative to dry dung/manure or wood burning. The population pressure in India has reduced the country's forests almost to nothing, causing severe fuel shortages in rural areas. Daily wood (and also water) gathering, primarily by women, has been a major effort that lasts for hours. To compensate for the fuel shortages, or a complete lack of fuel, cow and other animal (e.g., camel) manure is dried and used as fuel, and the resulting acrid smoke has led to endemic eye disease. Also, drying manure is a perfect smelly breeding ground for flies and a source of other public health problems. The same situation was (and still may be) typical for some rural areas of China (Inner Mongolia, Tibet) or in Haiti, Ethiopia, or sub-Saharan African countries (http:// http://www.habmigern2003.info/biogas/methane-digester.html) that do

Table 8.4   Caloric energy content and physical characteristics of biogas

Types of Gas	Biogas 1 Household Waste[1]	Biogas 2 Agrifood Industry[1]	Natural Gas	Syngas Municipal Solid Waste[2]
Composition	60% CH_4	68% CH_4	97.0% CH_4	18% H_2
	33% CO_2	26% CO_2	2.2% CO_2	24% CO
	1% N_2	1% N_2	0.4% N_2	6% CO_2
	0% O_2	0% O_2	0.4% other	3% CH_4
	6% H_2O	5% H_2O		49% N_2
				2% other
Energy content kWh/m^3	6.1	7.5	11.3	2.5[3]

[1] Naskeo (2009)
[2] Planet Power (2009)
[3] Yun et al. (2003)

not have forests. It has been estimated that India now has millions of small biogas generation units, often used for cooking in stoves instead of traditional small fires with wood.

VIII.4.3 Small and Medium Biogas Production Operations

India promotes decentralized energy production from renewable sources. However, other than hydropower, renewable sources are still lagging behind the national rural power needs. Until recently, burning wood or dried manure was the option for cooking and heating. Laundry has been commonly done by women in the nearest watercourse. Biogas production in small digesters is far safer and causes less nuisance and fewer health problems. Simple household digesters have also been introduced in China and several other developing countries.

In temperate and warm developing countries, a small-scale unit producing biogas from biomass can provide all the cooking gas a family needs. In addition, a small photovoltaic panel providing electricity and concentrated solar power for heating water supplies all the energy needed. Such "distributed" energy production systems are now appearing in several developing countries in Latin America and Asia. In these tropical and subtropical environments, maintaining mesophilic temperature in the digester may not require heating during most of the year. Modern biogas energy conversion can be installed at any location where plants can be grown and/or domestic animals are reared (Ravindranath and Hall, 2002). When such biomass is in short supply, for example, in small cities, it can be supplemented by leaves and similar vegetation (Jagadish et al., 1998). In fact, Jones, Nye, and Dale (2009) recommend adding vegetation and crop residue to the biomass because the nitrogen content of manure, with or without human fecal matter, is too high for optimal digestion. Conceivably, if wetlands are used for treatment in water reclamation plants in small communities (see Chapter VII), wetland vegetation and crop residues can be harvested, along with other clippings and leaves from gardens, parks, and urban forests, and used as biomass in biogas-to-methane conversion plants.

The size of the simple rural digester should be 10 to 15 times the daily volume of moist biomass daily input, providing 2 weeks average residence time of the waste in the digester (Jones, Nye, and Dale, 2009). Digesters range in size from around $1 m^3$ for a small household unit to as large as $2000 m^3$ for a large commercial installation. Figure 8.22 shows two possible configurations of small- to medium-sized digesters. The figure on the left is a single-tank batch feed digester consisting essentially of two drums. The upper inverted drum is mobile and stores the produced biogas. The figure on the right is a continuous feed two-stage digester similar to those used in typical municipal operations (Jones, Nye, and Dale, 2009).

Jones, Nye, and Dale (2009) estimated methane production from manure containing 8% solids as 0.5 to 1.5 m^3 per m^3 of reactor volume per day. Based on Jones, Nye, and Dale and on information in Tables 8.3 and 8.4, 1 kg of manure solids can produce 0.3 m^3 of biogas, which will contain 0.2 m^3 of methane. Small-scale biogas production units can provide 5 to 100 kW of gas power production, and medium-scale units can range from 1 to 10 MW.

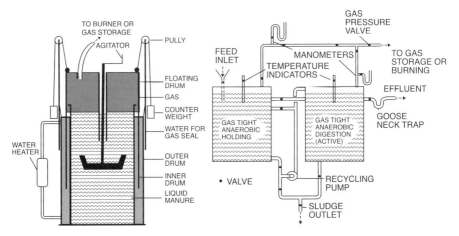

Figure 8.22 One- and two-stage smaller digesters for production of methane in farmsteads and small communities (picture credits: Jones, Nye, and Dale, 2009; Purdue University Cooperative Extension Service, West Lafayette, IN).

The mean residence time of solids in a high-rate digester is from 4 to 10 days, which is about 10 times the retention time in the aerated activated sludge units. However, anaerobic treatment units are simple, with minimum machinery and no energy requirements (they produce energy). Nevertheless, achieving concentrated influent by flow separation and eliminating all clean water inputs are the most fundamental requirements of the future designs. The separation should occur in the cluster (ecoblock) water reclamation facilities (see Chapter VII).

Normally, biogas should not have an objectionable rotten egg odor caused by hydrogen sulfide (H_2S), but if it contains too much of this odoriferous gas it could also cause severe corrosion. Therefore, the gas should be passed through a desulphurization filter containing lead filings or a mixture of wood chips and iron oxide to remove any hydrogen sulfide. "Sorb beads," manufactured by Mobil Oil, can also be used to remove hydrogen sulfide and water vapor. Desulphurization is also the first step in the fuel-reforming process to produce hydrogen for direct energy production (see the next section).

VIII.4.4 Anaerobic Upflow Reactor

While anaerobic digesters are widely popular, proven, and reliable old technology, their main disadvantage is the hydraulic residence time, which may be from 10 days with heating to 60 days without heating (Metcalf & Eddy, 2003). The anaerobic digesters also work best for biomass slurry with a high solids content of 5 to 10% (50–100 g/L) that would be difficult and energy demanding to pump longer distances. A 10% solids or thicker sludge mass does not flow by gravity, according to the common laws of Newtonian hydraulics, and, in many cases, would require vacuum or pressure flow sewers (Lettinga, 2009) and special pumps.

Anaerobic upflow reactors, also known as the upflow anaerobic sludge blanket (UASB) process, were introduced in the preceding chapter in Section VII.1.4, and shown in Figure 7.5. The hydraulic residence time (HRT) is much smaller than that for an anaerobic digester, yet they provide comparable sludge residence time (SRT). Also, the influent sludge concentrations into the anaerobic digesters must be high, which generally requires liquid sludge thickening. UASB reactors work well at a much smaller influent concentration, and do not require primary settling in most cases. The versatility of UASB reactors and their potential for biogas energy production allow them to be the first unit in the treatment process (Lettinga et al., 1980). The COD removal efficiencies are almost as good as those of the average activated sludge unit.

The best use of a UASB reactor is for treating concentrated used water such as black or brown water or liquid sludge slurry (without thickening); hence, sanitary sewers (existing or new) can be used to bring the used water/primary sludge mixture from cluster water reclamation plants into a centralized integrated resource recovery facility. The UASB process alone or preceded by a pre-acidification reactor will be the first step. In the future, a microbial electrolysis cell (see the next section) could be included in the sequence. If the influent contains high concentrations of mineral solids, a primary settler may be needed (Lettinga et al., 1980). Most of the nutrients in the treatment process will be incorporated into the supernatant (effluent), from which they could be subsequently recovered.

VIII.5 DIRECT ELECTRIC ENERGY PRODUCTION FROM BIOGAS AND USED WATER

The future (already proven and tested today) of biogas and dried solids (including dried manure in developing countries) conversion into energy is not inefficient combustion that produces carbon dioxide and other air pollution. The fundamental concept of fuel cell technology is to produce electricity directly and more efficiently without pollution, and ultimately without a carbon footprint. Two energy conversion concepts will be presented here.

The first process involves *hydrogen fuel cells*, in which electric energy is produced from an exothermic chemical reaction of hydrogen combining with oxygen to produce water, which is water electrolysis in reverse. Water electrolysis uses energy; recombination of hydrogen and oxygen releases energy, as typically shown in middle school chemistry laboratory experiments. Hydrogen is the most efficient energy source and is used by NASA to fuel rockets and space stations. This process is electrochemical. Fuel cells can also operate in "reverse," using electricity to create hydrogen and oxygen from water (World Energy Council, 2009) as is again done in the middle school experiments. Consequently, as documented in the MIT research presented in this chapter in Section VIII.4 (Trafton, 2008; Kanan and Nocera, 2008), production of hydrogen from water is an efficient way to store solar and wind energy. This opens the possibilities of hydrogen fuel cells becoming an ultimate reversible

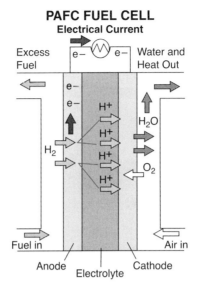

Figure 8.23   Hydrogen fuel cell (from U.S. Department of Energy) with phosphoric acid as an electrolyte.

small-to-large engine operating in the same way as (much larger) pump storage operated hydropower plants.

The second process is *microbial fuel cells*, in which microorganisms under anaerobic conditions oxidize organic matter by transferring electrons to electron acceptors other than oxygen (nitrates, sulphates), which is done outside of the cells (Logan, 2008). This flow of electrons can be harnessed to produce electric current.

VIII.5.1   Hydrogen Fuel Cells

New invented and reinvented concepts of energy recovery employ direct conversion of chemical energy into electricity in *hydrogen fuel cells* (Figure 8.23). Hydrogen fuel cells convert hydrogen directly into electric energy and heat with high efficiency. Hence, energy production by fuel cells is an ultra-clean process. Hydrogen is generated from water (Trafton 2008; Kanan and Nocera, 2008) or from hydrocarbons (natural gas, methyl alcohol, or biogas) or other chemical compounds rich in hydrogen (e.g., ammonia). Biogas containing methane (CH_4) is a good source of hydrogen. Fuel cells have a variety of applications: running automobiles, heating, and providing electricity to diverse facilities from homes to spaceships and space stations.

A hydrogen fuel cell produces clean electricity from hydrogen and oxygen, which react in the presence of an electrolyte. The reactants flow into the cell, and the reaction products flow out of it, while the electrolyte remains within it. The anode side receives hydrogen, and the cathode side receives oxygen from air, which generates electric potential, and electricity is produced by combining hydrogen and oxygen into water. Water is the only by-product. Fuel cells can operate virtually continuously, as long as the necessary flows are maintained. Fuel cells do not operate on a thermal cycle like combustion engines, and they are not constrained by thermodynamic limits.

Hydrogen fuel cells operate typically at a temperature of 25°C. However, energy is needed in the process of biogas (natural gas) reforming, which converts methane into hydrogen.

The advantages of fuel cells for production of energy and heat are (Barbir, 2005):

- Fuel cells have very high efficiency, more than any other engine or fossil fuel power plants.
- Fuel cells can be used for decentralized operations such as in cluster (ecoblock) resource recovery units.
- Fuel cells produce low or zero emissions. Even if methane (natural gas or biogas) is used, emissions are much lower than those from combustion of methane.
- Fuel cells are extremely simple.
- Fuel cells have no moving parts; hence, they are quiet.
- Fuel cell energy production power capacity can vary from watts to megawatts.

The U.S. DOE (2009) lists several types of fuel cells based on the type of the electrolyte. The one that appears to best fit the process of generating energy from biogas is the phosphoric acid fuel cell (PAFC), shown in Figure 8.23, with liquid phosphoric acid as an electrolyte (the acid is contained in a Teflon-bonded silicon carbide matrix) and porous carbon electrodes containing a platinum catalyst. This type of fuel cell is typically used for stationary power generation. PAFCs are more tolerant of impurities in fossil fuels, such as biogas, that have been reformed into hydrogen (see fuel reforming below) than some other types of fuel cells. They are 85% efficient when used for the co-generation of electricity and heat but less efficient at generating electricity alone (37%–42%) when heat is wasted. This is only slightly more efficient than electricity production by combustion-based power plants.

In 2010, the size of the cells ranged from 50 kW to 1 MW; hence, they are very suitable for use in small communities or ecoblocks up to 1000 people. A typical module is 250 kW, suitable for a large apartment house. For larger applications, several modules will be needed.

Under development (by NASA) are *regenerative fuel cells* that produce electricity from hydrogen and oxygen, generating heat and water, just like other fuel cells. However, regenerative fuel cells can reverse the process and by electrolysis generate hydrogen and oxygen from water by means of electricity from solar power or another renewable source. This excess hydrogen can be stored and reused as a fuel in the reverse process, when the renewable source is not active, which is an ideal and efficient way of storing solar and wind energy.

The conversion of biogas into electric energy by hydrogen fuel cells involves two processes: *fuel reforming*, in which biogas is converted into hydrogen and carbon dioxide, *and fuel cell conversion* of hydrogen to electricity (Figure 8.24). In the first process, fuel reforming, biogas is cleaned by removing sulfides in a desulphurization

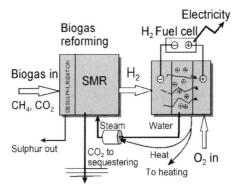

Figure 8.24   Hydrogen fuel cell for electricity production from biogas with biogas reforming.

process, which will produce a purified mixture of methane and carbon dioxide from biogas or natural gas. This process is followed by the fuel-reforming water-gas shift process of hydrocarbon conversion (Sammers, 2006; Halmann and Steinberg, 1999) into hydrogen, in which at high temperatures (700–1100°C) and in the presence of a metal-based catalyst (nickel), steam reacts with methane to yield carbon monoxide (a highly toxic gas) and hydrogen.

$$CH_4 + H_2O \rightarrow CO + 3H_2$$

In the second reaction, additional hydrogen can be recovered by a lower-temperature gas-shift reaction of the produced carbon monoxide to carbon dioxide after combining with steam in steam methane reforming (SMR). The reaction is summarized as:

$$CO + H_2O \rightarrow CO_2 + H_2$$

The first reaction is endothermic (consumes heat), and the second reaction is exothermic (produces heat). The product of the fuel reforming is hydrogen. The by-product is carbon dioxide that normally would be released into the atmosphere, and if natural gas (fossil fuel) is used it would contribute to global warming. As stated before, biogas conversion to energy is carbon neutral. However, because the produced CO_2 is far more concentrated in this process than in the air, Halmann and Steinberg (1999) suggest that carbon dioxide can be sequestered into aquifers (with caution, and not when the aquifer is pumped for potable water production), deep gas wells, or oceans, or chemically (e.g., by neutralizing elevated pH of water) or for producing algal biomass (see Section VII.6.2). CO_2 is heavier than air and would not normally escape back up to the atmosphere. Carbon sequestering by injecting underground is a well-established method when done on a large scale, but on a small scale it has not yet been proven (World Energy Council, 2009). Hence, in a vision of the future we will assume that carbon dioxide will be sequestered.

SMR is the most common method of producing commercial bulk hydrogen. In this process, biogas or natural gas is combined at a higher temperature with steam in the SMR reactor, and carbon monoxide combines with water steam to produce more hydrogen. It is also the least expensive method. However, there are still several problems that must be resolved by research, such as adverse interferences of sulfur and carbon monoxide residuals with membranes in the fuel cell. More than 95% of hydrogen produced today comes from the SMR process. A high temperature fuel cell (600-$700°C$) has been installed in the community center GlashusEtt in Hammarby Sjöstad (Chapter XI) converting biogas produced by the treatment plant directly into electricity.

Hydrogen Production by Solar Energy   Because steam is needed and the SMR reactor operates at higher temperatures, about 30% of energy in methane input is lost when converting it by reforming to hydrogen. Solar hydrogen production can remedy this problem, as described in an article by Bakos (2005). Chemically, the solar panels provide the needed energy in the endothermic processes of reforming methane to hydrogen. Chemically the process is the same as the SMR introduced in the preceding section. A mixture of carbon dioxide and methane from biogas is fed into a reactor heated by a solar concentrated heat array. The intermediate products of the first reaction (hydrogen and carbon monoxide) are fed into a water-gas shift reactor controlled at near atmospheric pressure, which produces more H_2 and CO_2. As previously stated, the first reaction is endothermic but heat is provided by solar energy instead of the electricity produced in the hydrogen fuel cell. In this way more energy can be produced.

VIII.5.2   Microbial Fuel Cells (MFC)

A microbial fuel cell (MFC) is a new bioelectrochemical process that produces electricity from the anaerobic oxidation of biodegradable organic substrates. Microbes in the anodic compartment produce electrons and protons from the oxidation of organic matter, with CO_2 and biomass as final products (Logan, 2008; Ahn and Logan, 2010). In MFC, microorganisms, as elsewhere, oxidize organic matter to obtain energy. In normal circumstances, the electrons are transferred to the terminal electron acceptors (TEA), such as oxygen, nitrate, and ferric ion, which may diffuse into the cells through the membrane and become reduced. However, some bacteria used in MFCs can transfer electrons exogenously (outside the cell) to a TEA, like an iron oxide.

Figure 8.25 shows the principle of MFCs schematically. In the chamber on the left, bacteria use biodegradable organics as an electron donor, oxidize them, and transfer the electrons to the "soup" in the reactor. The electrons pass through a load (e.g., a resistor) and reach the cathode in the chamber on the right, where they reduce oxygen and form water. The two chambers are separated by a membrane allowing the migration of protons produced in the anode chamber. This process requires special microbe exoelectrogens, most of which are bacteria. Microbiology researchers have

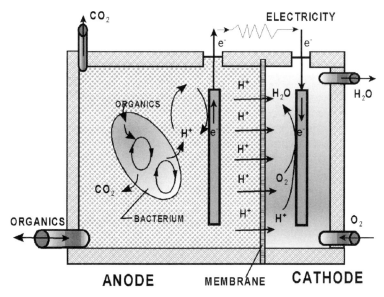

Figure 8.25 Microbiological fuel cell schematics. Adapted from Rabaey and Verstraete, 2005 with permission.

tried to find the best microbiological population, one that would not only produce electricity but also efficiently treat the wastewater. The mechanisms of exoelectrogenic pathways are still not well known.

Aelterman et al. (2006) used laboratory-scale MFCs to treat industrial and hospital wastewater. Ahn and Logan (2010) investigated domestic used (waste) water treatment at two different temperatures ($25°C$, and $30°C$) in batch and continuously fed laboratory reactors. The highest power recovery of 12.8 W/m^3 was achieved under continuous flow and higher temperatures. COD removal was relatively low, 25.8%. However, these removals were achieved at a very short hydraulic retention time (HRT) of only 13 minutes, which indicates that MFC can serve as pretreatment ahead of conventional biological treatment. A higher power density of 464 mW/m^2 (15.5 W/m^3), with total COD removal of 40–50%, was achieved by flowing the wastewater through the carbon cloth anode (Cheng et al., 2006). With two MFCs operating in series, removals of total COD were much higher, 93%, but the HRTs were also much longer, 18 hours in the first reactor and 15.5 hours in the second reactor. Such level of treatment would meet the effluent quality standards.

Another modification of MFC is to retrieve hydrogen from biodegradable organic matter rather than producing electricity. In Figure 7.8 in the preceding chapter, and in the preceding section, we described the process of anaerobic sludge digestion, which is descriptive of most anaerobic processes, including soluble BOD on one side and organic content of vegetation residues on the other side. Digestion is a three-step process consisting of: (1) hydrolysis and fermentation, (2) acetogenesis (production of acetate) and dehydrogenation (production of hydrogen), and (3) methane fermentation. A question can be asked as to why, under normal conditions in the digester, the second process cannot be carried to its logical completion—that is, breakdown of the organic acids into H_2, CO_2, and water. Logan (2008) pointed out that the second reaction is endothermic, and bacteria cannot fully convert acetates to methane. As a matter of fact, hydrogen is an inhibitor, and the breakdown of higher organic acids to acetate and hydrogen requires that hydrogen be in low concentrations (McCarty and Mosey, 1991). Only then, in the traditional digestion process, does the final fermentation reaction produce methane and carbon dioxide from acetate. The previous section outlined an endothermic chemical process of steam methane reforming that breaks methane into hydrogen and carbon dioxide.

Logan (2008) and his coworkers have experimentally proven that if a small amount of electricity, about 0.2 volts, is added to that produced at the anode in an MFC, and oxygen is not used at the cathode, hydrogen gas can be produced at the cathode. These units are called *microbial electrolysis cells (MECs)* or *bio-electrochemically assisted microbial reactors (BEAMRs)*. The extra current can be provided by the MFC and/or from a renewable source. The reactor also produces carbon dioxide, which is the same outcome as chemical methane reforming. The process is completely anaerobic, in contrast to MFC, which needs oxygen, and produces almost three times more energy than the electricity used to extract it. The MEC process may not need specific exoelectrogens; it may be accomplished by bacteria already present in the conventional anaerobic digestion process. These bacteria operate in a temperature range of 20 to 30°C.

MFCs and MECs are a promising technology but have not yet been used in practice. Numerous researchers have tested the feasibility of MFCs applied in wastewater (used water) treatment. One conclusion, which will be further elaborated, is that substrate of wastewater used in the process should not be diluted, and it does not make sense to separate organic solids by primary clarification. MFC energy in the laboratory is still less than that from the anaerobic digester. Laboratory measurements by Logan (2008) reached power densities on a volumetric basis of 115 W/m^3, which is 3.5 times less than the power density of an anaerobic digester. The goal of Logan's and his coworkers' research is to reach densities in the range of 83 to 415 W/m^3.

The introduction of MFCs and MECs and the increased emphasis on energy recovery and reduction of carbon footprint may necessitate rethinking of the conventional water reclamation process, which typically consists of pretreatment, physical treatment (settling and solids removal), activated sludge treatment with excess sludge removal and sludge preprocessing (thickening) and digestion (see also Chapter VII). This traditional process requires a lot of energy and emits significant amounts of GHGs (carbon dioxide, nitrous oxides, and methane). A new energy-generating process is emerging (Logan, 2008; Lettinga, 2009, Vestraete, Van de Caveye, and Diamantis, 2009). This process requires separation of used water flows into black and gray water flows, to minimize dilution and prevent entry of inert solids such as grit and finer particles from urban runoff, which carry toxic metals and other chemicals (see Chapter VII). In this process, anaerobic digestion or MEC (MFC) would be the first step. MEC cells are anaerobic, and so is the first compartment in the MFC reactor. Because anaerobic digestion (AD) as of the first decade of this century produces more energy than MFCs or MECs, AD in a form of digester or a UASB reactor is preferable as a first step.

VIII.5.3   Harnessing the Hydraulic Energy of Water/Used Water Systems

Water and wind are the two oldest clean energies used by human beings. The formula for water power is basic physics, simply that power is the capability to move a unit of mass vertically up or down by a unit of length in a unit of time. This translates to

$$P_w = \gamma Q H_t$$

where P_w is water power (watts), γ is specific density of water (9810 N/m^3), Q is flow (m^3/sec), and H_t (meters) is the total hydraulic head, which includes the elevation difference between the two points on the system minus the head losses due to friction or a change of momentum. Turbines generate mechanical energy from water power, while pumps convert mechanical energy into potential (pressure) and the kinetic energy of moving water. Structurally, pumps and turbines are very similar. Water turbines in connection with a generator can produce electric energy with an efficiency of about 75 to 80%, which is substantially more than the efficiency of combustion engines.

There are a number of points and instances in water systems where water energy is wasted and can be harvested. What is needed is more or less steady flow and excess unused elevation difference. Throughout history waterwheel mills were built on small creeks by impounding them, which provided storage and the head. The efficiency of old wind wheels, still pumping water from canals in the Netherlands, is much less than that of modern turbines.

Opportunities of recovering hydraulic energy from water/stormwater/used water systems are numerous and include:

- *Unused pressure differences in the pressurized water systems.* Pressurized water supply pipe networks are designed to operate at pressures from 2.5 to 10 atmospheres, which are approximately equal to 25 to 100 meters of the hydraulic pressure head (1 atmosphere = 10.3 meters of pressure head). If the pressure in the pipeline exceeds 10 atmospheres (10.3 bars), which may happen in hilly terrains, a pressure release valve is installed which wastes the water energy. The relation between the pressure inside the pipe and the pressure head is expressed by the centuries-old Bernoulli equation, featured in every basic fluids or hydraulic text. The hydraulic head difference between the high (1) and low (2) points is then calculated as

$$H_t = [\text{Elevation}\,(1) + p1/\gamma] - [\text{Elevation}\,(2) + p2/\gamma] - \text{losses}$$

where p is the pressure inside the pipe in newton/m^2.

Head losses are relatively small and include friction losses in the pipeline between the upper point 1 (before the pressure break) and the lower point (2). Losses in the turbine are accounted for in its efficiency. For example, if the pipeline pressure is to be reduced from 10 to 3 atmospheres and the water flow is 0.2 m³/sec, then the power-generating potential is

$$P_w = \gamma Q H_t = 9810 [\text{N/m}^3] \times 0.2 [\text{m}^3/\text{sec}] \times (10 - 3) \, [\text{atm}] \times 10.3 \, (\text{m/atm})$$
$$= 141,460 \, \text{W} = 141.46 \, \text{kW}$$

If a turbine and a generator with an overall efficiency of 75% are installed instead of the pressure release valve, they could generate in one day 141.46 (kW) × 24 (hrs) × 0.75 = 1811.28 kWh of electric energy and reduce GHG emissions by 1811.28 kWh × 0.62 kg of CO_2/kWh = 1062 kg CO_2/day or about one ton per day.

The same calculation can be performed for the energy recovery from the reject water of the highly pressurized reverse osmosis and nanofiltration systems (see Chapter V).

- *Drop shafts for conveyance of used water and stormwater flows into deep interceptors.* Sewer interceptors conveying used water or collecting, storing, and conveying combined sewer and/or sanitary sewer overflows (CSOs and SSOs) for treatment are sometimes drilled in rocks at great depths. For example, Milwaukee, Wisconsin, and Chicago, Illinois, deep tunnels (see Figure 1.18) are located at depths of more than 100 meters below the surface. A used water conveyance tunnel to the Changi Water Reclamation Plant in Singapore is 30 or more meters below the surface. Polluted water and sewage from these underground storage and conveyance facilities are pumped to the water reclamation facilities at a great expense of energy and money. A part of the energy could be again recovered by installing turbines and generators at the major drop shafts. The problem with recovery of energy from CSOs and SSOs is the intermittent nature of these overflows occurring only during rainfall, and their duration which is not long enough to justify energy recovery, based on cost, and technical difficulties with turbines run by water containing grit and other larger particles.
- *Other sources of energy.* The available and emerging technologies that water utilities can employ to capture the kinetic and potential energy of water, beyond the traditional storage of water behind dams, are numerous. They include run-of-the-river hydroelectric generation, or in some facilities tidal and wave energy. The pervasive incorporation of small-scale turbines into water facilities could offer significant potential in certain settings. Further, to match electricity demand and power supply during peak hours, it may be possible to utilize pumped storage facilities using reversible pumping and generation units.

VIII.6   SUMMARY AND A LOOK INTO THE FUTURE

This chapter investigated the water/used water nexus, including the possibilities and feasibility of reaching the net zero GHG footprint goal. The current solar and wind energy harvesting technologies and new research in the area of fuel cells (both electrochemical and microbiological) are extremely promising, and it can be concluded that the net zero GHG footprint is achievable. It is already being proven by several developments in China, the United Arab Emirates, the United Kingdom, and elsewhere, even in the US (Sonoma Mountain Village—see Chapter XI) . In Chapter V, we pointed out that with respect to water management, water conservation will bring real and significant savings of energy and, hence, reduction of GHG emissions. The GHG emission reductions are not as clear with high reuse and recycle, which may require energy-demanding technologies such as nanofiltration and reverse osmosis to produce water of acceptable quality for direct and indirect potable reuse.

However, the urban carbon footprint is not only caused by potable water, stormwater, and used water management, although it may be significant. Heating of buildings and using energy for warming water and for cooling are the biggest contributors to GHG emissions. There are several key sources of renewable energy: solar, wind, heat content of used water, ground and geothermal, the electrochemical potential of used water, and biogas from digesters.

VIII.6.1   A New Look at the Used Water Reclamation Processes

The majority of current and near-future urban sewerage systems aim to collect all water from the surface and from buildings and convey the used water and stormwater below the surface to large regional facilities. This was precipitated in the U.S. by the use of the simplistic economic cost analyses of the last century based on the economy of scale that led planners to "the bigger the better" conclusions, without considering the social costs of losing water from urban streams, effluent domination, and nutrient pollution. These linear systems use a lot of energy for conveyance (e.g., pumping in lift stations of sewers) and treatment by aerobic treatment process units. Furthermore, the wastewater arriving at the wastewater treatment plants (sometimes called water reclamation plants) is diluted by I-I inflows. With more stringent effluent quality requirements and with many cities running out of water, the tendency is towards using energy-demanding and increasingly complex treatment processes. Energy recovery is typically minimal and only includes sludge recovery of biogas from sludge digestion.

Visionaries such as Lettinga (2009), Barnard (2007), Asano and Tchobanoglous (Metcalf and Eddy, 2003, 2007), and Logan (2008) are calling for a change of the paradigm of used water reclamation, leading to integrated resource recovery. The previous section outlined the energy recovery potential from anaerobic unit processes that break down organic solids (both particulate and dissolved/colloidal) into biogas-containing methane and carbon dioxide. We have also suggested that production of algae fed by partially treated used water (e.g., with nutrients not removed) holds a promise of both producing biomass for biofuel and sequestering carbon. The previous discussion also introduced fuel cells (both microbial and electrochemical) that can convert biogas or organic solids directly into electricity or high-energy hydrogen fuel. These anaerobic recovery processes work best when the incoming used water is concentrated with organic particulate, colloidal and dissolved organic solids. The biological anaerobic digestion process also works for liquid (used water), slurry and so called non-newtonian organic fluids (sludge), and in solid but moist waste (landfill).

The change of the paradigm would include:

- Concentrating the used water flows by eliminating all clean water inflows such as surface runoff, groundwater, and roof downspout inflows, as well as rigorous water conservation. Underground conduits should be capable of transporting not only used water but also other organic solids such as shredded vegetation residues and food waste (frequently discharged in the U.S. by kitchen-sink disposals comminuting kitchen waste), as long as the slurry containing mostly organic particles is liquid enough so that the particles are dispersed and follow the hydraulic newtonian gravity flow without settling (e.g., flow velocity in the sewers should be more than 0.6–0.9 m/sec, which is a standard design parameter for conventional sewers).

- Black water with or without urine separation would be the best candidate for integrated cluster or regional resource recovery. Gray water can be reclaimed and treated in cluster (ecoblock) reclamation/reuse treatment units employing microfiltration and RO units followed by disinfection by UV radiation and adding ozone. Concentrate and sludge from these units would be sent to the ecoblock (cluster) biological resource recovery unit (see Chapter VII) or with the concentrated biodegradable used water and solids flow to the regional integrated resource recovery facility. The benefits of urine separation must be assessed because nutrients recovery, the main reason for urine separation, can also be effectively done in the integrated resource recovery facility (IRRF) described below. However, urine contains 50% of the phosphorus and more than 75% of the nitrogen load in less than 1% of the total flow; consequently, nutrient recovery from urine is more efficient and less costly than the recovery from the total flow, but it still would leave enough nutrients for the follow-up treatment and resource recovery.

- Consider anaerobic treatment in the form of an anaerobic digester or upflow anaerobic sludge blanket (UASB) unit as the first step in treatment and water and energy recovery. The product of the digestion would be an effluent laden with nutrients, biogas, and residual solids usable as soil conditioner after removal of excess moisture. Biogas could be converted by hydrogen fuel cells (HFC) into hydrogen and electricity. CO_2 produced in HFCs could be recycled and the excess sequestered. If the ongoing and future research proves the feasibility of direct large-scale and economical H_2 or electricity recovery by microbial fuel cells, the anaerobic step would be an MEC (BEAMR) for hydrogen production or an MFC for electricity. The BOD (COD) removal efficiency of well-designed and operated anaerobic units is more than 75%, and BOD removal efficiencies as high as over 90% can be achieved (Metcalf & Eddy, 2003).

- Organic solids can be converted to biofuel either by a chemical heat gasification process producing syngas (a mixture of carbon monoxide, carbon dioxide, and hydrogen) – pyrolysis -or by anaerobic digestion that produces methane biogas. Both carbon monoxide and methane can be reformed into hydrogen and electricity in a hydrogen fuel cell, and excess carbon dioxide can be sequestered.

- Aerobic/anaerobic posttreatment units requiring no or less energy, such as trickling filters or subsurface flow wetlands, could be used as a polishing step that could also convert sulfide to sulphate. Aerobic trickling filters can be combined with an anoxic unit for nitrogen removal. Vegetation from wetlands (two in a series) can be harvested, comminuted, and sent to the anaerobic reactor.

- Phosphorus and ammonium can be removed chemically, if needed, or as struvite in a special upflow fluidized bed reactor, or by algae in a algae growth reactor.

- Membrane filters followed by UV radiation (with or without ozone) would be the final step before discharge of the reclaimed water for irrigation and/or as ecological flow into a receiving water body.

- For indirect potable reuse, the effluent would have to receive additional treatment by reverse osmosis, mixed with a good-quality dilution water (e.g., treated stormwater), and stored for an extended period in a surface or underground basin or aquifer.

Daigger (2009, 2010) stated that dual distribution and source separation practices complement water reclamation and reuse by delivering "fit for purpose" water for various uses and separating the components of the typical waste stream to facilitate energy and nutrient recovery. Table 8.5 illustrates segregation of the typical residential waste stream into gray water (laundry, bath), black water (feces, kitchen), and potentially yellow water (urine). Gray water has the largest volume and is relatively less contaminated, facilitating low-energy treatment for reuse. Black water is relatively low in volume and high in concentration when grey water is removed, and

Table 8.5 Distribution of organic matter and nutrients in typical European wastewater (Henze and Ledin, 2001)

Source	BOD_5 (g/(person·day))	Total Nitrogen (g-N/(person·day))	Total Phosphorus (g-P/(person·day))	Potassium (g-K/(person·day))
Toilet Waste				
Feces	20	1.1	0.6	1.1
Urine	5	11.0	1.4	2.5
Kitchen	30	0.8	0.3	0.4
Bath/Laundry	5	1.1	0.3	0.4
Total	60	14.0	2.6	4.4

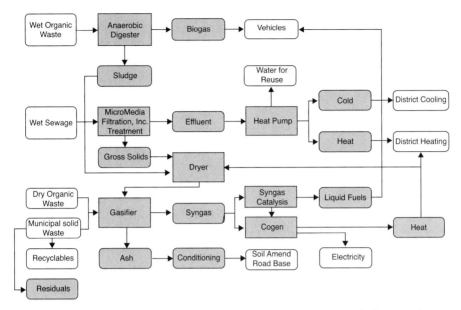

Figure 8.26 A conceptual flow chart of the decision-making process used when adopting an IRM design process (from O'Riordan et al., 2008).

can be treated directly by anaerobic processes for energy production. Yellow water is very low in volume (about 1 L/capita/day) and contains most of the nutrients. In short, source separation segregates the water, organic matter, and nutrient components for efficient recovery and reuse (see also Chapter V).

VIII.6.2 Integrated Resource Recovery Facilities

In this section three possible alternative integrated resource recovery facilities (IRRFs) are presented. *Alternative 1*, shown in Figure 8.26, is the integrated resource management (IRM) process used in development.

Figure 8.26 shows the integrated resources management concept used in development of sustainable communities in British Columbia, Canada, and described in O'Riordan et al. (2008). This system is focusing mainly on resource recovery similar to that used in Sweden. In IRM, decentralized aerobic/anoxic tertiary used water treatment produces heat, water, nutrients, and biosolids that can be reused rather than discharged into the environment. "Waste" heat energy from sewage is captured to heat homes and businesses, and the resulting chilled water is then used for cooling.

Organic solid waste solids are dried and used to create syngas or biofuels, together with the solids from sewage. These "new" carbon neutral energy sources displace fossil fuels, saving money and reducing GHGs, while increasing energy supply flexibility and security. IRM reduces carbon emissions by about 25% when compared to traditional wastewater treatment and energy recovery from solids (Corps et al.,

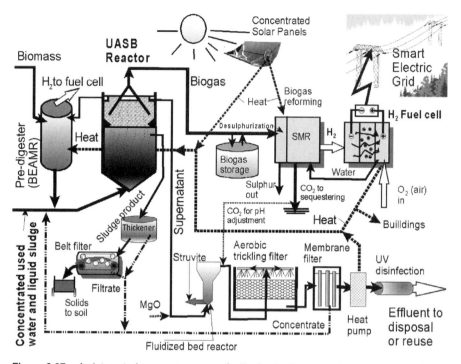

Figure 8.27   An integrated resource recovery facility for the Cities of the Future which reclaims water and produces electricity, heat, fertilizer, soil-conditioning solids, and hydrogen, by Novotny (2010). Includes basic concepts of Lettinga's (2009) Natural Biological Mineralization Process proposal.

2008). The processes included in the flow chart rely on combustion for biogas and syngas conversion to energy and electricity.

The next two alternatives present a vision of a future integrated resource recovery facility (IRRF) that converts resources contained in used water and organic solids into electricity, heat, hydrogen and biogas, nutrients, and organic solid residuals, and sequesters carbon (positive GHG benefit) while being a net clean energy producer. The proposed IRRFs contain unit processes that have already been invented and are either already used in practice or being tested with promising results in laboratories. Therefore, while it is a vision of the future, the vision and proposal are realistic. The idea and concepts of IRRF were presented by Novotny (2010) at the WEF/IWA Conference on the Cities of the Future.

The first IRRF (*Alternative 2*) in Figure 8.27 is a combination of the Natural Biological Mineralization Process proposed by Lettinga (2009) for biogas production, and also advocated by Logan (2008) for electricity and hydrogen production, with the hydrogen fuel cell. The illustrative integrated resource recovery process shown in the figure would begin with an upflow anaerobic sludge blanket (UASB) reactor unit. The retention time in the UASB rector is 7–10 hours (see Table 7.3), in contrast to 15 days or more in a conventional mesophilic high-rate sludge digestion process (Metcalf & Eddy, 2003). The UASB reactor would operate satisfactorily at temperatures greater than 20°C, as compared to more than 35°C needed in a mesophilic digester. The recommended inflow COD concentration should be between 0.5 and 5 g/L or greater, as long as inflow TSS is less than 8 g/L (Lettinga and Hullshoff Pol, 1991).

The liquid inflow could be mixed with outflow from the pre-digester decomposing biomass (decaying vegetation, food waste, wood chips) into acetates with suppressed methane fermentation. This reactor would require much shorter HRT than a conventional anaerobic digester. The process was identified by Lettinga and Hulshoff Pol (1991) as a pre-acidification or acidification, and the authors mentioned it as beneficial but not necessary. In view of Logan's (2008) discovery of direct hydrogen- or electricity-producing microbial cells and the general need for clean energy recovery, pre-acidification of the biomass could be an asset. Under the classic model, pre-acidification produces acetate and hydrogen without forming carbon dioxide, but hydrogen may be scavenged by hydrogen-scavenging methanogens (McCarty and Mosey, 1991), which can be prevented by lowering pH. Hydrogen was a useless by-product until this century, but today it is the best source of energy. Hydrogen production can be maximized, for example, by adding cathodes and anodes with electric current, essentially converting the pre-acidification of the biomass into a MEC–BEAMR reactor (Logan, 2008). The result would be production of hydrogen, preprocessing of organic particulate solids, and conversion of the solid biomass into soluble acetates that would then be converted into methane biogas with the incoming concentrated water and liquid in the subsequent UASB reactor. The hydrogen in the pre-acidification reactor from biomass could be conveyed directly to the fuel cell to produce electricity. A portion of the outflow from the UASB reactor would provide necessary moisture, nutrients, and microorganisms to the dry solids (arriving by trucks) in the BEAMR reactor. Another variant would be to comminute the dry solids at the cluster level, mix them with the concentrated used water, and send them as slurry to the regional resource recovery facility.

A second optional UASB reactor could be added to improve the treatment. A fluidized bed reactor precipitates and removes phosphorus and ammonium in the form of granules of struvite, which is facilitated by adding magnesium as magnesium hydroxide, magnesium chloride, or magnesium oxide (MgO) (Barnard, 2007; LaCorre et al., 2009; Britton et al., 2005). If magnesium chloride is added, pH could be increased by adding sodium hydroxide, but the salinity increase it would produce may not be desirable in some cases. pH adjustment to the range of 8.5 to 9.0 is needed for efficient (90%) struvite precipitation (Britton et al., 2005). Subsequently, carbon dioxide will be added after struvite precipitation to reduce pH to close to neutral. This CO_2 is sequestered and will not contribute to global climatic change.

Box 8.7 contains calculation of the biogas and energy production by the USAB reactor, which is expected to produce most of energy in the IRRF. However, this does not include the hydrogen produced in the BEAMR (pre-acidification) reactor.

BOX 8.7   CALCULATION OF METHANE AND ENERGY PRODUCTION IN A UASB INTEGRATED RESOURCE RECOVERY UNIT PROPOSED BY NOVOTNY (2010)

The unit serves a cluster with a population equivalent of 8000people. It receives concentrated black water effluents in the amount of 50 liters/capita-day (see Chapter V, Table 5.3). The concentrated waste composition is COD = 3000 mg/L, suspended solids TSS = 1000 mg/L. It is assumed that 75% of the solids are volatile.
Calculate the daily flow

$$Q = 8000 \text{ (inhabitants)} \times 0.05 \text{ (m}^3\text{/cap-day)} = 400 \text{ m}^3\text{/day}$$

Because methane can be expressed as COD, the COD mass balance of the UASB reactor is performed based on Metcalf & Eddy (2003)

$$\text{COD (influent)} = \text{COD (effluent)} + \text{COD(converted to biomass)}$$
$$+ \text{COD(methane)}$$

Hence $$\text{COD}_{\text{CH4}} = \text{COD}_{\text{infl}} - \text{COD}_{\text{eff}} - \text{COD}_{\text{VSS}}$$

In the mass balance the above components will be expressed as mass/day, which equals concentration times flow. Also note $(\text{mg/L}) = (\text{g/ m}^3)$. It is assumed that the reactor will operate with 90% COD removal efficiency. COD of 1 gram of sludge-type volatile solids is 1.42 grams of COD per one gram of VSS. Then

$$\text{COD}_{\text{infl}} = 3000 \text{ (g/m}^3) \times 400 \text{ (m}^3\text{/day)} = 1,200,000 \text{ (g/day)}$$
$$\text{COD}_{\text{eff}} = (1 - 0.90) \times 1,200,000 \text{ (g/day)} = 120,000 \text{ (g/day)}$$

The solids yield coefficient (from Table 8.5) is Y = 0.08 kg VSS/kg of COD removed. Then

$$\text{COD}_{\text{VSS}} = 1.42 \text{ (COD/VSS)} \times 0.08 \text{ (kg VSS/kg of COD)}$$
$$\times (1,200,000 - 120,000) = 122,688 \text{ (g/day)}$$

Then $\text{COD}_{\text{CH4}} = 1,200,000 - 120,000 - 122,688 = 957,312 \text{ g/day}$
From Table 8.5 (also Metcalf & Eddy, 2003), the methane gas equivalent of COD is 0.4 L of CH_4/g COD equivalent. Then the methane gas production will be approximately

$$CH_4 \text{ (produced)} = 0.4 \text{ (L } CH_4\text{/g COD)} \times 957,312 \text{ (g/day)} \times 0.001 \text{ (m}^3\text{/L)}$$
$$= 383 \text{ m}^3\text{/day}$$

From Table 8.5 the energy equivalent of methane at 25°C is 9.2 kWh/m^3 of CH_4. Assuming that the total energy conversion efficiency of the fuel cell is 85% of which 50% is electricity and 35% is heat, the produced energy will be

$$\text{Electric energy} = 0.5 \times 383 \text{ (m}^3\text{/day)} \times 9.2 \text{ (kWh/m}^3\text{)} = 1762 \text{ kWh/day}$$
$$\text{Heat energy} = 0.35 \times 383 \text{ (m}^3\text{/day)} \times 9.2 \text{ (kWh/m}^3\text{)} = 1233 \text{ kWh}$$

For this particular example, the corresponding electric and heat energy gain per person that could be counted towards the reduction of carbon footprint is

$$E = (1762 + 1233) \text{ (kWh/day)}/8000 \text{ (inhabitants)} = 0.38 \text{ kWh/cap-day.}$$

This estimate does not include energy derived from digested biomass (vegetation residues, food waste, or shredded wood).

Vestraete, van de Caveye and Diamantis (2009) presented similar but simpler alternatives of resource recovery from concentrated used water that also begin with the UASB-ST reactor. This reactor is an upflow septic tank in which used water entering the tank is in contact with the sludge blanket. The excess solids from the sludge blanket are diverted into a digestion chamber which is emptied about once a year. The effluent has to be post treated, the authors proposed membrane filters and reverse osmosis. The produced biogas can be collected but the gas production and energy yield is not as high as that from the UASB. This unit is suitable for cluster used water treatment and resource recovery.

If the one- or two-stage UASB process does not bring the effluent COD (BOD) to the acceptable concentration for disposal into receiving water bodies or reuse, then an aerobic polishing step may be necessary. In *Alternative 2* an aerobic trickling filter may require minimum energy. With a little pressure in the feed pipe, the rotating arm would turn by hydraulic inertia without an electric motor. The treatment process of water reclamation would be concluded by membrane filtration and ultraviolet disinfection. Because the supernatant leaving the UASB reactor will have a temperature greater than 20°C, heat can be extracted from the effluent by a heat pump and returned to the plant heating system. Excess heat will heat buildings.

Alternative 3 is the same as Alternative 2, except that the trickling filter would be replaced by an algal biomass–producing reactor (Figure 7.7), either following the struvite-producing upflow fluidized bed reactor or directly after the UASB reactor. By not including the struvite producing unit or including it after the algal bioreactor effluent, the biomass production would be greatly increased. The purpose of the algal biomass–producing reactor would be to:

1. Remove nutrients
2. Produce more biomass that, after separation by the algal production reactor, would be diverted back to the UASB reactor or even to the BEAMR reactor to produce more biogas

3. Sequester waste carbon dioxide that would be delivered from the SMR (or similar) biogas-reforming unit and/or any other source of waste CO_2 to:
   - Adjust the pH after struvite precipitation (*Alternative 2*)
   - Grow algae (*Alternative 3*)

The major output of the integrated resource recovery facility would be biogas with about 65% methane; solids for soil conditioning that may still contain a low nutrient content; hydrogen; electricity; struvite, which is granulated ammonium magnesium phosphate fertilizer; and heat. Upstream urine separation would not be needed but could contribute to a better nutrient recovery. Another benefit would be carbon sequestering credit that could be traded.

The anaerobic digestion (BEAMR) reactor unit for biomass should be heated by the heat provided by the fuel cell, by the concentrated solar power panels, and by the heat extracted from the effluent. Using solar power for heating the pre-digester and biogas reforming units will increase energy production by the system. When methane is converted to heat and energy, the process emits CO_2, which is also emitted by the aerobic trickling filter. Because this CO_2 is not produced from fossil fuel, the carbon dioxide emission would be carbon neutral. In the proposed IRRF, carbon dioxide would be sequestered by all three alternatives, with maximizing sequestering in *Alternative 3*. On the other hand, letting methane escape into the atmosphere would significantly contribute to global warming because methane is a GHG 25 times more potent than carbon dioxide. Biogas from integrated resource recovery facilities, after cleaning, may also be offered as a fuel for public transportation buses. However, in the future ecocities, all produced biogas might be converted to hydrogen gas, the ultra-clean fuel, and to electricity by a hydrogen fuel cell. The technology is already available and has been tested and implemented in pilot automobiles.

The residual solids after drying can be used as soil conditioning to increase organic content of soils, by which carbon would be sequestered. Alternatively, residual solids can be mixed with solid organic waste and gasified to produce syngas or converted by pyrolysis to charcoal and other products, which is a form of carbon sequestering.

If microbial fuel cell technology reaches efficiencies approaching the UASB units combined with electrochemical fuel cells, an MFC or MEC can replace anaerobic digestion and fuel cells in one unit or in tandem; that is, the MFC can produce electricity directly, while the MEC can provide hydrogen gas. The MFC and MEC can operate at near ambient temperatures without heating (Logan, 2008). Looking 15 to 25 years into the future, this process may become a reality. Excess biogas and/or hydrogen produced can be stored.

VIII.7   OVERALL ENERGY OUTLOOK—ANTICIPATING THE FUTURE

VIII.7.1   A Look into the Future 20 or More Years Ahead

In Section VIII.1.2, it was reported that the average carbon footprint of the larger urban area in the U.S. is about 8.5 tons CO_2 equivalent/capita-year. Table 8.6 contains

Table 8.6   Average statistics of energy use in 100 large U.S. cities (recalculated from and modified from Gleaser and Kahn, 2008)

Energy Use	CO_2 Emissions in Tons/Capita-Year	% of Total
Transportation by cars	4.091	47.0
Public transportation	0.388	4.4
Home heating by gas or oil	1.470	17.0
House electricity including that for cooling	2.751	31.6
Total	8.71	100

statistical data averages of GHG emissions for one hundred large U.S. cities, from Gleaser and Kahn (2008). These averages separate household energy use into transportation by vehicles, public transportation, household heating, and use of electricity. Heating is by natural gas or oil. Because data are from before 2005, information about hybrid or electric cars was not available.

Figure 8.28 shows that in 2007 the carbon emissions trend in the U.S., after 50 years of going up, reversed and started to decrease, 9% in 2007 alone, and continuing to decrease in 2008 and 2009 (Brown, 2009). The article, published by the Earth Policy Institute, noted a number of near-future trends that raise optimism:

- Automobile fuel-economy standards are becoming progressively stronger, supplemented by occasional government-subsidized (e.g., clunkers for cash) promotions for replacing low-mileage cars.
- Higher appliance efficiency standards (Energy Star program) are in place.

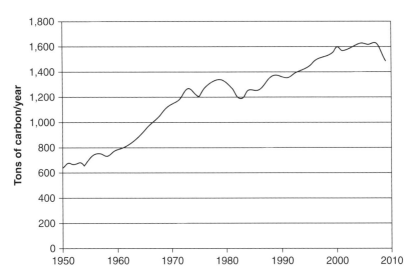

Figure 8.28   Carbon emissions in the U.S. started to decrease after 2007 (data source: U.S. EPA, DOE; replotted from Brown, 2009).

- There are financial incentives supporting large-scale development of wind, solar, and geothermal energy.
- Public awareness and movement to the green lifestyle are on the rise. Climate-conscious, cost-cutting Americans are altering their lifestyles to reduce energy use.
- The federal government—the largest U.S. energy consumer—is reducing vehicle fleet fuel use 30% by 2020, recycling at least 50% of waste by 2015, and buying environmentally responsible products.
- There is a virtual de facto moratorium on new coal plants; many plants are expected to close and be replaced by wind farms, natural gas plants, wood chip plants, or efficiency gains. Shifting to wind, solar, and geothermal energy drops carbon footprint to zero.
- U.S. solar cell installations are growing at 40% a year.
- Oil use and imports are both declining as the new fuel economy standards raise the fuel efficiency of new cars 42% and light trucks 25% by 2016.
- Forty-two percent of diesel fuel burned in the rail freight sector is used to haul coal; hence, falling coal use means falling diesel fuel use.
- The federal government provides incentives to development train networks run by electricity instead of diesel fuel.

Assumptions for the Future

Electric Energy Production

Electric energy production from sources that do not emit GHGs (nuclear, hydropower, and renewable wind and solar) will be increased from 30 to 60%. This will reduce GHG emissions attributable to power production by 43%, from 0.62 kg CO_2/kWh in the early 2000s to 0.35 kg CO_2/kWh, which the authors expect to be attainable in 2030. Most of the reduction could be realized by phasing out coal power plants and replacing them with renewable energy and nuclear power. Nuclear power has about 25 times the life-cycle footprint (construction, cement, transportation) of wind power, but is still far below fossil fuel power plants. This estimate is on the conservative side because the U.S. DOE estimates that wind energy potential alone will reach at least 20% of the total energy production in the U.S. Jacobson and Delucchi (2009), in *Scientific American*, presented a realistic plan by which all electric power would be produced by renewable sources that would include hydro, solar, and wind, with a target year of 2030.

Vehicular Traffic Assumptions

The majority of cars will be electric plug-ins or hybrids, which will double mileage and cut the GHG emissions from automobiles by half. It is anticipated that with better public transportation, fewer miles (kilometers) will be driven by drivers. Anticipated GHG emissions will be cut by 60%, or to about

1.5 GHG tons/capita-year. In the Cities of the future, extensive use of automobiles will be discouraged.

Public Transportation

Public transportation will be mostly by electric trains and light rail and buses powered by biogas, electricity, or green hydrogen. Because most buses will be powered by green fuels, GHG emissions from public transportation should be dramatically reduced to less than 0.1 GHG tons/capita-year.

Heating

GHG emissions from heating buildings can be reduced by 50–60% by switching to heat pumps deriving heat from air or, better, from used water or the ground. Passive energy house-heating measures (green roofs, good insulation, southern exposure) can also dramatically reduce heating needs. Comparing heating needs in Southern cities (Tampa, Orlando, New Orleans, Phoenix) with those in Northern cities (New York, Boston, Chicago, Portland), the difference in GHG emissions from heating represents about 2 tons of GHGs. On average, heating GHG emissions can be less than 0.3 GHG tons/capita-year in the South, and in Northern climatic conditions less than 1 GHG tons/capita-year.

Electricity demand

NSTC (2008) estimated that a 60–70% household electric energy use reduction can be achieved by more efficient appliances, new water-heating methods, water savings by conservation, reuse of rainwater, and reduction of cooling energy demand by cross ventilation, green roofs, insulation, and shading. Furthermore, the electric energy production from renewable sources, based on the assumption given above, would increase from 30% to 60% (still less than in France and most European countries), reducing the carbon footprint of power production to 0.35 kg CO_2/kWh. These measures would bring down the GHG equivalent urban emissions from 4.45 tons/capita-year to about 0.9 tons/capita-year.

With these measures the future potential of total GHG emissions in U.S. cities would be reduced by 2030 to about 3 tons GHG equivalent/capita-year, which is slightly more than it is today in Barcelona, Spain.

Water-Related Energy Savings

In Section VIII.1.3, the estimated energy equivalent of one cubic meter of *water service* (providing, transporting, treating, and disposal) is 2.26 kWh/m^3, with corresponding carbon emissions of 1.37 kg CO_2/m^3. Reducing water demand by water conservation (Chapter V) from 0.5 m^3/capita-day to 0.2 m^3/capita-day, making the same assumption of the reduction of GHG emissions by the power industry, could bring about a CO_2 reduction of 0.54 kg CO_2/capita-day or 0.2 ton CO_2/capita-year.

In Box 8.7 the contribution of *the integrated resources facility* to reduction of energy use and improving carbon footprint was estimated as 0.4 kWh/capita-day. Assuming the future carbon/energy conversion to be 0.35 kg CO_2/kWh, this would represent a reduction of the carbon footprint by 0.14 kg CO_2/capita-day or 0.05 ton CO_2/capita-year.

The effect of *extracting heat from used water* was calculated in Box 8.4 as 10.47 kWh/m^3. The future per capita water use is conservatively estimated as 0.2 m^3/capita-day; hence, the heat extracted from used water will have an energy value of 2.1 kWh/capita-day and a reduction of the carbon footprint by 0.73 kg CO_2/capita-day or 0.268 ton CO_2/capita-year.

Additional energy savings can be achieved by bringing surface runoff drainage to the surface, thus eliminating pumping from deep interceptors and by lift stations. Conservatively, such pumping and miscellaneous measures can save another 0.3 ton CO_2/capita-year.

In total, the minimum reduction of carbon footprint by the urban water sector could be almost 1 tons CO_2/capita-year, helping to reduce the total urban carbon footprint to 2 tons CO_2/capita-year.

Finding Savings of 2 ton CO_2/capita-year   The 2-ton CO_2/capita-year energy deficiency requires, with the future energy/carbon conversion of 0.35 kg CO_2/kWh, finding 15.6 kWh/capita-day savings or new renewable sources of energy. So far our look into the future did not consider cluster-wide and other local energy sources in the future ecocities. The additional energy savings and new sources of energy will come from:

1. More far-reaching reduction of carbon emission by private automobiles. The ecocities of today and the future will reduce driving by switching to walking, biking, and more public transportation. Public transportation stops should be within 15 minutes' walking distance from every home. Reducing the driving of an average driver from 20,000 km/year (12,437 mi/year) to 5000 km/year (3110 mi/year) and increasing mileage from 8.25 km/liter (20 mi/gal) of gasoline to 21 km/liter (50 mi/gal) (equaling current hybrid cars already on the market) could bring reduction of GHG emissions of 1 ton CO_2/capita-year.

2. Including photovoltaics and solar panels on every house, which would eliminate most of the house electric energy needs, even under the current technology. In warmer, sunny climatic conditions with the solar radiation rate of 4 kWh/m^2 and panel efficiency of 40% (double the current efficiency), a panel array on the roof of a family building with a size of 5 m^2 would provide 8 kWh of electricity per day on average (Section VIII.3.1). This would be equivalent to 1.0 ton CO_2/capita-year savings, assuming the 0.35 kg of CO_2/kWh conversion. The same panel would provide all the electricity needed for one person.

Note that Chinese government provides solar heat panel to many residence located in environmentally sensitive areas (Figure 8.9).

3. Installing small wind turbines throughout the city (See Dongtan in Chapter XI). A small 1-kW turbine operating 50% of the time could provide on average 12 kWh/day or 4380 kWh/year, which would provide enough renewable energy to reduce GHG emissions by 1.5 tons CO_2/year in 2030.

4. Energy recovery by burning flammable refuse and dried sludge solids in incinerators without carbon sequestering is carbon neutral.

Reaching the net zero carbon footprint is realistic, even in high-energy-consumption countries like the U.S.

VIII.7.2 Is Storage a Problem?

The most obvious argument against developing renewable sources of energy is that they are not available all the time; solar energy production needs sun or at least daylight, and wind energy needs winds in excess of some threshold wind velocity. In the U.S., most of the hydropower is already developed. Building-scale requires large, still expensive rechargeable batteries or production and storage of hydrogen by electrolysis during the times of excess energy production.

A study for California (Jacobson and Delucchi, 2009) found that the energy storage problems may not be as serious as they seem. On a larger regional scale, when

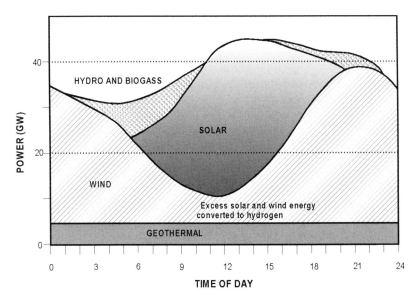

Figure 8.29 Symbiotic energy production by renewable sources to satisfy the future energy needs of California. Modified and replotted from Jacobson and Delucchi, 2009.

the sun does not shine it is usually a windy day, and if it rains, more water may be available to produce energy in hydropower plants. Coal-fired plants must be producing energy continuously; even during off-peak demand hours the fire in the furnace cannot be extinguished. On the other hand, biogas produced by digestion can be stored in aboveground storage tanks (not the best security risk) or in underground caverns; hydrogen fuel cell electric energy production can be started and shut down instantaneously. Hydropower from high dams is used mostly for peak power. Hence, if needed, biogas plants with or without hydrogen fuel cells can operate like peak power plants in places where fossil natural gas or imported oil is currently used for peak electricity production. Figure 8.29 shows the symbiotic cooperation of renewable sources in the California study. The plot and the analysis suggest that all sources of renewable energy are needed to complement each other and reduce the need for storage of wind- and sun-generated energy.

REFERENCES

Accent Energies (2009) "The Accent Heat Pump," http:/www.accentwater.com.au/page/accent_heat_pump.html

Aelterman, P., K. Rabaey, P. Clauwaert, and W. Verstraete (2006) "Microbial fuel cells for wastewater treatment," *Water Science and Technology* 54(8):9–15

Agrawal, L. K., Y. Ohashi, E. Mochida, H. Okui, Y. Ueki, H. Harada, and A. Ohashi (1997) "Treatment of raw sewage in a temperate climate using a UASB reactor and the hanging sponge cubes process," *Water Science and Technology* 36(6–7):433–440

Ahn, Y. and B. E. Logan (2010) "Effectiveness of domestic wastewater treatment using microbial fuel cells at ambient and mesophilic temperatures," *Bioscience Technology* 101, pp. 469–475

Allan, J. A. (1993) "Fortunately there are substitutes for water otherwise our hydro-political futures would be impossible," in *Priorities for water resources allocation and management*, ODA, London, pp. 13–26

Allan, J. A. (1994) "Overall perspectives on countries and regions," in *Water in the Arab World: Perspectives and Prognoses* (P. Rogers and P. Lydon, eds.), Harvard University Press, Cambridge, MA, pp 65–100

American Institute of Architects (2004) *The Architects Handbook of Professional Practice*, John Wiley & Sons, Hoboken, NJ

Bakos, J. (2005) "Renewable solar hydrogen production," *Environmental Science and Engineering*, May 2005, http://www.esemag.com/0505/solar.html (accessed October 2009)

Barbir, F. (2005) *PEM Fuel Cells: Theory and Practice*, Elsevier Academic Press, Burlington, MA

Barnard, J. (2007) *Elimination of Eutrophication Through Resource Recovery*, The 2007 Clarke Lecture, National Water Research Institute, Fountain Valley, CA

Britton, A. F.A. Kocj, D.S. Mavinic, A. Adnan, W.K. Oldham, and B. Udala (2005) Pilot scale struvite recovery from anaerobic digester supernatant at an enhanced biological phosphorus removal wastewater treatment plant, *J. Environ. Eng. Sci.*, 4(4):265–277

Brown, L. R. (2009) Plan B Updates: "U.S. Headed for a massive decline in carbon emissions," Earth Policy Institute, www.earthpolicy.org/index.php?/plan_b_updates/2009/update83 (accessed October 2009)

Burton, T., D. Sharpe, N. Jenkins, and E. Bossanye (2001) *Wind Energy Handbook*, John Wiley & Sons, Hoboken, NJ

Caton, J. A. (2000) "On the destruction of availability (exergy) due to combustion process with specific application to internal combustion engines," *Energy* 25(11):1097–1117

Cheng, S., H. Liu, and B. E. Logan (2006) "Increased power generation in a continuous flow MFC with advective flow through the porous anode and reduced electrode spacing," *Environmental Science and Technology* 40, pp. 2426–2432.

Consumer Reports (2009) "Alternative Energy," October, 2009

Corps C., S. Salter, W. P. Lucey, and J. O'Riordan (2008) *Resources from Waste: Integrated Resource Management - Phase I Study Report*, February 29, 2008. Prepared for: British Columbia Ministry of Community Services, Victoria, BC, Canada

Daigger, G. (2009) "Evolving urban water and residuals management paradigms: Water reclamation and reuse, decentralization, and resource recovery," *Water Environment Research* 81(8):898–823

Daigger, G. (2010) "Integrating Water and Resource Management for Improved Sustainability," in *Water Infrastructure for Sustainable Communities: China and The World* (X. Hao, V. Novotny, and V. Nelson, eds.), IWA Publishing, London, UK

Dodman, D. (2009) "Blaming cities for climate change? An analysis of urban greenhouse gas emissions inventories," *Environment and Urbanization* 21(1):185–2001

Energy Information Center (2008) *Residential Energy Prices: A Consumer Guide*, Department of Energy, Washington, DC, http://www.eia.doe.gov/neic/brochure/electricity/electricity.html

Energy Information Center (2009) *Average Retail Price of Electricity to Ultimate Customers by End-Use Sector, by State*, Department of Energy, Washington, DC, http://www.eia.doe.gov/cneaf/electricity/epm/table5_6_a.html

Friends of the Earth (2009) *Briefing: Pyrolysis, Gasification and Plasma*, Friends of the Earth, London, http:/www.foe.co.uk/resource/briefings/gasification_pyrolysis.pdf (accessed December, 2009)

Gevorkian, P. (2007) *Sustainable Energy System Engineering*, McGraw-Hill, New York

Glaeser, E. L. and M. E. Kahn (2008) *The Greenness of Cities: Carbon Dioxide Emissions and Urban Development*, Working Paper 14238, National Bureau of Economic Research, Cambridge, MA, http://www.nber.org/papers/w14238 (accessed March 2009)

Green, M. A. (2003) *Third Generation Photovoltaics Advanced Solar Energy Conversions*, Springer Verlag, Berlin, Heidelberg

Halmann, M. M. and M. Steinberg (1999) *Greenhouse Gas Carbon Dioxide Mitigation – Science and Technology*, CRC Press, Boca Raton, FL

Hau, E. (2006) *Wind Turbines – Fundamentals, Technologies, Applications, Economics*, 2nd ed., Springer Verlag, Berlin, Heidelberg

Henze, M. and A. Ledin (2001) Types, characteristic and quantities of classic, combined domestic wastewater. In *Decentralized Sanitation and Reuse: Concepts, Systems, and Implementation* (P. Lens, G. Seeman, and G. Lettinga, eds.), IWA Publishing, London, pp. 59–72

Henze, M. and Y. Comeau (2008) "Wastewater concentrations," in *Biological Wastewater Treatment - Principles, Modeling, Design* (M. Henze, M. C. M. van Looschrecht, G. A. Ekama, and D. Brdjanovic, eds.), IWA Publishing, London, UK

Hoekstra, A. Y. and P. Q. Hung (2005) "Globalisation of water resources: International virtual water flows in relation to crop trade," *Global Environmental Change* 15(1):45–56

Hoekstra, A.Y. and A. K. Chapagain (2007) "Water footprints of nations: Water use by people as a function of their consumption patterns," *Water Resources Management* 21:35–48

Jacobson, M. Z. and M. A. Delucchi (2009) "A path to sustainable energy by 2030," *Scientific American* 301(5):58–65

Jagadish, K. S., H. N. Chanakya, P. Rajabapiah, and V. A. Anand (1998) "Plug flow digestors for biogas generation from leaf digestors for biogas generation," *Biomass and Bioenergy* 14(5–6):415–423

Jensen, A. B. and C. Webb (1995) "Treatment of H_2S containing gases: A review of microbiological alternatives," *Enzyme and Microbial Technology* 17(1):2–10

Johnson, G. (2009) "Plugging into the Sun," *National Geographic* 216(3):38

Jones, D. D., J. C. Nye, and A. C. Dale (2009) *Methane Generation from Livestock Waste*, report published online, Purdue University Cooperative Extension Service, West Lafayette, Indiana, www.ces.purdue.edu/extmedia/AE/AE-105.html (accessed October 2009)

Kanan, M. W. and D. G. Nocera (2008) "In situ formation of an Oxygen-evolving catalyst in neutral water containing phosphate and Co^{2+}," *Science* 321(5892):1072–1076

Kibert, C. F. (2007) *Sustainable Construction: Green Building Design and Delivery*, 2nd ed., John Wiley & Sons, Hoboken, NJ

Kotter, D.K., S.D. Novack. W.D. Slafer, and P. Pinhero (2008) Solar nantena electromagnetic collectors, *Proc. of ES 2008, Energy Sustainability 2008*, ASME, August 10-14, Jacksonville, FL

Lettinga, G., A. F. M. Van Velsen, S. W. Hobma, and W. de Zeeuw (1980) "Use of upflow sludge blanket (USB) reactor for biological wastewater treatment, especially for anaerobic treatment," *Biotechnology and Bioengineering* 22; 699–734

Lettinga, G. and L. W. Hulshoff Pol (1991) "UASB – process design for various types of wastewater," *Water Science and Technology* 24(8):97–109

Lettinga, G. (2009) "Urgent need for integration of aerobic and anaerobic treatment," PowerPoint web presentation, www.leaf-water.org/files/symposium/2%20Gatze%20Lettinga.pdf (accessed September 2009)

Le Corre, K.S., E. Valsami-Jones, P. Hobbs, and S.A. Parson (2009) Phosphorus recovery from wastewater by struvite crystallization: A review, *Critical Reviews in Environmental Science andTtechnology* 39(6):433-477

Logan, B. E. (2008) *Microbial Fuel Cells*, Wiley-Interscience, Hoboken, NJ

Messenger, R. A. and J. Ventre (2004) *Photovoltaic System Engineering*, CRC Press, Boca Raton, FL

Metcalf & Eddy (2003) *Wastewater Engineering: Treatment and Reuse*, 4th ed. (revised by G. Tchobanoglous, F. L. Burton, and H. D. Stensel), McGraw-Hill, New York

McCarty, P. L., and F. E. Mosey (1991) "Modeling of anaerobic digestion processes (A discussion of concepts)," *Water Science and Technology* 24(8):17–33

Monsma, D., R. Nelson, and R. Bolger (2009) *Sustainable water systems: Step one – redefining the nation's infrastructure challenge*, A Report of the Aspen Institute's Dialogue on Sustainable Water Infrastructure in the U.S., Aspen Institute, Washington, DC

Naskeo (2009) "Biogas renewable energy," http://www.biogas-renewable-energy.info/index.html (accessed October 2009)

National Science and Technology Council (2008) *Federal Research and Development Agenda for Net-Zero Energy, High Performance Green Buildings*, Committee on Technology, Office of the President of the United States, Washington, DC

Novotny, V. (2010) "Urban water and energy use – From current U.S. to Cities of the Future," *Proceedings of the WEF/IWA International Conference, Cities of the Future*, Cambridge, MA, March 2010, Water Environment Federation, http://www.wef.org/Publications/page.aspx?id=3329

O'Riordan, J., W. P. Lucey, C. L. Baraclough, and C. G. Corps (2008) "Resources from waste: An integrated approach to managing municipal water and waste systems," *Industrial Biotechnology*, Fall 2008, pp. 238–245

Pacific Institute (2008) "Bottled Water and Energy: A Fact Sheet," http://pacinst.org/topics/water_and_sustainability/bottled_water/bottled_water_and_energy.html

Pasqualetti, M. J., P. Gipe, and R. W. Richter (2002) *Wind Power in View: Energy Landscapes in a Crowded World*, Academic Press, San Diego, CA

Patel, M. R. (1999) *Wind and Solar Power Systems*, CRC Press, Boca Raton, FL

Planet Power (2009) "Chemical Composition of Biomass Syngas, Coal Syngas, and Natural Gas," http://www.treepower.org/fuels/biomasssyngas.html (accessed December 2009)

Rabaey, K. and Verstraete, W. (2005) "Microbial fuel cells: novel biotechnology for energy generation," *Trends in Biotechnology* 23(6):291–298.

Ramlow, R. and B. Nusz (2006) *Solar Water Heating: A Comprehensive Guide to Solar Water and Space Heating*, New Society Publishers, Gabriola Island, BC, Canada

Ravindranath, N. H. and D. O. Hall (2002) *Biomass, Energy and Environment: A Developing Country Perspective from India*, Oxford University Press, Oxford, UK

Sammers, N., ed. (2006) *Fuel Cell Technology – Reaching Towards Commercialization*, Springer Verlag, London, UK

Sonoma Mountain Village (2008) *North America's First One Planet Living Community*, http://www.sonomamountainvillage.com/080410%20Short%20SAP.pdf

Stine, W. B. and M. Geyer (2001) *Power From The Sun*, http://www.powerfromthesun.net/book.htm (accessed September 2009)

Trafton, A. (2008) "Major discovery' from MIT primed to unleash solar revolution," *MIT News*, July 2008, http://web.mit.edu/newsoffice/2008/oxygen-0731.html?tmpl=component&print=1 (accessed September 2009)

U.S. Department of Energy (2006) Wind Power http://www.eere.energy.gov/de/wind_power.html (accessed September 2009)

U.S. Department of Energy (2009) Fuel Cell Technologies Program, EERE Information Center Program, http://www1.eere.energy.gov/hydrogenandfuelcells/fuellcells/fc-types.htlm (accessed October 2009)

U.S. GBC (2007) "A national green building research agenda," U.S. Green Building Council, http://www.usgbc.org/ShowFile.aspx?DocumentID=3402

Verstraete, W., P. Van de Caveye, and V. Diamantis (2009) Maximum use of resources present in domestic "used water, *Bioresource Technology* 100:5337-5545Wikipedia (2009) List of countries by carbon dioxide emissions per capita, http://wikipedia.org/wiki/List_of_countries_by_carbon_dioxide_emissions_per_capita (accessed October 2009)

World Energy Council (2009) "Focus – Fuel Cells – Fuel Cell Efficiency," http://www
.worldenergy.org/focus/fuel_cells/377.asp (accessed October 2009)

Yudelson, J. (2009) *Green Buildings through Integrated Design*, McGraw-Hill, New York

Yun, Y., S. W. Chung, and Y. D. Yoo (2003) "Syngas quality in gasification of high moisture
municipal solid wastes," *Preprints of Symposia, American Chemical Society, Division of
Fuel Chemistry* 48(2):823–824

IX

RESTORING URBAN STREAMS

IX.1 INTRODUCTION

IX.1.1 Rediscovering Urban Streams

In the water centric Cities of the Future, surface water bodies are/will be integral and dominant features of the landscape. The logical reasons are:

- Many cities have embarked on sewer separation as a solution for the point source CSO pollution problem, thus creating a dual system: one for sewage and other wastewater, and the second system for cleaner, yet still unacceptably polluted, urban runoff. Current and future regulations will require permits and further removal of pollutants from storm sewers. This is due to the fact that sewer separation preserves the end-of-pipe fourth paradigm system, leaving communities with the dilemma of installing costly, energy-demanding, and inefficient end-of-pipe stormwater treatment systems. It is not logical to put cleaner stormwater back underground. Even in densely built settings, surface best management practices incorporated into the landscape are available to prevent and/or control storm runoff and snowmelt pollution (see Chapter IV).
- Considering sewage and stormwater as a resource may require distributed cluster (ecoblock) stormwater/sewage reclamation systems that will:
  - Treat the polluted stormwater using landscape-based best management practices (Chapters III and IV), such as rain gardens, green roofs, pervious pavements, and biofilters
  - Reclaim water from sewage and stormwater in each ecoblock management unit by means of compacted automated water, energy, and nutrient resource

recovery plants that will produce effluents with a quality on par with, or better than, the water quality in good-quality receiving waters
- Use the produced reclaimed water to provide lost ecological (base) flow to existing and daylighted surface water bodies
- Hydraulically, surface channels with the same cross-sectional area as an underground conduit have a larger flow capacity and, with a managed floodplain/riparian zone, are far less vulnerable to the adverse flood impacts of extreme events. For example, underground stormwater conduits are typically designed to carry high flow (storm) with a recurrence interval of once in 5 to 10 years, while a surface channel with a riparian zone can safely handle flows of much larger magnitude (Figure 2.13).
- Surface water bodies, a desirable amenity for communities, provide leisure, recreation, and biking opportunities, recreational fishing, and in many instances boating and wading. Surface waters that have been lost by being buried underground and converted into sewers will regain great economic, social, and environmental benefits by being daylighted, which, by definition, will make them more sustainable.
- Reclaimed and other clean water (e.g., pumped seepage from basements and underground tunnels, foundation dewatering, clean water overflows from storage tanks, clean rainwater from roofs, and cooling water) should not be discharged into sewers, especially not into those carrying sewage flows that cause sanitary and combined sewer overflows (SSOs and CSOs). This water will constitute base flow of the preserved or daylighted streams.

It is clear that rediscovering urban streams, some of them out of sight for more than a century, will be difficult if the current regulations promoting fast conveyance drainage systems persist. However, as pointed out in the preceding chapters, retrofits are possible, desirable, and more economical than replacing the existing stormwater/wastewater underground infrastructure with the same century-old (or older) technologies of the second and third paradigm.

Landscape/urban development architects, beginning with Frederick Law Olmsted in the 19th century, for more than 150 years have recognized the high value of water bodies for the recreation, enjoyment, and health of the urban population (e.g., Dreiseitl and Grau, 2008). However, the contingent value of urban streams (see Chapters I and X) was suppressed during the third paradigm because of uncontrolled pollution of surface water bodies and the nuisance they created. As a result many streams were buried; wetlands and tidal bays with marshes were filled and disappeared from the surface (see Chapter I). The situation changed at the end of the last century, when in many developed countries point sources of pollution (urban sewerage with overflows of raw sewage, untreated industrial outfalls, concentrated animal operations such as feedlots and poultry farms with thousands of animals in a small place) were abated, and the water quality of the receiving waters greatly improved. Urban rivers today do not catch fire because of heavy pollution (see Chapter I) and, in most developed countries, the bad smell of anoxia does not drive people away from parts of the cities, as occurred frequently in London and many other places during the first 60 years of the last century. Consequently, after cleanup, people in the U.S. who left the cities for the suburbs have begun rediscovering urban rivers and have been slowly returning to central city living, with the real estate near the water resources at a high premium. Many cities are now calling themselves "Water Cities" and use this distinction in urban development. Chicago (Illinois) and Milwaukee (Wisconsin) on Lake Michigan, Toronto (Ontario) on Lake Ontario, Sydney (Australia), and Singapore in Asia have opened their architecture to their water bodies by preserving and restoring their waterfronts, including beaches, parks with bike and pedestrian paths, architectural monuments, museums and entertainment facilities (e.g., opera in Sydney, an art museum in Milwaukee designed by the renowned Spanish architect Calatrava, and museums in Chicago), restaurants and piers, and so on. Historically, Paris has had its Seine River, Prague has had the Vltava (Moldau) River, London is renowned for the Thames River, and Boston (Massachusetts) has the Charles River; Amsterdam, Bangkok, Venice, and several cities in China all have arch bridges and famous canals. These rivers and canals have been the lifelines of the cities, providing transportation for people and commerce, water and energy for industries developed on the shores, enjoyment and recreation opportunities for the people—but they also became a medium for disposing of wastes from the city, overwhelming their pollution-carrying capacity and, for some, leading to their demise.

Today, cities are also beginning to rediscover the high economic and social value of their water bodies, large and small, transecting the urban areas. Citizens and city governments have realized that streams buried by unrestricted developments during the third paradigm, and of poor quality, have no social or economic value. As a matter of fact, the areas above and around the buried streams in most cases have deteriorated to the point of social injustice and left disadvantaged populations in the dilapidated neighborhoods. This occurred in the watershed of Stony Brook in Boston, as well as in the watershed of the Cheonggyecheon River in Seoul, Korea (see Box 9.1).

BOX 9.1 DAYLIGHTING THE CHEONGGYECHEON RIVER IN SEOUL, DEMOCRATIC REPUBLIC OF KOREA

The name Cheonggyecheon means in Korean "clean water stream," and for centuries the river provided the Korean capital city with clean water from nearby mountains and washed its waste to the much larger Han River. But at the beginning of the 20th century, when the river received its current name, it was far from clean, and plans were made a hundred years ago by Japanese colonial occupying authorities to cover the river. These plans failed because of lack of funding and wars.

During the Korean War (1950–1953), refugees from the north settled on the Cheonggyecheon, and in the mid-1950s, the Cheonggyecheon was considered a

(continued)

Figure 9.1   The Cheonggyecheon River in Seoul, Korea, in the 1950s. Refugees from the Korean War built squatter shanties on both sides of the river. The river did not have enough low flow to dilute raw sewage discharges. Source: Public files of the Seoul Metropolitan Government.

symbol of poverty and filth, the legacy of a half-century of colonialism and war (Figure 9.1). Due to the development and upstream withdrawals for water supply of the burgeoning city, the river often had no base flow and no dilution of the untreated sewage discharged to it. The open sewer in the center of the city was also a major obstacle to the redevelopment of historic downtown Seoul. At a time of extreme economic hardship, during the postwar period in the 1950s, the only way of dealing with this problem was to put the stream underground (Preservation Institute, 2007). A 6-km (3.9-mile) four-lane elevated freeway, the Cheonggye Overpass, was built above the ten-lane road covering the stream/sewer between 1967 and 1971 (Figure 9.2). Hence, only the name remained and was given to the freeway. This retention of names of former rivers is ubiquitous, found in almost every city.

At the end of the last century, four decades after it was covered, the Cheong-gyecheon area became a shabby industrial area, filled with flea markets and tool, lighting, shoe, apparel, and used-book stores. The enormous concrete structure forming the Cheonggye Overpass, once a symbol of modernization and

Figure 9.2 Cheonggye Overpass, built over a 10-lane road covering the river in the 1980s. Alternate public transportation by electric buses and subway replaced the traffic capacity lost by the overpass removal.

industrialization, had become an eyesore, born out of Korea's drive in the 20th century to industrialize at all costs.

The former mayor of Seoul and later president of Korea, Lee Myung-bak, gave impetus to removing the freeway and restoring the stream. Mr. Lee's goal was to make Seoul an architectural hub of Northeast Asia by attracting tourism and investment from multinational companies and international organizations—and the "beautification" of the downtown was the main project to accomplish this goal. Despite some opposition from local businesses that were closed to allow the demolition, the project was supported by 80% of Seoul residents. The project started with planning in 2001, and the river restoration was competed in 2005 (Figures 9.3 and 9.4), at a total cost of U.S.$360 million. In 2007, Lee was elected president of Korea, and a *New York Times* article about the election began by saying:

"The man chosen as South Korea's next president in Wednesday's election owes much of his victory to a wildly successful project he completed as this city's mayor: the restoration in 2005 of a paved-over, four-mile stream in downtown Seoul, over which an ugly highway had been built during the growth-at-all-cost 1970s. The new stream became a Central Park-like gathering place here, tapped

(*continued*)

Figure 9.3 Recreated river in 2008 forms a 6-km-long urban oasis and cultural center in Seoul. Photo V. Novotny.

Figure 9.4 Detail showing stepping-stones, allowing people close contact with the river, and water cascade for aeration. Photo V. Novotny

into a growing national emphasis on quality of life and immediately made the mayor, Lee Myung-bak, a top presidential contender."

The total value of the social and economic benefits for the surrounding area has been reported as far exceeding the cost.

Sources: Lee, T.S. (2004); The Preservation Institute (2008); Lee I.K. (2007).

Realizing only economic and social gains without addressing the environment and pollution may not be sustainable, but the economic and social revitalization success of the daylighting (rediscovering) and restoration of urban rivers has shown the attractiveness of these projects. For example, after the partial point source pollution abatement in the fourth paradigm period, cities such as San Antonio in Texas (Figures 9.5 and 9.6) and Ghent in Belgium (Figures 9.7 and 9.8) modified and beautified their abandoned streams, and Ghent is now opening (daylighting) its buried water bodies—all of this with great economic benefits far surpassing the restoration cost. The most dramatic case of discovery and metamorphosis of a lost water body occurred in the early 2000s in Seoul, capital of the Republic of Korea (Figures 9.1 to 9.3 and Box 9.1). Examples of other stream restoration projects are listed in Table 9.1. Additional case studies are listed in Pinkham (2000) and France (2007).

Figure 9.5 The architecturally restored San Antonio River has brought great economic and social benefits to the downtown area of the city from tourism, conventions, downtown hotel and business development, visual enjoyment, and employment opportunities. Photo V. Novotny.

Figure 9.6 The beautified San Antonio River is a concrete impounded channel with very poor habitat and water quality (zero water environment benefits).

Figure 9.7 A partially covered historic canal in Ghent, Belgium. Photo V. Novotny.

Figure 9.8 Daylighted (uncovered) portion of the canal in Ghent. The canal receives CSOs and has relatively poor habitat. Photo V. Novotny.

However, the above three examples, no matter how architecturally attractive, are not sustainable based on the triple bottom line assessment. All three provide substantial economic benefits from increased tourism, higher values of real estate, and businesses surrounding the river and social benefits of enjoyment, a pleasant environment, employment, and so forth. But the San Antonio River is not ecologically or hydrologically functional; it is essentially an impounded concrete-lined channel that in some places suffers from hypoxia or even anoxia and cannot support a sustainable healthy fish population. Also, because of the relatively high dam that impounds the San Antonio River, there is a problem with fragmentation; fish and other organisms cannot pass the dam. Canals in Ghent also provide poor habitat, and in 2008 they suffered from pollution by CSOs that have not been fully abated. The Cheonggyecheon River is an artificial river without an upstream flow, which was lost by urbanization of the upper watershed. Flow is provided during dry periods by pumping 1.4 m^3/sec of water from the large Han River, located about 15 km downstream. This requires a lot of energy, which makes the project carbon (GHG) emissions positive. Also, almost no best management practices such as pervious pavements and other infiltration have been implemented in the watershed, resulting in frequent CSOs. A warning system has been installed along the river, asking people to leave the paths along the river because of the CSOs and high flow danger occurring at times when rainfall intensity exceeds 5 mm/hr. The Cheonggyecheon River is a highly sophisticated

Table 9.1 Examples of stream restoration and daylighting

River	Location	Type of Modification	Main Benefits
Providence River	Providence, Rhode Island (USA)	Daylighting the river in the downtown area, urban landscaping	River festivals, riverside restaurants, cafés, pedestrian paths, fishing, boating
Lincoln Creek	Milwaukee, Wisconsin (USA)	Renaturalization of 15 km (9.2 miles) of the creek by removing concrete lining, restoring wetland and habitat, flood storage, control of CSOs	Elimination of flooding for 2000 homes, restoration of flora and fauna, water quality improvement, recreational fishing, parks, neighborhood revitalization
Emscher River	Ruhr District, Germany	Formerly a raw effluent dominated surface sewer; raw sewage inputs and concrete lining were removed; renaturalization, brownfield remediation; connecting river with monuments of old heavy industrialization	Recreational and educational benefits; significant water quality improvement; the river is located in the formerly most industrialized part of Germany, and the entire watershed is being converted to a memorial to old industrialization
San Antonio River	San Antonio, Texas (USA)	Engineered concrete-lined impounded channel, in places aerated with cascades, very poor habitat	Very high economic benefits from tourism, conventions, downtown revitalization, social benefits of employment and aesthetics
Cheonggyecheon River	Seoul, Korea	Engineered recreation of a man-made river in historic downtown Seoul fed by pumping flow from another larger river, good habitat with functioning flora and fauna, receives CSOs	Large economic benefits of downtown commerce and business revitalization, tourism, and a cultural center, controlled flood and CSO conveyance, aesthetics
Strawberry Creek	Berkeley, California	Daylighted and restored creek from subsurface culvert; water quality improvements by control of illicit connections, recreating habitat, and introduction of flora and fauna	University of California campus beautification, education, water quality; the daylighting project continues to downtown Berkeley
Arcadia Creek	Kalamazoo, Michigan	Daylighted three blocks of concrete-lined channel (no renaturalization) and two blocks of open pond in densely populated downtown district, poor habitat	Flood control benefits and downtown beautification
Streams in Zurich	Zurich, Switzerland	20 kilometers of streams buried in the city have been daylighted and renaturalized, additional 15 km revitalized, base flow and habitat recreated	Daylighting highly desired by the population, brooks and creeks have been recreated even in dense parts of the city

hydraulic system with engineered CSO discharges, parallel sewer lines, and pipelines bringing flow to the river. Nevertheless, it is a sophisticated river creating project that is difficult to replicate elsewhere on the same scale. On the other hand, water quality in the river is relatively good, habitat has been established, and the river supports fish population.

IX.1.2 Definitions

BOX 9.2   RESTORATION, REHABILITATION, RECLAMATION, AND DAYLIGHTING (COMMITTEE ON RESTORATION OF AQUATIC ECOSYSTEMS, 1992; PINKHAM, 2000)

Restoration is the process of returning an ecosystem to a close approximation of its former condition. The restoration process reestablishes the general structure, function, and dynamic self-sustaining behavior of the system. However, it may not be possible to recreate the system exactly because the surroundings and stresses may not be the same as in the predisturbance period.

Reclamation is a process designed to adapt a resource to serve a new or altered use. This could mean a process to convert a resource into a productive use. For example, a restored stream can provide water for irrigation or other nonpotable uses, including wading and other types of recreation. Other examples are reclaiming urban brownfields, removing freeways and reclaiming surfaces covered by pavements for ecological uses, and converting a mining excavation pit into a lake.

Rehabilitation puts a severely disturbed and/or partially irreversibly modified and damaged resource back into good working order. It is often used to indicate improvements primarily of a visual nature or to an ecological status less that that of a natural system. The ecologic potential of rehabilitation must be determined by a study (Use Attainability Analysis).

Daylighting restores to the open air some or all of the flow of a previously covered or converted and buried river, creek, or natural stormwater drainage. Daylighting reestablishes a waterway in its old channel wherever feasible, or in a new channel threaded between the buildings, streets and roads, parking lots, or playing fields. Some daylighting projects also recreate wetlands, ponds, or estuaries.

An *ecotone* is a transitional zone between aquatic ecosystems and built urban habitat. It contains built floodplain and ecological (natural) riparian zones.

The fundamental goal of restoration is to return the ecosystem to a condition that approximates its condition prior to disturbance (Cairns, 1988; Committee on Restoration of Aquatic Ecosystems, 1992; Interagency Task Force, 1998; Dunster and Dunster, 1996). The terms "restoration," "reclamation," and "rehabilitation" are often used interchangeably, but their meanings are different (see Box 9.2). Daylighting is a more recent term used to describe bringing buried and highly modified underground culverted streams to the surface.

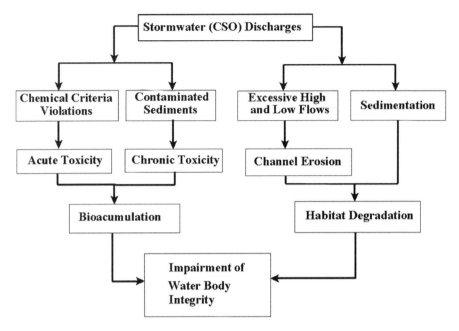

Figure 9.9 The impact of stormwater discharges on integrity of water body includes toxicity caused by pollutant discharges and physical habitat degradation.

IX.2 ADVERSE IMPACTS OF URBANIZATION TO BE REMEDIED

IX.2.1 Types of Pollution

"Pollution" was defined in Chapter II as adverse anthropogenic stream modification and/or discharge impairing the integrity of the receiving water body. The integrity of urban streams has been compromised by pollution (Figure 9.9). Restoration/rehabilitation and daylighting remedy physical impediments, including legacy pollution in sediments; therefore, pollution by pollutant discharges from effluents, CSOs and SSO, and contaminated stormwater must be remedied first or concurrently with restoration/rehabilitation. Daylighting implicitly includes restoration because underground streams are contained in culverts or sewers.

The adverse effects of urbanization to be remedied before or during the restoration are:

Pollutant Discharges

- Untreated or insufficiently treated effluent (point source) discharges, including excessively warm cooling water discharges from industries and power utilities
- Other legal point sources such as combined sewer overflows (CSOs), and sources such as spills from sanitary sewers (SSOs) and treatment plant

bypasses, which are not allowed and will result in legal action in the U.S. if occurences are repeated

- Pollutant discharges in stormwater
- Legacy pollution in sediments
- Development along the riverbank and shoreline served by direct septic tank discharges

Habitat Degradation

- Caused by increased flows and channel/stream bank erosion (Figure 4.2)
- Channel lining and cutting trees along the water body
- Constricting channel by dikes, levees, vertical masonry, or concrete river banks
- Loss of riparian ecotones and floodplains, i.e., development all the way to the river or shoreline bank
- Decrease or loss of base flow
- Substrate degradation including embeddedness (area of the bottom containing clay and silt which makes poor feeding and spawning medium for fish and benthic vertebrate and invertebrate organisms)
- Habitat loss (see Figures 4.2, 4.3, and 9.10)
- Burying the stream

Figure 9.10 Fragmentation by culverts and concrete lining. Shallow channels with no pools are uninhabitable by aquatic biota, except by some attached growth slimes. Photo V. Novotny.

Other Adverse Impacts

- Flow diversions and withdrawals
- Infrastructure such as drop structures (Figure 4.3) and bridges or culverts impassable for aquatic biota (Figure 9.10) and causing longitudinal ecological fragmentation
- Dams, often filled with contaminated sediments, causing longitudinal fragmentation and loss of migration
- Drainage and loss of wetlands
- Narrowing or loss of riparian ecotones

Effect of hydrologic changes. Because of imperviousness of urban watersheds, the hydrology of the watershed and the frequency of low and high flows in urban streams have changed (Figures 4.1 and 9.11). Imperviousness reduces infiltration and recharge of groundwater aquifers that provide base flow to the streams, decreases the time of concentration, and increases the volume of high surface runoff events.

Hence, the high flows increase, and low flows decrease or even disappear during the dry season. Because the channel's morphological characteristics/parameters are in equilibrium with the hydrological frequency and magnitude characteristics of flows, a change of hydrology has a tendency to modify the cross section of the channel by bank erosion (Figure 4.2), resulting in wider channels with less or no low flow during dry periods without effluent discharges. Bank erosion adversely affects the bank habitat and refuge for fish. As a result, the biological integrity of the urban stream is downgraded by urbanization, resulting in bank erosion.

Stream restoration projects often are labor-intensive but infrastructure-inexpensive (Figure 9.12); therefore, use of funds for such projects has the additional benefit of creating numerous jobs, potentially in urban and rural areas with high

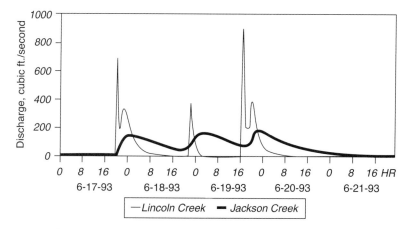

Figure 9.11 Comparison of flow following a storm in the rural Jackson Creek and the fully urbanized Lincoln Creek in Milwaukee, Wisconsin. Both creeks have about the same drainage area at the points of flow measurements. Source: Wisconsin Department of Natural Resources.

Figure 9.12   Daylighted and restored Strawberry Creek on the campus of University of California, Berkeley. Blocks from the removed culvert were used to stabilize the banks.

unemployment rates. They are most effective when stream restoration (with or without daylighting), flood control, and neighborhood revitalization are considered concurrently. Without restoration benefits, flood control of urban streams is often economically inefficient and would represent a massive transfer of payments from the city population to a few beneficiaries residing in the floodplain. Adding ecological restoration has a much wider appeal and greater benefits (Novotny et al., 2001a, b). However, water body restoration efforts are ineffective if point source control and remedies to repair hydrology are not implemented first. For example, some restoration projects partially failed because of the lack of base flow or due to excessive flushness of the water body, causing continuing stream bank erosion.

Stream restoration also improves the loading capacity, LC, of the water body for many pollutants and may have an effect on the size of the point source allowable waste load. For example, removing a dam significantly increases the aeration capability of the stream and reduces the chances of infestation of the water body by algae (Hajda and Novotny, 1996). Remedying legacy pollution in sediments,

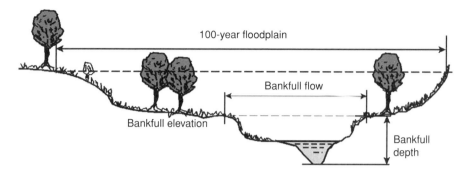

Figure 9.13 Typical cross section of a natural or renaturalized stream.

especially when removal of highly contaminated sediments or sediment capping may be necessary, will also require restoration (Novotny, 2003). In such cases, it may be cheaper and more equitable to enhance the LC of the water body and improve habitat than to require unreasonable additional treatment of discharges from point sources that may not even help.

The water body—a stream or a lake—cannot be separated from its surroundings, the riparian lands. A natural stream and its surrounding form a stream corridor that includes the water body, floodplain (which may include riparian wetlands), oxbow lakes (abandoned channels), forests, and buffers, as shown in Figures 9.13 and 9.14

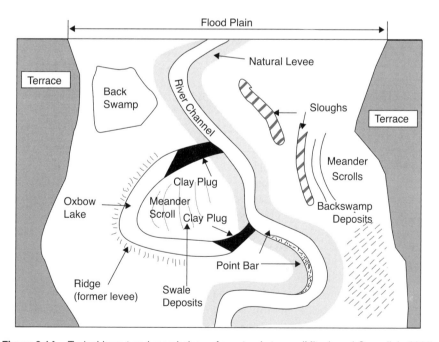

Figure 9.14 Typical layout and morphology of a natural stream (Mitsch and Gosselink, 2000; based on Leopold et al., 1992).

(Rosgen, 1996). An extensive manual on stream corridor restoration was prepared by the Interagency Task Force (1998), and a chapter is included in Novotny (2003). The stream and stream corridors evolve in harmony with and in response to the surrounding ecosystem in a dynamic equilibrium with their surroundings and outside stresses. These changes and effects of stresses are best detected by biological and habitat monitoring (Karr and Chu, 1999).

Water body restoration is an important and concluding part of the watershed/water body restoration effort. Water body restoration is not possible if the external sources of pollution are not taken care of first and controlled. Novotny (2003) pointed out that the process of water body/water restoration begins with the Use Attainability Analysis defining the ecologic potential of the water body, continues with the Total Maximum Daily Load TMDL) process, and concludes with water body/floodplain restoration.

IX.2.2   Determining Main Impact Stressors to Be Fixed by Restoration

The United States Clean Water Act set forth the national goal of "restoring and maintaining the chemical, physical and biological integrity of the nation's waters." A similar goal of attaining a good ecological status was set forth in the Water Framework Directive by the European Community. Integrity was defined as a condition of a water body able to support balanced aquatic life, resembling as closely as possible its natural state. The concept of an "Index of Biological Integrity" (IBI) was developed and published by Karr et al. (1986). It is a method of quantifying the ecological impact of human-induced alterations in stream ecosystems, using fish and macroinvertebrate organisms as indicators. Several variants of the IBI have been developed in the U.S. in the last 20 years and adopted by the U.S. EPA (Barbour et al., 1999) and, in regionally modified forms, by many states.

A fish IBI is constructed from field-measured component metrics that include parameters related to species richness and composition, trophic composition, and organism abundance and condition. It is based upon the premise that fish respond to environmental stressors in a species- or guild-specific manner. Metrics are scaled relative to covariation with natural factors (e.g., stream size or geographical distinctions), and when properly calibrated, allow for the calculation of a "rating" that describes the stream's ecological health relative to best case, or non-impacted ecoregional reference. Thus IBIs can provide a "biological response signature" for monitoring compliance with antipollution regulations (Yoder and Rankin, 1999). Two other "companion" indices are also used in the integrity evaluation of water bodies to evaluate the habitat and macroinvertebrate community. For example, the Invertebrate Community Index (ICI) and the Quantitative Habitat Evaluation Index (QHEI) are used in the Midwestern U.S. All three indices, IBI, ICI, and QHEI, use metric scoring to arrive by summation at the value of the overall index.

Since 1990 the U.S. Environmental Protection Agency and state agencies have conducted extensive monitoring of the physical, landscape, and chemical parameters of watercourses, along with using the metrics of the IBI (fish and macroinvertebrates). Since the early 1990s there has been considerable interest among scientists in developing models for correlating the biotic integrity indices with various stressors—for example, degree of imperviousness or urbanization in the watershed (Wang et al., 2000; Novotny et al., 2005). However, very few water bodies, streams, and impoundments (both man-made reservoirs and natural lakes) are being degraded by only one stressor. Furthermore, an increase in one stress may affect one group of organisms negatively but some other group (e.g., species with pollution tolerance) positively. The sequence of shifts in fish population from salmonid to carp-dominated fish populations, continuing to fish disappearance at very high levels of stress, is well known. The multiple effects of stress and the shifting composition, diversity, and density of the fish, macroinvertebrate, and algae populations are not gradual and may have several intermediate semiresilient states typical for a group of sites. This grouping results in clustering of the sites according to the composition of organisms, which is in IBIs determined by the metrics. There is an almost an infinite number of combinations of relationships among multiple stressors affecting the biotic population and impacting the magnitude of the metrics. A cluster is a group of sites exhibiting similar multiparametric responses to a multiplicity of stressors.

Manolakos, Virani, and Novotny (2007) and Bedoya, Novotny, and Manolakos (2009) analyzed the impact of chemical (pollutants), physical (habitat) and landscape (e.g., land use, watershed characteristics, fragmentation) stressors on the biotic integrity of hundreds of streams in Ohio and other states. The measure of the biotic integrity was the multimetric Index of Biotic Integrity (IBI) that is enumerated for fish and benthic macroinvertebrates. These indices have been developed using advanced data mining and modeling methodology described in the above papers. Division of the sites into similar clusters was done by self-organizing mapping (SOM), by unsupervised artificial neural nets (ANN) (Kohonen, 1990), followed by the Canonical Correspondence Analysis (CCA) identifying the parameters with the greatest influence on the magnitude of the IBI in each cluster. Such parameters were called Cluster Dominating Parameters. The research described in the aforementioned articles related 35 or more variables to the magnitude of the fish IBI, which has 12 metrics (see also Novotny et al., 2009).

Through the SOMs of fish metrics, it was possible to divide the sites in Ohio, Maryland, and Wisconsin intro three clusters reflecting the quality of the fish community. The Ohio SOM is shown in Figure 9.15. The overall fish IBIs in the clusters indicated that Cluster I had "superior" fish composition, Cluster II was intermediate, and Cluster III was inferior. However, the IBI ranges within each cluster showed an overlap because the overall IBI is a summation of scoring of metrics, and the same IBI is achieved by many variants of metric scores. Because each neuron of SOM contains several physical monitoring sites, it was possible to locate the clusters regionally and put them on the map (Figure 9.16). It can be seen that most Cluster III sites are located in the northwestern, highly agricultural corner of the state (monocultural corn growing) and around the Cleveland–Akron industrial area and the Columbus–Dayton and Cincinnati urban zones. By further clustering analysis, it was possible to separate Cluster III into agricultural sites primarily impacted by habitat parameters such as embeddedness and urban sites that were also adversely affected by concentration of chemical pollutants (Bedoya, Novotny, and Manolakos,

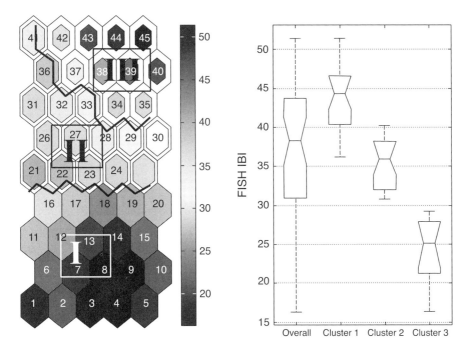

Figure 9.15 SOM of IBIs based on the IBI metrics for Ohio, indicating three clusters of the IBI metrics. Cluster I neurons are the best; those of Cluster III are the worst (from Manolakos et al., 2007).

2009). The best Cluster I sites are in the hilly, more pristine eastern and southern parts of the state, but the east also contains coal-mining areas identified as Cluster III sites.

The subsequent nonlinear CCA then linked the stressors to the SOM and quantitatively ranked the stressors as to their impact on IBIs and their metrics (Figure 9.17) (Virani et al., 2005). CCA is a powerful method for the multivariate exploration of large-scale data (Legendre and Legendre, 1998). CCA is a weighted average ordination technique providing simultaneous ordering of sites and species, rapid and simple computation, and very good performance when species have nonlinear and unimodal relationships to environmental gradients (Palmer, 1993). CCA also identified the Cluster Dominating Parameters (CDP). It can be seen that Cluster I (superior) is most sensitive to habitat quality parameters and forested land (including riparian land), while Cluster III (inferior) is most impacted by pollutants and intensive land use by human beings.

SOM and the subsequent CCA identified the following parameters as having the greatest impact on the fish IBI and its metrics in Ohio and Maryland: substrate, embeddedness, cover, channelization, riparian quality, gradient, riffle, hardness, and total suspended solids. If the vectors are close to each other in both directions, the parameters are cross-correlated. Cross-correlated parameters are pool, riffle, and

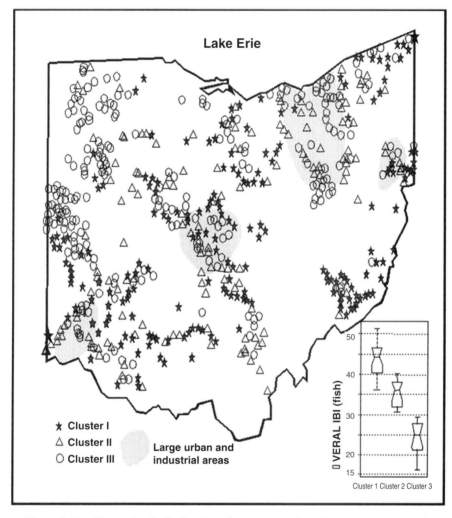

Figure 9.16 Cluster site distribution throughout the state (from Manolakos et al., 2007).

gradient, also embeddedness and substrate quality. Weak impact can be seen for percent agriculture, percent urban development, percent forested wetlands, BOD, TKN, iron and sulphates. In Figure 9.17, the magnitude of the impact is identified by the length of the parameter impact vector. The three variables that consistently had the greatest impact on the fish IBI were embeddedness (the fraction of the bottom area covered by clay and silt-sizes sediments), substrate quality, pool quality, and channelization. This indicates that impoundments and slow-moving streams have a significant impact on biotic integrity. BOD, conductivity, arsenic concentration, hardness, and TKN were the top chemicals but with less impact than the habitat parameters.

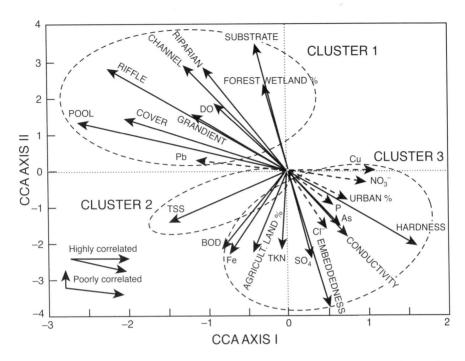

Figure 9.17 Canonical Correspondence Analysis of top 25 input parameters superimposed on the cluster domains identifying the Cluster Dominating Parameters (from Novotny et al., 2009).

Such analyses also identify, cluster by cluster, the optimum ecologic potential of the urban water bodies, which, in most cases, cannot be restored to the original pre-development natural conditions due to irreversible modifications of the watershed, stream corridor, and channel. For one thing, such streams may be effluent dominated during low flows, instead of fed by natural springs (Novotny, 2007); they can be irreversibly constricted by city flood protection measures or by irremovable infrastructure.

IX.2.3 Effluent Dominated and Effluent Dependent Urban Water Bodies

The changed hydrology of urban streams and of current and even future wastewater and sewerage management must be recognized. It has been stated several times throughout this book that restoration of hydrology to a condition as close as possible to the predevelopment status is the goal of restoration efforts and of water/stormwater/wastewater management in the new and retrofitted ecocities.

In the last century, regional sewerage and water supply planning agencies and utilities used the argument of the economy of scale and of the possible adverse effects of the uncontrolled water-sewage-water cycle (Novotny, 2007; Imhoff, 1931; Imhoff and Imhoff, 1993) to promote and develop large regional sewer systems that transfer water and wastewater over long distances, often to another watershed, which dramatically changed the hydrology of the impacted surface waters. Urban waters affected by these long-distance withdrawals and concentrated discharges became flow deficient after withdrawal, and after the water body received the effluent discharge, often far downstream, the water body then became effluent dominated. The following definition assists in understanding the issue of effluent domination (Novotny, 2007; Pickus et al., 2002; Gensemer et al., 2008):

- *An effluent dominated water body is a water body that predominantly contains wastewater effluents during all or a part of a year.* For water quality analyses and assessment purposes, this definition implies that the design low flow is predominantly made of upstream effluent. Effluent dominated streams do not have to be naturally ephemeral.
- *An effluent dependent water body is generally an ephemeral (either naturally or due to excessive upstream withdrawals) stream where aquatic biota and other uses of the water body are or can be sustained by treated effluents creating perennial flows.*

Figure 2.10 in Chapter II shows the closed-loop hydrologic cycle. It is replotted here without showing the loop to explain the hydrologic representation of the effluent dependent and dominated reaches of a river flowing through a city/metropolitan area. Typically, upstream of the city, water is withdrawn for water supply from reservoirs built upstream that store water, and it is then delivered to a water treatment plant for processing to potable water. In many areas water is also withdrawn from rivers for other purposes such as irrigation, industrial production, thermal power plant cooling, and so on. Water not returned to the watercourse is called consumptive use, which includes losses by evapotranspiration in the irrigation systems, evaporation in cooling systems, and water transferred into other watersheds in products and other interbasin transfers. Irrigation systems usually withdraw more water than is lost by evapotranspiration to keep the salt content in soils in check; otherwise soil fertility would be lost. This results in relatively polluted irrigation return flow being discharged from the irrigated field back into the water body from which irrigation water was taken.

The water supply system for Beijing receives most of the water for the 15 million people from mountains 100 kilometers northeast of the city and stores it in the large Miyun Reservoir. Almost all water stored in the reservoir is delivered to the city, leaving the river downstream without flow. A similar situation can be found in Phoenix, Arizona, and many other localities throughout the U.S. and other countries.

Understanding the categorization of effluent dominated/dependent water bodies is necessary in planning the restoration process. In the U.S., the CWA and the ensuing Water Quality Standards Regulation (40 CFR 131) do not give agencies or dischargers a "carte blanche" to exclude effluent dominated water bodies from attaining the integrity goals of the Act and allow them to take all flow. A Use Attainability Analysis (UAA) is needed if a downgrade or any change of the designated use is contemplated (40 CFR 131.10; U.S. EPA, 1994). In a large majority of cases, UAAs have not been prepared, and such water bodies remain downgraded. Typically, these water bodies are water quality limited and have been put on the CWA Section 303(d) TMDL action list (see Committee to Assess the Scientific Basis of TMDL, 2002); however, the TMDL process, as stated before, is only capable of dealing with the pollutant discharges, and not with adverse stream morphology and hydrology modifications caused by urbanization that fall under the definition of "pollution" in the CWA but are not discharges of pollutants. Furthermore, the TMDL process is not designed to change the designated use or associated standards.

The Water Quality Standards Regulation (40 CFR 131.10) provides six reasons that can be used in petitioning (through a UAA) for a downgrade of the statutory water uses defined in the CWA, that is, maintaining integrity of aquatic life and providing for primary recreation, considering also uses for water supply and navigation (U.S. EPA, 1994; Novotny et al., 1997). However, § 131.10(g)(2) allows the removal of an aquatic life use that is not an existing use (i.e., the water body has been impaired), where "natural, ephemeral, intermittent or low flow conditions prevent the attainment of the use, *unless these conditions may be compensated for by the discharge of a sufficient volume of effluent discharge without violating state water conservation requirements to enable uses to be met.*" (Empasis added.) This rule has been interpreted to imply (U.S. EPA Region 8, 2003) "**where an effluent discharge creates a perennial flow, the resulting aquatic community is to be fully protected**," which means that the integrity goals specified in Section 101 of CWA should be attained. Integrity has been defined as a stream condition suitable for balanced aquatic life (Karr et al., 1986) and contact and noncontact recreation. Effluent dominated streams present a challenge but also a potential for restoration. Furthermore, these streams represent a limit of restoration and the emerging successes (e.g., in the Southwest or Midwest U.S. or in Japan) provide proof of the feasibility of full restoration of these apparently most stressed water bodies.

Hundreds of effluent dominated/dependent streams can be found in the U.S. and thousands abroad. The categories may include:

1. Ephemeral streams in arid regions were made perennial by effluent discharges (the Santa Ana River in Orange County, California; Las Vegas Wash between Las Vegas and the Colorado River in Arizona; the Santa Fe River in New Mexico; the Los Angeles River in California; the Salt/Gila River around Phoenix in Arizona). Some ephemeral streams in the U.S. Southwest might have been perennial before the settlers moved in *en masse*, approximately one hundred years ago.

2. Perennial rivers where flow was diminished by upstream withdrawals for water supply and irrigation and which became effluent dominated by large effluent discharges from municipal or regional treatment plants (e.g., the South Platte River in Colorado; the Trinity River in Texas; the Chattahoochee River in Atlanta, Georgia; the Charles River upstream of Boston, Massachusetts).

3. Smaller perennial rivers in humid regions that were overwhelmed by treated or untreated effluent discharges from sewer outfalls and regional treatment plants

receiving wastewater from outside the original watershed (the Des Plaines River in the Chicago metropolitan area, Figure 1.14; the Emscher River in Germany).

In Arizona, more than three-fourths of effluent discharges are into formerly (upstream) ephemeral riverbeds. It has to be expected that the effluent dominated rivers, even with a high degree of treatment, will be different from their natural counterparts because of more nutrient enrichment and the presence of residual chemicals not present in the natural flows. An extensive research effort by Pima County (Pima County Wastewater Management Department, 2007), which includes metropolitan Phoenix, Arizona, revealed that discharging treated effluents into an ephemeral river bed, if properly designed and carried out, can revitalize not only the aquatic flora and fauna but also the riparian ecosystem surrounding the river. Figures 9.19 and 9.20 show examples of the changes to the ephemeral Gila River in the Southwestern U.S. All upstream flow was withdrawn for water supply of the Phoenix metropolis and upstream irrigation.

The same research, however, pointed out that the type of treatment is crucial. Originally, the biggest problem was disinfection of the effluent, which left large concentrations of residual chlorine in the river. In the late 1970s, this led the Metropolitan Water Reclamation District of Greater Chicago and other dischargers into the

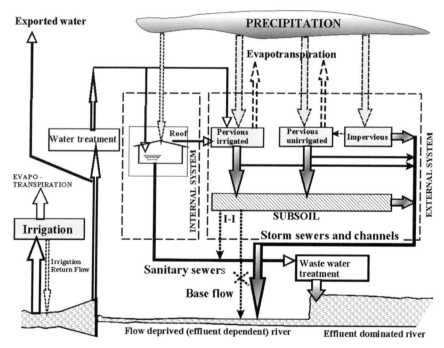

Figure 9.18 Representation of the open linear hydrology of urban areas creating effluent dominated and dependent water bodies and stressed urban basins; also included as Figure 2.10 in Chapter II (adapted from Grimmond et al., 1986; Mitchell et al., 1996; Heaney, 2007).

Figure 9.19 Ephemeral Gila River in Arizona before effluent discharge from Phoenix WWTP (Photo R. Meyerhoff, CDM).

effluent dominated Lower Des Plaines River to petition for removing the bacterial standard (hence stopping chlorination) for this stressed water body and also for all waterways throughout the Chicago metropolitan area. The petition was granted, and these water bodies then received special use designation that waived the bacteriological standard. Stopping chlorination (without dechlorination) resulted in the return of some fish and other organisms in the Lower Des Plaines River. However, other uses such as recreation were not possible. The Use Attainability Analysis of the river recommended reinstating the standard for *Escherichia coli*, providing limited protection to accidental swimmers (Novotny et al., 2003, 2007). Table 9.2 shows the effects of treatment on the macroinvertebrate community in the Southwestern U.S. ephemeral streams receiving treated municipal effluents that converted the downstream reaches into perennial rivers. The table shows that chlorinated effluents are generally toxic, and that dechlorination improves the aquatic community. Because the streams were ephemeral above the discharges, repopulation of perennial reaches from upstream was unlikely, and the streams acted similarly to ecologically discontinued/fragmented water bodies. Today and in the future cities, chlorination would not be used, being replaced, as shown in the preceding chapters, by UV radiation with or without ozone addition.

Figure 9.20 Gila River in Arizona downstream from the Phoenix effluent WWTP discharge (Photo R. Meyerhoff, CDM).

Table 9.2   Treatment level vs. macroinvertebrate community in Southwestern effluent dominated streams

Treatment Level		TaxaRichness	Percent Clean Water Taxa
Higher-quality effluent	Chlorination with no dechlorination	Significant decline below discharge	None present
	Chlorination with dechlorination; nitrification with denitrification	Increase or decrease below discharge	None present to low percentage (circa 10%)
	Chlorination with dechlorination; nitrification with denitrification; filtration	Decrease below discharge	Consistently present (17–99%) but few taxa

Source: Pima County waterwater management Department (2002)

The Pima County program (Pima County Wastewater Management Department, 2007) also documented that the water quality of the perennial effluent dominated reaches was improving with the distance from the outfall, and that the riparian vegetation and habitat had also dramatically improved.

This predicament of effluent domination must be incorporated into the planning and design of the surface water bodies in the ecocities. The linear hydrology of urban streams in the cities—whereby most or all water is taken upstream, leaving the stream water short—or of naturally ephemeral urban streams will be improved when distributed resource recovery (water and energy reclamation) from sewage and wastewater is implemented, and reclaimed ecological flow is discharged into the effluent limited reaches. Consequently, the effluent's purity, from water and resources recovery units, must (1) be at a level commensurate with the quality of receiving water bodies capable of supporting a balanced indigenous aquatic biota, (2) meet all chemical and physical water quality standards to protect the aquatic life, and (3) provide conditions for primary (at least wading) and secondary recreation. Hence, logically, restoration, including daylighting, will be focused on providing habitat and hydrology (adequate base flow) to support the aquatic life and provide conditions for further improving the residual pollution of the receiving water body.

IX.3   WATER BODY RESTORATION IN THE CONTEXT OF FUTURE WATER CENTRIC (ECO)CITIES

IX.3.1   Goals

Urban surface water systems are the main means of connectivity of urban ecosystems. They are natural corridors for migration of urban animals and fish as well as movement of pedestrians and bicyclers. A healthy urban water body also requires healthy riparian zones, including floodplains. Urban rivers, streams, lakes, and reservoirs also include groundwater systems, surface and underground drainage infrastructure, estuaries, and wetlands. In existing urban areas, these water systems have been modified, often to the point that the affected urban water bodies have lost hydrological and ecological functionality. Often, these modifications have adverse effects on the resilience of the urban watershed to extreme events. Chapter I and several figures in this chapter show urban water bodies that were covered or converted to lined and straightened flood conveyance channels, receiving overflows from the sewer systems—channels with no riparian and benthic habitat. Navigable water bodies are obviously impacted by barge and boat traffic as well as by pollution.

In general, most of the examples of restored/recreated rivers introduced in the first section of this chapter had a goal of increasing the visual attractiveness of the urban area, bringing people to the river banks for walking, sightseeing, cafés, restaurants, limited fishing, shops, and in San Antonio and Ghent, for boat rides. It was implicitly assumed that water quality was acceptable as long as it would not cause a nuisance, and that if a nuisance occurred (e.g., anoxia and emanation of hydrogen sulfide), it could be remedied by such architectural measures as stream aeration by cascade waterfalls or fountains.

In the context of the future sustainable ecocities, restoration of urban streams should provide, in a balanced way, the trinity of environmental, social, and economic benefits (see Chapter III) in an intergenerational context, as well as contribute to the reduction of carbon emissions. Hence, the urban surface water bodies are not just visual assets of the community that might spur downtown or community development. They are a lifeline of the development, serving multiple purposes such as:

- Receiving residual treated reused and/or excess reclaimed water, and excess clean stormwater.
- Serving as a source of water for reuse:
  - For buildings (e.g., flushing toilets)
  - Landscape and green roof irrigation
  - Cooling
  - Street and sewer cleaning
- Reducing and controlling stream bank erosion.
- Accepting other clean water inputs (cooling water, water from dewatering tunnels, deep basements and foundations, roof water, groundwater and upstream stream flows buried underground when they reached populated urban zones) that would otherwise discharge into sanitary and combined sewers.
  - Eliminating clean water inputs into sanitary and combined sewers saves energy by reducing pumping mixed wastewater in the lift stations of the sanitary or combined sewer system.
  - Reducing clean water inputs reduces the treatment capacity and energy use in the treatment plant (traditional regional systems) or resource recovery units (distributed systems).
- Providing recreation such as individual and tourist boating, swimming, recreational fishing, and enjoyment.
- Revitalizing neighborhoods and areas surrounding the water bodies and contributing to the solution of environmental injustice.
- Natural, created (e.g., man-made wetlands and ponds) and restored/daylighted water bodies attenuate residual pollution from surrounding inhabited residential, industrial, and commercial areas and roads and highways, instead of treating polluted runoff in hard infrastructure treatment plants.
- Sequestering carbon in the ecotones.
- In combination with landscape best management practices, surface streams are more efficient conduits of floodwater than underground drainage.
- Providing habitat with a condition for a balanced aquatic life.
- Providing an on-site source of water for various needs, instead of using delivered treated water brought from long distances.

In the new developments, existing water bodies must be preserved as close to their natural state as possible, or restored with an ecological limitation in mind to prevent excessive pollution and impairment of the integrity of the water body. As was shown in Chapter III, and also emphasized in this chapter, the entire urban landscape/watershed represents one system. Other nonaquatic green areas (parks, playgrounds, green boulevards) could and should be connected to the main ecosystem artery, which is the urban stream and its corridor. Many urban water bodies, not just the restored ones, must also be managed, including supplemental aeration, fish restocking, sediment capping, or alum treatment (impounded water bodies to prevent and manage cyanobacteria blooms). Sometimes, warning systems may be installed to inform the population about near-future flooding or potential bacterial contamination of swimming areas from overflows (if present). Stream restoration and management technologies are extensively covered in Novotny (2003) and the Interagency Task Force manual (1998).

IX.3.2 Regionalized versus Cluster-Based Distributed Systems

Chapter II presented the new cluster (ecoblock) partially cyclic distributed drainage system concept that dramatically contrasts with conventional linear regional systems. In the regional linear system (Figure 9.21), most of the water for the city or

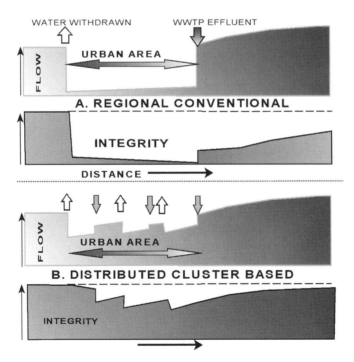

Figure 9.21 Impact of conventional linear once-through flow urban systems and distributed cluster (ecoblocks) based systems on dry-weather flow and integrity of receiving streams.

urban region is withdrawn upstream of the city, sometimes from a source outside the watershed, treated, and transported via pipeline to the users; the wastewater is then conveyed by the underground sewer system to one or a few treatment plants and then discharged into the receiving water body. This leaves the stream between the point of withdrawal and point of discharge short of flow. In some regions of the U.S. (e.g., the Southwest) or arid water-short countries, so much water can be withdrawn that streams have no flow. Even in the relatively humid state of Massachusetts, the Ipswich and Sudbury Rivers have no flow every other year due to municipal withdrawals. Similarly, the river between the Miyun Reservoir that supplies water to Beijing in China and the streams in the Beijing metropolitan area are pools of standing water for most of the year, or dry beds. The impact of effluent discharges then converts the streams into effluent dominated water bodies. These streams have no or insufficient base flow and are overwhelmed by high flows during storms. Consequently, the integrity of the water body between the withdrawal and effluent discharge and downstream from the discharge is severely impaired, and it takes a long time (distance) for the river to recover.

The distributed cluster (ecoblock) system in the ecocities (Figure 2.17, see also Chapter X) will provide the following benefits:

- Because of water conservation, reuse, and storage within the urban area, ecological flow can be provided.
- The resource recovery/water reclamation cluster-based and regional integrated resource recovery units will distribute the residual highly treated reclaimed water gradually into the receiving water body by means of surface buffered (e.g., by ponds and wetlands) and subsurface (infiltrated) flows. The impact on integrity (expressed by water quality and biotic integrity measures) will be gradual and in most cases far more beneficial than adverse. Downstream recovery sections will be short or nonexistent.
- The water bodies, because of the high-quality residual discharges and buffering by surface and systems and/or soil filtration, will attain the designated uses of aquatic life protection and propagation and recreation, and will be suitable for other uses such as landscape irrigation and additional ecological flow.

Figure 9.21 shows the difference between a traditional (fourth paradigm) urban water/wastewater system and the cluster-based distributed system during dry periods. The former situation results in a dramatic disruption of the hydrological and ecological integrity of the stream system that may not fully recover, or recovery will occur only after it joins a downstream larger, better-quality water body or a large reservoir. The latter situation represents a far more favorable situation. For example, the effluent dominated Trinity River draining the Dallas–Ft. Worth metropolitan area (population 6 million) reaches its maximum recovery after it passes the large Livingston Reservoir, 400 km downstream of Dallas–Ft. Worth, that provides approximately six months of retention time for the river flow. The lake is the major source of raw potable water for the water supply of Houston (population 2.2 million).

A system analysis approach could define the best locations of the distributed inflows of relatively clean effluents from the water reclamation and resource recovery. The stream will also receive most if not all clean water inputs (see Section 9.3.3). Generally, the measures of integrity outlined in Section IX.2.2 are:

- Chemical Integrity – Degree of compliance with standing water quality standards
- Biological Integrity – Indices of Biotic Integrity (Barbour et al., 1999; Novotny, 2003):
  - Fish diversity, health, and abundance (Ohio EPA, 1999)
  - Benthic macroinvertebrates integrity indices (Barbour et al., 1999)
- Habitat integrity index

IX.3.3 New Developments and Retrofitting Older Cities

New cities will evolve either as annexes of new developments on idle land adjoining the old historic center or in completely new locations (see Tianjin ecocity in Chapter XI) or by retrofitting old developments such as brownfields of abandoned industries, railroad yards, or even airports (e.g., Denver) or by gradually changing the living neighborhoods. In a new ecocity development, the existing natural riverine and lake systems should be preserved, but sometimes they are enhanced by additional man-made canals to allow individual developments to have their own access to boat piers (e.g., New Wuhan and Dongtan in China or communities in Florida) or for boat tours for tourists. In older developments, the stream system may be recreated because the original natural stream and lake system was lost and is underground. Some new sustainable communities are built on brownfields left by abandoned industries, warehouses, or ports (e.g., Hammarby Sjöstad—see Chapter XI). Hence, water body restoration is one component of restoring and retrofitting the entire watershed, including the current sewers.

According to the Clean Water Act, both previous stream modification and pollutant discharges constitute pollution. However, initiating restoration is a difficult and tedious process due to legacy building regulations and ordinances that are making current unsustainable drainage and sewerage difficult to change, and the old existing underground infrastructure such as water pipes, electricity and natural gas conduits, sewers, subways, and the like. The body of ordinances and approval processes may prevent implementation of new innovative concepts.

Obviously, retrofitting with daylighted streams in high-density urban environments will be far more costly than preserving and sometime enhancing the existing natural water bodies. In some cases, large cities and suburbs have developed and implemented ordinances that effectively restrict floodplain development (for example, Milwaukee, Wisconsin), which makes retrofitting easier. However, even with such regulations, it has to be realized that floodplains have been changed by urbanization. Areas developed decades ago outside the floodplain could be frequently flooded today (see Box 9.2 in this section). Many restoration projects are actually triggered by the increased frequency of flooding of populated neighborhoods. On the other hand, after the point source cleanup, the most frequent and important parameters that impair the integrity of urban streams are habitat parameters and siltation caused by increased sediment inputs from construction, stream bank erosion, and other erosion.

Although most restoration/daylighting projects are unique, given the history of past restoration projects combined with the demands of the new ecocity concepts, restoration projects can be categorized, and unique features within each category can be identified. It should be emphasized that mandatory point and nonpoint source cleanup must be done first, or concurrently with the restoration project. Restoration should not be a substitute for point source cleanup; however, landscape-based (soft) best management practices can be a viable alternative to the high-cost post-treatment of residual pollution of treatment plant effluents and the control of urban runoff pollution by end-of-pipe (hard) treatment infrastructures.

1. *Restoration to remedy damages to the existing water body caused by changed hydrology of the urbanized watershed*   This most common restoration repairs the damages caused by increased urban runoff and flooding. In the past these streams were candidates for hydraulic fast conveyance modifications and channelization. These streams may be receiving polluted urban runoff from storm sewer outlets and combined sewer overflows; they may have legacy pollution in slow sections, caused by contaminated sediments, and active erosion in fast sections.

Flood control typically gives the main impetus for restoration; however, the current regulations and public resistance do not allow channelization and stream lining. Residents often demand restoration of channelized and fragmented streams, as exemplified by the citizens living in the Lincoln Creek watershed in Milwaukee, Wisconsin (Box 9.2). Restoration of streams includes the following activities:

- Removal of hard lining installed in the past.
- Calculating the two- and hundred-year recurrence flows. Two-year flow is recommended for establishing the main channel capacity in the design (Leopold et al., 1992), and hundred-year flow determines the extent of the floodplain (see Figure 9.13).
- Installing best management practices in the watershed that would evaporate, infiltrate, treat, and attenuate urban runoff (see Chapters III and IV) such as green roofs, rain gardens, pervious pavements, infiltration basins and tree boxes, dry and wet storage ponds and wetlands, biofilters, and so forth. In most cases these practices will fully control, and could infiltrate, urban runoff flows caused by frequent rainfalls (e.g., up to two- to five-year recurrence interval), significantly reduce or eliminate CSOs, and enhance base flow. Creation of surface decorative drainage should be included (Dreiseitl and Grau, 2008).
- Collecting all clean water inputs into sewers and diverting them directly into surface drainage and into restored stream.
- Implementing stream bank stabilization by bank armoring, fascines, and revegetation by bioengineering properly anchored to the banks; in some cases stone

(riprap) or concrete pavement blocks allowing vegetation to take hold and develop could be installed (Novotny, 2003).

• Increasing resiliency against extreme events caused by rare storms and other cyclones, by recalculating the flow capacity of bridges and other bottlenecks and either removing redundant nonfunctioning bridges and other channel obstructions or enlarging the bridge underpass and channel capacity.

• Recalculating the design flow for the extreme flows and outlining the floodplain, in light of the above watershed measures. Floodplains, which between extreme events serve a multitude of purposes (parks, biking, walks, playgrounds, and sport fields), should be designed to include off-line and in-line flood storage in dry and wet storage basins. In the U.S., the typical extreme design flow is the one that has 1% of probability of occurring in any year.

The restored stream must be ecologically and hydrologically functional. Restored urban streams commonly do not have enough base flow. It must be provided by collecting all clean water inputs from the watershed. See Section IX.3, and also consult and apply the 150-year-old F. L. Olmsted Emerald Necklace concept (Figure 2.5). Olmsted designed the river as a sequence of pools and free-flowing channels, starting with a large pond and wetland that provided base flow and habitat for many organisms populating the river. For ecological base flow generation, see Chapters IV and X.

Two examples of stream restoration are featured here. Lincoln Creek in Milwaukee, Wisconsin (Box 9.2 and Figures 9.22 and 9.23), and The Kallang River in Bishan Park of Singapore (Figures 9.24 and 9.25). Both rivers were converted to concrete-lined lifeless channels. The differences between the concepts of restoration are minute, but the reasons are not the same. Channelization and lining of both rivers were implemented in the past to control flooding using the *fast conveyance* solution. In the Lincoln Creek watershed the grass-roots effort by the citizens stopped the continuation of the fast conveyance conversion, and a decision was made to provide flood protection and water quality improvements by restoration of ecology and habitat. Subsequently grants were received from the U.S. Environmental Protection Agency and the state of Wisconsin for the restoration project.

In Singapore, the need for restoration was partially driven by the conversion of Marina Bay (see Chapter VI) into a freshwater reservoir from which water is pumped into water supply reservoirs, some located in the central watershed of the headwaters of the Kallang River, which, along with the Singapore River, is a tributary of the Marina Bay. In this case a large-scale central watershed restoration was needed, which also included renaturalization of the river and cleansing biotopes in Bishan Park as a barrier to pollution that could come from the watershed. Extensive BMPs are being implemented throughout the watershed. The 65-hectare Bishan Park, created on a fill, is being improved. The park receives 3 million visitors each year (Public Utilities Board, 2009; Dreiseitl and Grau, 2008). The lined Kallang River is shown in Figure 9.24, and the architectural rendition of the future park is in Figure 9.25. The new Bishan Park will be home to diverse wildlife, with existing water bodies (ponds, creeks) restored and developed into water features integrated with the river.

Figure 9.22 Restored middle section of Lincoln Creek in Milwaukee. Note bioengineered bank armoring and riffle and pool sequence. Photo courtesy of Milwaukee Metropolitan Sewerage District (MMSD).

BOX 9.3   RESTORATION OF LINCOLN CREEK IN MILWAUKEE, WI

The Lincoln Creek restoration project is one of the largest in the U.S. The 14.8-km (9.2-mile) creek is an urban water body located entirely in the north part of Milwaukee County. Seventy years ago, the lower part of the 54-km^2 (21-sq-mile) watershed of Lincoln Creek was a desirable residential area, while the upper half was rural. In the second half of the last century, the channel was constricted by urbanization that is now encompassing the entire watershed. As a result peak flood flows have significantly increased, with a concurrent enlargement of the floodplain that has encroached into residential zones. Consequently, 2000 homes in the lower part of the watershed were located in the floodplain and flooding was frequent. The response of planners between 1940 and 1985 was to channelize the creek and line it with concrete (Figures 1.17, 2.12 left, and 9.10).

In mid-1980, NGOs and citizens resisted the completion of fast conveyance channelization and demanded renaturalization/restoration. In the late 1980s, the political attitude changed and the Milwaukee Metropolitan Sewerage District (MMSD) began to renaturalize the creek. Concurrently, the CSOs were eliminated by diverting them into Milwaukee's deep tunnel underground storage (see Chapter 1), and some best management practices were installed to control urban runoff. The project includes numerous elements to improve water quality, restore fish and wildlife habitat, enhance community value and pride, expand education, and preserve historical features such as the architecturally valuable stone bridges built in the 1930s during the Depression. The specific components are (see also Figures 9.22 and 9.23):

- Removal of 3.2 km (2 mi) of concrete lining
- 14.2 km (8.85 mi) of bioengineered channel
- Two off-line flood control facilities with total storage of 283,500 m^3 (75million gallons)
- Five enlarged vehicular and pedestrian bridges to remove bottlenecks
- Wetland restoration and nature educational area with native prairie and wetland planting

The project was finished in 2002 at a cost of U.S. $115 million, and fish and aquatic biota have returned. The restoration and integrated management of Lincoln Creek in Milwaukee, Wisconsin, and subsequently several other water urban rivers have been impressive with several social and environmental benefits such as:

- Providing flood protection to the homes from floods of up to 1% probability of occurrence (hundred-year recurrence interval). As a result homeowners' cost of extra flood insurance was eliminated, which also increased the value of the homes
- Enhanced recreation opportunities and community value
- Return of fish population and improved ecology of the creek and surrounding ecotones

The project, being one of the largest in the country, is a laboratory. The ecologic system is still fragile, and in the first years after completion it partially failed because the restoration was not based on the total hydrologic balance. The creek is lacking sufficient base flow, which was reduced by dropping grouwater table caused by urbanization within the watershed, and the pollutants present in the stormwater runoff entering the creek have not been fully controlled (e.g., salt from road de-icing operations and toxic compounds in urban runoff). Consequently, the biota and oxygen levels crashed one year after the completion due to excessive growth of algae *Cladophora* stimulated by these deficiencies.

Source: MMSD, Milwaukee, WI, and CDM.

Figure 9.23   Lower portion of Lincoln Creek looking upstream. The white line shows approximately the original cross section. The right bank of the floodplain was excavated to provide flood storage with a riparian wetland. Photo V. Novotny.

Figure 9.24   The Kallang River in Bishan Park in Singapore in 2009. Photo CDM.

Figure 9.25   The future of the Kallang River in Bishan Park. Courtesy public files of the Public Utilities Board (Singapore).

2. Restoration to remove fragmentation of the stream channel   Fragmentation involves small dams, culverts impassable by aquatic organisms (e.g., they generate fast shallow supercritical flows or are lined with concrete), weirs, and channel drops. In the U.S., many dams were built on streams for various purposes more than one hundred years ago. In Europe, building stream weirs (low head dams) dates back several centuries. These were the structures providing hydraulic head for hydropower plants, mills, and navigation. Sediment behind these dams has accumulated in the last 150 years. It came from urban and rural diffuse sources, wastewater dischargers, and later from combined sewer overflows. In many cases this sediment was contaminated by toxic pollutants and exhibits high sediment oxygen demand (SOD). Because of the sediment accumulation and the demise of small mills using hydropower, these impoundments ceased to function, and the sediments contain legacy pollutants.

After the dam is fully or partially removed, or the culvert is modified to allow passage of aquatic organisms, or the drop structure is eliminated, the stream channel is restored as described in Section IX.1 above. In many cases, the old channel has been covered by sediments; hence, the site must be surveyed and a new channel excavated and landscaped. Thousands of small dams built in the New England region of the U.S. have no function today other than some aesthetic value to a few riparian property owners. Removal of these structures should be considered (see Novotny, 2003) if:

- They cause longitudinal ecologic fragmentation and impede important migratory routes of anadromous (e.g., salmon) and catadromous (e.g., eel) fish.

- They have no function other than accumulation of sediments.
- The impoundment has poor quality caused by excessive algal and cyanobacteria populations, resulting in low dissolved oxygen in the impoundment and massive loading of organic algal matter into slow downstream reaches such as estuaries.
- The dam has deteriorated to a point of danger to the downstream population, and cost of repair would far exceed the benefit of keeping the dam in place.

Dam removal and stream restoration to its original function may be legally difficult if the water body is used for active navigation, which in the U.S., due to the Commerce Clause of the Constitution, is a protected use (Novotny et al., 2007). Box 9.3 describes removal of the North Avenue Dam in Milwaukee. The state of Wisconsin recognized that such dams impair the integrity of the state water bodies, prevent spawning and migration, and, generally, are a cause of water quality deterioration. The state has a successful program of dam removal and stream restoration.

If the dam cannot be removed, fish migration passage can be partially restored by a fish ladder, which contains cascading pools, allowing fish to move upstream and downstream through the pools. Smaller organisms can migrate in fish ladders only in the downstream direction, with the cascading water. Anoxia or hypoxia can be controlled by aeration (Novotny, 2003), by cascades (Figure 9.4), or by sidestream elevated pool aeration (SEPA), invented and installed by the Metropolitan Water Reclamation District of Greater Chicago on Chicago's waterways (Butts et al., 1999; Novotny, 2003).

Restoration of the Milwaukee River has had significant beneficial economic, environmental, and social impacts on the city. Thirty years ago, downtown Milwaukee was dilapidated, and most businesses and people had left for the suburbs. Today, in addition to lakefront beautification, the river between the city center and the end of the former North Avenue Dam impoundment is vibrant with river walks, cafés and restaurants, boating regatta, condominiums, sports events, and so forth. The downtown has become a place to come for arts, entertainment, recreation, and city living.

BOX 9.4   REMOVAL OF NORTH AVENUE DAM IN DOWNTOWN MILWAUKEE, WI

This small dam about 10 meters (30 ft) high, located less than 1 km (0.6 mi) north of downtown Milwaukee, was built in the second half of the 19th century (Figure 9.26). Its original use was to provide navigation head for a canal between Lake Michigan and a tributary of the Mississippi River, which was never built. Additional uses of the dam, in the first half of the 20th century, included power production and swimming. After more than one hundred years of

Figure 9.26   North Avenue impoundment in Milwaukee before the dam removal in 1992. Photo public domain archives of MMSD.

Figure 9.27   Dam removal, sediment capping, and stream bank armoring of the North Avenue impoundment in 1995. Photo V. Novotny.

(continued)

existence, the dam became a water quality problem. Because of excessive algal problems (eutrophication), turbidity (Secchi disc depth) in the impoundment decreased during the summer to 0.25 meters, the dissolved oxygen of flow typically dropped by 2 mg/L during the time of passage of the flow through the impoundment, and often anoxic conditions developed in the impoundment. Almost the entire volume of the impoundment was filled with sediments, with the water depth in some sections being less than 1 meter (3 ft).

Figure 9.28   Restored Milwaukee River about 3.5 km north of the city center. Water quality, biotic integrity, and habitat substrate quality have dramatically improved; fish have returned; and anadromous fish migrate during the spawning season (October) from Lake Michigan. Photo taken in 2002, courtesy MMSD.

In 1990 the dam was opened for repair and left open. In a few months a new channel eroded in 10 meters of clayey-organic sediments (Figure 9.27), and the exposed mudflats were experimentally revegetated. Although the accumulated sediments had a very high content of toxic metals (lead, zinc, and others), PAHs, and PCBs, the new riverbed sediments in the eroded channel were more sandy and far less toxic than the accumulated riverbank sediments that stayed. A study was conducted to determine the remediation of the sediments. The final solution was to remove sediment from a few hot spots, cover the remaining exposed floodplain sediment with a clean layer of soil, and revegetate with wetland plants. The dam was partially removed in 1995, and a free-flowing channel was created connecting Lake Michigan with the 30-km-long segment of the river. In subsequent years, salmon began to migrate upstream (Figure 9.28), and game fish have returned in high numbers, to the delight of fishing enthusiasts in the Milwaukee area. The dam removal also contributed to improved water quality in the Milwaukee downtown and to its remarkable revitalization.

3. Remediation of legacy pollution of contaminated sediments of ur-ban impoundments and fragmentation of migration Some rivers' sedi-ments (e.g., the Hudson River in the state of New York, the Sheboygan River and Cedar Creek in Wisconsin), contaminated by high concentrations of PCBs and other toxic pollutants, have become a major problem. The PCB contamination propagates and biomagnifies through the food web and has been found in the tissues of or-ganisms. Some of these sites were declared hazardous contaminated sites requiring cleanup. The most common methods are sediment removal and safe disposal or sed-iment capping.

Extreme caution must be taken when sediment is dredged because of resuspen-sion of the contaminants and the possibility of their moving downstream. In extreme cases, underwater dredging is not feasible, and the river must be diverted from the contaminated site for the site to be dredged; sediment then has to be removed to suit-able landfill sites, and the channel restored. Such drastic rehabilitation/restoration measures are small-scale. The Hudson River (NY) rehabilitation is a large-scale action.

Box 9.4 describes sediment remediation and stream restoration of Cedar Creek in Cedarburg, Wisconsin.

BOX 9.5   RESTORATION OF CEDAR CREEK

Cedar Creek transects the tourist city of Cedarburg in southeastern Wisconsin. Years ago two manufacturers (already relocated from the city) discharged PCBs from their plants into the creek through storm sewers. Over the years the PCB concentrations in two impoundments of the creek reached alarming levels of thou-sands of mg of PCBs per kg of sediment. The PCB contamination was carried by sediment and algae developed in the ponds downstream, all the way to the

(continued)

Figure 9.29   Cedar Creek bypass pipes are shown. After the pond created by the dam (center of the figure) was drained, sediments were dredged, and a new substrate (sand and gravel) was placed on the bottom of the pond.

Figure 9.30   Cedar Creek (Rock Pond) after restoration. Photos by V. Novotny.

Milwaukee River and Lake Michigan, 40 kilometers (25 miles) downstream. The sources of PCBs were identified, and the industries agreed to pay the cost of the Cedar Creek cleanup, rehabilitation, and restoration.

In the mid-1990s, the entire creek flow was conveyed around the contaminated site by several large pumps, the sediment was dredged and disposed of in the licensed landfills, the dredged bottom was covered by sand, and banks were restored (Figures 9.29 and 9.30).

4. Daylighting previously buried streams converted into storm sewers and underground culverts   During the third paradigm, urban streams were buried and converted into combined sewers (see Chapter I). However, forced by the fourth paradigm regulations and practices, many cities separated the sanitary sewage inputs and converted these underground conduits into storm sewers receiving (polluted) urban runoff and clean flows from upstream tributaries and from the city. Sewer separation has always been a costly construction activity. In retrospect, and for future polluted-versus-cleaner flow separation projects, it is clear that daylighting is preferable to building storm sewers. Sewer separation is not sustainable, has a very disruptive impact on neighborhoods during construction, and confers no social benefit thereafter. It only has a marginal environmental benefit, and brings no economic neighborhood revitalization benefits.

Category 4 daylighting involves opening an underground conduit (storm sewer and/or culvert) that does not carry sewage or wastewater and creating a surface natural-looking channel for:

1. Clean water flows from the watershed area outside the urban developed zones, from upstream creeks, springs, and small rivers
2. Clean water flows from the watershed from various sources such as basement and foundation dewatering, roof runoff, treated urban runoff, overflow from wells and public fountains, cooling water, and drainage water from tunnels
3. Residual discharges (ecologic flow) from resource recovery and water reclamation units

Daylighting of former streams that have been turned into storm sewers and underground culverts is relatively straightforward (see the examples in Section IX.1) and creates large economic and social revitalization benefits. However, in built environments it has many challenges (Pinkham, 2000), which are *social* (resistance of local public works and sewerage utilities, objections of nearby residents, access to riparian properties), *institutional* (new waterway ownership and maintenance, open channel liability, potential new regulations restrictive to future riparian homeowner, lack of leadership), and *technical* (underground utilities will be affected; building bridges; soil suitability; groundwater table and contamination; location of streamside paths, picnic areas, green vegetation, etc.).

Designers must decide whether there are conditions allowing restoration to restore the stream as close as possible to its original state, or whether a new channel and stream corridor should be created. Considering the fact that burying streams was done to provide space for development, "recreating" streams may be a more appropriate description of the process than "restoring." Examples of stream daylighting in Zurich, Switzerland (Conradin and Buchli, 2008); Beijing, China; and Seoul, Korea, document (see Section IX.1) that streams can be daylighted even in densely built-up environments, and/or when they have been built over with roads and freeways.

Channel design reflects local conditions and irreversible modifications of the urban area. In many instances, the channel is threaded into existing streets and parks. The following design issues are considered (Pinkham, 2000):

- Pollution inputs into the new channel must be eliminated or controlled to the point where the created water body would comply with water quality and biotic integrity standards and criteria.

- Establish the original configuration of the stream, which can be retrieved from old maps, aerial photos, and historic documents.

- Select the design of the channel cross section and sinuosity, including floodplain and pool and riffle sequence. In most cases, the flow capacity of the daylighted channel will be greater than the capacity of the underground conduit it is replacing.

- Identify points where new bridges or culverts will be built. These structures should be passable by aquatic organisms in both directions.

- Consider safety. The channel velocity and depth during high flows should not create dangerous fast and deep flows. Vertical concrete, masonry, or steel sheet pile embankment walls delineating deep channels should have fences.

- Ideally, the flow and hydraulic configuration of the channel should be as close as possible to the natural channel cross section (Figure 9.13) and flow patterns; otherwise, the forces of the flow will carve a new more stable channel. In densely populated zones, new urban stream landscape designs may be required, but these still should provide good habitat and conditions for aquatic biota and recreation. Decide which native species of plants and grasses would be compatible with the site.

- Provide base flow by directing all clean water flows and residual effluent flows from the resource recovery units in the watershed into the new channel; identify withdrawal points where water from the new stream would be withdrawn for reuse in the cluster.

Figures 9.31 and 9.32 show the daylighted Zhuan River in Beijing, China. In addition to daylighting and landscaping the stream channel, the project also included implementation of best management practices (e.g., pervious pavements) throughout the revitalized neighborhoods.

Figure 9.31 The Zhuan River in Beijing, China, before restoration. Photo courtesy of the Beijing Hydraulic Research Institute.

Figure 9.32 The Zhuan River, after daylighting and development of the area, has become a tourist attraction. Photo courtesy of the Beijing Hydraulic Research Institute.

5.   Daylighting of streams converted into combined sewers   In Zurich and in Tokyo, streams were daylighted while the combined sewers containing the original stream flow were left in place. Hence, the objectives of daylighting are (1) reduction of clean water inputs and urban runoff into combined sewers, (2) neighborhood beautification and, (3) in the ecocity context, providing an ecologically and hydrologically functioning surface conduit to maximize the social, environmental, and economic benefits. Box 9.5 presents the daylighting of a small stream in Tokyo, Japan, the drivers for restoration, and the final outcome. This project was created for the enjoyment of the population and to be a recipient of the highly treated tertiary effluent that otherwise would be discharged into an underground conduit. It also helps to reduce the CSO from the sewer below. In a sense, it is a monument to the lost river.

BOX 9.6   DAYLIGHTING OF KITAZAWA RIVER IN TOKYO

The story of the Kitazawa stream in Tokyo, shown in Figure 9.33, stands out as an example of social interaction between the public and the sewerage agency, and demonstrates the social and educational benefits. The story of the river is presented in Figure 9.34. In highly urbanized Tokyo, before the 1964 Olympic Games, most homes had no flushing toilets, and the community had no sewers. The night soil from outhouses was collected daily by city or private haulers. Thus,

Figure 9.33   View of the Kitazawa Creek in Tokyo. Photo V. Novotny.

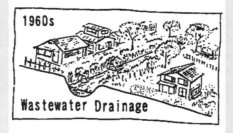

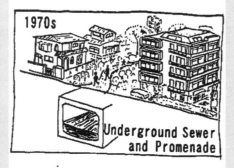

Figure 9.34   Schoolchildren's art describing the metamorphosis of the Kitazawa River in Tokyo, shown as a poster in the Kitazawa promenade.

(*continued*)

the small Kitazawa River was relatively clean and used by the population for enjoyment, and sometimes for laundry and fishing. After 1964, flushing toilets, washing machines, and dishwashers became common, but, because of the absence of sewers, house sewage outlets were directly connected to the stream, as in urban areas of Europe and the U.S. 50 or more years before. The stream became a smelly open sewer, and the city put the river into a culvert, essentially a combined sewer, and built a park/promenade above the underground conduit. However, the attitudes of people in Tokyo in the 1980s were different from those of their European counterparts at the beginning of the 20th century. People protested and demanded that the stream be brought back. The Tokyo Sewage Works Agency decided to create an artificial stream several kilometers long on the top of the culvert. Flow was fully provided by the effluent from a nearby tertiary treatment plant—in essence, a water reclamation plant. The creek has been planted, and fish have been restocked.

The dominant triple bottom line benefits of the Kitazawa River are more social than environmental. The creation of the small river in the boulevard helped to revitalize the neighborhood (as all stream daylighting projects do), but the river has no purpose other than public enjoyment and flood conveyance, which, however, could be a valid reason in an urban densely populated area with a disadvantaged population. It also serves for environmental education, as shown on the story of the stream created by the schoolchildren and displayed in the park surrounding the rivulet (Figure 9.34).

Because some modifications of streams by urbanization may be irreversible, restored urban streams need to be managed.

In Zurich, the primary technical objective of stream daylighting was to create natural-looking surface conduits for clean water that would otherwise be carried by combined sewers and contribute to CSOs. The city has numerous small mountain streams that were culverted upon entry into the populated urban zones, and many were converted into combined sewers, as was common in the first 70 years of the last century, and during the third paradigm. The consequence was delapidation of urban landscape, loss of recreation, and disappearance of natural flora and fauna (Conradin and Buchli, 2008). In addition to the streams from mountains and upstream rural areas, numerous groundwater infiltration-inflow (I-I) inputs were also increasing flow in the sewer and presented heavy clean water loads on the wastewater treatment plant (WWTP). About one-third of the dry-weather flow to the WWTP was clean I-Is. Instead of sewer separation, the city has chosen to daylight the streams and convey the I-I and upstream clean water inputs into this newly recreated surface system. The program started in 1988, and since its inception:

- Twenty kilometers (12.4 miles) of streams have been daylighted or revitalized.
- Of the estimated 800 L/sec (26.3 cfs) total extraneous water in the sewer system, approximately 300 L/sec (10.6 cfs) are diverted into new streams.

Additional extraneous clean water inputs diminished when the old leaking sewers were replaced by watertight pipes.

• New brooks have become important assets of the urban landscape.

• The population enjoys the new landscape enhancement, and the streams are especially popular with children (Conradin and Buchli, 2008).

The principle of the system is shown in Figure 9.35. The conventional sewerage concept, which was over a century old, intercepted the upstream creeks with a grit separator, from which the flow entered the combined sewer. Throughout the sewer system one or more flow dividers separate large flow into the flow that is sent to the treatment plant (typically four to six times the dry-weather flow) and the flow that is sent into the CSO. The clean water inputs could significantly increase the frequency of CSOs, and in some instances (e.g., in the sewer system of the Mestre community of Venice in Italy) CSOs have become permanent discharges. Dry-weather flow inputs are also undesirable when they enter separated storm sewers because they may be polluted (e.g., by cross-connections to sanitary sewers) and result in beach closings and other pollution problems. The city's response is sometimes to connect

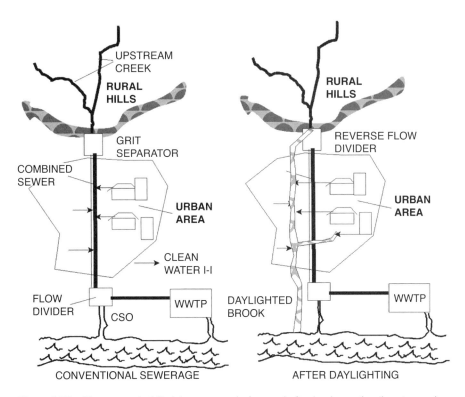

Figure 9.35   The concept of Zurich sewerage before and after implementing the stream daylighting for clean water and I-I flows. Adapted and redrawn from Conradin and Buchli, 2008.

dry-weather flows from storm sewers into sanitary sewers, which defies the purpose of separation. In such situations, putresible solids accumulate in the storm sewers and, with other polluted solids accumulated on the surface, cause a severe first flush pollution effect from storm sewers.

The Zurich system of daylighted streams is designed to accept clean water flows that would be about six times greater than the dry-weather flow, and the flows of greater magnitude would be redirected back into the rehabilitated (watertight) combined sewer system. In this case the CSOs are not fully controlled, but they are significantly reduced.

The sustainability of the program was investigated by Hugentobler and Brändli-Ströh (1997), who pointed out the following triple bottom line benefits (also quoted in Conradin and Buchli, 2008):

- *Social system:* Increased human interaction and networking in the neighborhoods where citizens participated in planning, implementation, and interdisciplinary cooperation with various city government departments
- *Human (individual) system:* Provided opportunities for beautification of the immediate living environment such as landscape planting and improvement, aesthetically pleasing features of the landscape, stimulation and opportunities, especially for children, to play and creatively use the immediate environment as well as educate themselves about the urban natural systems
- *Biological system:* Increased diversity of flora and fauna, leading to better sustainability of neighborhoods
- *In the chemical/physical system:* The savings achieved by reduced energy requirements of the sewerage system achieved by less pumping and treatment energy cost and consequent reduction of GHG emissions are significant and positive. A 2000 study in Zurich estimated annual energy cost for delivery and treatment per each liter/second (0.26 gallons/sec) of continuous flow in the sewerage system as U.S. $5000; hence, the savings achieved by clean water removal from sewers may be substantial. However, the system does not solve the problem with flooding, due to the small capacity of the surface channels, and risk of water pollution by toxic substances in urban (untreated) runoff entering the channels may reduce the sustainability considerations.

IX.4   SUMMARY AND CONCLUSIONS

The goal of this chapter was to show and document the feasibility of revival of urban streams that during the third paradigm were abandoned, lost their functionality, were modified into lifeless concrete channels, became open sewers and, in the ultimate demise, were buried underground and converted into underground sewers. Their demise was paralleled in the U.S., and to some degree also in Europe and

Asia, by decay of neighborhoods, abandonment by the middle class, and social environmental injustice. This was in correlation with the state of wastewater treatment technology, which, at that time, was absent or unable to treat waste to a degree that would make it suitable for reuse as surface flow. However, this lack of technology, under the philosophy of the third paradigm (emphasizing economic development and the growth of the industrial age at the expense of environmental and social sustainability) was tolerated by the population and generally accepted by urban planners and governments.

At the beginning of this millennium, the status of water/stormwater/wastewater treatment and reuse technologies made a quantum leap forward, and understanding of the functions of the urban landscape put the urban water bodies at the focus of the revitalization of urban areas that will make them sustainable, energy conscious, and better for living. It can be safely said that the new sustainable ecocity concepts cannot be implemented without considering treated effluents as a resource and keeping them on the surface for reuse and for providing ecological flow to sustain balanced aquatic life.

The conclusions and recommendations of this chapter are:

1. The situation of urban streams within the city, which are flow (effluent) limited or buried underground, is not sustainable.

2. In the U.S., attaining and preserving the integrity of surface water bodies are required by the Clean Water Act, in the European Community by the Water Framework Directive, and elsewhere by similar pollution control acts. Complying with these legislative mandates, enhancing the quality of the aquatic biological community, and providing habitat for urban wildlife are the most important environmental benefits of water body restoration and daylighting.

3. Urban streams provide connectivity of urban ecologic systems (green areas, parks, riparian zones/ecotones, larger water bodies, estuary and coastal zones).

4. Urban streams serve multiple purposes, including flood control, which may be one of the most important drivers for restoration. Past fast conveyance stream modification are inappropriate. Storage-based solutions with habitat restoration are needed to correct the past abuses of the hydrological and ecological functioning of urban streams.

5. Some past injuries left in legacy sediment contamination may be costly to remedy, but this is required by law.

6. Daylighting requires removal of overlying land use features such as roads, pavements, and parking lots. It also requires building or enlarging the flow capacity of bridges and utility overpasses. Typically, ordinances preclude building over buried streams. In most cases the cost of removing the impervious surfaces is far less than the benefits gained by daylighting.

7. With the current high level of treatment and resource recovery technology, even effluent dominated water bodies can support a balanced aquatic life and be safe for reuse and recreation.

8. Resource recovery and stream restoration work best in distributed systems with cluster-based resource recovery units reclaiming water for providing ecological flow and other on-site uses, such as irrigation or reducing domestic and commercial potable water use in buildings.

9. Stream restoration is a comprehensive plan that includes water body rehabilitation; renewal, greening, and enhancement of riparian flood zones (ecotones); urban runoff and CSO controls throughout the watershed by implementing modern best management practices; change of drainage from fast conveyance to storage-oriented; and replacing or reducing storm sewering by rain gardens and decorative infiltration.

10. Each cubic meter of water saved by water conservation and reuse within the city saves more than one cubic meter of produced potable water (because of leaks and other loses that may be significant) with reduction of cost and energy use (GHG emissions) in the water delivery system. The saved water may stay and enhance the flow in the water body. Saving cost is an economic benefit, while reducing energy and GHG emissions is both an economic and social benefit.

11. All clean water from the watershed should be discharged into the revitalized urban streams, and not into sewers. As with the reasoning above, each cubic meter of clean water discharged into surface stream systems will reduce cost, energy use, and GHG emissions from pumping the sewage and treatment into the WWTP, and will reduce CSOs and SSOs, which are significant environmental benefits.

12. Stream restoration and daylighting bring large economic benefits by increasing the quality of the neighborhoods, revitalizing businesses, often attracting tourists, and increasing the tax base of the community. The triple bottom line-life time assessment economic gains will be larger than the cost.

13. On the social side, high-quality urban streams, with their environmental integrity maintained, are the greatest architectural assets of surrounding neighborhoods. They provide a sense of enjoyment, relaxation, opportunities for recreation, and nature education. They bring communities together.

14. After restoration, urban streams cannot be left alone, and must be managed. The risks of injury to the ecosystem cannot be fully eliminated; clean natural and highly treated reclaimed base flow and connectivity have to be provided to increase resilience.

15. With daylighting and water body restoration in the cluster-distributed comprehensive water/stormwater/sewage management system, large interceptors, sewage conveyance and storage tunnels and storm and combined sewers may become obsolete.

16. Daylighted waters should include stepping-stones and other access features wherever possible and be surrounded by paths, picnic areas, and lookouts for people to enjoy (see Chapters III and IV).

REFERENCES

Barbour, M. T., J. Gerritsen, B. D. Snyder, and J. B. Stribling (1999) *Rapid Bioassessment Protocols for Use in Streams and Wadeable Rivers: Periphyton, Benthic Macroinvertebrates, and Fish*, 2nd ed., EPS-841-B–99/002, U.S. Environmental Protection Agency, Washington, DC

Bedoya, D., V. Novotny, and E. Manolakos (2009) "Instream and offstream environmental conditions and stream biotic integrity. Importance of scale and site similarities for learning and prediction purposes," accepted for publication in *Ecological Modeling*

Butts, T. A., D. B. Shackleford, and T. R. Bergerhouse (1999) *Evaluation of Reaeration Efficiencies of Sidestream Elevated Pool Aeration (SEPA) Stations*, Illinois State Water Survey, Champaign, IL

Cairns, R. F. (1988) "Increasing diversity by restoring damaged ecosystems," in *Biodiversity* (E. O. Wilson, ed.), National Academy Press, Washington, DC, pp. 333–343

Committee on Restoration of Aquatic Ecosystems (1992) *Restoration of Aquatic Ecosystems*, National Research Council, National Academy Press, Washington, DC

Committee to Assess the Scientific Basis of TMDL (2002) *Assessing the TMDL Approach to Water Quality Management*, National Academy Press, Washington, DC

Conradin, F. and R. Buchli (2008) "Separating clean water inputs from underground pipes," in *Handbook of Regenerative Landscaping* (R. L. France, ed.), CRC Press, Taylor & Francis, Boca Raton, FL, pp. 47–60

Dreiseitl, H. and D. Grau (2008) *Recent Waterscapes: Planning, Building and Designing with Water*, Birkhäuser, Basel, Boston, Berlin

Dunster, J. and K. Dunster (1996) *Dictionary of Natural Resources*, University of British Columbia, Vancouver, BC, Canada

France, R. L. (2007) *Handbook of Regenerative Landscaping*, CRC Press, Taylor & Francis, Boca Raton, FL

Gensemer, R. W., R. D. Meyerhoff, K. Ramage, and E. Curley (2008) *Relevance of Ambient Water Quality Criteria for Ephemeral and Effluent-dependent Water Courses of the Arid Western United States*, Society of Environmental Toxicology and Chemistry, Pensacola, FL

Grimmond, C. S. B., T. R. Oke, and D. G. Steyn (1986) "Urban water balance 1. A model for daily totals," *Water Resources Research* 22(10): 1397–1403

Hajda, P. and V. Novotny (1996) "Modeling the effect of urban and upstream nonpoint sources on eutrophication of the Milwaukee River," *Water Science and Technology* 33(4–5): 153–158

Heaney, J. P., L. Wright, and D. Sample (2000) "Sustainable Urban Water Management," Chapter 3 in *Innovative Urban Wet-Weather Flow Management Systems* (R. Field, J. P. Heaney, and R. Pitt, eds.), TECHNOMIC Publishing Co., Lancaster, PA, pp. 75–120

Hugentobler, M. and M. Brändli-Ströh (1997) "Sustainable urban development. A concentrated framework and its application," *Journal of Urban Technology* 4(2): 85–103

Imhoff, K. (1931) "Possibilities and limits of the water-sewage-water-cycle," *Engineering News Report*, May 28

Imhoff, K. and K. R. Imhoff (1993) *Taschenbuch der Stadtentwässerung (Pocket Book of Urban Drainage)*, 28th ed., R. Oldenburg Verlag, Munich, Germany

Interagency Task Force (1998) *Stream Corridor Restoration*, USDA, U.S. EPA, TVA, *et al.*, Washington, DC

Karr, J. R. and E. W. Chu (1999) *Restoring Life in Running Waters*, Island Press, Washington, DC

Karr, J. R., K. D. Fausch, P. L. Angermeier, P. R. Yant, and I. J. Schlosser (1986) *Assessing Biological Integrity in Running Waters: A Method and its Rationale*, Special Publication 5, Illinois Natural History Survey, Champaign, IL

Kohonen, T. (1990) "The self-organizing map," *Proceedings of the IEEE* 78, pp. 1464–1480

Lee, I. K (2007) "Cheong Gye Cheon Restoration project," http://www.irbdirekt.de/daten/iconda/CIB7972.pdf

Lee, T. S. (2004) "Buried treasure: Cheong Gye Cheon restoration project," *Civil Engineering* 74(1): 31–41

Legendre, P. and L. Legendre (1998) *Numerical Ecology*, Elsevier Science BV, Amsterdam

Leopold, L. B., M. G. Wolman, and J. P. Miller (1992) *Fluvial Processes in Geomorphology*, Dover Publications, Inc., New York

Manolakos, E., H. Virani, and V. Novotny (2007) "Extracting Knowledge on the Links between Water Body Stressors and Biotic Integrity," *Water Research* 41, pp. 4041–4050

Mitchell, V.G., R.G. Mein and T.A. McMahon (1996) Evaluating the resource potential of stormwater and wastewater: An Australian perspective; Proc. 7th International Conference on Urban Storm Drainage, Hanover, Germany, p. 1293-1298

Mitsch, W. J. and J. G. Gosselink (2000) *Wetlands*, John Wiley & Sons, Hoboken, NJ

Novotny, V. (2003) *Water Quality: Diffuse Pollution and Watershed Management*, 2nd ed., John Wiley & Sons, Hoboken, NJ

Novotny, V. (2007) "Effluent dominated water bodies – their reclamation and reuse to achieve sustainability," in *Cities of the Future: Towards Integrated Sustainable Water and Landscape Management* (V. Novotny and P. Brown, eds.), IWA Publishing Co., London, UK, pp. 191–214

Novotny, V. et al. (1997) *Use Attainability Analysis: A Comprehensive UAA Technical Reference,* Water Environment Research Foundation, Alexandria, VA

Novotny, V., D. Clark, R. J. Griffin, and D. Booth (2001a) "Risk based urban watershed management under conflicting objectives," *Water Science and Technology* 43(5): 69–78

Novotny, V., D. Clark, R. Griffin, A. Bartosova, and D. Booth (2001b) *Risk Based Urban Watershed Management under Conflicting Objectives*, U.S. EPA, NSF, USDASTAR Watershed Program, Final Report, Marquette University, Milwaukee, WI

Novotny, V. *et al.* (2003) *Lower Des Plaines River Use Attainability Analysis*, Report submitted to the Illinois Environmental Protection Agency by AquaNova International, Ltd., and Hey and Associates, Springfield, IL

Novotny, V., A. Bartošová, N. O'Reilly, and T. Ehlinger (2005) "Unlocking the relationship of biotic integrity of impaired waters to anthropogenic stresses," *Water Research* 39:184–198

Novotny, V., N. O'Reilly, T. Ehlinger, T. Frevert, and S. Twait (2007) "A River is Reborn: The Use Attainability Analysis for the Lower Des Plaines River, Illinois," *Water Environment Research* 79(1): 68–80

Novotny, V., D. Bedoya, H. Virani, and E. Manolakos (2009) "Linking Indices of Biotic Integrity to Environmental and Land Use Variables - Multimetric Clustering and Predictive Models," *Water Sci. & Technol.*, 59(1):1–8.

Ohio EPA (1999) "Association between Nutrients Habitat, and the Aquatic Biota in Ohio Rivers and Streams," *Ohio EPA Technical Bulletin* MAS/1999-1-1, Ohio EPA, Columbus, OH

Palmer, M. W. (1993) "Putting things in even better order: The advantages of canonical correspondence analysis," *Ecology* 74(8), p. 2215

Pickus, J. M., W. B. Samuels, and D. E. Amstutz (2002) "Applying GIS to identify effluent dominated waters in California," ESRI Conference, gis.esri.com/library/userconf/proc02/pap0393/p0393.htm

Pima County Wastewater Management Department (2007) *Arid West Water Quality Research Project, Final Report*, submitted to ES Environmental Protection Agency, Region 9, Phoenix, AZ

Pinkham, R. (2000) "Daylighting: New Life for Buried Streams," Rocky Mountain Institute, Snowmass, Colorado, available online: http://www.rmi.org/images/PDFs/Water/W00-32_Daylighting.pdf

Preservation Institute (2007) "Seoul South Korea Cheonggye Freeway," http://www.preservenet.com/freeways/FreewaysCheonggye.html

Public Utilities Board (2009) "Kallang River – Bishan Park Project," Public Utilities Board, Singapore, http://www.pub.gov.sg/abcwaters/AdditionalInformation/Pages/KallangRiver.aspx, accessed December 2009

Rosgen, D. (1996) *Applied River Morphology*, Wildland Hydrology, Pagosa Springs, CO

U.S. Environmental Protection Agency (1994) *Water Quality Standards Handbook*, 2nd ed., EPA-823-b-94-005A, Office of Water, Washington, DC

US Environmental Protection Agency - Region 8 (2003) EPA Region 8 - Effluent-dependent streams and the Net Environmental Benefit Concept, Discussion with the Colorado 309 Workgroup, April 28, 2003, San Francisco Irani, H., E. Manolakos, and V. Novotny (2005) *Self Organizing Feature Maps Combined with Ecological Ordination Techniques for Effective Watershed Management*, Technical Report No. 3, Center for Urban Environmental Studies, Northeastern University, Boston, MA

Wang, L., J. Lyons, P. Kanehl, R. Bannerman, and E. Emmons (2000) "Watershed urbanization and changes in fish communities in southeastern Wisconsin streams," *Journal of the American Water Resources Association* 6(59): 1173–1189

Yoder. C. O. and E. T. Rankin (1999) "Biological criteria for water resources management," in *Measures of Environmental Performance and Ecosystem Condition*, National Academy Press, Washington, DC, pp. 227–259

X

PLANNING AND MANAGEMENT OF SUSTAINABLE FUTURE COMMUNITIES[1]

X.1 INTEGRATED PLANNING AND MANAGEMENT

X.1.1 Introduction

How to plan and manage sustainable communities? At the time of writing this study, the state of the art of planning sustainable Cities of the Future was evolving and was discussed at conferences using buzz words such as "sustainability," "reuse," "reclamation," "desalination," "low impact," "carbon neutrality,, "water neutrality," and "decentralization." Questions were asked: How can existing communities be retrofitted to become sustainable? What are the criteria of sustainability? Is complete water recycling a goal for a community that currently has, and, in the foreseeable future will have, adequate water supply, sewerage, and disposal of pollutants? What is the difference between a sustainable city and an ecocity? Can a city be sustainable by practicing water and energy conservation but not extensive water reuse and without developing renewable sources of energy? Can a city located in a desert or in an extremely water-short area be sustainable because of the low demand on drinking water resources (that may not even exist at the location), by virtue of developing and applying the most modern water reclamation and recycle technologies? We have already explained that these cities may have high energy demand and would have to rely to a maximum extent on developing renewable sources of energy (see Chapter XI). Can a Low Impact Development (LID) community be sustainable? These and many other questions will have to be answered by means of a comprehensive and integrated systems analysis during the planning process. Developing a sustainable

---

[1]W.P. Lucey and C.L. Barraclough (Aqua-Tex Scientific Consulting, LTD), Glen Daigger (CH2M-Hill), and P.R. Brown (CDM) contributed to this chapter.

community, either as a new development or by a retrofit, is not easy and requires an exceedingly complex degree of cooperation among planners, citizens, government officials, and all other stakeholders.

At the end of the 20th century, the goal of the United Nations was to provide adequate water supply and sanitation to all who did not have them. The drive towards sustainability and ecocities (see Chapter II for ecocity definition) emerged at the end of the last millennium because of the realization of the consequences of business as usual in cities under the major stresses of (1) population increases and migration, (2) threats of adverse impacts of global climatic changes, and (3) increasing water shortages in many highly populated regions of the world. The International Water Association, the Water Environment Federation, the American Planning Association, and many other professional associations made the "Cities of the Future" movement a major initiative in the first decade of this millennium, which has resulted in a number of international and national conferences on this topic. LEED (Leadership in Energy and Environmental Development) gold or platinum certification (see Chapter II, Section II.4.2) has become a measure of acceptance for new buildings and subdivisions in developed countries. However, at the same time, we have to realize that the worldwide goals of adequate water supply and sanitation of the last decade of the 20th century have not been met in many developing countries with growing megapolis developments; the problems of poor public health and inadequate water supply may be worsening as their populations rise. Concurrently, the infrastructure for water conveyance, used water treatment, and waste disposal in developed countries has been crumbling and needs to be upgraded or replaced. Should it be done in the old, clearly unsustainable, and energy-demanding way, at great expense? Because of the increasing restrictions of effluent and ambient water quality criteria, the cost of replacing the old infrastructure with a more expensive and more energy-demanding new infrastructure has become prohibitive.

Leaders and members of professional associations in many disciplines (environmental engineering and science, urban planning, architecture, urban and suburban ecology) have realized that the current infrastructure and urban planning paradigm have become impediments to achieving sustainable urban development and living, and to addressing the impacts of global climatic change because of their reliance on fast surface and underground conveyance of water and wastewater; on regional water and wastewater management systems; and on energy overuse for sustaining living processes, commerce, transportation, and use of other resources in the cities. A paradigm shift from current unsustainable urban development and living to sustainable future ecocities is needed. This paradigm shift in thinking has helped shape a working definition of Cities of the Future (COF) (see Chapter II, Section II.5) for both existing and new urban areas.

Rigorous analysis of the performance of current and alternative urban systems is required to give weight to the paradigm shift toward more sustainable urban designs. This analysis must take the form of an illustration of the footprint of various urban systems, and tools that allow easy analysis of system options in future scenarios. This book is a guide to the development of such an analysis.

X.1.2 Footprints

A "footprint" is a quantitative measure showing the appropriation of natural resources by human beings (Hoekstra and Chapagain, 2007). Footprints can be local as included, for example, in the LEED (USGBC, 2005) or OPL (WWF (2008)) criteria, or regional to global. The major categories of footprints related to urban metabolism have been identified in the literature: the social footprint of GHG effect on climate; the water footprint; the economical footprint of affordability; and the ecological footprint. The One Planet Living (OPL) sustainability criteria (see Chapter II, Section II.4.2) provide a more stringent yardstick for certification of urban sustainability and, as of the date of publishing this work, only a small handful of medium-size new developments throughout the world can claim OPL compliance (see the next chapter). The OPL evaluation parameters (see Section II.4.2) include zero emissions (pollution and GHG), near zero solid waste, use of local materials, minimizing life cycle and virtual value of nonrenewable resources, protection of natural resources, supporting local culture and heritage, and providing good public health amenities.

Civil and environmental engineers and urban planners organize their evaluation of sustainability along the triple bottom line sustainability assessment lines categorized as (1) social, (2) ecological, and (3) economical. Sustainability also implies intergenerational (sustainability) equity. These assessments consider the lifeline horizon and extent of the infrastructure component benefits and costs and their environmental and social impacts.

Carbon (Social) Footprint   *The push for carbon neutrality* could be considered as self-preservation by the global society from the worldwide effects of climatic changes that are expected if nothing or little is done to reduce emissions of GHGs. Today pressing news about the deleterious effects of increasing concentrations of CO_2 and other GHGs gives added impetus to the need to make our cities more sustainable. Global warming solutions cut across all the major systems of the city: energy provision for buildings; energy use by transportation systems and the "discovery" by transportation engineers that land use and urban design decisions can reduce mileage (kilometers) traveled by cars and other traffic—a major contributor to transportation carbon emissions; and the interrelationship (nexus) of water and energy. Energy is needed by the water/wastewater industry (the greatest energy cost is for transporting water and used water) as well as by the energy industry, particularly for nuclear power plants. Global warming solutions are also assisted by taking a more eco-friendly approach to development.

The water-energy nexus is also a premise of global sustainability. In the area of water management, achieving this goal implies water (energy) conservation, reuse of used water and use of stormwater, development and use of renewable energy, reduction in energy use in urban and suburban transportation and building infrastructure, and reliance on local and sustainable agriculture. Figure 8.4 in Chapter VIII shows a possible path towards achieving the net zero GHG emissions suggested by the U.S. president's National Science and Technology Council. According to the council, the current scientific research indicates that 60 to 70% of energy reductions could be achieved with more efficient appliances, such as better water and space heaters and heat pumps, and with significant reduction of water demand. Further, 30 to 40% of energy could be produced by renewable sources, including wind, solar energy, biogas production, and heat recovered from used water or extracted from ground and groundwater. Used water has to be considered as a resource, providing clean water by water reclamation, clean hydrogen and electric energy from biogas and microbial fuel cells, nutrients for fertilizers, and soil-conditioning solids for agriculture.

Ecological Footprint *Global scale/regional ecological footprint* has been defined as the total area of productive land and water required to produce, on a continuous basis, all the resources consumed by the city and to assimilate all the wastes produced by its population, wherever on earth the land may be located (Rees, 1996; 1997).

Rees (1997) and Wackernagel and Rees (1996) calculated the ecological footprint of a "typical North American city" as being 4.8 ha/person, which, if multiplied by the expected population 20-30 years in the future, will be 3 to 4 times the available productive land on earth. If things are let go as usual this disproportionality of resource use may lead to great societal disruptions. Hence the COTF goal must be to implement cyclic metabolisms and reduce the this ecological footprint.

Local ecological footprints are to some degree different than global footprints focusing on sustainability of resources to provide viability to the city. The local/subregional ecological footprints were divided into those considering (a) urban waterways and impoundments, (b) water corridors and urban open green space; and (3) urban hydrology, including surface and subsurface water resources and drainage.

Urban Waterways This is the most dominant component of the ecological footprint of the COF. Previous practices buried urban streams used as sewers or built culverts out of sight because of severe pollution, decades or a century ago. In COF, water conservation and treatment will provide ecological flow (lacking today because of overuse) to surface water bodies. Current and future used water reclamation technologies will enhance water quality to support aquatic life, water supply, and recreation. Therefore, it is important to daylight lost streams and water bodies as well as to restore and protect existing urban lakes and streams.

Responsible nutrient management. Many water bodies, not just the urban, are severely affected by eutrophication, which in some cases has led to a hypertrophic status characterized by massive algal blooms of cyanobacteria. These resilient microorganisms greatly impair beneficial uses of water bodies such as fish and wildlife propagation, recreation, and water supply. These problems are caused by excessive nutrient (nitrogen and phosphorus) inputs both from urban and rural point and nonpoint sources. To make matters worse, the world is running out of phosphorus, which is needed to grow crops. Efficient and responsible nutrient management and phosphorus recovery is a COF goal and a measurable footprint as well as the vulnerability of the water body and its watershed to the adverse effects of bad nutrient management and overuse.

Restoration of Ecological Corridors and Urban Water Bodies   Urban ecology, consisting of green areas, water bodies, and ecotones (which separate nature from built habitat), has to provide connectivity and passage to the urban biota and people. The opposite of connectivity is fragmentation, which impedes a healthy ecology and survival during times of stress. Ecological corridors along the urban surface water bodies also provide resiliency to extreme meteorological events such as floods (Chapter III). The width and ecological health of the water body riparian zones could be an important footprint.

Restoring Hydrology   Past urbanization has dramatically changed the hydrology of our cities by reducing infiltration and groundwater recharge, and increasing flooding. This has led not only to water shortages but also to dangerous subsidence of buildings, monuments, and other infrastructure in many communities—including Venice (Italy), Mexico City, Philadelphia, and Boston—and increasing vulnerability to catastrophic flooding. Unrestricted development and climatic changes will also increase the portions of urban areas in floodplains. Restoring hydrology as close as possible to the natural water cycle should be a goal and also one of the measures of progress towards sustainability (Chapter II). The most obvious parameter of this footprint will be the percent imperviousness of the watershed and of the riparian corridor, degree of channelization/sewering of the of ephemeral and first order streams.

Economic Footprint   The COF will implement advanced technologies of water and energy conservation in water-short areas based on reuse and tapping sources that were not accessible in the past, such as seawater (high salinity) and deep brackish aquifers. Extracting, transporting, and treating water from these sources and relying on high-cost recycling is expensive, uses more energy, and has limits. Some COFs are being planned and built in areas that have very little or no freshwater, with a goal of maximum reuse, including direct and indirect potable reuse. Current and future technologies will allow treatment of reclaimed water to achieve the quality of bottled drinking water; however, economic cost and affordability must be considered. Public health requirements limit the amount of reuse that can be safely achieved in each city, and how many times a particle of water can pass through the reclamation-reuse cycle is not known, especially when considering accumulation of conservative trace and emerging pollutants in the cycle.

Water Footprint   In Chapter V, the water footprint was discussed by introducing water use disparities among countries; the U.S. leads the water use demand. The water footprint shows the extent of water use in relation to consumption by people. Hoekstra and Chapagain (2007) defined the water footprint of a country as the volume of water needed for the production of the goods and services consumed by the inhabitants of the country. The internal water footprint is the volume of water used from domestic water resources; the external water footprint is the volume of water used in other countries to produce the goods and services imported and consumed by the inhabitants of the country. The water footprint is often equated with the virtual water content of products consumed, which, on a per capita basis, far exceeds the population's actual water consumption. The water footprint is a new concept introduced at the beginning of this millennium in order to have a consumption-based indicator of water use, in addition to the traditional production-sector-based indicators. For example, rice consumption has the largest global water footprint (2291 m^3 of water per 1 ton of rice) (Hoekstra and Chapagain, 2007). The water footprint is being developed analogously to the ecological footprint of cities, which was introduced in the 1990s (Rees, 1992), represented by the total area of agricultural and forest land needed to provide food and raw materials to sustain life in the city.

X.2 URBAN PLANNING

Urban planners organize and expand the sustainability assessment in the following categories:

- Land use/urban form
- Water, used water, and stormwater
- Solid waste
- Transportation
- Ecology/nature
- Energy
- Social/cultural relationships
- Economic base/jobs/income inequality, and environmental justice
- Health
- Food

All of the above categories relate to water. Water and the water bodies that provide it for many beneficial purposes are the lifeline of cities, and throughout history, until the Industrial Revolution, sources of water were protected and forcefully defended by city dwellers, especially in water-short areas.

Urban planners also recognized the close relationship of water resources and water to transportation. Building highways has an impact on hydrology, but, traffic emissions, road deterioration, and use of de-icing chemicals degrade the water quality of urban streams, lakes, and groundwater even more. The salt content of groundwater in cities and near highways is now linearly increasing, with no sign of leveling off. For example, in the northern metropolitan area of Minneapolis, Minnesota, only about 25% of applied salt on the streets of this metropolitan area leaves the watershed by the Mississippi River. The rest remains in lakes, where salt causes lasting chemical stratification, and in groundwater (Novotny, Sander, Mohseni, and Stefan, 2009). The water-energy nexus was extensively covered in Chapter VIII. The impact of urbanization on receiving waters was also discussed in Chapter IX.

X.2.1 Ecocity Parameters and Demographics—Population Density Matters

At the end of the last century, the ideas of smart growth development and Low Impact Development (LID) were promoted by urban planners and urban environmental engineers.

Smart Growth Development According to Wikipedia (2009) definitions, **smart growth** is an urban planning and transportation theory that concentrates growth in the center of a city, to avoid urban sprawl, and advocates compact, transit-oriented, walkable, bicycle-friendly land use, including neighborhood schools, complete streets, and mixed-use development with a range of housing choices.

Smart growth values long-range, regional considerations of sustainability over a short-term focus. Its goals are to achieve a unique sense of community and place; expand the range of transportation, employment, and housing choices; equitably distribute the costs and benefits of development; preserve and enhance natural and cultural resources; and promote public health.

The components of smart growth are:

Compact neighborhoods that attract more people and business. Creating such neighborhoods is a critical element of reducing urban sprawl and protecting the climate. Such a tactic includes adopting redevelopment strategies and zoning policies that channel housing and job growth into urban centers and neighborhood business districts, to create compact, walkable, and bike- and transit-friendly hubs. This sometimes requires local governmental bodies to implement code changes that allow increased height and density downtown, and regulations that not only eliminate minimum parking requirements for new development but establish a maximum number of allowed spaces. Other topics that fall under this concept are:

- Mixed-use development
- Inclusion of affordable housing
- Restrictions or limitations on suburban design forms (e.g., detached houses on individual lots, strip malls, and surface parking lots)
- Inclusion of parks and recreation areas

A transit-oriented development (TOD) is a residential or commercial area designed to maximize access to public transport, and mixed-use/compact neighborhoods tend to use transit at all times of the day. Many cities striving to implement better TOD strategies seek to secure funding to create new public transportation infrastructure and improve existing services. Other measures might include regional cooperation to increase efficiency and expand services, and moving buses and trains more frequently through high-use areas.

Pedestrian- and bicycle-friendly design. Biking and walking instead of driving can reduce emissions, save money on fuel and maintenance, and foster a healthier population. Pedestrian- and bicycle-friendly improvements include bike lanes on main streets, an urban bike-trail system, bike parking, pedestrian crossings, and associated master plans. The most pedestrian- and bike-friendly variant of smart growth and new urbanism is new pedestrianism, because motor vehicles are on a separate grid.

Other features:

- Preserving open space and critical habitat
- Reusing land, and protecting water supplies and air quality
- Transparent, predictable, fair and cost-effective rules for development

Low Impact Development As defined and described in Chapters III and IV, LID uses site design techniques to create a hydrologically functional landscape. This, in turn, reduces the need to build infrastructure to convey stormwater runoff away from a property. According to the National Association of Home Builders Research Center, LID can decrease the costs of development by reducing infrastructure construction and stormwater management costs.

Population Density Effects It can be seen that the Smart Growth Development concepts are those that would be closely applicable to the sustainable ecocities; as a matter of fact, it can be postulated that future sustainable cities will conform to the concepts of smart growth development. However, it also appears that the OPL criteria for ecocities are more stringent and already contain the key smart growth criteria. Furthermore, the current calls for GHG emissions and water neutrality are not fully incorporated into the smart growth development concepts.

Novotny and Novotny (2009, see also the next chapter) analyzed seven ecocities throughout the world that were either being built or were in the concluding pre-realization phases of their planning. These cities were Hammarby Sjöstad in Sweden; Tianjin, Dongtan, and Qingdao in China; Masdar in the United Arab Emirates; and Sonoma Mountain Village and Treasure Island in California. The result of the analysis is included in Table 10.1, and a more detailed description of the communities is given in the next chapter. Table 10.1 shows that these progressive cities varied in design and performance. The hydrology and water movement ranged from a linear system in Hammarby Sjöstad to almost closed systems in Qingdao and Masdar. All strived to meet or exceed the OPL criteria, but only three (Masdar, Qingdao, and Sonoma Mountain Village) could claim to meet the OPL specifications, and only the Masdar and Qingdao projects could claim GHG emissions neutrality.

With the exceptions of Sonoma Mountain Vollage, the smallest development, and the Quingdao ecoblock, the development with the highest population density, the density of the developments varied between 117 to 170 people/ha. From the presentations and findings in the literature, it was evident that all the design teams used some kind of a proprietary model that balances the population and its energy use based on the probability of walking and biking instead of driving, energy insulation of buildings and exposure to sun, renewable energy sources, and other determinants for GHG emissions from urban areas. Three sites, Dongtan, Tianjin, and Treasure Island were designed by an English architectural and development consultancy, Arup.

Table 10.1 Summary analysis of the parameters of the seven ecocities

City	Population Total	Population Density (#/ha)	Water Use (L/capita-day)	% Water Recycled	Water System	% Energy Savings Renewable	Green Area (m²/capita)	Cost (U.S.$/unit)*
Hammarby Sjöstad	30,000	133	100	0	Linear	50	40	200,000
Dongtan	500,000 (80,000)++	160	200	43	Linear	100	100	~40,000
Qingdao	1500+	430–515	160	85	Closed loop	100	~15	?
Tianjin	350,000 (50,000)++	117	160	60	Partially closed	15	15	60,000–70,000
Masdar	50,000	135	160	80	Closed loop	100	<10	1 million
Treasure Island	13,500	170	264	25	Mostly Linear	60	75	550,000
Sonoma Valley	5000	62	185	22	Linear	100	20	525,000

Source: Novotny and Novotny (2009)

*Based on average 2.5 members per household, +Qingdao ecoblock, ++Phase I

Figure 10.1 An architect's conception of the harbor entrance into Dongtan from the Yangtze River, showing a lagoon, canals with water taxis and wind turbines (Art picture by Arup).

The population density issues arose during the competition among architects for the contract for the Dongtan ecocity (Figure 10.1), to be located on the alluvial Chongming Island in the mouth of the Yangtze River. The island, located 40 kilometers from Shanghai, was sparsely populated by farming communities in the early 1990s, and had extensive wetland ecology. Most of the submissions by firms were in the mode of LID, to provide low-density housing on the entire buildable land of the island. However, Arup realized that building low-impact spread-out American-style subdivisions with low-rise condominiums and single-family homes scattered across the island, with lawns and parks in between, would lead to a dead end and environmental disaster on Chongming Island. For one thing, low-impact spread-out developments would need automobiles and/or uneconomical public transportation, which would lead to more adverse global warming effects, traffic pollution, and congestion. Arup decided that, to be a sustainable ecocity, Dongtan, instead of having 50,000 people on the entire buildable portion of the 1200-km^2 island, needed a lot more people on a smaller area. If population density is low, then public transportation and energy and water reclamation/recovery are not economical (McGray, 2007; Urban Agent, 2008).

Arup's engineers most likely analyzed and plotted the population density relation to energy consumption. They could have found that when the population density reached 120 residents per hectare—which was approximately the density of Hammarby Sjöstad in Stockholm or Copenhagen in Denmark—such communities used more biking, and walking and public transportation were most economical. Furthermore, heating and cooling energy recovery and savings made more sense. When the population density is increased—for example to 290 people/ha, as in Singapore, or up to 720 people/ha as in Hong Kong—further energy savings are negligible. Hence, Arup proposed a Venice-type water centric city, with canals, lakes, and electric water taxis, that had the density of 150 residents per hectare (McGray, 2007) (see Table 10.1). Two authors (V. Novotny and E. V. Novotny) of this book compiled and

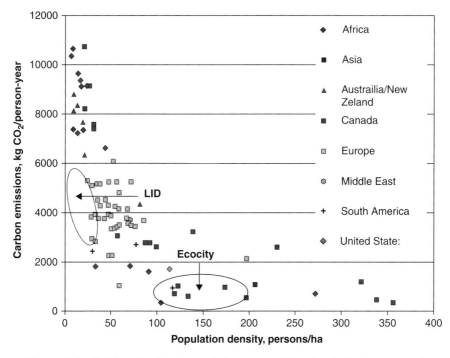

Figure 10.2   Relationship of carbon emissions to population density (various sources).

developed a plot of urban carbon emissions from private and public transportation, heating, and electricity (including cooling), which is presented in Figure 10.2. The plot does not include industrial uses and emissions. Because the difference in the emissions between the urban sprawl–type communities and the ecocities is dramatic, and the pressure to reduce carbon emissions is immense, population density matters. This also fits with the concepts of smart livable developments and city conversions. It should be noted that the development of Dontang is on hold, but the concepts are sound.

The literature indicates that low-density "American style" suburban areas with one oversized house on 0.4 ha (1 acre) of land are the most wasteful regarding energy use and efficiency (Newman, 2006). The fact that medium-density development is the optimal design refutes, to some degree, the utility of the "low impact" subdivisions, which in most cases have an objective of minimizing stormwater impacts and discharges, and generally result in low-density developments with high energy and automobile uses. Ecocities achieve the same environmental impact—i.e., minimization of pollution by urban stormwater—but these effects are achieved by judicious landscape attenuation, dramatic reduction of vehicular traffic inside the city, and extensive reuse of stormwater. In contrast, LID practices rely mostly on infiltration, ponds and wetlands, reuse is typically minimal, and the mode of transportation is less relevant.

With the exception of shantytowns in developing countries, the high-density developments, even though the carbon impact is less than that for very low-density suburbs, still require a lot of energy, mainly for commercial activities such as office lighting and heating, subways and other public transportation, tourists (hotels), pumping water to high elevations, long-distance transfer of water from far away water resources, pumping and treatment of wastewater, and other energy uses that would be less in medium-density, smaller developments. Medium-density, smaller developments (fewer than 0.5 million people) are also more amenable to low-energy transportation, such as electric light rail, buses, and water taxis, and to installation of the renewable energy sources. Installation of renewable energy would be inefficient on per area basis in high-density developments.

X.3 INTEGRATED RESOURCES MANAGEMENT (IRM)

X.3.1 Sustainability

In Chapter II the triple bottom line – life cycle assessment (TBL-LCA) was introduced as a measure to evaluate the sustainability of the development or retrofit of the urban area. Sustainable development based on the TBL and LCA triple bottom line assessment and evaluation will balance, in an intergenerational context, **social equity**, which recognizes the needs of all members of the society (e.g., public health, reduction of GHG emissions, recreation and leisure time); **economic development**, ensuring economic growth and employment, better life for society today and in the future; and **protection and enhancement of the quality of the environment**. But we have also pointed out that the TBL components are not equal, and that if the current and near-past generations have damaged the environment or distorted social equity for the sake of development, a new triple bottom line relationship must be achieved to restore the balance for future generations.

Traditional linear development design models almost always result in landscape disintegration and the loss of ecosystem function. While the TBL method is intended to ensure that the economic outcomes of development do not ignore either social equity or ecological concerns, we have seen few TBL developments that truly take a "valuation-based" approach to creating a regenerative, adaptive outcome. Thus, traditional TBL design practice produces an economically viable outcome that is sufficiently profitable to provide adequate funds for environmental and social enhancement. The traditional design approach is, thus, a linear process that can be characterized by various trade-offs, which tends to oversimplify causal factors (Hegg, 2009). In contrast, regenerative adaptive design (RAD) involves a multidirectional and systems thinking approach in order to identify creative solutions to a problem or process. Simply put, the conventional design process views nature as a liability, whereas the RAD design process views the environment as an asset. Related to this approach is the fact that the RAD design process recognizes the importance of preserving, enhancing, and maintaining healthy/functional ecosystems as a part of development design, since these systems are characteristically resilient and resistant to external shocks. The RAD design process also implicitly recognizes the value of the services that are provided by functioning ecosystems.

Chapter II outlined the land use planning for the IRM/RAD urban communities and the steps to achieve the sustainability goals of the healthy and functioning urban ecology. Another rationale for developing an IRM/RAD-based development design approach is that such a method should contribute to addressing three global challenges:

- The increasing real cost of carbon-based energy and need to reduce carbon emissions
- The potential or real water shortages
- The effects of a changing climate on ecosystems

Although ecosystem valuation methods are imperfect, the valuation exercise can be extremely beneficial as a means by which ecological costs and values can be adequately represented in decision-making processes. Historically, in the context of urban development, where the building of homes and cities is *"not unity with nature, but conquest . . . and that is built upon economics"* (McHarg, 1969), ecosystem valuation can help decision makers to evaluate the long-term consequences of ecosystem degradation/loss as a result of economic objectives. Thus, ecosystem valuation is a means to help achieve decision-making goals regarding ecosystems, but it is not an end in itself. Since the historical design process focuses on buildings and the economic/social components of development, nature used to be (and sometimes still is) treated as a valueless commodity. The result is urban sprawl that has long-term economic, social, and environmental costs. However, these traditional developments are now being viewed as economic, social, and environmental risks, as communities worldwide now face an unprecedented changing climate. In the extreme situation after the economic recession of the real estate market in 2008–2009, many urban sprawl–type communities in the U.S. ended in bankruptcy, and many properties were abandoned by the owners and foreclosed on by the lenders. One of the central issues to be addressed is the need for a new integrated design model that incorporates TBL principles and addresses the need for city-scale sustainable water infrastructure within the context of the water-energy nexus.

The approach advanced in this book is based upon an alternative design principle: An integrated design should be based upon maximizing economic, ecological, and social equity value *and restoring ecosystem function*. Therefore, the design process is an iterative one that models the value created when a design incorporates ecological regeneration, is adaptive, enhances social capital, and is demonstrably profitable—. The development is profitable *because* the ecological and social elements have been enhanced and are inherently more resilient to change. In addition, the development of infrastructure is based upon a number of fundamental principles termed "integrated resource management" (IRM). The core concepts of IRM are (Corps et al., 2008; O'Riordan et al., 2008):

- There are no wastes, only resources.
- Optimization of resource value requires an integration of water and energy (the water-energy nexus), in addition to ecological and social resilience.

Daigger (2009), Novotny (2008) and others agree that the current "linear" approach to urban water management—sometimes called the *take, make, waste* approach in the sustainability literature—when applied more broadly to natural resources use and global climatic change, has become increasingly unsustainable. The most obvious effects discussed in this book are growing water shortages caused by population increase, pollution, and overuse of resources throughout the world. Concerns are also growing about the ecologic status of the water bodies impacted by urban development, resource consumption, and the dispersion of nutrients resulting in severe algal blooms.

Daigger (2009) has suggested the triple bottom line goals to achieve sustainable urban water and resource management listed in Table 10.2. The objectives listed in Table 10.2 must be met, but future urban water and resource management systems must also address the broader goals listed in Table 10.3. Utilities providing these services must be viewed as providing sufficient value for system users to be willing to adequately fund them to achieve long-term fiscal sustainability. Broader environmental goals related to water quantity and resource utilization must be met, and approaches to providing services to all must be developed and implemented. Accomplishing these goals will require transformations in approaches to urban water and resource management. Monsma et al. (2009) have outlined the key principles, as follows:

- The traditional definition of water infrastructure must evolve to embrace a broader, more holistic definition of sustainable water infrastructure that includes both traditional man-made water and wastewater infrastructure and natural watershed systems.

Table 10.2   Triple bottom line urban water and resource management sustainability goals (from Daigger, 2009)

Sustainability Area	Goal
Economic	Financially stable utilities with the ability to maintain the infrastructure
	Valorization of the resources in used water and considering the value of resource and energy recovery in economic analyses of the system
Environmental	Locally sustainable water supply (recharge exceeds net withdrawal)
	Energy neutral (or positive if possible), with minimal chemical consumption
	Responsible nutrient management that minimizes dispersal to the aquatic environment
	Providing ecological flow to restored and/or preserved urban water bodies
Social	Provide access to clean water and appropriate sanitation for all
	Minimize and adapt to adverse impacts of global climatic changes

Table 10.3   Elements of sustainable urban water management systems (From Monsma et al., 2009)

Transparency	Security & emergency	Advanced procurement	Network
Public outreach &	preparedness	& project delivery	optimization
stakeholder	Conservation & water	methods	Regulatory
involvement	efficiency	Modernized plant	optimization
Good governance	Environmental	operations	Workforce
Full cost pricing	stewardship	Management of	management
Allocation of cost	Energy management	environmental	Affordability
of development	Climate change	impacts	Research and
Asset management	mitigation &	Watershed & regional	technological
	adaptation	optimization	and
			managerial
			innovation

- This definition of sustainable water infrastructure should be embraced by all public and private entities involved in water management, and these same entities have a shared role in ensuring that their decisions consider and integrate a set of criteria to include environmental, economic, and social considerations (the sustainable path).

- A watershed-based management approach is required for drinking water, wastewater, and stormwater services to ensure integrated, sustainable management of water resources.

Monsma et al. further identified the elements of sustainable urban water management systems, as listed in Table 10.3. These objectives, goals, key principles, and elements form the basis for developing sustainable urban water and resource management systems.

Switching from concepts described by the terms "waste", "refuse" and "wastewater" to those characterized by "resource recovery" or "used water reclamation" cannot be done under the linear system scenario, even when the name of the utility is changed from wastewater treatment to water reclamation. In the prevailing current linear system, water is taken from upstream sources, delivered to the urban area by underground conduits, used and polluted, then delivered by underground conduits to a regional wastewater treatment facility many kilometers downstream from the points of potential reuse, and finally the receiving water body is overwhelmed by the effluent discharge, often creating an effluent dominated water body. Traditional simple economic cost analysis for water systems, based on economy of scale dogma, was leading planners to build large regional facilities and (in the 1970s, after the passage of the Clean Water Act in the U.S. and equivalent legislature elsewhere) to abandon smaller community-based treatment plants that were deemed uneconomical and inefficient.

Daigger (2009) proposed a hybrid system in which water/stormwater/used water management can be accomplished by means of a system that incorporates a remnant of a linear system but is mainly a distributed system, which was described, in an expanded version, in Table 2.4 in Chapter II. This hybrid system concept can be further expanded into home (building) scale, cluster or ecoblock scale, and regional scale measures and into management leading to sustainability. This creates a portfolio or a toolkit of measures that planners and developers can adopt, adapt, and expand on a site-specific basis. Malmqvist et al. (2006) published guidelines on how to develop strategies towards sustainable cities.

X.4 CLUSTERS AND ECOBLOCKS—DISTRIBUTED SYSTEMS

X.4.1 The Need to Decentralize Urban Water/Stormwater/Used Water Management

The water centric COTFs need water for various uses, including for the healthy ecology of the urban waters, in- and on-water recreation, water supply, often navigation, and so forth. As documented in the preceding chapters, in many communities functioning surface water bodies have disappeared either by being converted into sewers and buried or by having insufficient flow because of excessive upstream withdrawals.

In Chapter II, Sections II.4 and II.5, we have provided an introduction to the concept of clusters, ecoblocks, and ecocities. The design and configuration of resource recovery were developed in Chapters VI to VIII. The integrated resource management cluster (IRMC) was defined as a semiautonomous water management/drainage unit that receives water; implements water conservation inside the structural components of the cluster and throughout the cluster; reclaims sewage for reuse, such as flushing, irrigation and providing ecological flow to restored, existing, or daylighted streams; recovers heat energy from used water; and possibly recovers biogas from organic solids. (The term "ecoblock" was used for the same purpose, but the preposition "eco" may create some degree of ambiguity because it could refer to either economic management or management focusing on ecology.) Without recovery, many components, such as organic carbon and nutrients, would become pollution. Standard traditional treatment creates great energy demand and also emits GHGs. In a distributed cluster system, multiple potable and used water reclamation facilities are located throughout the regional service area. Such facilities should be located within the cluster so that perishable reclaimed resources (e.g., heat) can be effectively used, and the cost of transmitting the reclaimed resource (water) is not excessive. The IRMC treatment and recovery unit can be incorporated into the neighborhood as shown in Figure 10.3.

However, it has also become clear that not all functions of resource recovery may be suitable for recovery in clusters. The integrated resource recovery facility (IRRF) proposed in Chapter VIII for recovery of biogas, hydrogen, heat, electricity, and organic solids could more efficiently serve several clusters insteaad of being installed in each cluster, and it could be situated on the premises of current municipal water reclamation plants but not far downstream from the population centers. The preferable site should be near the center of the community, to minimize the cost of transmission (e.g., the Tianjin Sino-Singapore city project in China – see Chapter XI).

Figure 10.3 Dockside Green, Victoria, British Columbia, on-site tertiary wastewater treatment, which provides disinfected reclaimed water for use throughout the site, including toilet flushing, irrigation, and top-up water for the central waterway. (Courtesy W.P. Lucey, Aqua-Tex Scientific Consulting, Ltd.).

This facility would accept liquid sludge and organic solids and/or concentrated black water. Gray water could then be reclaimed and reused in the IRMC (cluster) recovery facilities

A cluster may include:

- A large high-rise building (or buildings) identified as a community (e.g., Battery Park in New York)
- A subdivision
- Several city blocks forming a small watershed
- A small community, or a community within a large metropolitan area
- A large recreational area (e.g., large hotels and seaside communities)

An IRMC facility includes a **water reclamation plant (WRP)** and **energy recovery units (ERU)**, which could be installed in most clusters at the **points of reuse**. A point of reuse is a reclaimed water outlet, such as one used for toilet flushing in buildings, street washing, landscape irrigation, groundwater recharge, and providing ecological base flow for urban streams (see Chapters VII and VIII). To minimize the energy losses, it is necessary that the cluster water and energy reclamation units be located in or near the cluster they serve. WRPs and ERUs can be located underground in commercial shopping areas or in basements of large commercial buildings (Figure 10.3).

Interconnectivity. Although the clusters are semiautonomous in water, sewage, and energy recovery management, they should be interconnected to increase resiliency against the failure of a cluster operating system, namely its WRP. In the case of failure, there should be an option to store the untreated used water and send it to the nearest cluster plant that has available capacity or to the regional IRRF. Consequently an **online real-time optimization and control cyber infrastructure** will have to be developed.

Storage. IRM requires storage of captured and reclaimed water, solids, biogas, produced nutrient fertilizers, and other products. In a sense an IRRF is more a clean industrial plant than a waste-processing and disposal facility.

Storage of captured and reclaimed water can occur:

- On the surface in ponds, wetlands, and restored water bodies
- In shallow aquifers or subsurface zones
- In man-made subsurface basins (in rainwater collection cisterns, in the basements of buildings, under plazas and parking lots, under parks and other multiuse open areas)
- In deeper aquifers for indirect potable reuse

The storage sites need to be identified and storage capacity calculated by dynamic hydrologic and water quality models. Biogas and, eventually, hydrogen can be stored in surface or subsurface tanks similar to those of natural gas. The storage sites need to be indentified and storage capacity calculated by dynamic hydrologic, water qyality, and energy models.

X.4.2 Distribution of Resource Recovery, Reclamation and Management Tasks

Water centric developments recognize the ecological value of surface water resources first. In this approach the ecological integrity of the water resources and riparian and flood zones is preserved or restored, using integrated resource management, which also considers the impact on GHG emissions. It has been pointed out that integrated resources management of water centric cities will consider:

1. Water conservation (green development)
2. Distributed stormwater management using best management practices of rainwater harvesting, infiltration and storage of reclaimed water and excess flows, and surface drainage
3. Distributed used water treatment, generating water for reuse in buildings, landscape irrigation, and ecological flow of existing or restored water bodies

4. Using landscape and landscape components (ponds, wetlands, grass filters, etc.) for attenuation of diffuse pollution and post-treatment of effluents recovered for reuse
5. Heat and energy recovery
6. Nutrient recovery
7. Biogas and hydrogen production
8. Electricity production
9. Developing other renewable energy sources such as growing biomass in algal ponds for carbon sequestering and additional biogas and oganic solids production.

This involves many tasks and activities that will be performed locally, in the clusters, and regionally (centrally).

Decentralized building or city block management

- Stormwater management and rainwater harvesting (Chapter VI)
  - Permeable pavements
  - Green roofs
  - Rain gardens
  - Bioretention
  - Storage of rainwater in cisterns for irrigation and in-house use (toilets) and discharging excess water into receiving waters, recharging shallow aquifers, or sending it to cluster water management system
- Water conservation (Chapter V)
  - EcoStar-certified plumbing (shower heads, faucets)
  - Xeriscape
  - Additional local sources of water (basement and foundation dewatering sump pumps), recharging shallow aquifers with excess pumpage from basements and foundation, and sending excess to cluster water management system
- Local (household) energy recovery (Chapter VIII)
  - Use of photovoltaic (electricity) and concentrated solar energy panels (hot water) and potential storage of energy by converting it to hydrogen or selling the excess electricity to the grid
  - Heat generation by air-to-air or water/ground-to-air or ground/water-to-water heat pumps
  - Heat recovery from gray water (preheating hot water)
- Source separation
  - Black and gray water separation and sending to cluster or regional resource recovery facility
  - Urine (yellow water) separation and sending to cluster or regional IRRF

Cluster (ecoblock) management

- Water reclamation and reuse (Chapter VII)
  - Cluster (ecoblock) used water treatment and reclamation
  - Biomass production from black water either by subsurface flow wetlands or algal ponds. Biomass transferred to the regional IRRF for biogas and electricity production and solids recovery.
  - Raw sludge separation and organic waste preparation and transfer to regional IRRF
  - Tapping into sewers for treatment and on-site reuse, including ecologic flow
  - Gray water treatment and recycle
  - Surface and/or underground storage of reclaimed water and recycle
- Stormwater management
  - Pervious pavements, road, and street bioswales
  - Surface drainage (conversion from underground to surface) operation and maintenance
- Management of ecological flow and quality in receiving water bodies
  - Ponds and wetlands
  - Ecotones
- Energy recovery and production
  - Cluster wind and solar energy harvesting
  - Heat recovery from used water
  - Electricity production by microbial fuel cells

Regional (utility) management

- Water supply (grid) management
  - Development and management of freshwater sources, including harvested and stored rainwater and stormwater and pumped groundwater
  - Desalination of water from saline sources (brackish groundwater and seawater)
- Integrated resource recovery facility
  - Management of deliveries of concentrated used water and sludge from the decentralized (cluster) water reclamation units
  - Management of collection and deliveries of solid organic solids (food waste, vegetation residues, wood shredding) and leachate from landfill for digestion and energy recovery
  - Large-scale biogas and energy production from organic solids, sludge, landfill leachate, and concentrated used water
  - Nutrient recovery (struvite production)
  - Heat recovery and use for heating reactors and buildings

- • Management and digestion of organic solids from sources other than used water and sludge (vegetation residues, food waste, fuel crops)
  - • Algal farm management for sequestering carbon and residual nutrients and additional biomass production
  - • Commercial distribution of products (water, biogas, hydrogen, struvite, agricultural solids) and selling electricity
- • Ecological services
  - • Operation, protection, and management of freshwater reservoirs providing water supply, (limited) recreation, and buffer zones
  - • Stream daylighting, restoration, and ecological management (fish restocking)
  - • Recreation management
  - • Education about sustainability, global climatic change effects, and ecology; tourist promotion and guidance
- • Real-time control (RTC) of the entire system
  - • Online monitoring and near-future forecasting of the system network performance, cluster management coordination
  - • System component malfunction management
  - • Management of the entire system during extreme events
  - • Statistics and optimization modeling of the system, and online reporting
- • Financial management
  - • Collection of fees and distributing revenues for cluster management
  - • Recovering a portion of municipal taxes attributable to desirability of the developments
  - • Grants and subsidies
  - • Selling energy, nutrients, heat, and biosolids

The above listing is a portfolio from which to select the best and optimal management, not necessarily a list of all tasks. The list may expand in the future. It is clear the distributed nature of cluster management will need both regional and cluster (distributed) management. For example, the distributed water reclamation and reuse facilities will have a high degree of automation and online monitoring connected by cyber-infrastructure to a central management unit; hence, real-time control will be crucial. In essence, the regional management and oversight may persist in a modified form, but a computerized center may replace a lot of the current hard infrastructure, such as tunnels, extremely large treatment plants, and resulting overloaded effluent dominated receiving waters.

Decentralized stormwater management has benefits other than water reclamation and energy generation. Most of the best management practices (BMPs) attenuate peak flows and volumes and, hence, reduce flooding. Therefore, the role of regional utility management is also emergency and disaster prevention management, during extreme events on both sides of the spectrum, flooding and drought. This will require in the RTC-mode system wide storage monitoring, continuous forecasting, and management plans for emergencies, system failures, and—in an extreme situation—illegal terrorist actions and sabotage.

By recharging shallow aquifers and/or by releasing water from surface storage, needed base (ecologic) flow is provided to restored urban streams. Rain and stormwater capture also has been historically (see Chapter I), and still is today, a significant source of potable water. For example, the Tianjin ecocity in China will rely heavily on precipitation as a source of potable water for its 500,000 population. The local sources of water are brackish (a source after desalination) and/or polluted (Chapter XI).

X.4.3 Cluster Creation and Size

The size of the cluster is obviously variable. For example, the U.S. Berkeley-designed Qingdao ecocity layout is divided into 3-hectare uniform ecoblocks with 1500 to 1800 inhabitants (see Chapter XI). The ecoblock has both semiautonomous distributed and centralized water/energy reclamation and reuse management. Masdar ecocity in the UAE has centralized management for approximately 50,000 inhabitants and another 40,000 commuters. Hammarby Sjöstad in Sweden, which will have a population of 30,000, is also centralized and closely tied to Stockholm refuse management. Tianjin with its 500,000 population is organized into several ecoblocks of various sizes but most of the water management is centralized.

Many factors will determine the size and configuration of the cluster, including the persistence of the tradition of large centralized utilities. Questions are being asked about what to do with the current heavily centralized infrastructure and whether or not the repair of the current crumbling infrastructure should be the solution. Obviously, the size of the cluster, which is a semiautonomous operation and management unit, will depend on cost and revenues, the morphological features of the area, proximity and inclusion of a water body, cost of transporting water and energy, location of public transportation stations, and other parameters. However, the benefits of the decentralized hybrid and cyclic system over the traditional regional linear system are overwhelming.

Viability of the Cluster Hydrology and Ecology The cluster can be organized around a first- or second-order surface water body, which also determines its surface drainage. Hence, a water centric cluster could be a low-order natural or created watershed. The daylighted and/or restored water body is the centerpiece that will also include the nature areas surrounding it, that is, floodplain parks, nature trails, and preserved forests (See Chapter III). The cluster water management, in addition to homes and commercial establishments, will provide base flow to the water bodies, which, in turn, will provide reclaimed water for some uses within the cluster (e.g., irrigation). Figures 10.4 and 10.5 schematically illustrate the water cluster's water management in respect to the cluster surface water body (a stream or lake with a tributary). In the first figure, rainwater is harvested and storm runoff is conveyed and infiltrated into a shallow groundwater aquifer. In the second figure, the hydrology of the cluster watershed is restored, and a first-order stream is created from base

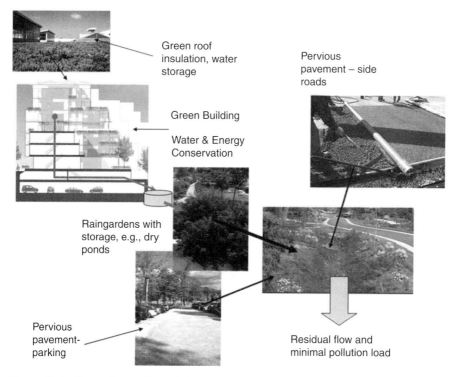

Green roof insulation, water storage

Pervious pavement – side roads

Green Building

Water & Energy Conservation

Raingardens with storage, e.g., dry ponds

Pervious pavement-parking

Residual flow and minimal pollution load

Figure 10.4 Recreating cluster hydrology to a condition approaching the predevelopment status, by means of storage and infiltration of rainfall and stormwater. Other best management practices are available.

flow provided by the discharge from the shallow aquifer and by highly treated effluent from the cluster water reclamation facility. A viable first-order water body can be recreated even in close vicinity to the habitat (Figures 10.6 and 10.7).

Both the Dockside Green development in Victoria, British Columbia (Canada) and Zhiangijawo in the Tianjin region of China, featured in Figures 10.6 and 10.7, are examples of cluster-type developments related to water but not necessarily focusing on fully closed-loop water management. Dockside Green is a 6.5-ha mixed-use development adjacent to the Upper Harbour and downtown Victoria. This site, built on a former brownfield, is now a setting filled with green spaces traversed by waterways and walking trails. It also has a vehicle-sharing program, mini-transit system, boat launch, pedestrian and bike trail, amphitheater, public art, and dock facilities.

Zhiangijawo is a new 28-hectare start-up community located near the city center. The site has a historic irrigation canal, with decentralized surface stormwater management practices that include biofilters and bioswales, channels, and other best management practices that treat water and allow it to infiltrate. This also prevents the intrusion of brackish water from the sea. Because the Tianjin region is relatively dry during winter, gray water from the settlement is treated and provides the necessary

Downstream flow

Figure 10.5   Creating a viable urban first-order water body. Connectivity to wetland and ponds is important for the biota of the water body. The wetland could be a part of the treatment train. The cluster's water reclamation facility must control nutrients (also to be removed in the wetland).

ecological flow, in addition to water stored in the shallow aquifer (Grau, 2009; Dreiseitl and Grau, 2009).

The above two sites are examples of forming a cluster with a population between 1000 and 10,000 that, to some degree, may still rely on the regional system for supplying water, energy, and disposal of used water and trash. Both developments are considering or have already installed renewable energy–producing technology. At this point in time, they are more than halfway to the ecocity designation.

X.4.4   Types of Water/Energy Reclamations and Creation of a Sustainable Urban Area

In Chapter VII we covered several possible water reclamation systems and technologies, and Chapter VIII covered the water-energy nexus. In Chapter VIII we also suggested a possible future IRRF that would be central to the water reclamation, energy and nutrient recovery, and heat-producing and organic soil–conditioning solids production—that is, an income-generating and carbon-sequestering facility (Figure 8.27). It could also be used in a scaled-down version as a cluster resource recovery facility for black water streams. It was also shown that, to reduce the GHG footprint, a maximum water conservation effort is most energy efficient in situations

Figure 10.6 The Dockside Green development in Victoria, British Columbia, during the last phase of stream creation. The original site was a reclaimed brownfield. Note wind turbines and solar panels on the building on the right of the picture. Ccourtesy W. P. Lucey, Aqua-Tex Scientific Consulting, Ltd., Victoria, BC.

where water availability is not an issue. Furthermore, a full closure of the cycle will require more energy to be produced by renewable carbon emission–free sources (wind and solar). Hence, the reclamation and reuse trains of new developments or city renewals can be divided into several potential categories:

I. Water conservation train and portfolio – Hammarby Sjöstad model of linear partial recycle systems (goal water demand as low as 100 L/cap-day)
   - Local and Cluster Level
     Water supply from the grid and potentially other local sources (local well)
     Restoration and daylighting of urban streams
     Maximum implementation of water conservation by water efficient water fixtures, no irrigation use and toilet flushing with potable water
     Rainwater harvesting and potential withdrawal of water from local surface water bodies and storage for irrigation and toilet flushing
     Used water separation (black, gray, and, potentially, yellow)
     Treating gray and yellow waters locally and sending the black water to the regional IRRF

Figure 10.7   Recreated stream in Zhiangijawo in the Tianjin region of China. The original stream was a polluted canal. Note the stepping-stones to provide closeness to the water for inhabitants (including children). Courtesy Herbert Dreiseitl (Atelier Dreiseitl, Berlin, Singapore).

Used water treatment by low energy–demanding treatment unit processes (wetlands, trickling filter followed by membrane filtration and disinfection) for partial reuse (public toilets) and ecologic flow

Sludge (from all recovery units) and comminuted food and other solid organic waste conveyance to the regional IRRF

Urban runoff minimization by infiltration (pervious pavements), storage, and reuse for irrigation; conveyance on surface

Heat recovery from used water

Local and cluster renewable energy production

Bike and pedestrian paths

Maintenance and care for ecotones surrounding the restored and existing surface waters

- Regional Level

IRRF and regional renewable energy generation

Public transportation

Planning and real-time control of the entire system, including also emergency preparedness and response

II. Maximum water conservation and reuse – Qingdao and Masdar total cycle and reuse (goal water demand as low as 50 L/cap-day)

- Local and Cluster Level

  Maximum water conservation and minimum water delivery from the grid and/or desalination (50 L/cap-day)

  Rainwater harvesting for supplementing the potable water cycle

  Black, yellow, and gray water separation, treatment, and reuse

  Potable and nonpotable water supply pipe systems

  Urban runoff minimization by infiltration (pervious pavements) and reuse, after storage, for irrigation and ecologic flow

  Xeriscape

  Surface and underground (basin or shallow aquifer) storage of excess rainwater and treated urban runoff for supplementing the potable water cycle, and reclaimed used water for nonpotable uses

  Additional management as outlined for the Category I communities

- Regional Level

  Same as Category I communities

  Selecting the train from the portfolio is site-specific and related to local water reuse and discharge laws and regulations. The above two scenarios are more relevant to the new developments on brownfield or demolished city sections (urban renewals) or developments on low-quality arid nonagricultural lands. Converting forested areas is undesirable and typically leads to LID low-density developments. Agricultural land should remain in agricultural production.

III. Lower-density urban zones – Seattle mode (goal water demand 200 L/cap-day or less)

Converting typical low-density U.S. suburbs rapidly into ecocities with net zero GHG impact is difficult, and a gradual return of population to the restored and retrofitted city or satellite city settlements may take a generation. Such low-density developments may be smaller (fewer than 5000 to 20,000 inhabitants) so that clusterization and cluster management may not be a priority. Furthermore, in older suburban communities, distributed used water treatment by septic tanks and subsurface soil disposal might have been practiced but abandoned or might be on the verge of collapse. Existing low-density suburbs can be converted to LID developments with low or net zero GHG effect by:

- Switching to hybrid and electric plug-in vehicles and carpooling, getting rid of gas guzzlers
- Homestead water and energy conservation similar to that proposed for ecocities
- Encouraging xeriscape and/or irrigation using harvested and stored rainfall; using gray water for irrigation on the home scale is not recommended at this time but may be possible in the future

- Installation of renewable wind turbines throughout the community and/or solar energy panels on each house
- Eliminating curb and gutter drainage with storm sewers and installing LID best management practices (bioswales, rain gardens, infiltration, rainwater harvesting)
- Continuing implementation of stormwater wetlands and ponds in highly impervious zones (freeways, shopping centers, village centers)
- Eliminating old malfunctioning septic systems and developing cluster-distributed used water conveyance and water reclamation facilities
- Restoration and maintenance of nature areas and natural water bodies (there may be plenty of instances where developers damaged the water bodies, including putting them into concrete channels and culverts)
- Encouraging neighborhood shopping instead of suburban malls, building bike paths and pedestrian walkways
- Public transportation
  Chapters III and IV showed a Street Edge Alternative neighborhood in Seattle, and an example of a conversion of a low-density suburban zone into an exemplary LID development.

IV. Historic densely populated cities mode – Singapore, Chicago, Battery Park (in New York City), Malmö (Sweden), and Ghent (Belgium) (goal water demand 100 – 200 L/cap-day)

Conversion of historic cities to sustainability and water/stormwater/used water decentralization may be a challenge. Tradition, sense of land ownership, and the NIMBY (not in my back yard) attitude may have to be addressed by education and presenting examples. Traditional densely populated cities typically have a regional linear used water conveyance system, with a wastewater treatment plant and landfills located at some distance from the city, underground stormwater conveyance in combined and/or separate storm sewer systems plagued by overflows (CSOs and SSOs), most of the original smaller natural streams buried in culverts or converted into sewers, and filled lakes and bays. Traditional remedies include underground tunnels capturing overflows (reducing CSO frequency from more than 50 times in a year to a still unacceptable and beach-closing 10 to 15 overflows in a year), and satellite CSO treating facilities using settling and disinfection. However, a recent article in the *New York Times* (Duhigg, 2009) noted that more than a third of U.S. municipal systems in 2009, thirty-seven years after the passage of the Clean Water Act, have violated the water quality laws. The cities named in the article included New York, San Diego, Houston, Phoenix, Philadelphia, San José and San Francisco. Most of the violations occurred because of CSO. Cities located in water-short areas may have chronic water shortages due to overuse and expansion, and may be implementing desalination and/or transfer of water from increasing distances.

The measures that could retrofit the communities and bring them towards sustainability are:

- Identify and delineate stormwater/used water management clusters (high-rise buildings, commercial clusters, residential communities).
- If the linear system is completed, the first focus should be elimination of CSOs and SSOs by:
  - Implementing pervious pavement on side streets, back alleys, and parking lots
  - Green roofs
  - Infiltration of stormwater
- Identify buried former streams, canals, and filled wetlands and designate those that could be daylighted and made a backbone of the gradually surfaced stormwater drainage, essentially separating storm drainage from combined sewers and eliminating storm sewers (Malmö, Sweden, see Stahre, 2009) (Figure 10.8, Ghent).
- Continue revitalization of cities' waterfronts (Figure 10.9, Singapore River and Marina Bay).
- Consider black and gray water separation and urine collection from public toilets.
- Identify places where satellite water reclamation plants could be located that would connect with the existing sanitary sewers to extract used water for treatment and reuse for irrigation (no nutrient removal) or for ecologic

Figure 10.8 Restored canal in Ghent, Belgium. Daylighting and water quality improvements are changing this historic community of 600,000 people into a Venice type city with canals and rivers attracting tourists (photo V. Novotny)

Figure 10.9   The Singapore River. The former polluted brackish estuary was converted by the Marina Barrage (see Chapter VI, Section VI.3.6) and extensive and thorough urban stormwater cleanup into a catchment freshwater reservoir that also provides water supply. The riverbanks are lined by many restaurants and parks and provide recreation (Photo V. Novotny).

(nutrient removed) base flow of the restored/daylighted streams. The satellite water reclamation treatment plants could also treat the cluster-produced used black and gray water and send separated urine and sludge to the regional IRRF.

- Recover heat from used water.
- In the more futuristic design, the cluster resource recovery unit could also produce renewable electricity; hence, the anaerobic unit process could be the first unit to treat the separated black water.
- Install photovoltaic and concentrated solar energy panels and turbines on the roofs of the buildings and in open spaces.
- Improve public transportation.

V. Conversion of large cities (megalopoli) in developing countries to sustainability (goal: keeping water demand below 100 L/cap-day and low GHG footprint)

Currently these cities may not have adequate sanitation, especially in shantytowns; water use is low, typically less than 50 L/cap-day; and potable water, if available, may not be of acceptable quality because of low pressure in the pipeline systems and water contamination. High energy use of electricity and for transportation is not a problem, and per capita energy use is low (Figure 10.2). Some families living in shantytowns in countries such as India rely on outdoor cooking. Surprisingly, many families in poor neighborhoods in some Latin and South American countries may have a photovoltaic solar panel providing basic electricity for TVs and recharging batteries for light. These communities cannot afford and should not build Western-type linear regional sewerage with expensive water reclamation plants and, as the living standard may improve, they should avoid past mistakes of the Western world.

- The first task is to provide an adequate and healthy water supply. Yamada (2009) pointed out that developing and maintaining a traditional underground pipeline system is very difficult in depressed economies of the third world, and that bottled water in recyclable jars provided by the public utility may be the first solution, combined with standpipes with safe drinking water. The utility will thus raise funds for improving the pipeline distribution system, rather than private water suppliers charging high prices for bottled water of dubious quality. Beijing, China, still has problems with drinking water contamination; consequently, each household and hotel room boils water for personal hygiene and uses bottled water for drinking. In South Africa, each person has a right to 25 liters per day of safe drinking water that is provided free of charge by the government. Citizens pay a fee for excess water use.

- Instead of using communal latrines, the communities may develop periodical (daily or weekly) cluster-wide collection of excreta (a widespread practice in Japan) from households latrines or flushless toilets. The fecal matter would be digested with other organic solids in a community (cluster) biogas plant (Chapter VIII) and produce biogas that could be distributed to families either by a pipeline or in recyclable bottles for home cooking, to replace burning wood or charcoal, which is leading to deforestation in India and African countries. The organic digested solids could be used in community gardening, which could be encouraged by education and NGO outreach.

- Education by NGOs and encouragement by religious leaders should help convince communities to keep urban water bodies clean. Typically, gray water should be collected by surface channels and treated by wetlands before being discharged into the receiving water bodies or reused. In Belo Horizonte, Brazil, the pollution of local rivers is so bad that an entire river is treated by flotation (Figure 10.10). The SWITCH program sponsored by UNESCO is implementing a community-based cleanup of local streams by wetlands. Polishing ponds can be stocked with fish.

Water/stormwater/used water or solids management in developing countries will, in the foreseeable future, rely on low-cost measures and best management practices that, if selected and applied properly, can be very effective. Unfortunately, the governments of most developing countries have poor records in the environmental public health domain, and most efforts are driven by outside and local NGOs and some foreign aid programs (for example, Japan is very active in Southeast Asia).

Figure 10.10   A flotation treatment plant (energy demanding) treating low flows of a local urban stream in Belo Horizonte, Brazil. Photo V. Novotny.

NGOs such as Engineers Without Borders from the U.S. are also actively involved with local populations in developing countries in building local community (cluster-size) water supply systems that will provide safe drinking water. In India and through South and Southeast Asia, religious authorities may be important. A well-known (in India and throughout Asia) Western-educated (PhD from University of California, Berkeley) environmental scientist, philosopher, university professor, and religious leader, Guru Agrawal, in his keynote address to the 2001 IWA Diffuse Pollution Conference in Milwaukee stated (Agrawal and Trivedi, 2001):

What conserved India's, China's and South/ South-East Asia's environment was not government regulations or economic measures, but pro-environment religious and cultural traditions. It is undeniable that conditions today are much different than they were in the past. But I believe the human psyche has not changed and even today religion in developed countries is stronger than government regulations, and cultural traditions are stronger than scientific/ economic rationale. Else why would some 100 million Indians congregate at Allahabad to have a dip in Gangaji during January-February, 2001, some 30 million on a single day, entirely at their own cost, with no economic incentives or subsidies? Or why do 20 million Moslems from all over the world congregate in Mecca each year on a particular day? Should we not exploit cultural traditions for conservation of environment, including abatement and control of pollution?

Conventional Western-type formal or informal education or media-blitz cannot convince or convert the still traditional ethnic communities of developing countries. Taking the support of ethnic cultural traditions can, as shown by the success of efforts in rejuvenating rivulets of arid Alwar through community action. The success of voluntary cultural groups in managing the stream reaches taken up by them, without any government support or financial resources, but exploiting cultural respect for the water-reach also points in the same direction. In contrast government regulations and their enforcement often get corrupted and drawn in to nepotism, bureaucratic egos and vested interests, particularly so in developing countries. Case studies in India show how bureaucratic nepotism can obliterate any chances of meaningful action in developing country-situations, particularly when implementation depends on government or foreign funds. Concluding this argument, I recommend reducing the role of government regulation and maximizing the role of generating community pressures as shown by moving from the present relative stress on government regulations to a dominant and controlling reliance on community pressure for successful abatement and control of pollution in developing countries.

X.5 SYSTEM ANALYSIS AND MODELING OF SUSTAINABLE CITIES

X.5.1 Complexity of the System and Modeling

Designing the new sustainable cities or retrofitting the older communities to become sustainable requires a complex system analysis and virtual computer modeling. A macroscale study in which the impacts of microscale changes, new developments, and retrofits will be assessed is needed and should be a part of every design. The leading consultants in this field and some universities have developed, or are working on development of, complex multifaceted urban models that would hierarchically incorporate the urban systems and would assess their sustainability to various degrees. An *integrated analysis* approach being developed by CDM[2] enables the designer to analyze relationships and interdependencies between the different subsystems to be explored in the decision-making context. The key subsystems are

- Building infrastructure
  - Green buildings
  - High-performance buildings
  - Population density (high-rise, medium, or low height)
  - Carbon emissions
- Transportation
  - Public transportation by electric trains and buses
  - Automobile

---

[2]CDM's Neysadurai Global Technical Centre for Integrated Water Resources and Urban Planning, Singapore

- Walking, biking
- Fossil fuel use and carbon emissions
- Water systems
  - Linear regional vs. closed-loop cluster-wide
  - Surface vs. subsurface drainage
  - Nutrient management
  - Carbon and methane production and emissions
- Energy systems
  - Energy from the grid
  - Locally produced solar and wind energy
  - Energy from used water, organic solids, and combustible trash
  - Energy use in buildings
  - Energy used by water and water infrastructure (water reclamation and reuse plants, pumping)
  - Energy recovery (hydraulic energy)
- Natural systems
  - Interconnectivity (fragmentation)
  - Hydrology and water flow availability or shortages
  - Pollution
  - Ecotones and floodplains
- Goods and solid waste
  - Virtual water and energy consumption
  - Life-cycle benefits and costs

To organize the thinking about quantifying the sustainability performance of different urban designs, the following elements of an overall urban system are categorized as systems, activities, and layers. Figure 10.11 represents the hierarchy of the system.

- *Systems* are the main entities that one wants to model or track in the process. They are characterized by components that can be defined as stocks and flows.
- *Activities* take place within the system. They are either desired results from the system or by-products of its performance. Generally, activities drive the flows in the system, and the levels can be controlled and/or monitored.
- *Layers* sit atop the overall system structure and represent key performance measures that we are interested in analyzing and evaluating, such as the economic, environmental, energy, and physical aspects.
- *Technologies* get applied to the system components. They alter the rates (i.e., flows) and/or change the system structure.

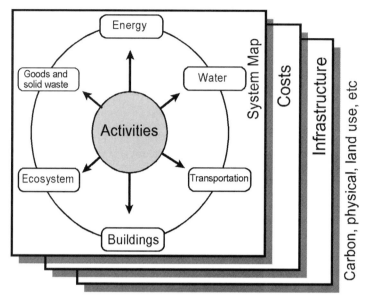

Figure 10.11 The hierarchical representation of urban systems (*Source:* CDM).

Table 10.4 and Figure 10.12 provide examples of systems, activities, and layers. The model by CDM's Neysadurai Global Technical Centre for Integrated Water Resources and Urban Planning in Singapore is to be developed for conditions scalable for site, cluster, or ecocity applications. This model, like several other models, is developed in the GIS (geographic information system) environment. GIS is a system

Table 10.4 Systems, layers, and activities examples

List of Systems	List of Activities	List of Layers
Buildings	Population/occupancy	Costs
Transportation	People movement	Energy and carbon emissions
Energy	Energy consumption	Infrastructure
Water	Water consumption	Physical sites: topography,
Ecosystems, natural systems	Economic, rental revenues	water/rain/sunlight, wind,
Energy	Goods movement	geothermal
Goods and solid waste	Used water and solids disposal	Land use: usage, zones, etc.
	Water reclamation and recycle	
	Energy recovery and renewable energy production	
	Nutrient management	

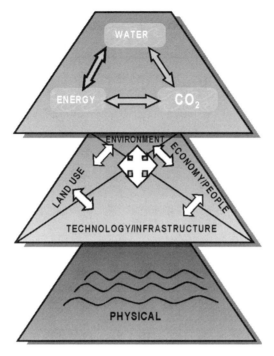

Figure 10.12 Model layers schematics.

of hardware and software used for the storage, retrieval, mapping, and analysis of geographic data. In GIS, descriptive attributes in tabular form are associated with spatial features. Spatial data and associated attributes in the same coordinate system can then be layered together for mapping and analysis. The GIS software is connected to many internal and external models that can analyze and simulate various spatial dynamic phenomena and processes.

Recent advances in the availability of spatial (i.e., GIS) and temporal (i.e., sensor) data are expanding the scope of models that can be developed. For example, in the U.S., LIDAR (Light Detection and Ranging) can provide physical elevation data at vertical accuracy of 15 cm, which can be used to delineate catchments in urban areas (Maidment, 2002) and fed into lumped parameter hydrological models (Hellweger and Maidment, 1999; Bedient et al., 2008). However, except for some coarse-scale efforts (e.g., Mitchell and Diaper, 2006), most urban models do not consider the complete urban water cycle. This is a problem, because the different "layers" of the urban water cycle (drinking water distribution, stormwater and wastewater collection, surface water, groundwater) intersect. On the positive side, all the different layers of the water cycle are relatively well understood, and models exist. The art of estimating sewer flow in well-characterized urban areas is impressive. Of course, groundwater hydrology is not as well understood in the highly heterogeneous urban environment, which contains numerous preferential flow paths. The urban model to be developed could be based on existing GIS technology and approaches (Hellweger and Maidment, 1999; Maidment, 2002) and existing hydrologic models.

In addition to mechanistic models—such as basin hydrology, pollution loadings and impacts on ecosystems, energy production and recovery, treatment plant effluent-influent relationship for various treatment pollutant attenuation, in the flow through the selected treatment and recovery trains, the customized GIS interfaces will also rely on specific available, reputable "expert models" for buildings, water, and transportation systems.

X.5.2 Triple Bottom Line (TBL) Assessment

In the preceding sections, we categorized the water management of future communities as occurring on homestead, cluster, and regional scales. Changing to the sustainability (fifth) paradigm will require urban planners, city officials, developers, and the public to be persuaded, in addition to regulations and, initially, subsidies encouraging such developments, because the benefits will go to the entire society, not just to the people living in the COTFs. The latter incentives will be related to the necessary society-wide goals of reducing GHG emissions (Kyoto Protocol and Copenhagen Agreement) and improving the ecology of urban receiving waters and other natural resources to meet the water quality regulations embedded in law. However, the best argument is the economic viability of these developments and retrofits. Simply stated, urban planners, developers, and the public will get excited about sustainable development (and implement it) if it achieves better profits or cost savings in comparison with traditional developments. Homeowners and commercial establishments will experience more satisfaction with the development, which can then be translated into higher willingness to pay for the units in the development, high values of homes, and, for the community, a higher property tax level. Subsidized housing for lower-income people should always be a part of the development.

The triple bottom line assessment over the life cycle (see also Chapter II) is an extrapolation of the traditional cost/benefits analysis of public and private projects by including quantitative assessment of (1) environmental/ecological protection and enhancement, (2) social equity, and (3) economics. To evaluate resiliency to extreme events, a TBL analysis should consider: (1) flood-causing precipitation, (2) water shortages, and (3) extreme pollution, also related to global warming. The carbon emissions footprint is mostly a societal impact related to preservation of society and its resources.

The costs and benefits are tangible and intangible, and TBL analysis makes an effort to include both. Valorization of intangible benefits can be based, for example, on the results of willingness to pay (WTP) surveys. For example, one survey found that people are willing to pay up to 2% more for a property if it is built according to "green buildings" principles (Lippiant, 2007). Novotny et al. (2007) calculated WTP for restoration and flood control benefits of restored urban streams. The energy and hydrological benefits of green roofs have been estimated by Carter and Rasmussen (2006). The benefits of energy savings and of water reclamation and reuse are tangible and known and can be directly calculated. The following are examples of the tangible benefits of IRM of the COFs:

- Increased value of homes and revenues to the community
- The value of electric energy and heat produced by the IRRF or cluster energy recovery unit and from selling the excess energy to the regional or municipal grid
- Selling biogas and hydrogen to transportation companies
- Savings on fuel
- The economic value of businesses and employment by riverside commercial establishments
- Urban restoration economic effects
- Sales of recovered fertilizers
- Savings on decreased water demand
- Savings on elimination of subsurface storm sewers, and rental fees for the use of excess capacity of existing sewers (e.g., from telephone and cable companies)
- Savings on pumping energy cost for transmitting water. Used water and stormwater as a result of water conservation and local reuse, and harnessing energy from excess pressure in pipelines and reject water conduits
- Boat launching and excursion fees, and fees for recreational and boat habitat uses of restored water bodies (e.g., Ghent, Belgium)
- Fees for solid waste collection
- Fees for reclaimed water (e.g., irrigation of private golf courses and gardens)
- Savings on waste discharge fees and profits from selling "cap and trade" energy credits (due to carbon neutrality or net sequestering) in countries that have implemented nationwide payments

It is very difficult to use traditional economic methods for enumerating benefits for projects that involve pollution abatement, carbon and other footprints, watershed management, aesthetic amenities, convenience of transportation, the value of enjoyment, and satisfaction with living in a "green community," recreational fishing, and so forth. One obvious reason is the fact that a great part of the benefits is in the category of intangibles. The second reason is the fact that, regarding the benefits of proximity to healthy water and nature, the funds collected from the beneficiaries (homeowners, commercial establishments, public agencies) will be in the form of fees and taxes paid by the citizen groups that may include both direct beneficiaries (e.g., riparian owners of properties surrounding the impaired water body) and citizens who use the water body for enjoyment only occasionally, i.e., payment wil be made by the entire community which could be a city, an entire utility, or a state. A program of recovering the benefits via public funds can be successful only if the citizens involved are *willing to pay* for the benefits. Otherwise it will be an unfunded mandate by the government.

The benefits of daylighting and restoration of urban streams are also very high but often intangible, and information is still scarce. The daylighting of a river in downtown Seoul, Korea (see Chapter IX) has been estimated as bringing revitalization and tourism benefits many times higher (about U.S. $100 billion overt the life cycle of 50 years) than the cost of bringing the historic Cheonggyecheon River back to the surface (about U.S. $180 million) (Lee, 2004).

The lifeline costs and benefits are typically estimated for urban developments over a period of 50 years, of which 10 years could be gradual linear implementation, and the remaining 40 years are then a period of full accumulation of the operation, maintenance, and repair (OMR) costs and benefits. The tangible cost and benefits are converted to their present values, which involve discounting of future benefits and costs to their present value. Classic cost/benefit analysis is covered in most textbooks on engineering economy.

Barriers. Clune and Braden (2007) present a model for financing "green development" focusing on stormwater projects. "Green development" generally encompasses many different strategies employed to reduce the environmental impacts of urbanization. Successful green development may integrate multiple, complementary components designed not only to moderate stormwater flows, but also to reduce energy use and transportation requirements while incorporating natural aesthetics. However, many of the costs of poor stormwater management take the form of environmental externalities, whereby one community imposes the costs of its environmental activities on other, perhaps distant, groups of people. There is substantial evidence showing that a large share of the costs of environmentally unsound stormwater development is borne locally by owners, neighbors, and their immediate municipal organizations. In fact, the financial and other motivating benefits of green development are well known at the municipal level, and many communities already try in various ways to encourage on-site management of stormwater. Clune and Braden also outlined and categorized the barriers to implementation. As the market for development based on green design principles expands, the nature and severity of the barriers to adoption will change. The barriers intertwine. A useful way to think about them is in terms of the ways they interact to influence the supply of green development, buyer preferences, and market prices.

As Clune and Braden observed, regulatory barriers are important to recognize and address. Inflexible zoning and building codes protect municipalities and their employees while deterring innovations that might reduce overall costs to the community. Technical certification of on-site technologies could help to alleviate the liability concerns. The adoption of performance-based standards in lieu of design standards adds flexibility, but reliable technical guidance for developers and regulators is an essential ingredient, and communities must commit to ongoing monitoring of system performance. The development of model zoning and building codes may help communities to adapt their current regulations. Finally, development fees and stormwater management fees that reward on-site management measures engage the economic interest of both developers and property owners in the cause of stormwater management following principles of green development.

The range of pre-existing incentives for better green development and COF practices makes the inquiry into barriers all the more important: not only to discover why present incentives have been slow to induce solutions, but also to formulate better solutions. In the presence of positive incentives, policy should focus on removing the impediments that prevent market actors from employing green design strategies.

Willingness to Pay The following discussion on payments and realization of benefits is taken from Novotny (2003). The willingness to pay is not the same as the damage (cost) related to pollution nor the value of monetary benefits of water quality improvement or stream restoration, but it is related to them. Using flood control as an example, if a project benefit includes improvement in flood control, the tangible benefit would be a reduction of the flood damages caused by urbanization. In a typical urban watershed, barring some catastrophic and rare floods, such as Hurricane Katrina, actual flood damages are incurred only by a small number of homeowners because most properties are located outside the floodplain. Obviously, this hypothesis does not apply to all urban areas, especially those in developing countries where floodplain development is common. An extreme example can be found in Bangladesh, where a large area of this populous country is a floodplain. In a great majority of cases of urban flooding, the project cost for flood control alone far exceeds the monetary (tangible) benefits of reducing flood damage for citizens living in the floodplain. Typical "flood damages" such as sewage backups into basements represent a different category of damages that may be unrelated to flooding of urban streams. Yet, a great portion of the population living outside the floodplain would be willing to contribute to a project that would reduce flooding.

The same is true for fish and wildlife. The tangible benefits of improving the fishing or recreation opportunities of impaired water bodies are relatively small and, standing alone and without regulations, would not warrant the cost of a project that would restore the integrity of the water body. Yet, many citizens of urban communities feel attached to their water bodies, use them for recreation, picnicking, and fishing, and would be willing to pay a certain sum of money to preserve and restore their integrity (Novotny et al., 2000; 2001). The book by Dreiseitl and Grau (2009) has numerous examples documenting peoples' attachment to and enjoyment of water and waterscapes. In the final outcome, the funds citizens living in the watershed would be willing to pay for restoring the integrity of the water body and its corridor and for flood control may be more than sufficient to carry out the project. However, the way in which the funds are collected may also be important. Generally, raising general tax revenues is the least popular form of generating funds for the public projects of water body restoration and flood control. User fees (e.g., stormwater utility fees) clearly earmarked for these purposes and scaled according to the potential benefit or damage (e.g., fees related to the amount of imperviousness of the property, charging riparian owners of a lake a higher fee, imposing boat-launching fees) are preferable and more equitable. There may be less willingness to pay for flood control alone than for multi-objective watershed management that would also include water body integrity restoration and preservation (Novotny et al., 2000).

The third example of intangible benefits is the extra price people would be willing to pay for living in a sustainable "green" community. From the "Woodlands" development in Texas (www.thewoodlands.com), which dates back almost 40 years, to the present ecocities, people have consistently been willing to pay more for the opportunity to live in such developments. There are also far more opportunities for other economic gains (boat rides, restaurants, business location, recreation) in water centric green cities.

Willingness to pay for something valuable but intangible is found in surveys of the population (stakeholders) who, for example, reside in the watershed or in the region of interest. In the surveys reported by Novotny et al. (2000, 2001), contingent valuation was used. "Contingent valuation" (CV) is one of the methods used to find out the value of nonmarket commodities such as ecological enhancement, water quality improvement, improvement of aesthetic assets, and flood control benefits for the population residing outside the floodplain. Sometimes, these intangible benefits are derived for "nonuse values," that is, what is the willingness to pay (WTP) for an improvement of a resource that is not used because of poor quality, or, alternatively, how much compensation would a person be willing to accept (WTA) to allow deterioration of a resource from a point of use to nonuse—or how much more to pay for "green" renewable power, in comparison to power produced by GHG-emitting fossil fuel power plants? For living in a "green" building?

The contingent valuation method involves structured surveys of population/ stakeholders affected by the plan, or, in cases of wider analysis of the benefits, of a regional or state population sample. Typically, a reliable sample involves several hundred to one thousand respondents, with a ratio of number of respondents to the total number of people contacted of about 50 to 70%. The principle is simple: If one wants to find the value of an intangible benefit, just ask the stakeholder about it. From the questions, find out how much the people are willing to pay for the environmental benefit such as improvement or return of fishing, providing contact recreation opportunities, reducing flood damages and inconveniences, green power, improved public transportation—and/or find out what they are willing to accept as compensation to tolerate the cost of nonuse of the resource. The steps of developing the CV method were presented by Hanley, Shogren, and White (1997) as:

- *Setting up a hypothetical market.* In order to value environment in a nonmarket situation, offer bidding vehicles such as property taxes, income taxes, utility fees, user fees, etc.
- *Obtaining bids.* Bids are obtained in a phone survey or by mail. Phone surveys by experienced interviewers provide the most efficient method and have a higher response. In the survey, participants are asked to state their maximum willingness to pay or minimum acceptable compensation for the increase or decrease of environmental quality or other intangible benefits.
- *Estimating mean WTP or WTA and bid curves.*
- *Extrapolating from the surveyed population sample to the total population, and obtaining the benefits.*

The survey is prepared by using scientific communication methodologies, yet the questions must be simple. A survey preparation phase involving small focus groups is needed in advance of the survey. The communication to surveyed citizens and, later, to the public can be based on adaptation of the heuristic-systematic model of information processing, using Ajzen's Theory of Planned Behavior (Ajzen, 1988; Fishbein and Ajzen, 1975). The modified model is described in Griffin, Dunwoody and Neuwirth (1999). A survey example was published by Giese, Griffin, and Clark (2000).

The model proposes seven factors: (1) individual characteristics, (2) perceived risk characteristics, (3) affective responses to the risk, (4) felt social pressures to possess relevant information, (5) information sufficiency, (6) one's personal capacity to learn, and (7) beliefs about the usefulness of information in various channels; this factor will introduce the extent to which a person will seek information in both routine and nonroutine channels and the extent to which he/she will spend time and effort analyzing the risk information critically.

Also, policy makers are interested in stable responses because they will translate into stable support for remedial actions over time. For this reason, two wave surveys, one year apart, could be conducted. The Theory of Planned Behavior (Ajzen, 1988) predicts the variance of WTP and helps to explain cognitive, attitudinal, and social normative reasons for WTP. The answers and cognitive behavior of respondents are investigated by considering numerous factors such as upstream (source) versus downstream (impact) location, living or owning real estate inside or outside the floodplain, other demographic parameters including the standard of living, and, finally, the measured and calculated flood prevention and ecological benefits.

Revenues and payments for services. Current payments and revenues are heavily based on recovering the cost of the capital outlays financed in the public sector by bonds and in private sectors by investments loans and payments for OMR. Public utilities do not add profit, and for private utilities the profits are and should be regulated.

The water/stormwater/used water/solids management in the COF cannot be based on the current model of a regional utility covering only repayment of the interest paid on bonds for capital outlays and OMR cost by collecting fees from users. Under the new paradigm, used water and solids are resource commodities, and the utilities managing water, used water, and collected solids for recycle will be producing and selling valuable products. For example, a precedent of struvite recovery is the production and commercial sales of Milorganite fertilizer from dried and pasteurized sludge by the Milwaukee (Wisconsin) Metropolitan Sewerage District, which has been a commercial success for decades. Similarly, all treated effluent in Tucson, Arizona, is sold for irrigation to private golf courses and municipal parks. These are notable examples but also exceptions in the current management paradigm. The utilities of a City of the Future will be selling fresh and reclaimed new water, electricity to the municipal or national grid, fertilizer to agricultural operators or citizens for community gardening, biogas or hydrogen to the municipal fleet of buses and other vehicles, and heat to the citizens and businesses; they will receive fees for accepting organic solids for biogas production, or leachate from landfills. It is quite likely these revenues could pay, over the long run, for a substantial percentage or all of the cost of the water/stormwater/used water development and operation and, ideally, make a profit.

Some of these services can be done by private enterprises, and all will create jobs. The utility can also rent the unused capacity of existing sewers to communication and cable companies as is being done in the Tokyo–Yokohama (Japan) megalopolis.

The purpose of collecting payments based on the stakeholders' willingness to pay for the benefits of green development, better public transportation, reduced or net zero carbon footprint, water conservation, improved water quality and flood control, and the cost of the projects (whichever is smaller) is also to provide funds, including subsidies, to developers and ecocity management agencies for reducing carbon footprint, installing renewable energy in the clusters and, carrying out the water body restoration by the cities and public utilities. However, the cities will be responsible for collecting fees for abatement of pollution generated by the industries city. Note that water pollution emissions from sustainable COTF will not be an issue; due to water conservation and reuse, pollution emissions from sustainable cities will be minimal and certainly will meet the water and air quality standards. Public transportation users pay fees for the use; however, in the U.S. and elsewhere throughout the developed world, public transportation is heavily subsidized, but often the subsidies pay plenty for freeways and far less for electric trains (a malady of the U.S. public transportation policy). With cap and trade regulations in the U.S. and with regulations already implemented by countries that signed the Kyoto Protocol, there will be a strong incentive to reduce GHG emissions through fees paid by the emitters. However, subsidies to homeowners and communities for installing solar and wind power–producing devices and insulating homes will most likely persist into the foreseeable future.

It is very likely that in the process of building ecocities and converting current communities into sustainable cities, the financial and valorized intangible benefits will be of such magnitude that current barriers will be broken making the new paradigm a financial success.

Under the *cap and trade regulations*, municipalities, utilities, and homeowners will be rewarded by "selling" their GHG emissions quota and other pollution discharge allotments to those exceeding their allotment (see Novotny, 2003). This approach, called "bubble trading," has worked in the U.S. quite well in reducing the acid rain–forming substances and other pollution from the stack effluents of power plants and industries. Under this scenario, a regulating authority determines the maximum emissions from a region and, in the first go-by, distributes the allotment of the maximum emissions among the large dischargers, i.e., municipalities and industries. It would also be fair if agricultural conglomerates were included, because they may discharge methane. The large emitters can then negotiate among themselves, and those for which the cost of control would be high can negotiate with and buy the allotment from those that would be willing to reduce more than the amount of their allotment. The fact that the new sustainable cities could be net zero GHG emitters could bring income to the municipality, providing additional subsidies to homeowners and commercial establishments in the city.

In the United States, air emissions trading began in the 1970s, followed in the 1980s by trades involving lead gasoline and chlorofluorocarbons, and in the 1990s by acid rain emissions trading. The success of the acid rain emission trading programs will lead to the expansion of environmental trading programs.

Today, several environmental trading programs exist in the United States. Two models for trading dominate: so-called "cap and trade" programs and "mitigation" (sometimes called "offset") programs. The cap and trade model is based on pollution limits (caps) that are established through a regulatory permitting program by the state pollution control agencies. Permit holders are allowed to discharge only a certain volume of a pollutant and must offset any discharge above the cap by purchasing "credits." Credits are typically purchased from someone who is discharging less than their permit allows and thus has unused pollution "credits." In this system, while a certain volume of pollution is "free of charge" for all polluters (just as it is in traditional permitting programs), pollution above the cap has a cost, and "unused pollution" below the cap has value (Bowditch, 2007). In Europe and other countries any pollution discharged into the environment is subjected to a fee which is collected by the state agencies and is used for subsidies for or even assuming the full cost of installing pollution controls.

X.6 INSTITUTIONS

The long and complex institutional history of water does not easily accommodate the transition to fully integrated, holistic governance and resource management approaches. Because the individual components of our present-day water infrastructure (i.e., potable water, wastewater, stormwater, flood protection, and navigation) each evolved in response to related but separate needs and drivers, each has been supported by long-established institutional and legal frameworks that traditionally had little reason to fully coordinate activities or collaborate in accomplishing complex, multibenefit system objectives.

Furthermore, while there has been a relatively universal approach to the engineering and science of water technologies as they pertain to levees, reservoirs, pipelines, pumping facilities, treatment plants, distribution and collection systems, and disposal facilities, there is a wide and diverse range of legal and institutional frameworks involving both public and private sector actors that have developed to plan, design, finance, construct, manage, and maintain those facilities.

The significance of water resources as a fundamental organizing force in human history is well documented. Contemporary theories regarding the history of civilization have identified the development of large-scale irrigation systems as a primary driver in the formation of state-level authorities and governance. The emergence of organized empires from small independent city-states has been attributed by some to the ability of a strong centralized government to provide large-scale irrigation systems and to control watersheds for economic advantage. The civilizations of the Nile, Tigris and Euphrates, Indus, and Hwang-ho (Yellow River) valleys are all examples of the economic benefits created by central control over the irrigation systems of entire watersheds.

While the institutions and governance surrounding the agricultural and economic uses of surface water reach back to the emergence of advanced civilization itself, the institutional history of public health and sanitation appears to emerge in response to the new challenges created by the increased population density which led to increase the water supply. In this context, the health and sanitation aspects of water management have always seemed to lag behind the development and exploitation of water resources.

All of these examples illustrate the long and deep relationship that exists between institutions charged with the development and control of urban water infrastructure and the fundamental function and duty of government to provide for the well-being of citizens and communities. Consequently, the process of integrating and reorienting these institutions towards new objectives and purposes requires extensive effort and collaboration. The shift towards a new paradigm for urban water management represents a far more complex institutional transformation than will be required to transform the technology needed to make it happen.

Historically, the legal and institutional mechanisms governing water and its uses can be organized under three broad headings:

1. Those that emerged in response to economic drivers like the need for agricultural irrigation, basic water supply, goods movement and navigation, power generation, and flood protection
2. Those that developed in response to the human health and sanitation issues that accompanied growing industrial activity, increased population size and density
3. Those that were created to protect the natural environment from the deleterious effects of the prior two

In this study (Chapter VIII), we have emphasized the water-energy nexus, and electricity will be one of the major "products" generated by the integrated resource recovery facilities. Therefore it is paramount that the new water utilities closely collaborate with power grid utilities and establish a mechanism for renewable energy production on their premises and selling the excess to the grid.

X.6.1 Institutions for Integrated Resource Management

Integrated resources planning was initially developed in the electrical power industry in the late 1980s to ensure the implementation of conservation and demand-side management in planning for future power supply development. By the early 1990s the water industry incorporated this approach for the same reasons—ensuring that conservation, improved water use efficiency, and the development of local resources were coordinated with the expansion of large-scale water supply development, particularly in the Southwestern United States.

Federal or Central Government Role In the United States, the federal government's role in water resources management began in 1866, when Congress directed the U.S. Army Corps of Engineers to remove obstructions to navigation on the upper Mississippi and upper Missouri Rivers.

In general, the first category of institutions is **regulators**. These institutions are already in place in most countries; for example the U.S. Environmental Protection Agency executes federal policies, and state pollution control agencies (departments of environmental protection, natural resources, etc.) are responsible for intrastate pollution control. The ministries of environment (e.g., the Ministry of Environment and Water Resources in Singapore) in most countries have the same regulatory role. These institutions carry out the legislative policy mandates; specify standards and criteria; provide oversight, arbitration, and research; provide and distribute grants when federal and state interests in water quality remediation are involved, and perform other activities. They may designate uses of the water bodies, specify the standards, and execute water body abatement plans such as the Total Maximum Daily Load (TMDL) and Use Attainability Analysis (UAA) in the U.S. They may provide oversight of water management and pollution abatement agencies. These agencies should not manage water resources or pollution abatement. Financing for these agencies should be derived from state or federal sources and should not use funds (taxes) obtained from cities, dischargers, and/or users.

However, in the U.S., despite the Clean Water Act's established goal "to restore and maintain the chemical, physical, and biological integrity of the Nation's waters," its primary regulatory authority is focused on the discharge of pollutants (U.S. EPA, 2002). The Act provides no regulatory mechanism to support, for example, adequate base flows, or to prevent excessive stormwater peak flows, although addressing these issues is critical to achieving the Act's goals. Hopefully, the situation will change in the next 10 years with the new amendments to the Act. Problems with the present regulatory system were covered by Bowditch (2007).

Additional central government institutions are departments or ministries that are responsible for energy, transportation, housing, and (suburban) agriculture.

Large Municipal- and Watershed-Scale Utilities   The regional utility and large watershed organizations should have authority to raise revenues and implement management, which would be more efficient than fragmented local and state organizations that are lacking financial resources and rely mostly on the government. Many regional wastewater utilities that could potentially assume the role of the central management agencies may have the authority to regulate and control wastewater discharges and collect fees, but do not have the authority to conduct in-stream water quality management (wetland restoration, low flow augmentation) or to collect user fees from the beneficiaries of improved water quality. Present municipal sewerage utility agencies belong to this category. Singapore's Public Utility Board (PUB) is an effective city/state agency that has broad power. Similar regional agencies in the U.S., such as the Metropolitan Water Reclamation District of Greater Chicago or the Milwaukee Metropolitan Sewerage District, can also assess and collect fees for pollution abatement and stream/watershed restoration, but not for water supply. The Massachusetts Water Resources Authority has broader powers to provide water supply, collect and treat sewage, implement CSO controls, and conduct stream restoration (e.g., rehabilitation of the Muddy River in the Emerald Necklace in Boston–Brookline, Massachusetts).

As an alternative to municipal sewerage districts, which in most cases would be responsible for urban point and diffuse pollution abatement and which utilize taxes for financing, cities and larger urban areas have recently created separate stormwater utilities financed by user fees. Users are people and organizations that generate used water and stormwater runoff discharged into the utility drainage system. Ideally, water supply, point and diffuse source pollution, water reclamation, and energy production in the COTFs should not be fragmented among several organizations. Agricultural drainage or irrigation districts or cooperatives may constitute equivalent agencies in the agricultural sector and will be important when suburban sustainable agriculture is a part of the overall system.

The hybrid system of the COTFs may include a central utility/management agency, and cluster and watershed management units. A watershed may include several clusters. A regional watershed water management agency (the Watershed Management Unit) could be responsible for issuing discharge permits. Under the Clean Water Act regulations in the U.S., this would require that the regional agency itself file an application with a state regulating agency for a comprehensive regional effluent and stormwater discharge permit. The waste load allocation in the first round of permit negotiations could be distributed equitably (which in most cases would be uniformly), but not necessarily most economically, among dischargers. In the second phase, the agency could referee the negotiations among the polluters on the "most economical" distribution of the waste loads. The dischargers may choose to accept the load as it is and not participate in the permit trade-off process. This pollution discharge permit trading could lead to a more efficient and more economical (less costly) solution. The same principle is applied to the "bubble" approach for GHG emissions.

Discharge permit trading (the bubble approach to pollution discharge and GHG emissions permitting) can only be accomplished if a regional watershed-wide authority is a "referee" in the trading process, and the same authority will then make sure the final emission loads agreed upon by the polluters will not exceed the assimilative capacity of the environment. Subsequently, in a continuous process, the agency, in cooperation with the state pollution control authority, should have the authority to reissue the permits based on the assimilative capacity (TMDL) of the receiving water bodies for effluent discharges or on the size of the "bubble" for GHG emissions.

The central utility agency could be responsible for:

- Providing water and operating the regional water supply
- Planning and financing the water supply/stormwater/used water and solids facilities
- Construction of the facilities
- Operating and managing a systemwide water supply grid that provides water to the clusters
- Fiscal issues such as collecting fees from users and selling the products (i.e., electricity to the regional power grid; water, fertilizers,and solids to agricultural

enterprises and homeowners/gardeners) through an established commercial distribution system

- Applying for grants and subsidies from the central government and, in developing countries, from international funds (World Bank, European Community, Asian Development Bank, United Nations Environmental Programme, etc.)
- Operating the central integrated resource recovery facility (IRRF) and systemwide interceptors connecting clusters with the IRRF; for a fee, the IRRF may also accept leachate and biogas from landfills, septage from on-site septic systems or excreta from flushless toilets, organic solids (commercial food waste, vegetation residues), and produce renewable electricity on the IRRF premises
- Developing cyber-infrastructure and collecting data on the performance of the entire system, including surface and subsurface water quality and flows
- Developing and operating real-time control (RTC) of the system
- Providing central data collection and statistics
- Systemwide planning and financing of stream restoration and daylighting
- Developing an emergency system to handle systemwide and cluster emergencies, including population movement during extreme events such as flooding, systemwide optimization of used water management during overloads or failures in the clusters, and providing water during failure of the water supply system caused by toxic spills or sabotage
- Coordinating with other central agencies regarding public transportation, landscape development, and zoning

The role of the central utility will shift from the current function of collection and management of the total flows and treatment of stormwater and wastewater (using the old term implies that wastewater is not reused) by heavy infrastructure such as large interceptors, overflows, tunnels, and central wastewater treatment facilities to systemwide planning, management, real-time control, and commercialization of the resource products.

The **management unit of the integrated resource management cluster** may be responsible for:

- Cluster management of water supply
- Tapping into sanitary sewers to withdraw flows of used water for cluster water reclamation and reuse
- Used water separation and/or sending black water and sludge to the central IRRF
- Used water reclamation and reuse supervision and control; regular maintenance
- On-site biogas and electricity production
- Recovering heat from used water
- Distributed stormwater BMPs installation and management, reuse of stormwater

- Renewable electricity production in the cluster
- Providing ecological flow to receiving water bodies and reclaimed water for irrigation
- Education of homeowners and other stakeholders and serving as their liaison with the central utility agency
- Liaison with watershed management associations
- Coordinating with local solid waste management companies (municipal or private) regarding separation and recycle of solid wastes.

Solid waste management, reuse and disposal. Although the focus of this work is on water and energy management, the last item of responsibility, solid waste management, cannot be overlooked. COTF management puts an increased emphasis on recycle and reuse; hence, a system of cooperation with municipal public or private solid waste collection and disposal agencies must be developed, as well as a new sustainable reuse system to virtually eliminate waste from landfills. COTFs are cities that reduce the landfill solid waste input by 99% or more. In Hammarby Sjöstad, in Sweden, an underground solid waste disposal system that separates solid waste into categories has been developed on a source, block, and citywide scale (Figure 10.13).

Figure 10.13 Solid waste disposal site in Hammarby Sjöstad in Stockholm, Sweden. Each disposal chute is connected to the underground system for handling different types of solid waste. Photo courtesy Malena Karlsson, GlashusEtt, Stockhom, Sweden.

Generally:

Combustibles such as furniture and wood are incinerated to produce electricity or could be converted to biogas (Syngas).

Newspapers, catalogues, glass, tires, metals, etc. are sent for recycling.

Food waste and vegetation residuals, including shredded wood, should be accepted by the central IRRF to be digested with sludge to produce biogas and electricity.

Electrical waste (light bulbs, batteries, fluorescent lights, old TVs and computers) are sent for recycling.

Construction waste (asphalt, concrete) can be recycled.

Paint, solvents, and other toxic waste are handled at the hazardous waste collection sites and either chemically reused or incinerated at very high temperatures, converting them into CO_2. In COTFs, consideration should be given to sequestering the carbon.

Stockholm, where Hammarby Sjöstad is located, uses a sophisticated automated waste disposal system that conveys the source-based waste into underground tanks, separated by category, from which the waste is emptied by vacuum into large collection vehicles and delivered for processing.

A watershed management agency will be established if the basin area of the central water body includes several clusters, or if the water body is a distinct natural asset (which it should be). If the entire first-or second order water body (stream or lake) is contained within the cluster, the cluster unit could also manage the water body. Watershed agencies may evolve from successful river and lake management associations. Today, many watershed and lake associations are far more comprehensive than a typical NGO. In Europe, China, Japan, Australia, and elsewhere, these associations are government agencies responsible for comprehensive planning, and financing of all river-connected or discharging entities, including multipurpose reservoirs. In Germany the river basin water management agencies are given authority over wastewater treatment, stormwater disposal, and water quality management as well (see Chapter IX, also Novotny, 2003). However, these activities have been separated from the economic activities that cause the pollution. In the U.S., watershed management is fragmented between the federal agencies (U.S. Army Corps of Engineers, U.S. Bureau of Reclamation) and the watershed agencies in some states such as Florida's watershed management districts. In some other states watershed associations (e.g., the Charles River Watershed Association in Massachusetts) are powerful NGOs that assume some of the functions of a watershed management agency, such as planning, monitoring, and public outreach. Hence, the activities performed by watershed management agencies will vary from country to country and state to state, but generally will include:

- Local (watershed) water quality and stream corridor or lake riparian zone planning
- Water quality monitoring and data collection

- Oversight of the central and cluster water management agencies and generating proposals for improvements
- Advocacy and lobbying regarding water supply and water resources on behalf of citizens in the watershed; representing, sponsoring and coordinating citizens' groups that have vested interests in good environmental quality
- Maintenance of streams and stream corridors and riparian zones of lakes, including pedestrian and bike paths, camping and picnicking sites, beautification
- Collecting fees for the use of water bodies, such as boat launching or swimming
- Applying for and receiving grants for stakeholder and community projects and education

X.6.2   Enhanced Private Sector

The private sector has always played a role in the development of the public water infrastructure. A report by the Committee on Privatization of Water Services in the United States describes the history of private water service ownership and the current structure of the industry.

Historically, water services were initially delivered by private providers in many cities in the United States, such as Boston, New York, and Philadelphia. As these and other larger U.S. cities grew, water services became a core function of local government. This trend accelerated largely because of a legislative change after World War I, when Congress exempted interest payments on municipal bonds from federal income tax. This ensured that municipalities could issue bonds at lower interest rates than those for taxable bonds.

The U.S. water industry today is highly diversified. As of 1999, there were nearly 54,000 community water systems in the U.S. (the U.S. Environmental Protection Agency defines community water systems as systems serving more than 25 people, regardless of ownership). The vast majority of these systems serve small populations. In fact, 85% of U.S. community water systems serve only 10% of the population.

Investor-owned water supply utilities (i.e., "private utilities") accounted for about 14% of total water revenues and for about 11% of total water system assets in the U.S. in 1995. Investor ownership of wastewater utilities is more limited than investor ownership of water supply utilities, in part because of extensive federal funding of wastewater treatment plants, which began after World War II. Investor-owned water supply and wastewater utilities are subject to state economic regulation that oversees rates charged, evaluates infrastructure investments, and controls profits. In contrast, private contract arrangements under public ownership are not subject to this regulation. The private sector has favored public–private relationships that are not subject to state economic regulation. According to the National Association of Water Companies (NAWC), the proportion of water services in the U.S. provided by private water companies, whether measured by customers served or volume of water handled, has remained close to 15% since World War II (Committee on Privatization of Water Services in the United States, 2002).

X.6.3 Achieving Multibenefit System Objectives

The planning, design, and implementation of sustainable urban water infrastructure have led to the active involvement of many community stakeholders throughout the process. Since the triple bottom line approach stresses that environmental, social, and economic benefits be part of every project, the need for input and direction from the potential beneficiaries and stakeholders throughout the community is critical. Obtaining the full involvement of a broad cross section of community stakeholders in the analysis and evaluation of alternatives and their active participation in decision making is a necessary but frequently underutilized approach.

In order to achieve more sustainable solutions, a necessary step is the introduction of decision-making processes that are fundamentally more democratic in terms of setting goals and objectives, and evaluating the trade-offs necessary to arrive at multibenefit solutions. The stakeholders in the process can be divided into two groups: (1) internal governmental and agency interests that are already responsible for the management of urban water infrastructure and their counterparts in other utility functions (e.g., transportation, parks and recreation, and education), and (2) the community of interests representing citizens and corporations that support and receive services from utility and infrastructure agencies.

The SWITCH program (Howe and Van der Steen, 2008) developed a multiobjective approach to planning and developing the COTF systems (Figure 10.14). They emphasized the differences between traditional systems engineering projects and projects tackling global problems like COTF. The main points that distinguish the multiobjective approach are:

- There are multiple clients and multiple potential solutions.
- Time frames are long term with no real ability to see if the expected outcomes are realized.
- The project must take account of interactions and solutions across multiple scales.
- Projects must optimize the component parts as well as the "bigger picture."

In addition, TBL sustainability assessments must balance the economical, social, and ecological goals. In the COTF candidate cities, "learning alliances" platforms have been developed as the vehicle to transition research into action through local engagement in the decision-making process. These alliances build on existing formal and informal networks and are designed to optimize relationships, break down horizontal and vertical barriers to learning, and share a common desire to address an underlying problem. They will also be willing to share or develop common approaches—visions, strategies, and tools—for achieving these goals (Hove and Van der Steen, 2008). Citywide learning alliance platforms have been developed in Birmingham, UK; Hamburg, Germany; Lodz, Poland; Zaragoza, Spain; Alexandria, Egypt; Beijing and Chongqing, China; Belo Horizonte, Brazil; Accra, Ghana; Tel Aviv, Israel; and Lima, Peru.

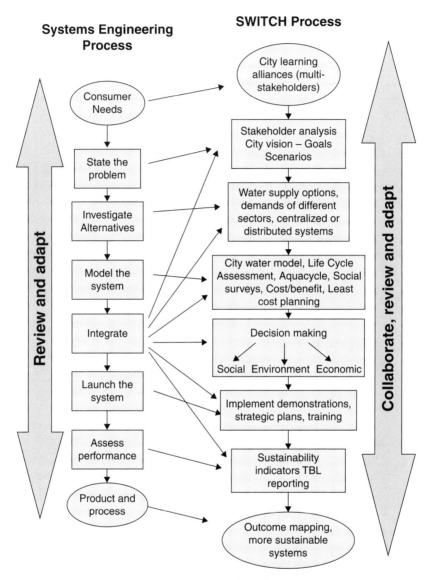

Figure 10.14 Comparison between traditional and SWITCH sustainable city planning concepts (From Howe and Van der Steen, 2008).

Learning alliances are a relatively new concept. These groups of individuals or organizations have a shared interest in innovations and are a method to speed up the process of identification, adaptation, and uptake of innovations. These platforms bring together a wide range of public, private, and NGO partners with capabilities and mandates in implementation, regulation, policy, legislation, research, learning, activism, documentation, and dissemination. This approach fits in well with the literature on sustainable development, which stresses that new ways of working are needed because the stakes and/or uncertainty are high. Hove and Van der Steen highlighted

> *the contribution of all the stakeholders . . . is not merely a matter of broader democratic participation. . . . These new problems are in many ways different from those of research science, professional practice or industrial development. . . . Quality depends on open dialogue between all those affected. This we call an "extended peer community" consisting not merely of persons with some form . . . of institutional accreditation, but rather of all those with a desire to participate in the resolution of the issue (Funtowicz et al., 1999).*

In this process the engineer or scientist acts as a type of "facilitator" in deliberative, participatory processes, relinquishing the idea of control, resolving trade-offs, helping to set directions and visions for the future. The expert must take a longer-term, broader-scale, systems perspective, to engage a range of stakeholders in consultative decision processes, and to practice personal quality assurance through self-critical and reflective practice. This paradigm moves the technical expert away from propounding the right answer and toward offering a range of technological options. It calls on specialists to withhold judgment, to be explicit about their own values and perspectives—which often form the unexamined foundation to professional engineering practice—and to accept that a wide range of stakeholder perspectives and priorities will determine the eventual decision.

REFERENCES

Agrawal, G. D. and R. C. Trivedi (2001) "Management of Diffuse Water-pollution in Developing Countries - Goals and Policies," Keynote address, IWA Conference on Diffuse Pollution and Watershed Management, Milwaukee, WI

Ajzen, I. (1988) *Attitudes, Personality and Behavior.* Open Univeristy Press, Milton Keynes, UK

Bedient, P. B., W. C. Huber, and B. E. Vieux (2008) *Hydrology and Floodplain Analysis*, 4th ed., Prentice Hall, Upper Saddle River, New Jersey

Bowditch, K. (2007) "Restoring the Charles River watershed using flow trading," in *Cities of the Future: Towards Integrated Sustainable Water and Landscape Management* (V. Novotny and P. Brown, eds.), IWA Publishing, London, UK

Carter, T. and T. Rasmussen (2006) "Hydrologic Behavior of Vegetated Roofs," *Journal of the American Water Resources Association* 42 (5):1261–1274

Clune, W. H. and J. B. Braden (2007) "Financial, economic, and institutional barriers to 'green' urban development: the case of stormwater," in *Cities of the Future: Towards Integrated Sustainable Water and Landscape Management* (V. Novotny and P. Brown, eds.), IWA Publishing, London, UK

Committee on Privatization of Water Services in the United States (2002) *Privatization of Water Services in the United States: Assessment of Issues and Experiences*, National Research Council, National Academies Press, Washington, DC

Corps C., S. Salter, W. P. Lucey, and J. O'Riordan (2008) *Resources from Waste: Integrated Resource Management*, Phase I Study Report, February 29, 2008, prepared for British Columbia Ministry of Community Services, Victoria, British Columbia

Daigger, G. (2009) "Evolving Urban Water and Residuals Management Paradigms: Water Reclamation and Reuse, Decentralization, Resource Recovery," *Water Environment Research* 81 (8):809–823

Dreiseitl, H. and D. Grau, eds. (2009) *Recent Waterscapes: Planning, Building and Designing with Water*, 3rd ed., Birkhäuser Verlag, Basel, Boston

Duhigg, C. (2009) "As sewer fill, waste poisons waterways," New York *Times*, November 22, 2009

Fishbein, M. and I. Ajzen (1975) *Belief, Attitude, Intention, and Behavior: An Introduction to Theory and Research*, Addison-Wesley, Reading, MA

Funtowicz, S. O., J. Martinez-Alier, G. Munda, and J. R. Ravetz (1999) "Information Tools for Environmental Policy Under Conditions of Complexity," *Environmental Issues*, Series No. 9, European Environment Agency, Copenhagen

Giese, J. K., R. J. Griffin, and D. Clark (2000) *Survey of Attitudes and Willingness to Pay for Flood Control and Water Body Restoration*, Technical Report No. 5, Project sponsored by the U.S. EPA, USDA, NSF STAR Watershed program, Institute for Urban Environmental Risk Management, Marquette University, Milwaukee, WI

Grau, D. (2009) "New water culture for Tianjin," in *Recent Waterscapes: Planning, Building and Designing with Water* (H. Dreiseitl and D. Grau, eds.), Birkhäuser Verlag, Basel, Boston

Griffin, R. J., S. Dunwoody, and K. Neuwirth (1999) "Proposed model of the relationship of risk information seeking and processing to the development of preventive behaviors," *Environmental Research*, Section A 80, pp. S230–S245

Hanley, N., J. F. Shogren, and B. White (1997) Environmental Economics: In Theory and Practice, Oxford University Press, UK

Hegg, D. (2009) "Valuing ecological systems and services and community design - implications for the private market and local governments," master of science thesis, University of Victoria, Victoria, BC

Hellweger, F. L. and D. R. Maidment (1999) "Definition and Connection of Hydrologic Elements Using Geographic Data," *Journal of Hydrologic Engineering* 4 (1):10–18

Hoeksttra, A. Y. and A. K Chapagain (2997) "Water footprints of nations: Water use by people as a function of their consumption patterns," *Water Resources Management* 21, pp. 35–48

Howe, C. and P. van der Steen (2008) *SWITCH – A Systems Approach to Urban Water Management*, INCOSE, UNESCO-IHE Institute for Water Education, Delft, The Netherlands

Lee, T. S. (2004) "Buried treasure: Cheonggyecheon restoration project," *Civil Engineering* 74 (1):31–41

Lippiant, B. C. (2007) *BEES 4.0: Building for Environmental and Economic Sustainability, Technical Manual and User Guide*, NISTIR 7423, National Institute of Standards and Technology, U.S. Department of Commerce, Washington, DC

Maidment, D. R. (2002) *Arc Hydro: GIS for Water Resources*, ESRI Press, Redlands, California

Malmqvist, P-A., G. Heinicke, E. Kärrman, T. A. Stenström, and G. Svensson (2006) *Strategic Planning of Sustainable Urban Water Management*, IWA Publishing, London, UK

McGray, D. (2007) "Pop-up cities: China builds a bright green metropolis," *Wired Magazine*, April 24, 2007, http://www.wired.com/wired/archive/15.05/feat_popup.html, accessed November 2009

McHarg, I. L. (1969) *Design With Nature*, Natural History Press, Garden City, NY

Mitchell, V. G., C. Diaper (2006) "Simulating the urban water and contaminant cycle," *Environmental Modelling & Software* 21, pp. 129–134

Monsma, D., R. Nelson, and R. Bolger (2009) *Sustainable water systems: Step one – redefining the nation's infrastructure challenge*, A Report of the Aspen Institute's Dialogue on Sustainable Water Infrastructure in the U.S., Aspen Institute, Washington, DC

Newman, P. (2006) "The environmental impact of cities," *Environment and Urbanization* 18 (2):275–295

Novotny, E. V., A. R. Sander, O. Mohseni, and H. G. Stefan (2009) "Chloride ion transport and mass balance in a metropolitan area using road salt," *Water Resources Research* 45 (12), doi:10.1029/2009WR008141

Novotny, V. (2003) WATER QUALITY: Diffuse Pollution and Watershed Management, John Wiley & Sons, Hoboken, NJ

Novotny, V. (2008) Sustainable urban water management, In *Water & Urban Development Paradigms* (J. Feyen, K. Shannon, and M. Neville, eds.), pp. 19–31, CRC Press, Boca Raton, FL

Novotny, V., D. Clark, R. Griffin, and D. Booth (2000) "Risk based urban watershed management under conflicting objectives," *Water Science and Technology* 43 (5):69–78

Novotny, V., D. Clark, R. Griffin, and A. Bartosova (2001) "Balancing flood control and ecological preservation/restoration of urban watersheds," in *Source Control Measures for Stormwater Runoff* (J. Marsalek, W. E. Watt, F. Sieker, and E. Zemann, eds.), Kluver-Academic Publishing, Dordrecht, Netherlands

Novotny, V. et al. (2007) *Developing Risk Propagation Model for Estimating Ecological Responses of Streams to Anthropogenic Watershed Stresses and Stream Modifications*, Final Report, STAR R 83-0885-010, Submitted to U.S. Environmental Protection Agency, Washington, DC, www.coe.neu.edu/environment

Novotny, V. and E. V. Novotny (2009) "Water centric ecocities – Towards macroscale assessment of sustainability," *Water Practice and Technology* 4 (4)

O'Riordan, J., W. P. Lucey, C. L. Barraclough, and C. G. Corps (September, 2008) "Resources from waste: An integrated approach to managing municipal water and waste systems," *Industrial Biotechnology* 4 (3):238–245

Rees, W. E. (1992) "Ecological footprints and appropriated carrying capacity: What urban economics leaves out," *Environment and Urbanization* 4 (2):121–130

Stahre, P. (2008) *Blue green fingerprints in the City of Malmö, Sweden – Malmö's way towards sustainable urban drainage*, VASYO, Malmö, Sweden

Urban Agent (2008) "Sustainable City Race, Part 3: Dongtan," Agents of Urbanism, April 16, 2008, http://agentsofurbanism.com/2008/04/16/sustainable-city-race-part-3-dongtan, accessed March 2009

U.S. Environmental Protection Agency (2002) Federal Water Pollution Control Act (33 U.S.C. 1251 *Et Seq.*), As Amended Through P.l. 107-303, http://www.epa.gov/r5water/cwa.htm

Wikipedia (2009) *Smart growth*, http://en.wikipedia.org/wiki/Smart_growth, accessed November 2009

Yamada, K. (2009) "A new strategy for water supply system in local city or town of developing country," in *Proceedings 2009 International Conference on Sustainable Water Infrastructure for Cities and Villages of the Future,* Beijing University of Civil Engineering and Architecture, Beijing, November 6–9, 2009

XI

ECOCITIES: EVALUATION AND SYNTHESIS[1]

XI.1 INTRODUCTION

This chapter presents case studies of seven ecocities already designed and in various stages of development. The concepts of the cities were elaborated in detail in the preceding chapters. New developments claiming to be sustainable and/or ecocities are sprouting up in several countries. In Chapter III we presented four developments: (1) Westergasfabriek Park in Amsterdam, (2) Staten Island Blue Belt in New York, (3) Augustenborg Neighborhood in Malmö, Sweden, and (4) Western Harbor, also in Malmö. Throughout the book we have described several components of the water/stormwater/used water systems of Dongtan, Tianjin and Qingdao in China, Masdar in the UAE, and Dockside Green in British Columbia, Canada. In China, another large (300,000+) new cities are being developed based on water conservation/ecology concepts, in CaoFeiDian on the Pacific coast, about 200 km from Beijing (Ma, 2009), Harbin, Shenyang, Beijing, Chengdu, cluster of cities in the Pearl River Delta, and others to come in the near future in Portugal and the UK.

In 1999, the Swedish Parliament adopted sixteen objectives relating to the quality of Sweden's environment, and most of them are to be achieved by the year 2020 (Environmental Objectives Council, 2006). The aim is to pass on to the next generation a society in which all major environmental problems have been solved. Among these sixteen goals, three are particularly relevant for the urban wastewater sector: Zero Eutrophication, A Non-Toxic Environment, and A Good Built Environment. Other goals of concern for the urban wastewater sector are Reduced Climate Impact, Natural Acidification Only, and Good-Quality Groundwater (Malmqvist and Heinicke, 2007).

[1] This chapter is coauthored by Eric V. Novotny with a contribution from W.P. Lucey and C. Baraclough and CH2M-Hill Masdar team.

539

Zero Eutrophication   The main goal is: *"Nutrient levels in soil and water must not be such that they adversely affect human health, the conditions for biological diversity or the possibility of varied use of land and water."* Discharges of phosphorus and nitrogen to the receiving water are a main task for all Swedish wastewater treatment plants (WWTPs). Most of them are designed for phosphorus and nitrogen removal (nitrogen removal is required only if more than 10,000 persons are connected to the water reclamation (treatment) plant.).

A Non-Toxic Environment   The main goal is: *"The environment must be free from manmade or extracted compounds and metals that represent a threat to human health or biological diversity."* Of concern to the wastewater sector is the discharge of heavy metals and organic hazardous substances to the receiving water and to soil in case the sludge is used in agriculture.

A Good Built Environment   The main goal is: *"Cities, towns and other built-up areas must provide a good, healthy living environment and contribute to a good regional and global environment. Natural and cultural assets must be protected and developed. Buildings and amenities must be located and designed in accordance with sound environmental principles and in such a way as to promote sustainable management of land, water and other resources."*

Among the short-term goals are:

- "To promote more efficient energy use, use of renewable energy resources and development of production plants for district heating, solar energy, biofuels and wind power."
- "The quantity of household waste landfilled, excluding mining waste, will be reduced by at least 50% by 2005 compared with 1994, at the same time as the total quantity of waste generated does not increase." The first part of the goal has been achieved, the second not (Environmental Objectives Council, 2006).
- "By 2010 at least 35% of food waste from households, restaurants, caterers and retail premises will be recovered by means of biological treatment. This target relates to food waste separated at source for both home composting and centralized treatment."
- "By 2015 at least 60% of the contents of phosphorus in wastewater will be brought back to productive land, of which at least half will be brought back to agricultural soils."

China is very serious about bringing ecocity concepts into urban planning and development of the new urban areas, and so are British Columbia (Canada), the United Kingdom, Sweden, Germany, the Netherlands, and several cities in the U.S. (for example, San Francisco, Chicago, Philadelphia). It has been pointed out that China must build, in the next 25–30 years, urban settlements for about 300 million people. In 2003, the Chinese government proposed building "a conservation-oriented and environmentally-friendly society." The Chinese Ministry of Environmental

Protection issued "Guidelines on building ecological provinces, ecological cities and ecological country" in May 2003 (revised in 2007). The response has been an upsurge in building sustainable cities. However, how to develop actual plans has not been clearly defined (Ma, 2009). Therefore, some renowned ecocity projects in China are planned, developed, and built with foreign partners; for example, Tianjin is a Sino-Singapore joint venture, Dongtan was done with British cooperation, and CaoFeiDian is being done in cooperation with Sweden. Several U.S. universities are also involved in planning and developing Chinese sustainable cities (e.g., University of California, Berkeley; Massachusetts Institute of Technology; and Harvard University).

There are many commonalities to characterize the sustainable (eco) cities that have been updated from the original 20-year-old definition given by University of California, Berkeley, professor Richard Register (1987), and quoted in Chapter II. The following descriptions of the eight developments, plus the abovementioned additional developments in the Netherlands, Sweden, and British Columbia, Canada, all have these concepts in common:

- Sustainable new (eco) cities are built on lands that might not be suitable for development using traditional criteria. The land is either a brownfield or an arid land. One of the COTF requirements is not converting prime agriculture land or forest into an urban area.
- They are not like Low Impact Development (LID) cities common in the U.S. Typically, they are medium-density multiple-family mixed-use developments which; however, incorporate drainage concepts common to LID (e.g., pervious pavements and rain gardens).
- They are very frugal with water use.
- They use less energy; some developments have a net zero carbon footprint.
- They recycle and eliminate most or all refuse from landfills.
- Vehicular traffic within the city is restricted and/or discouraged; people in the cities walk or use bikes and have convenient public transportation.

This book has also extensively covered Singapore as the world hub of the newest developments and implementation of the sustainable infrastructure for water reclamation and reuse. Singapore is approaching sustainability, but in spite of its progressive and groundbreaking water management, it is not yet an ecocity. The city-state is in the final phase of heavy water/stormwater/used water infrastructure investments for water reuse and recycle; it is now focusing on softer approaches such as water conservation that includes labeling of appliances based on their water-saving capability (virtual water), collecting and treating urban runoff, stream restoration, and implementing best management practices throughout the watersheds. The per capita water use, according to the statistics of the Ministry of Environment and Water Resources, is about 150 liters/capita-day (approaching the ecocity level), and the programs of reducing water use and reuse continue. Singapore has an excellent public transportation system and, because of its small size, the citizens do not drive cars excessively. The city is rich with beautiful parks, and one can walk everywhere; hence, the majority of One Planet Living (OPL) criteria would be satisfied. However, the water system relies heavily on massive infrastructure and energy-demanding used water conveyance (in a tunnel), pumping, and conventional activated sludge treatment and recycle relying on reverse osmosis without fully compensating energy use by production of renewable solar and wind energy. Because of its tropical climatic conditions, use of electricity is high (about 8000 kWh/capita-year) and almost all electricity is produced by fossil fuel–powered plants (oil). Singapore is capable of developing renewable energy sources at a relatively fast pace.

Such a system, relying on heavy infrastructure, will be difficult to reproduce in most large cities in the world, and it may not be suitable for those located in developing countries. Nevertheless, technology and concepts developed in Singapore are highly applicable to the Cities of the Future, and the city-state is now in the final phases of becoming an ecocity. The last lap could be run in a relatively short time.

The following sections provide more focused and detailed information on eight developments that have the elements of ecological development. The case studies of seven cities were first analyzed and reported by Novotny and Novotny (2009). The description of the Dockside Green development in British Columbia was provided W. P. Lucey and C. Barraclough of Aqua-Tex Scientific Consulting Ltd.

XI.2 CASE STUDIES

XI.2.1 Hammarby Sjöstad, Sweden

Hammarby Sjöstad means "a city surrounding Hammarby Lake." The city was conceived in the early 1990s and was planned originally as part of Stockholm's bid for the 2004 Summer Olympic Games (which were awarded to Athens, Greece). About 200 ha (480 acres) of old industrial and port brownfields were converted into a modern, sustainable neighborhood. The development has a strong emphasis on water, ecology, and environmental sustainability (Figure 11.1). Once the city is fully built, in 2015, there will be 11,000 residential units for more than 25,000 people. A total of 35,000 people is estimated to live and work in the area. The following information is taken and adapted from information in published materials about the city (GlashusETT, 2007).

Located on Lake Hammarby Sjö, the waterside environment shaped the project's infrastructure, planning, and building design into a modern mixed-use urban space. The scheme has attracted international acclaim for the quality of habitat it created, and convinced many that carbon-neutral development does not require lifestyle changes. The development is linked with Stockholm's center, including adoption of the contemporary inner city street dimensions, block lengths, building heights, and density. Its mix of uses provides a quality neighborhood.

The use of glass as a core material maximizes sunlight and views of the water and green spaces. The scheme successfully connects the historic landscape with aquatic areas. which act as stormwater drainage and encourage biodiversity, creation of new

Figure 11.1   Hammarby Sjöstad with a central canal (Courtesy Malena Karlsson, GlashusEtt).

habitats, informal amenity areas, and formal areas of public open space (Figure 11.2). Sustainability is also enhanced through the use of green roofs, solar panels, and eco-friendly construction products. The city has a fully integrated underground sanitary (separated) waste collection system conveying wastewater to the local district treatment and heat recovery plant.

The development has its own ecosystem, known as the Hammarby Model, which also includes a wastewater plant. The "GlashusEtt", Hammarby Sjöstad's environmental information center, disseminates knowledge to residents and visitors via study trips, exhibitions, and demonstrations of new environmental technologies. Public reports by the center (GlashusEtt, 2007, 2009) provide most of the information.

In the early 1880s the area was a popular park, providing enjoyment of nature for the inhabitants of Stockholm. However, in the late 1800s a large bay, a part of the original area, was filled for a planned port. The natural elements of the area were also partially destroyed during the construction of a highway transecting the area. The port was never built, and the original and reclaimed (filled) area was made available for storage depots and industries. However, until 1998 most of the buildings were temporary shantytown structures (Vestbro, 2005).

The industrial operation left the soil contaminated, and the site became a brownfield. Hence, the first step in preparing the land for the Hammarby Sjöstad ecocity development was the monitoring and decontamination or removal of highly contaminated soil.

The project planners worked with various companies to change the brownfield area into a livable, sustainable habitat to make sure all aspects of clean energy were considered and included. Great emphasis was placed on the importance of

Figure 11.2   Residential area with a surface stormwater channel (Courtesy Malena Karlsson, GlashusEtt).

collaboration and synergistic thinking between these agencies, each of which had responsibility for a different segment of the system.

Hammarby Sjöstad is a full-scale, living proof that use of clean energy and energy-saving solutions does not have to increase project costs. Table 11.1 presents the basic characteristics and parameters of the city.

Environmental Goals   The overall environmental goal of the development is to preserve the existing natural areas as much as possible and create new parks and green areas within the city.

- The city will have at least 15 m^2 of green courtyard and 25 to 30 m^2 of open courtyard and park space available to each inhabitant of the city. Park area should be available within 300 meters of every apartment building.
- At least 15% of each courtyard should be sunlit for 4–5 hours on sunny days during vernal and fall equinoxes.
- Development of the green public areas shall be compensated by creating biotopes benefiting the biological diversity in the immediate area.
- Natural areas shall be protected from development.

Table 11.1 **Characterstics and parameters of Hammarby Sjöstad**

Location	Stockholm, Sweden
Area of development	200 ha (480 acres)
Population served	25,000–35,000 when fully developed in 2015
Population density	133 inhabitants/ha (56/acre)
Project team	
Key partners	Exploateringskontoret Stockholm Stad and app. 20 different proprietors
Lead planners	Stadsbyggnadskontoret
Architect	Stadsbyggnadskontoret in cooperation with architects from the app. 20 other architectural and consulting companies
Water and wastewater	Stockholm Energi, Stockholm Water, and SKAFAB (the city's waste recycling company)
Contact web site	www.hammarbysjostad.se
Project cost	20 billion Swedish krones (appr. U.S. $2.4 billion)
Type of drainage	
Sanitary	Subsurface (sewers) connected to a centralized on-site experimental treatment plant
Storm runoff and snowmelt	Local surface channels and green roofs; stormwater from streets with more than 8000 vehicle/day traffic treated by local BMPs (infiltration, storage, sedimentation)
Renewable energy	Solar cells, solar panels
	Heat extraction from treated wastewater (also converted to cooling)
	Buildings green architecture
	Heat extraction from incineration of combustible solids
	Biogas production by digestion from organic solid residuals
Water conservation	Outside source, in-house water-saving fixtures (low-flushing toilets, low-flow dishwashers, showers; potential gray water reuse)
Wastewater system and management	Linear and centralized (no water reclamation from the central treatment plant), heat recovered by heat pumps
Transportation	Light rail and (free) ferry to Stockholm
	Car pools
Recreation, leisure, sports	Extensive network of foot and bicycle paths, cross-country skiing, downhill ski slope
	Sports arena and a cultural center
Green areas and nature	Extensive, interconnected, natural and man-made; see below

Transportation Hammarby Sjöstad offers the following low-energy transportation alternatives to minimize the energy-demanding use of private cars: (1) light rail connection with Stockholm center, (2) ferry to Stockholm, and (3) car pools. Eighty percent of residents' and workers' trips in the city will be by public transportation, by bicycle, or on foot. Private car users are limited to 0.7 parking lots per household.

Approximately one-third of the town's residents are members of the car pool, whereby two or three people share a car. Most people use the car pool to shop at the supermarket over the weekend. Light rail provides commuting possibilities to residents who work in Stockholm center, which is about 15 to 20 minutes away by all means of transport. The water ferry is free between the city and the nearest subway stop. There are also several bus lines in the city.

Integrated Planning   The goal of the integrated planning was to create a residential environment based on sustainable resource use, where energy consumption and waste production would be minimized, and resource and energy savings maximized. This is accomplished by:

- Heat energy extracted from treated used water by heat pumps and used for heating and cooling
- Heat energy extracted from water solids by incineration
- Biogas from digestion of organic sludge and solids
- Biosolids for soil conditioning from digested sludge and other organic solids
- Solid waste recovery and recycle
- Surface stormwater management and treatment

Table 11.2 lists the components of integrated resource recovery management. The goal is to reduce the per capita water use to 100 liters/capita-day by conservation measures, which is about one-half of the current average water use in Sweden.

Table 11.2   Components of integrated management (GlashusEtt, 2007)

Energy	Water and Sewage	Solid Waste
—Combustible waste is converted into heating and electricity in the city incinerator. —Biodegradable waste is converted into biofuel and subsequently into heat and electricity. —Solar cells convert solar energy into electricity. Solar panels use solar energy to heat water. —Good heat insulation, southern exposure, solar panels, and building materials reduce the energy demand.	—Water consumption is reduced through the eco-friendly installations, low-flush toilets, and air mixer taps. —A pilot used water treatment plant was built to treat separated sanitary used water and to research treatment technologies. —Digestion extracts biogas from the sewage sludge. —Digested solids are used for fertilization. —Rainwater is drained via surface paths to the lake. —Local BMPs are used for treatment of polluted street runoff.	—An automated waste disposal system with three deposit chutes, a block-based system of recycling rooms, and an area-based environmental station sorts and disposes the waste. —Organic waste is converted/digested into biosolids and used as fertilizer. —Combustible waste is converted into electricity and heating. —All recyclable materials are sent for recycling. —Hazardous waste is incinerated or recycled.

The area has an experimental on-site centralized wastewater treatment and re-source recovery treatment plant (no water is reclaimed currently from the plant for reuse), officially opened in 2003. The plant, receiving only sanitary sewage flows, reduces the nitrogen level in the effluent to below a standard of 6 mg/L and recovers 95% of phosphorus for reuse on agricultural lands. The phosphorus concentration in the effluent is expected to be below 0.15 mg/L.

Landscape Architecture The street dimensions, block lengths, building heights, density, and usage mix were designed to take advantage of water views, parks, and sunlight. Restricted building depths, setbacks, balconies and terraces, large glass areas, and green roofs are the main features.

Landscape architecture planning is crucial in the implementation of surface storm drainage. Stormwater from the developed area is infiltrated, routed on the surface in channels into three surface canals transecting the city.

A green avenue links the city district's public green spaces, creating a green corridor running through the southern part of the city. The parks are also linked to the nature conservancy and forests. Most predevelopment natural areas have been preserved, and new nature areas around the shoreline (former brownfield areas) were recreated.

Solid Waste Recycling Solid waste management is conducted on three levels: (1) domestic, building source–based, (2) block-based, and (3) area-based:

> *At the source*, solid waste is separated into combustibles, food waste, and paper (newspapers, catalogues, etc.). Wastes are deposited into three color-marked chutes or bins (Figure 10.13).
>
> *The block-level depository room* receives other solid waste such as glass, plastic and metal packaging, bulky items (e.g., furniture), electrical and electronic waste (light bulbs, small batteries, fluorescent lights), and textiles, which are deposited in special recycling depository rooms.
>
> *The area-based depository facilities* receive potentially toxic wastes—such as paint, solvents, and large batteries—that cannot be deposited with the other block-level waste or poured into household drains. These wastes are separated and handled at the hazardous waste collection location.

Combustible wastes are recycled as heat and converted into electricity in an incinerator located in South Stockholm. Food waste is composted into soil along with the sludge residuals after sludge digestion and methane extraction. The biosolids are currently used in the surrounding forest, and the application will be expanded also to farmland.

Newspapers and similar items are delivered to paper-recycling enterprises. Metals are recycled; some other discarded bulky items (e.g., furniture) are incinerated. Noncombustible and nonreusable solid waste is disposed of in landfills. Hazardous waste is either incinerated or recycled.

The city uses a sophisticated automated waste disposal system that conveys the source-based waste into underground tanks, separated for each type of waste, from which the waste is emptied by vacuum into large collection vehicles and delivered for processing.

Health and Social Well-Being   The city strives to be a healthy place, to stimulate the body and soul by providing ample opportunities for exercise, sports, and culture. It has numerous foot and bicycle paths, a slalom ski slope, a sports hall, and a nature reserve. Cultural outlets include a social and cultural center and a library. It offers tuition for students and adults to engage in art classes.

Summary   The Swedish traditional urban and suburban housing model is different from the prevalent model in the U.S. Swedish suburbs mainly contain large blocks of apartment houses, not the detached single-family units typical for U.S. suburban developments. Also, the central city developments prefer a high-density habitat that provides better conditions for local services and lively streets. This policy of higher density resulted in more households per 10,000 inhabitants than anywhere else in the world (Vestbro, 2005). At the time of planning Hammarby Sjöstad, virtually all major political parties in Sweden supported the traditional high-density development concepts. This resulted in the planners' opting for a compromise between the suburban and urban concepts. The average population density in Hammarby Sjöstad is 133 inhabitants/ha, which is in between the typical suburban density in Sweden of 34 inhabitants/ha and that in the central city, which ranges between 163 and 273 inhabitants/ha. It was pointed out in Chapter X that higher-density developments are more environmentally friendly and have a smaller carbon footprint than typical suburban developments. A "compacted" city with good transportation and other services such as recreation, shopping, and the like, reduces the demand for private car ownership.

The ecocity is still based on the linear water utilization model, and water reclamation and reuse are not included.

The Hammarby Sjöstad goal, model, and reality document that energy and water use can be halved in comparison to standard Swedish urban settings, even when considering the fact that typical Hammarby apartments are larger and more illuminated (with light provided by oversized windows) than a typical flat in the Stockholm area (Vestbro, 2005).

The city development promotes a sustainable lifestyle and serves as a laboratory for sustainable development. In this sense, Hammarby Sjöstad is the first city built on ecological principles that broke the barrier preventing sustainable urban development. It uses modernistic and advanced 20th-century principles and concepts. It is a true lower impact development without the drawbacks typical for some other "low impact" developments that result in low-density developments in rural suburban areas. Hammarby does not incorporate some key 21st-century principles such as closed-loop or water/stormwater/wastewater management or wind power. However, because Sweden is rich with water resources and has low population density, the closed water loop criterion of sustainability may not be as critical as in ecocity development in water-deficient areas, and wind power can be relatively easily installed to further reduce energy use from the city fossil fuel power plant. The environmental, social, and economic benefits are balanced.

XI.2.2 Dongtan, China

Introduction Dongtan, planned near Shanghai, China, is apparently one of the first comprehensive conceptual ecocity design and has changed the direction in which ecocities are going. Phase I, a demonstration settlement for 5000–10,000 inhabitants, was to be ready for opening at the time of the Shanghai International World Expo in 2010. The Exposition follows traditional World Fairs held at the beginning of each decade throughout the world. Because of the theme of the exposition, "Better City–Better Life," the Dongtan development was implicitly assumed to be one of the main attractions for visitors, and was to represent Shanghai, a megacity with 19 million inhabitants (2009), as a 21st-century major economic and cultural center (Langellier and Pedroletti, 2006).

Dongtan was planned to be at the eastern tip of Chongming Island at the mouth of the Yangtze River (Figure 11.3), in the middle of a designated nature reserve with outstanding biodiversity. Chongming is the third-largest island in China (at 1200 km^2 or 120,000 ha), and its principal land use has been agriculture, rice farming (Figure 11.4). Chongming Island contains vast environmentally sensitive wetlands that are home to migratory birds flying all the way from Siberia to Australia. The natural resource value is very high (SIIC, 2003). The location of the proposed city on the island is about 40 kilometers from downtown Shanghai, and in 2008 was accessible only by ferry. A bridge and tunnel, opened in 2009, shorten the commute from Shanghai from 3 hours to about 45 minutes, but it also shortens the commute to Shanghai to less than 1 hour, meaning, if the city is built in one form or another, many people living in the city will be commuting to Shanghai and not developing the local economy.

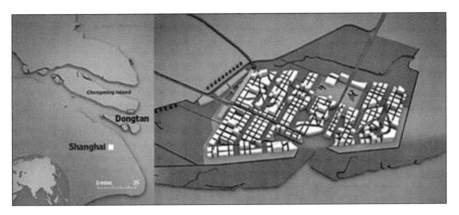

Figure 11.3 Dongtan Phase I (South Village) Location (Source: Arup).

Figure 11.4 The site of Dongtan, Chongming Island, comprises large naturalistic wetlands and agriculture.

The contract for the project was awarded to the Shanghai Industrial Investment Corporation (SIIC), a state-owned developer, in 1999. SIIC appointed Arup, a British-based urban planning consultancy, to design Dongtan. The contract was signed in London in November 1999 at 10 Downing Street (the British prime minister's office) in the presence of the former prime minister, Tony Blair, and the visiting Chinese president, Hu Jintao. Construction was supposed to begin in 2008 and the demonstration Phase I development for about 10,000 inhabitants finished by the time of the World Exhibition. The final population, in about 2050, was planned to be 500,000. The commitment to British–Chinese cooperation for the development of ecocities in China was reconfirmed by the former prime minister, Gordon Brown.

However, the realization of the Dongtan vision has stalled, mainly for political reasons, including the inability of the SIIC to obtain the necessary permits after a primary government mover for the project on the Chinese side was deposed in 2006. The outlook is uncertain, and the start-up of construction has been indefinitely postponed (Moore, 2008; Anon, 2009a). Nevertheless, because of the pioneering nature of the Dongtan concepts and the possibility that it might be built later, either at the original location or somewhere else in China, it has been included in this book.

How Big, and How Many People? Arup's approach is pioneering and initiated the new paradigm of ecocity building that was then adapted by other ecocity developments. However, although the city design is extremely water centric, the main architectural functions of the canals and lagoons within the development are aesthetics, recreation, and transportation. In other words, Arup architects decided to create another Venice (Italy) with all the characteristics of a Venice type of city (Figures 11.3 and 10.1), including parking only outside the city, but based on the Chinese thousand-year-old tradition of several historic water cities of their own. It is also necessary to consider the flooding potential, exacerbated by global warming, because the alluvial Chongming Island has very low elevation and is geologically unstable, like the 128 wetland alluvial islets forming Venice. The ecological and hydrological functions and benefits of the water bodies have not been fully addressed on a macroscale, but implicitly assumed and incorporated into a potpourri of various reuse, recycle, and infiltration measures incorporated in the landscape design. Arup has not used a comprehensive hydrological and ecological model to assess and design the overall macroscale impact of the city (Stanley Yip, Arup-China Director of Planning, personal communication).

Because the foundation conditions on the island did not allow high-rises, Arup settled on a range of four- to eight-story buildings across the city. The combination of such buidings allowed for passive energy savings from sun and ocean breezes, as well as for implementation of solar panels, voltaics, and small wind turbines (Figure 11.5). The result was that the population could be increased to about 500,000 on less than 10% of the island area, thus leaving ample space for wetlands, nature, and sustainable agriculture.

Characteristics The water centric nature of the city is inspired less by Venice than by the thousand-year-old Chinese traditions of their own water cities located in the Yangtze River delta and elsewhere along the Yangtze River (e.g., WuXi), with

Figure 11.5 Architectural rendering of the East Village and the lagoon, showing the medium-height (4–8 story) buildings with solar panels and vertical wind turbines surrounding the lake (Picture provided by Arup).

canals, ponds, arched bridges, and water transporation, although the concepts are very similar. Arup's landscape architect, Alejandro Gutierrez, created flood cells within the city, so that if Dongtan were hit by a once-in-a-century storm, the seawater would stay in a single cell. At the water's edge, instead of a high levee, a gentle hill would recede into a wide wetland basin—a park, bird habitat, and natural storm barrier would be created.

Regarding energy, the designers located the energy (electricity and heat) producing plant in the center of the city to efficiently distribute heat to the buildings. As an energy source, the city would use rice husks, abundantly produced by the agriculture on the island. In addition, a big wind farm, conversion of waste sludge into biogas, and numerous smaller contributions to the grid—including photovoltaic panels and small wind turbines—were planned. Dongtan would receive 100% of its energy from renewable sources within 20 years after Phase I.

The city's sustainability influence would also extend to suburban agricture, which would be mostly organic and use treated effluent for irrigation. Most of the food would be produced on the island. The features of the city are summarized in Table 11.3, as follows (Head and Lawrence, 2008; Arup, 2008; Urban Agent, 2008):

Energy:

- Energy demand in Dongtan would be substantially lower than in comparable conventional new cities. Dongtan ecocity aims to have:
  - 60% smaller ecological footprint

Table 11.3   Dongtan characteristics

Project	Dongtan New City
Construction start date	Indefinitely postponed
Anticipated completion phases	Final (500,000 pop.) by 2050
Location	Chongming Island, 40 km from Shanghai
Island size	1200 km² (120,000 ha)
Dongtan city size	86 km² (8600 ha)
Connection to Shanghai	Bridge and tunnel completed in 2009
Travel time to Shanghai	45 minutes
Project built-up size	30 km² (3000 ha)
Target total population	500,000
Population density	160 people/ha
Developer	Shanghai Industrial Investment Corporation (SIIC)
Design and master plan	Arup, Shanghai (Stanley Yip Cho-Tat, Director)
Financing	HSBC and Sustainable Development Capital LLP
Cost estimate	$1.3 billion for Phase II 80,000 population
Water/wastewater management	Centralized, partially closed cycle (water reclaimed from wastewater will be reused for toilet flushing and irrigation)
Arup web site	http://www.arup.com/eastasia/project.cfm?pageid= 7047

- 66% reduction in energy demand
- 40% energy from bio-energy
- 100% renewable energy for buildings and on-site transport
- If Dongtan is completed, the energy used within the city would not add to the level of greenhouse gases in the atmosphere. This would be accomplished by:
  - Energy in the form of electricity, heat, and fuel would be provided entirely by renewable means.
  - In buildings, this would be achieved by specifying high thermal performance and using energy-efficient equipment and appliances to encourage building users to save energy.
  - Transportation energy demand would be reduced by eliminating the need for a high proportion of motorized travel, and judicious choice of energy-efficient vehicles.
- Energy supply would be via a local grid, and electricity and heat would be supplied by:
  - A combined heat and power (CHP) plant located in the center of the city that runs on biomass of rice husks, which are the waste products of local rice mills.
  - A wind farm.
  - Biogas extracted from the digestion of municipal solid waste and sewage.
  - Electricity would also be generated in buildings using photovoltaic cells and micro wind turbines.
- Some of the electricity generated would be used to charge the batteries of electrically powered vehicles or to produce hydrogen for vehicle fuel cells.
- A key feature of energy management in Dongtan would be the level of information provided to consumers to encourage them to conserve energy by means such as smart metering and financial incentives. A visitors' center located close to the energy center would explain how cities can be sustainable in energy terms.

Resource and Water/Waste Management:

- Two water networks would provide water throughout the city: one that supplies drinking water to kitchens and another that supplies reclaimed treated wastewater for toilet flushing and landscape and farm irrigation.
- Approximate water use for the city (80,000 population) was estimated as 16,5000 m^3/day, or 200 L/cap-day, of which 43% would be reclaimed.
- The design aims to collect 100% of all waste within the city and to recover up to 90% of collected waste.
- Waste is considered to be a resource; most of the city's waste would be recycled, and organic waste would be used as biomass for energy production.
- Waste to landfill would be reduced 83%. There would be no landfills in the city.

Ecological Management of Wetlands:

- The delicate nature of the Dongtan wetlands and the adjacent sites for migrating birds and wildlife has been one of the driving factors of the city's design.
- The existing wetlands would be enhanced by returning agricultural land to a wetland state, to create a "buffer zone" between the city and the protected coastal wetlands.
- Only around 40% of the land area of the Dongtan site would be dedicated to built-up urban areas, and the city's design aims to prevent pollutants (light, sound, emissions, and water discharges) from reaching the adjacent wetland areas.

Sustainability:

- A combination of traditional and innovative building technologies would reduce energy requirements of buildings by around 66%, saving 350,000 tons of CO_2 per year for the start-up area.
- All housing would be within seven minutes' walk of public transport and have easy access to social infrastructure such as hospitals, schools, and work.
- Although some may choose to commute to Shanghai for work, there would be local employment for the majority of people who live in Dongtan, across all social and economic demographics. By means of effective policy incentives, companies would be attracted to Dongtan, and people would choose to live and work in the city.
- Dongtan would produce sufficient electricity and heat for its own use, entirely from renewable sources. Within the city, there would be practically no emissions from vehicles—vehicles would be battery- or fuel cell–powered.
- Farmland within the Dongtan site would use organic farming methods to grow food for the inhabitants of the city; nutrients and soil conditioning would be used together with processed city waste.
- The development of techniques that increase the organic production of vegetable crops would mean that no more farmland would be required than is available within the boundaries of the site.

Buildings and Architecture:

- Where possible, labor and materials would be obtained locally to reduce the transport and energy costs associated with construction.
- A combination of traditional and innovative building technologies would reduce the energy requirements of buildings by up to 70%.
- Public transportation with reduced air and noise pollution would enable buildings to be naturally ventilated, and in turn reduce the demand on energy.
- Buildings with green roofs would improve insulation and water filtration and provide potential storage for irrigation or waste disposal.

- The compact city design would reduce infrastructure costs and improve the amenity and energy efficiency of public transportation systems as well.
- The original three villages would be retained and form the historic city center.
- The city would contain 20% affordable housing.
- No building would be higher than eight stories.
- Dongtan would be a water centric city where lakes and canals would be the focus of the city, used for enjoyment, recreation, and transportation.

Transportation:

- Dongtan would be a city linked by a combination of bicycle paths, pedestrian routes, and varied modes of public transportation, including buses and water taxis.
  - All housing is to be within a seven-minute walk of public transportation; walking and bicycling would be promoted.
  - Public transportation by hydrogen fuel–cell buses and solar-powered water taxis would be provided.
  - Businesses, schools, hospitals, and other public facilities should also be easily accessible.
  - Vehicular parking would be allowed only outside the city.
- Improved accessibility in Dongtan would reduce travel distances by 1.9 million kilometers, reducing CO_2 emissions by 400,000 tons per year.
- Canals, lakes, and marinas would permeate the city, providing a variety of recreation and transport opportunities.
- Visitors would park their cars outside the city and use public transportation within the city.
- Public transportation with reduced air and noise pollution would enable buildings to be naturally ventilated, and in turn reduce the demand on energy.

The Current Situation and the Future *Barriers.* Arup's original plan envisioned that in 2010 about 5000 to 10,000 people would live in the demonstration site, at the time Shanghai hosted the World Expo. Apparently this did not happen; the project is suspended, but not canceled. A new bridge/tunnel, opened in 2009, connects the island site with the Shanghai mainland. Obviously, the major barrier to the realization of Dongtan is the political situation in Shanghai, and now there is potential pressure to develop Chongming Island as a traditional suburban car-based community. Another barrier could be the fragile nature of the island wetland ecology and its low elevation, with the potential of increased tidal and typhoon flooding.

Arup envisioned Dongtan as a vibrant city with green "corridors" of public space ensuring a high quality of life for residents. The city was designed to attract employment locally across all social and economic demographics, in the hope that people would choose to live and work there.

"Dongtan is designed to be a beautiful and truly sustainable city with a minimal ecological footprint. The goal is to use Dongtan as a template for future urban design. As China is planning to build no less than 400 new cities in the next twenty years, Dongtan's success is of crucial importance." (World Business Council of Sustainable Development, quoted in Urban Agent, 2008)

XI.2.3   Qingdao (China) Ecoblock and Ecocity

Ecoblocks   The ecoblock concepts were developed by the team of Dean Harrison Fraker of the University of California College of Environmental Design specifically for urban developments in China. China's current rapid pace of urban development consists of so called *super blocks* (Fraker, 2006, 2008). A super block is a typical high-rise residential development in China, usually 100–200 ha (240–480 acres) in an area with 2000 to 10,000 residential units housing 6000 to 30,000 people. China is now building 10–15 super blocks a day. A super block is a traditional unsustainable development relying heavily on municipal services for power, potable water, wastewater and stormwater conveyance (sewers), centralized wastewater treatment, and solid waste collection and disposal.

In contrast, an *ecoblock* is a city block much smaller than a super block. It is self-sustained and semi-independent in its water and energy needs. It generates its own energy from renewable sources, harvests rainwater, produces its own water, and processes and reclaims its wastewater. It is a module that can be repeated many times to form an ecocity or a part of thereof. Unlike the IRMC (water centric cluster) it is not connected to, nor does it rely on, natural or restored water bodies, with the exception of constructed wetlands that treat wastewater. In water-rich urban areas (e.g., Dongtan), landscape architects may include water bodies for enjoyment and recreation. However, the size of the ecoblock is relatively small, and conceivably it could be shaped and fitted into the local topography and natural environment. Fundamentals of the ecoblock and the history of its development and application in China are described in Fraker (2006).

A typical standardized ecoblock has 600 units on 3.5 hectares and would house 1500–1800 residents. A layout of the ecoblock is shown in Figure 11.6. The Qingdao ecoblock includes several 5- to 7-story townhouses, six 12-story tower blocks, and four 24-story tower blocks arranged around a green courtyard. Limited underground and on-street parking is intended to encourage walking, biking, and public transportation. The maximum time to reach public transportation by walking should be less than 15 minutes.

Ecoblock Characteristics   **Energy:** The components and concepts of energy reduction and self-sufficiency in the ecoblock are (Fraker, 2008):

1. The best techniques for energy conservation, such as:
   - Insulation
   - Passive solar energy capture by windows

Figure 11.6   Plan and view of the ecoblock module (Courtesy: H. Fraker, U.C. Berkeley, and Arup).

- Natural ventilation and daylighting
- Energy-efficient home appliances and lights
These technologies can reduce the energy demand of a unit by as much as 40%.

2. Renewable energy sources:
   - Vertical axis wind turbines on the tops of tall buildings (30%)
   - Building-integrated roof and canopy photovoltaics on the lower buildings, which would both generate electric energy and provide shading (21%)
   - Building-integrated solar water heaters (3%)
   - Bioconversion of sewage sludge, kitchen solid waste, and organic yard waste, by a two-phase anaerobic digestion process, into methane running a backup generator (6%)

3. Shared plug-in hybrid cars

Water   *Dual closed-loop water, stormwater, and wastewater recycling and reuse system.* A conceptual plot of a dual system proposed for the Qingdao ecoblock was shown in Figures 6.5 and 7.17. Figurs 7.17 was modified to remove direct potable reuse. Reducing water demand and instituting water reclamation would be achieved by:

1. Increased efficiency of water use, providing a 35% reduction of typical water use (160 liters/capita-day), includes:
   - Xeriscape of the green areas, reducing need for irrigation water
   - Using reclaimed water for irrigation
   - Low-flow fixtures in the units (dishwashers, washing machines, faucets)
   - Toilet flushing with reclaimed water
2. Recycling water in a dual system, one for gray water and the other for potable water, would provide 50% reduction.
   - Gray water from bathroom sinks, showers, and washing machines would be collected; conveyed for physical treatment by settling, aeration, reverse osmosis, and UV disinfection; and then returned into the potable water cycle. Sludge would be pumped into the biogas digester.
   - The gray water cycle would be supplemented by make-up potable water from the municipal system, representing about 15% of the total water use, to replace the losses by evaporation and to control buildup of nonremovable pollutants in the system (e.g., pharmaceutical residues and inorganic solids such as salt). Potable water would be pumped into hot and cold water tanks and distributed to the tenants.
   - Black wastewater from toilets and kitchen sinks would be conveyed to biological treatment by batch reactors and then discharged into wetlands. After subsurface flow wetland treatment it would be collected into large communal reservoirs, where it would be mixed with rainwater and thereafter reused, after UV disinfection, for landscape irrigation, toilet flushing, and washing machines.
   - Impervious surfaces would be limited mostly to roofs. Rainwater would be harvested and directed to cisterns from which water would be pumped after UV disinfection for toilet flushing and laundry. Pervious pavement would be used on all streets and paths to enhance groundwater recharge. Groundwater would also be pumped and either added to the cisterns with harvested rainwater or directly used for irrigation.

Assembling an Ecocity   The Qingdao ecocity would consist of 16 ecoblocks (Figures 11.7 and 11.8). The city would be connected to the public transportation station and the main road. The commercial center would be located around the station. The city would have central surface and subsurface reservoirs and energy recovery (digestion) units. Used and locally treated black water would be conveyed to the reservoirs via the constructed wetlands. Rainwater collected from roofs and pavements would be treated by grass biofilters (swale) and conveyed to the reservoirs.

The system includes proven technologies; however, the creators of the ecocity, in their video of the Qingdao plan (Green Dragon Media Project, 2008), pointed out

Figure 11.7 Architectural rendering of the ecoblocks in the Qingdao ecocity (Courtesy Professor Harrison Fraker, University of California, Berkeley).

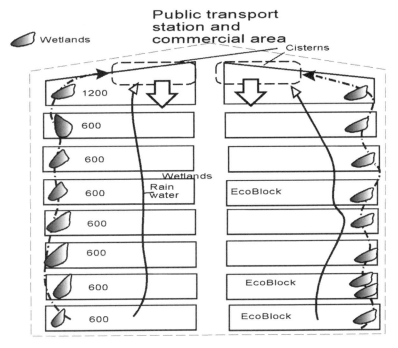

Figure 11.8 Schematic of the Qingdao ecocity, consisting of 16 ecoblocks. The number in each ecoblock represents the number of units. Based on the Green Dragon Media Project (2008) video.

that the new ecological and sustainable urban concepts would provide incentives for new and better technologies for water saving and reuse as well as renewable energy that would be less costly. It was estimated that the capital cost of such developments would be about 15% more than the conventional super block development in China.

Barriers. Schlaikjer (2007) discussed the social and economic barriers to the ecoblock and ecocity developments in China, acknowledging that:

1. Clean technologies are still expensive.
2. It would be difficult for foreign and Chinese firms to collaborate, which would be necessary in order to develop and reinforce responsible business practices.
3. The idea of a "gated community," though having historic precedent in the "forbidden city" (Fraker, 2006), does not always appeal to local cultural attitudes and norms.
4. The incentives to build and operate "green" cities are not always evident at the onset.

Similarly to Dongtan, the construction phase of the Qingdao ecocity is uncertain but the concepts of the ecoblock decentralized water/used water management is sound and was applied in Tianjin (Personal communication of Prof. Fraker to the primary author) and most likely other Chinese cities. It is applicable for retrofitting the medium to high density urban zones typical of central and east European cities and also many megalopolis of the developing world.

XI.2.4   Tianjin (China)

Overview   On October 28, 2008, a groundbreaking ceremony attended by Chinese and Singapore government officials and media celebrated the start of construction of one of the first ecocities in China, the Tianjin ecocity. The site of the new city development is about 150 kilometers southeast of Beijing and 40 kilometers from historic Tianjin City (population about 12.3 million in 2009), which is the center of the Tianjin region and the largest port city in northeast China (Figure 11.9). The city is part of a huge regional development of the Tianjin–Binhai New Area (TBNA) that would include several industrial parks, manufacturing (e.g., Airbus airplane production), several science and commercial centers, a port, and recreational tourist zones. The building of the ecocity would be based on an international treaty signed by Premier Wen Jiabao of the China State Council and Singaporean Prime Minister Lee Hsien Loong one year earlier. The Sino-Singapore Tianjin ecocity would use Singaporean advanced experience and leadership in sustainable urban planning, environmental protection, resource conservation, circular economy, ecological construction, renewable energy utilization, reclaimed water usage, and sustainable development. The overall goals of the projects should comply with "three harmonies," namely the harmony between (1) people and people (social), (2) people and the economy, and (3) people and the environment, which would be the same goal as balancing the triple bottom line (TBL) component of sustainability. Both governments consider the

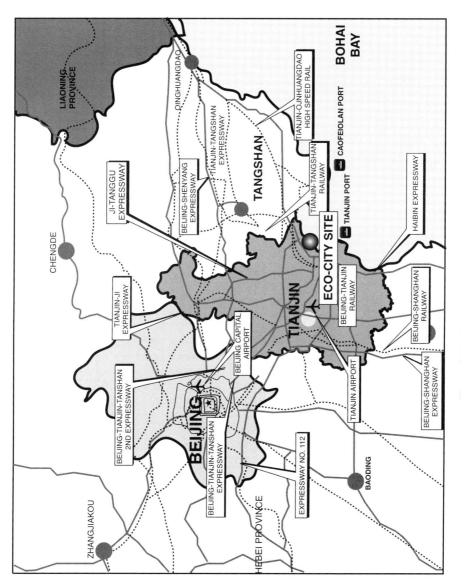

Figure 11.9 Location of the Tianjin ecocity

Table 11.4   Characteristics and parameters of the Tianjin ecocity

Location	Tianjin region 150 km southeast of Beijing
Area of development	30 km² Phase I, 4 km²
Population served	350,000 Phase I, 30,000
Population density	117 inhabitants/ha (49/acre)
Project team	
Key partners	Governments of China and Singapore
Lead planners	China Academy of Urban Planning and Design; Tianjin
Architect	Urban Planning and Design Institute; Singapore Urban
Developers	Development Authority
Water and wastewater	Sino-Singapore Tianjin Eco City Investment and
	Development Co., Ltd; Arup
	Arup, China; Siemens
Contact web site	www.tianjinecocity.gov.sg
Project cost	50 billion yuan (U.S. $9.7 billion)
Type of drainage	
Sanitary	Subsurface (sewers) connected to a centralized on-site
Storm runoff and	treatment plant; distributed system considered for future
snowmelt	expansions
	Local surface channels, rainwater capture, and recycling
Renewable energy	Solar cells, solar panels, wind energy capture, 15% energy
	use from renewable sources
	Buildings green architecture
	Biogas production by digestion from organic solid residuals
Water conservation	Water reuse, in-house water-saving fixtures (low-flushing
	toilets, low-flow washing machines, showers; potential
	gray water reuse)
Wastewater system and	Closed system but centralized in the first phase Extensive
management	water reclamation in the treatment plants for landscape
	irrigation and in-house use
	Ecoblocks considered
Transportation	Light rail to Binhai and Tianjin City, 90% of all trips by
	public transportation, on foot, or by bicycle
	No restriction on the use of private cars; the incentive to
	reduce car use: excellent and easily accessible public
	transporation, walkways and bike paths linking homes,
	shops, and public facilities
Recreation, leisure,	Extensive network of foot and bicycle paths, cultural center
sports	
Green areas and nature	Extensive, interconnected, natural and man-made; extensive
	water-based recreation, proximity of large recreation zone

Tianjin ecocity as a model for other cities in China and the world. The parameters of the Sino-Singapore Tianjin ecocity are summarized in Table 11.4.

After the signing of the treaty in 2007, the "Master Plan of Sino-Singapore Tianjin Eco-city" was developed by the design groups from the China Academy of Urban Planning and Design, the Tianjin Urban Planning & Design Institute, and the Urban Redevelopment Authority of Singapore. In order to expedite the project, the site selection had to minimize the land acquisition process and legal procedures, save productive land and water, enable resource recycling, and encourage independent innovation. Consequently, the project has intentionally been placed in a water-short area with salty land, scarce vegetation, desert, unfavorable natural conditions, and fragile ecology.

The sources of the information on the Tianjin ecocity are the web sites listed in the References section. plans for the city and its water management draw heavily on the Singapore experience, where such technologies have already been successfully implemented on a large scale. Figure 11.10 shows the architectural rendering of the Tianjin ecocity. It can be seen that the city would be divided into (eco) blocks. The smallest block unit has an area of 400 × 400 meters (16 ha or 38.4 acres). Harrison Fraker (Clean Energy, 2008) has testified that Tianjin would also include the Qing-dao ecoblocks, which are self-contained smaller ecoblocks with 600 units that have their own water reclamation and energy recovery program (see Qingdao ecoblocks in preceding section).

The city will have residential areas; protected historic-cultural districts; urban surface water bodies specified in the plan, including rivers, lakes, reservoirs, dykes, and

Figure 11.10 Architectural rendering of the Tianjin ecocity located on the Jiyun River. The city will be developed on both sides of the river. This is the original concept of the city in 2009. In 2010 planners were changing the city configuration.

wetlands; urban infrastructure influencing the overall urban development; and transportation facilities such as railway, light railway, and subway.

The city will feature an "eco-valley," which is the main north–south green connector in the city. The city site will retain a large ecological wetland, set aside habitat for birds' migration, and preserve the former watercourse of the Jiyun Canal to guarantee the smooth connection of Jiyun County Natural Reserve in the north to Binhai Bay Corridor and to form a regional ecological network with rivers as its arteries. It will be a water centric ecocity emphasizing the proximity and aesthetic functions of surface water bodies.

Overall Goals   The selection of the site was based on the requirement that no agricultural or other natural land would be used. The site consists of wasteland such as salt pan (deserted beach). One-third of the area consists of polluted water bodies, including a 270-ha waste pond. The polluted sites are being decontaminated and cleaned up. Another goal would be to restore the water quality of the Jiyun River, which transects the future city (Figure 11.10).

The environmental goals for the city were formulated using 26 key performance indicators based on the Chinese and Singapore national standards. The most important indicators are (www.tianjinecocity.gov.sg):

- Ambient air quality: at least Grade II China national standards
- Tap water quality: potable
- GHG emissions/unit of GDP: ≤ 150 tons of carbon/U.S. 1 million GDP
- Proportion of green buildings: 100%
- Transportation: 90% of all trips in the form of green trips (nonmotorized transport, cycling, and walking) and public transportation
- Proportion of affordable housing: 20% subsidized
- Usage of renewable energy: 15% of the total energy use
- Water use: not to exceed 160 liters/capita-day
- Sources of water: 50% from desalination (high energy use) and rainwater and 50% recycled water (see Figure 6.6)
- Employment generated: 50% of the ecocity residents to be employed within the ecocity

Integrated Resource Recovery System

Water Use and Reuse   Figure 6.6 in Chapter VI is a representation of the water and energy recovery system. The primary sources of water are desalinated water and rainwater. These sources would constitute more than 50% of water used in the city. An extensive system of rainfall collection and sewage reuse would be established, relying heavily on the landscape. The city will have centralized used water treatment and recycling, and would develop and utilize nonconventional water resources such as recycled water and desalinated seawater, in multiple channels, to improve the use proportion of nonconventional water resources. It would implement a reasonable and scientifically based water supply infrastructure that would reduce the need for conventional water resources. It will intensify the ecological rehabilitation and reconstruction of the surface water systems, which will be used also as a source of water in the recycle system, collect and use rainfall in the rainy season, strengthen groundwater resource conservation, and construct a favorable aquatic eco-environment connected to the Jiyun River.

Energy The city will rely on a mix of renewable and conventional energy sources that would be linked together. The plan envisions that at least 15% of energy would be from renewable sources. Traditional energy sources would be "clean coal" and other high-quality fuels. However, some may argue that from the standpoint of GHG emissions, there is no such thing as "clean coal." To reduce the influence on the environment, the plan forbids the use of high polluting fuels such as non-clean coal and other low-quality fuels. The proportion of clean energies would reach 100%. All buildings will use green energy conservation technologies and will be built according to the green building standards.

Heat pump technology will reclaim heat and electricity from wastewater and electricity, and will also be generated extensively by solar panels, wind turbines, geothermal energy, and methane generated by the anaerobic digester. During the initial phases, the ecocity willd draw waste heat from a nearby major power plant.

Culture, Leisure, Education The city would provide ample opportunities for land- and water-based recreation: beaches, boating, walking, and biking. There are several coastal wetland nature areas that provide habitat for water fowl.

A university would be planned for the city, which would focus on environmental science and technology. The Tianjin–Binhai New Area would have many large research centers and other institutions of higher leaning.

Summary *Barriers.* The system will be modeled after the Singapore water reclamation closed system, which is not distributed.

The plan features a partially closed hydrologic cycle system with extensive reuse of reclaimed water from the central treatment plant. The city planners intend to incorporate Qingdao's "ecoblocks," which are almost fully carbon-neutral (Harrison Fraker, personal communication). Most likely in the final outcome the city may be a hybrid between a centralized resource (water and solid waste) and energy management system and a decentralized rainwater-harvesting and local renewable energy–generating system.

The price tag of $9.7 billion for the city may change, as new elements may be incorporated in the future. Considering the average family size in China is fewer than three people, and subtracting about 20% for commercial, transportation, government, and education infrastructure, the cost of one flat would be about $60,000–70,000, which would be on the high side but still reasonable in Chinese economic conditions,

Figure 11.11   Artistic representations of Masdar City (Courtesy Shamma Al Muhari, Masdar).

especially in the higher income TBNA development region. Twenty percent of housing units will be subsidized.

The city development is on a wasteland with no resources except coastal wetlands that would be preserved. The site also has meager freshwater resources, and make-up water is only available from desalination and rainwater and eventually from the cleaned up and created water bodies. In the Tianjin area, rainfall is low and occurs only during two to three months in an average year. The city will be a habitat for many people working in theTianjin–Binhai New Area, but it will also provide many employment opportunities for its residents within the city.

XI.2.5   Masdar (UAE)[1]

Overview   Masdar City in the Emirate of Abu Dhabi, capital of the United Arab Emirates (Figure 11.11 and 11.12), is about 17 kilometers from Abu Dhabi island and close to the Abu Dhabi International Airport. It is also about 1 hour driving distance from Dubai. The city development is seed funded by the government of Abu Dhabi through Masdar, a wholly owned subsidiary of the Mubadala Development Company, and was designed by the British architectural firm Fosters & Partners. CH2M-Hill is the main strategic partner contractor for water, used water, reuse, energy, information communication technologies, public realm, and solid waste management. With expansion carefully planned, the land surrounding the city's built environment may contain waste management facilities, energy farms, research fields,

[1] Masdar section was revised and prepared by CH2M-Hill Masdar team.

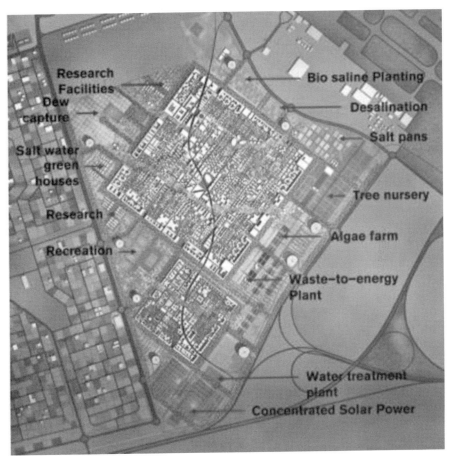

Figure 11.12 Layout of city with space allocation outside of the populated areas (Courtesy Masdar)

plantations and dense green spaces. The city is being built and designed to address the "One Planet Living" (OPL) ten principles (see Section II.4.2) (WWF, 2008).

At full build out, the city may be home to around 50,000 people, as well as hundreds of businesses in an area of 6 km^2 (1500 acres). 40,000 people are expected to commute to work in the city with a total living and working population of around 90,000 people (Bioregional, 2008; Hartman, Knell and Witherspoon, 2010). Basic characteristics of the development are summarized in Table 11.5 and the locations of many of the utilities around the city are displayed in Figure 11.12.

Open Spaces Green and open spaces will be located throughout the city and along the outside of the populated areas. Recreational fields are provided between the large and small populated centers. Native plant species will be used throughout the city in order to reduce irrigation needs. Open spaces will also be located throughout

Table 11.5   Characteristics and parameters of the Masdar City development

Location	Abu Dhabi, UAE
Area of development	700 ha or 7 km^2
Population served	50,000 residential, 49,000 commuters
Population density	140 people/ha (333 people/acre)
Project team	
Key partners	Foster + Partners, CH2M HILL, WSP, ETA, Transsolar, QS Cyril Sweett cost consultancy, Systematica
Lead planners	Foster + Partners, CH2M HILL
Architect	Foster + Partners
Contact web site	http://www.masdarcity.com/
Type of drainage	
Sanitary	Subsurface connected to a centralized treatment plant & black and grey water reuse
Stormwater	Rainwater harvesting, stormwater reuse
Renewable energy	Primarily photovoltaic ground and roof mounted with solar thermal, waste-to-energy, and geothermal as other possible sources
Water conservation	Most of water supplied to be recycled for irrigation, toilet flushing, district cooling and other uses. Gray water will also be collected and recycled. Low-flow and water-monitoring systems will be installed in all condominiums and offices.
Used water system and management	Centralized, 80% water recycled
Transportation	Car-free city. Every resident will live within a short walk of a transportation hub. A driverless four passenger rapid transportation system is being pilot tested and other options are being considered. Compact design encourages walking and biking. Light rail system connects city to Abu Dhabi.
Recreation, leisure, sports	Network of foot and bike paths, recreational center with sport fields
Green areas and nature	Green areas surround the city, with parks and community squares located throughout.
Project cost	U.S. $22 billion

the city, including three large pathways (two through the large rectangular populated area and one through the smaller area) to allow the daytime and nighttime winds to flow through the city. These pathways will be oriented to facilitate wind flow through the city to help promote airflow.

Energy   Masdar City aims to be one of the world's most sustainable urban developments in the world powered by renewable energy. Many factors would need to be implemented to reach this goal. These factors include energy reduction throughout the city through the implementation of advanced technology to reduce the city energy demand, city and street orientation, and the use of renewable energy sources to provide all of the energy needs of the city. Through the implementation of the most advanced technologies, the power requirements of the city will be reduced to 230 MW of power, instead of the usual 800 MW for a city of similar size (WSP, 2009). Narrow streets, shaded walkways, and orienting the city northeast would minimize the amount of direct sunlight on building sides and windows, reducing the need for air conditioning, as shown in Figures 11.13 and 11.14. All other electricity and

Figure 11.13 Artist's representation of a shaded street in the city (Courtesy Sharma Al Muhari, Masdar).

Figure 11.14 Artist's representation of a public square the city (Courtesy Sharma Al Muhari, Masdar).

cooling needs is provided by renewable energy generated on site (BioRegional, 2008; Hartman, Knell and Witherspoon, 2010).

The low carbon emission goals do not just apply to the city after implementation, but also during construction. A 10 MW photovoltaic power plant is providing clean electricity to support the construction process, In order to offset part of CO_2 emissions generated during construction, much of the plant's electricity is pumped back into the grid (Palca, 2009). There will be large photovoltaic arrays and panels throughout the city and a concentrated solar power (CSP). As well, there are plans for a waste-to-energy plant. In addition to these power sources, there are discussions regarding drawing clean energy from renewable power plants located outside Masdar City. The 10-MW solar farm is generating 17 million kWh per year. Built across 22 hectares by Abu Dhabi-based Enviromena, the plant was connected to the Abu Dhabi power grid in April 2009 and is supplied by 50 percent thin film photovoltaic modules and 50 percent polycrystalline photovoltaic modules. The CSP plant will be used to produce electric power by converting the sun's energy into high-temperature heat using various mirror configurations. This in turn will generate steam that will then be channeled through a conventional generator to provide electricity (Crampsie, 2008).

In addition to solar energy, geothermal and waste heat may also be used. Also both sewage and municipal solid waste may be used to provide energy. The waste-to-energy strategy involves treatment of both the organic fraction of the waste and the residual fraction to produce energy rich gases. Organics may be converted to biogas through anaerobic digestion. The residual fraction may be converted to syngas through pyroloysis and/or gasification. The produced gases in both cases may be used to produce electric energy.

The city itself will be a global clean-technology cluster and home not only to sales, marketing, demonstration facilities, headquarters and R&D facilities of firms ranging from start-ups to global multinationals, but also to the International Renewable Energy Agency (IRENA) and the Masdar Institute of Science and Technology, a graduate-level research-focused university focused on renewable energy and clean technologies and developed in cooperation with the Massachusetts Institute of Technology.

Transportation   Masdar City will be free of cars on the street level. Instead of private automobiles, a variety of alternatives will be used and test piloted, including a Personal Rapid Transit (PRT) system. This system could become the world's first large-scale personalized electric transport system powered by solar energy. It will work on the principle of small electric driver-less cabs carrying up to four passengers at a time. Every resident will be just a short walk from a transportation station where they can request one of the PRT vehicles. Once inside, a passenger can select any PRT station in the city at the touch of a button. (WSP, 2009).

In addition to the PRT system, the city is compact with short distances transportation links encouraging walking or biking to destinations. The shaded walkways and narrow streets will create a pedestrian-friendly environment in the context of Abu Dhabi's extreme climate. It also articulates the tightly planned, compact nature of traditional walled cities (Foster + Partners, 2007).

To accommodate the commuters who will travel in and out of the city every day, there is a Light Rail Transit (LRT). The LRT is an overland train that runs from Abu Dhabi city centre to the international airport stopping at Raha Beach, a popular resort development just outside Masdar City, and Masdar City itself (WSP, 2009). In addition to the light rail, parking structures are located at the city edge for travelers driving to the city. These parking structures are near transportation stations for easy access to the city.

Water Use   Masdar City does not have any viable freshwater sources and the sea water from the Arabian (Persian) Gulf has very high salinity (40,000 mg/L). Rainwater sources are very limited, the area is essentially a desert. The local groundwater salinity is twice that of the gulf waters (Hartman, Knell and Witherspoon, 2010). It should be pointed out that all Arabic states on the gulf use desalination for their water supply and discharge high salinity brine reject into the gulf. Potable water sources for the city are currently being evaluated. Each building is designed to have three unique water streams to enter and exit the unit (Hartman, Knell and Witherspoon, 2010). Potable water enters the building for personal consumption and bathing. Reclaimed used water is used for toilet flushing. Spent graywater and blackwater are conveyed to the water reclamation plant for treatment and reuse for toilet flushing, district cooling, and irrigation. The water reclamation plant is centrally located due to the large central demand for each stream. In addition to treated used water, landscape irrigation will also use graywater (Bioregional, 2008). In some cases the irrigation water will also be reused. For a typical city of Masdar City's size, consumption of desalinization water is expected to be around 20,000 m^3/day (5.3 MGD). In order to minimize water consumption, passive demand control strategies and water reuse approaches will be used to reduce water consumption significantly from typical city usage. Individual water consumption is expected to be much lower than typical city usage. This is a dramatic reduction from the current water use in Abu Dhabi which is currently ranging from 55- to 353 L/person-day, of which a significant volume is used for irrigating lawns of villas (Hartman, Knell, and Witherspoon, 2010). In order to reach this goal, reduction in water leakage from pipes is needed and installation of the most modern water saving fixtures and appliances. Masdar City will also adopt water efficiency measures that minimize consumption through a combined approach of technology and cultural change. Typically, an estimated 20% of water is lost through pipe leakages. Through the implementation of advanced technologies and systems, Masdar City expects to cut the losses down to less than 1%.

Other tools are used to reduce the amount of water consumed by each individual. These tools include installing water consumption tracking devices throughout the buildings to notify residents and office managers of how much water they are using and its associated carbon footprint impacts.

Masdar City will use a plethora of water management principles in order to treat all parts of the water cycle and use them as a water source. As many as nine water conveyance systems will be employed in 12 different ways and treated at three treatment levels. The variety of water sources being explored include groundwater, seawater, surface runoff, rainwater harvesting, dew/fog capture, gray and black water reuse and resource (nutrient) recovery from urine streams (CH2M-Hill, 2008).

Waste Management   The ultimate goal of waste management within Masdar City is to reach a state where waste is recycled, reused, converted to energy and/or reduced to zero at City buildout. More realistically the city plans to divert most of the waste from landfills by full built out (Crampsie, 2008). An intensive recycling program is implemented along with nutrient recovery to be used in the creation of soils in landscaping as well as waste to power scheme (Palca, 2009). Both sludge and organic garbage wastes will be used to create power by production of biogas from anaerobic digestion of organic solids.

Summary and Current Status   Masdar City seeks to be one of the most sustainable cities in the world, with the long-term goal of being carbon neutral at City

Figure 11.15   Masdar city headquarters (Courtesy Sharma Al Muhari, Masdar).

build out. In order to sustain this achievement, the city will control growth instead of sprawl, implement low rise high density developments and use sustainable methods of transportation while following the 10 principles of "One Planet Living" (OPL), independently verified by the World Wildlife Fund (WWF, 2008). The project began in 2008. The first buildings were set to be occupied in the second half of 2010. Not only will the city be used as a model for sustainable development, but also the majority of the people living inside of the city will work in the renewable energy and sustainable technology business. Located at Masdar City, Masdar Institute is researching and teaching renewable energy and clean technology practices. The total cost of the Masdar City project is expected to be $22 billion and is expected that at full build out it will house as many as 50,000 residents, with another 40,000 people commuting to the city for work in hundreds of businesses. The project broke ground in 2008 with the first building phase of Masdar City (the university/technology center and surrounding neighborhood) set for completion in 2013. The entire city will be built over more than a decade. Figure 11.15 show headquarters (city hall) if Masdar with a waterscape pool.

   The City's real estate business model requires careful spending and development growth to meet fiscal goals of a sustainable city. The final costs will be in line with the City's fiscal goals whilst being one of the most sustainable cities in the world

XI.2.6   Treasure Island (California, U.S.)

Overview   Treasure Island is a man-made island built by dredging sediments from the San Francisco Bay, constructed to host the 1939 Golden Gate International

Exposition. It was originally scheduled to become an airport after the exposition, but instead was acquired by the U.S. Navy for a naval base. In 1990 Treasure Island supported a population of more than 4500 people and a daily employee population of almost 2000. In 1997 the naval base was closed as part of the Base Realignment and Closure III (BRAC) program, and redevelopment plans have been developed to transform Treasure Island and the nearby Yerba Buena Island into one of the most sustainable cities in the United States.

The majority of the information on this redevelopment was obtained from the Treasure Island Development Authority development plan (TIDA, TICD, 2006). By 2018 the Treasure Island and Yerba Buena Island development will be an entirely new built community of 6000 homes supporting 13,500 residents, a retail-focused town center including 21,800 m^2 (235,000 sq ft) of retail space, hotels with a total of 420 hotel rooms, adaptive reuse of historic structures, a marina district including ferry transport to San Francisco, a range of essential services, and an extensive open space program (Figure 11.16). No official references regarding the cost are available, but the total cost has been unofficially estimated to be around $3 billion.

Currently the island is still owned by the U.S. Navy. Negotiations are ongoing regarding the purchase price for the City of San Francisco. In addition to purchasing the land, the city will have to deal with various contamination sites that are present throughout Treasure Island. Soil contamination is related to fuel storage and fueling operations, previous fire training activities, above- and below-ground storage areas, ammunition storage, and petroleum pipelines. The U.S. Navy is partly responsible for hazardous materials remediation. The city has been working with CH2M HILL and Geomatrix to monitor the Navy's cleanup work to date. The parameters of the city are in Table 11.6.

Other important issues that need to be addressed during the construction of the Treasure Island development relate to seismic conditions and traffic. Under current land conditions, Treasure Island is expected to perform poorly in a major earthquake

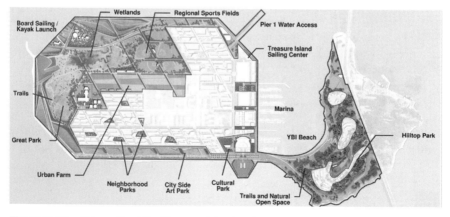

Figure 11.16 Schematic of the Treasure Island project with open spaces, recreation, parks, and urban farms (Sylvan, 2008).

Table 11.6 Characteristics and parameters of the Treasure Island development (TICD, 2006)

Location	San Francisco, CA
Area of development	1.8 km² (180 hectares, 450 acres)
Population served	13,500
Populations density	150 people per hectare of built area
Project team	
Key partners	Treasure Island Community Development (TICD), Treasure Island Development Authority (TIDA), City of San Francisco
Lead planners	SMWM, SOM
Architect	SMWM, SOM with the help of 18 other architecture and consulting firms
Type of drainage	
Sanitary	Subsurface connected to a centralized treatment plant
Stormwater	Green roofs, xeriscape, and gravity pipes for excess runoff to a centralized wetland area for treatment
Renewable energy	Photovoltaics, small vertical axis wind turbines, solar hot water heating, biogas power generation from WWTP
	Peak energy use 17.4 MW
	5% renewable
Water conservation	Low-flow fixtures (faucets, toilets, showers, dishwashers, washing machines), recycling 25% of water for flushing toilets, irrigation, boat washing, etc.
	Water use 264 L/cap-day (70 gpcd)
Used water system and management	Centralized; 25% of the water is recycled
Percent solid waste diverted from landfill	95%
Transportation	100% of the population within a 15-minute walk to the transit hub, ferry transportation to San Francisco, bus service to San Francisco and Oakland from transit hub, car-share program, extensive network of bike paths
Recreation, leisure, sports	Network of foot and bike paths, recreational center with sports fields, marina access, neighborhood parks, and on-island organic farm
Green areas and nature	Extensive, covering 56% of the 180 hectares
Affordable housing, %	30
Probable cost	$3 billion

event, resulting in possible soil liquefaction and lateral spreading. Traffic is another issue that needs to be addressed. Currently access to Treasure Island and Yerba Buena Island is only possible via the Bay Bridge. The high volume of traffic on the Bay Bridge and the design of connecting ramps to the two islands mean that vehicular traffic access could be constrained in the future. For an increase of population on Treasure Island from 4500 to 13,500, regulators will allow only a 5% increase in traffic on the Bay Bridge.

Construction is phased over 10 years. Initial estimates were to have construction begin in 2009 and continue until 2018 in four phases; currently it is scheduled to start in 2010, with the first homes available in 2013. Phase I will center on island-wide infrastructure improvements, including but not limited to: seismic stabilization, utility distribution systems (water sewer and storm sewer), environmental remediation, and deconstruction. This phase is expected to take four years. The following three phases are expected to last two years each and include the phased construction of homes, retail space, and open spaces, starting near the transit center during Phase II and expanding to the rest of the islands in Phases III and IV.

Many groups are involved in the design of this project, including the following government agencies: the San Francisco Department of the Environment, the Treasure Island Community Development (TICD) team, the Treasure Island Development Authority (TIDA), and the San Francisco Public Utilities Commission. Private companies involved include Skidmore, Owings & Merrill (SOM) Architects, Arup, BFK, BVC Architects, CMG Landscape Architects, Concept Marine Associates, CH2M HILL, Concord Group, Engeo, Geomatrix, Hornberger + Worstell Architects, Korve Engineering, Nelson/Nygaard, SMWM, Treadwell & Rollo, Tom Leader Studio, and William McDonough + Partners.

Characteristics

Open Spaces   Extensive open spaces are provided, covering approximately 120 ha (300 acres) for the entire project, including 55% of Treasure Island. Because of the population density in the residential areas, a substantial amount of land is free to be used for the entire population of Treasure Island. There will be a number of neighborhood parks spaced among the residential areas for access and use by the residential occupants. Trees will be retained or planted throughout the island to reduce the heat island effect and to sequester carbon to offset emissions. Throughout the island native or noninvasive, climate-appropriate, and low-maintenance plants will be used in consultation with restoration ecologists and landscape architects.

At the center of the island an organic farm will be created. This will allow for the production of local foods, opportunities for training and job creation, as well as a place to use composted organic wastes from the residential areas. Other areas throughout the development will include public plazas, a sailing center, a 400-slip marina, wetlands for stormwater treatment, connections to the Bay trails, sports, and recreational areas, and an art park.

Energy   A large portion of the energy management is to reduce power demand and energy consumption. Many design criteria will be used in order to reduce the power demand throughout the island, including: appropriate building orientation at 35° from due south to optimize solar exposure and create wind protection, natural ventilation, high-performance glazing, maximum use of daylighting, integrated lighting and energy controls, specification of Energy Star–certified appliances, centralized heating and cooling, and solar hot water for residential areas.

A central utility plant for heating and cooling is planned to reduce energy consumption by 20% below the projected consumption, for certain buildings in the central core of the development. This central plant will use a distribution heat pump loop with heat pumps in each building and heat exchanges to either reject or absorb heat from the Bay, depending on the season.

Energy production on the island will be gained from the sun, wind, biogas, and tides. In order to harness energy from the sun, roof-mounted PV cells will be used. Solar panels will cover 70% of the rooftops, generating 30 million kilowatt-hours of electricity per year. Solar power will also be used for water-heating systems that can support up to 80% of the hot water needs.

To harness wind power, large-scale wind turbines and small-scale vertical turbines will be placed on top of buildings. Other energy solutions being considered include the installation of tide-driven turbines on the floor of the Golden Gate channel and a biogas generator at the used water treatment plant. The biogas generator could provide half the power and heat needed for the treatment plant.

Local energy production will only be enough to provide 50% of the power needs of the community; hence, the remaining energy will be provided by the city grid during periods of low solar energy availability. In the middle of the day, when solar output is at a maximum, more energy will be created on the island than needed, and the extra energy will be exported out of the island to the grid. The overall goal of the development is to reduce the per capita carbon emission in the city by 60%, from 3.5 to 1.3 tons (Ward, 2008).

Transportation   The main goal of the transportation design is to reduce car use and promote public transportation, walking, and biking. The transportation network throughout the island is oriented first around walking and biking, and provides integration into the regional transportation system via ferry and bus. Ninety percent of the residents will live within 1.2 km from retail services and within a 15-minute walk from an intermodal transit hub (Biello, 2008). An on-island transit system will also be provided, with a small fleet of electric or alternative-fuel shuttles. The transit terminal will provide transportation to San Francisco through a bus and ferry system.

Car use will be limited by a fee and pricing system. Parking management will be based on the policy that all auto users incur a parking charge. A congestion pricing program will be applied to people who choose to use their car to get to and from the island during peak travel periods. Ramp metering will also be used to limit the number of vehicles that can leave the island during periods of bridge congestion.

Water Use and Treatment   Potable water will be imported from San Francisco. Low-flow faucets, shower heads, and toilets with sensors and controls, along with low-water-use appliances including dishwashers and side-loading washing machines will be installed in all residential units to reduce water consumption. All of the water use practices will provide a reduction in water use from the existing 380–450 liters per capita per day to 265 liters/capita-day (70 gallons), a 30% reduction. The water system is essentially linear with some 25% reuse.

After the water is used it will be sent to the local WWTP. Treatment will include influent screening, a combined primary/secondary treatment process (either membrane bio-reactors or sequencing batch reactors), aerobic sludge digestion, clarification, sludge dewatering, disinfection, and odor control. After the water is treated to meet secondary standards, 75% of the water will be discharged into the Bay, and the 25% will be treated to tertiary levels using coagulation, filtration, and disinfection and then recycled for irrigating the farm, as well as for flushing toilets in commercial buildings and washing boats in the marina. When the recycled water is taken into consideration, only 834,000 m^3 (218 million gallons) of water will be used every year. This value is equivalent to the amount of water produced by 51 cm of rain (average annual rainfall) falling on the island (Ward, 2008).

Stormwater management will center on xeriscape, permeable surfaces and pavements, green roofs, and routing excess runoff to be treated in a wetland. The impermeable area of Treasure Island will shrink from 64 to 39% through these practices (Ward, 2008). Once the excess runoff is collected, it will be routed to a constructed treatment wetland. The treatment capacity will be to treat 0.5 cm/hour, which includes 80–90% of storms in the Bay area. The hydraulic residence time in the wetland will be at least 48 hours. Stormwater in excess of the treatment flow will be collected and discharged into the Bay directly.

Solid Waste   The Treasure Island development plans to divert 75% of the solid waste from landfills, with an overall goal of 100% diversion by 2020 (TICD, 2006). Organic waste will be composted with an on-site aerobic digester capable of processing 6 tons per day of compost, to be used on the island's urban farm and community gardens. Separate bins will be used for compostables, recyclables, and general waste in public areas as well as residential units.

Summary   The Treasure Island development, including both Treasure and Yerba Buena Islands, will combine high-density residential areas with large open community parks, neighborhood areas, and a community organic farm. Renewable energy will provide 50% of the power required by the island by using technology such as PV cells, wind turbines, and biogas digesters, along with building designs that reduce energy consumption. Walking and biking will be promoted through extensive bike and pedestrian paths, close proximity of residential areas to the transportation depot, and commercial areas designed to fit the needs of the community. Stormwater will be treated with a centralized wetland, and 25% of the wastewater will be recycled for irrigation and commercial use. With the installation of low-flow appliances and fixtures, total water use will be reduced by 20%; however, the water use will still be very high and not commensurate with other ecocity developments. Energy produced in the island central power plant will be mostly derived from conventional fuels, and only 5% from renewable energy sources. On-site composting of waste to be used on the island and an extensive recycling program are expected to reduce trash exports from the island by 100% by 2020.

Figure 11.17 Layout of the Sonoma Mountain Village development (Courtesy Sonoma Mountain Village, SOMO).

XI.2.7 Sonoma Mountain Village (California, U.S.)

Overview Sonoma Mountain Village is located 64 kilometers (40 miles) north of San Francisco in the city of Rohnert Park, California, USA (Figure 11.17). It is being created in an area formerly occupied by an industrial park and striving to become the first North American community to be certified by the One Planet Communities criteria and the fourth in the world (Peters, 2009). The construction and design of this village are being led by BioRegional and Codding Enterprises at a total cost of $1 billion.

The construction of the Sonoma Mountain Village community started with models available for viewing in 2011–2012. Homeowners are expected to begin moving in by late 2011–2012, with the completion of the entire project expected after 2020 (Sonoma Mountain Village, 2010). In total, 1900 homes will be constructed on 0.8 km^2 of land with a mix of 900 apartments and condominiums and 1000 single-family homes. These homes will vary between single-family homes, row houses, affordable-by-design homes (smaller square footage), townhouses, multifamily condos, lofts, flats, and luxury homes, ranging from 56 to 420 m^2 (600 to 4500 sq ft) with prices from $300,000 to $3 million (Sonoma Mountain Village, 2010). The total population after construction is complete is expected to be around 5000 (Sonoma Mountain Village, 2008).

The community will include 500,000 square feet of commercial, retail, and office space to serve the needs of the neighborhoods and surrounding communities (Figure 11.18). Currently 21 businesses are located in the community, as well as 27 sustainability-oriented and socially relevant technology start-up companies (Sonoma Mountain Village, 2010). Table 11.7 presents the basic characteristics and parameters of the city.

Following the 10 principles of the One Planet Living concept (see Chapter II), the ecocity principles for Sonoma Mountain Village pertain not only to the final product, but also to the construction process. A number of measures are being conducted

Figure 11.18   Artist's rendition of public areas (Courtesy Sonoma Mountain Village SOMO).

Table 11.7   Characteristics of the Sonoma Mountain Village development

Location	Sonoma Village, CA, USA
Area of development	0.8 Km^2 (80 ha or 200 acres)
Population served	5000 people
populations density	62 people/ha (25 people/acre)
Project team	
Key partners	BioRegional, Codding Enterprises
Lead planners	Codding Enterprises
Architect	Farrell-Faber & Associates, Fisher Town Design, KEMA Green, Scott Architectural Graphics, MBH Architects, WIX Architecture
Contact web site	http://www.sonomamountainvillage.com/
Type of drainage	
Sanitary	Subsurface connected to a centralized treatment plant
Stormwater	Rain gardens, biofiltration, swales, pervious pavements in alleyways, construction of stream to transport runoff out of village
Renewable energy	Photovoltaic arrays on building tops
Water conservation	Rainwater harvesting, water reuse, and use of low-flow devices including ET irrigation technology
Used water system and management	Centralized
Transportation	Promote biking and walking within the city, car share/carpool programs, rail transport to nearby cities
Recreation, leisure, sports	Network of foot and bike paths, sports fields
Green areas and nature	Green areas throughout city
Project cost	$1 billion

to ensure that energy use and damage to the environment are minimized during the building phase, as they will be when the city is complete. During construction, vehicle access is constrained to existing roads and new asphalt roads, storm drains are protected with filter strips and settling areas as needed, and any significant vehicle use off roads is preceded by soil stabilization with gravel and the use of additional silt fences and earth dikes. All asphalt and concrete removed from previous construction are reused on-site. Stockpiling of these materials will require appropriate containment areas to prevent oils and concrete dust from mobilizing. Temporary seeding and mulching are used to stabilize bare soil throughout the project. Silt fences, sediment traps, basins, and biofilters are used.

Open Spaces   Open spaces, parks, and community areas are located throughout the 0.8-km^2 land area, including over 10 hectares of parks, many kilometers of trails for walking and bicycling, dog parks, and an international all-weather soccer field (Sonoma Mountain Village, 2009). Landscaping will include groupings of plant species native to California and species adapted to the local climate. Throughout the development, turf areas will be limited. Trees will also be planted along the streets and chosen for their heartiness, shade, and beauty. Residents will have access to community gardens, fruit trees, and a year-round farmers' market (Peters, 2009). In addition to the local farmers' market, 65% of all food consumed by the community will come from within 300 miles, with up to 25% coming from within 50 miles, promoting locally grown sustainable farming practices (Sonoma Mountain Village, 2008). In addition to all of the green spaces located on ground level, green roofs will be used throughout the community. In all, 10% of the land is set aside for habitat and 20% of the land for green spaces, with a total of 50% of the project area acquiring conservation easements using pollinator gardens on green roofs, native flowers, trees, and grasses throughout the community (Sonoma Mountain Village, 2010).

Energy   The energy plan in the village community is focusing on solar energy and energy conservation. A $7.5-million, 1.14-MW, 5845-photovoltaic panel solar array was mounted on the roof of an existing building (Figure 8.3) within the community in 2006 (Peters, 2009). This array is being used to power the construction of the development and will then be used to help power the community. When the community is finished, the solar power output is expected to quadruple, with excess energy routed to the utility grid.

The energy efficiency of the designed buildings, when compared with the state of California's current energy code, uses 50% less energy. The use of ground source heat pumps, ultra-efficient lighting and appliances, super-insulated walls, floors, and roofs, along with solar hot water preheating systems is expected to accomplish this goal (Sonoma Mountain Village, 2008). By 2020 the energy use in buildings will have zero carbon equivalent emissions, while average California homes have CO_2 equivalent energy emissions of around 8240 tons per year.

Transportation   The transportation goals in the community are walking and biking as the primary transportation methods. Every resident will be no more than a five-minute walk from groceries, restaurants, day care, and other amenities offering local, sustainable, and fair trade products and services. These services are located in the town square at the center of the community (Peters, 2009).

Narrow tree-lined streets, paths, and convenient bicycle parking will be available throughout the village. Free bikes, electric vehicles that connect to the smart grid, a biofuel filling station, plug-in hybrid car-share programs, and carpool concierge services will reduce car traffic throughout the village. A commuter rail is also available for transportation from and to nearby cities, with a station located within 10 minutes of the village (Peters, 2009). Overall, the goal of the community is an 82% reduction of greenhouse gas emissions from traveling to, from, and within the village (Sonoma Mountain Village, 2008). A typical California resident emits annually 22,140 tons of equivalent CO_2, whereas it is estimated that the people located inside this development will emit only 3940 tonnes (4343 tons) annually for transportation (Sonoma Mountain Village, 2010).

Water Use   The goal for water used within Sonoma Mountain Village is a reduction in water consumption by 60% from the general norm for single-family homes in

Table 11.8   Water saving, reclamation, and reuse in the Sonoma Mountain Village development (Codding Enterprises, 2007)

Municipal Drinking Water Supply	Reclaimed Water	Rainwater	Gray Water
All contact uses in buildings Toilet flushing in most existing commercial buildings Residential toilet flushing per Title 22 Private backyard irrigation	Toilet flushing in new commercial buildings All common area irrigation Colling tower Fire hydrants	Habitat maintenance Groundwater recharge Common area irrigation Cooling tower	Small-scale private backyard subsurface irrigation

the region (Codding Enterprises, 2007). This will be accomplished through water reduction devises, education, rainwater harvesting, and reuse of water. The municipal drinking water supply will be used inside all buildings and for irrigation in private backyards. Reclaimed water will be used for irrigation of all public parks, medians, and street trees, along with irrigation of all common areas, private front yards, and fire hydrants (Sonoma Mountain, 2010). Reclaimed stormwater will be used for habitat maintenance, for groundwater recharge, and as a supplemental irrigation supply for all landscape areas (Water Balance, 2006). Habitat-protected bioswales will act as wetlands connected to a 15,100-m^3 (4-MG) underground reservoir, from which water will be recycled for irrigation purposes (Kraemer, 2008). The savings, reclamation, and reuse components of the water system are presented in Table 11.8.

The Water Plan for the village (Codding Enterprises, 2007) estimates average daily water use for the village as 1186.5 m^3/day, of which 31% is for irrigation (with reclaimed water), 60.5% for residential water demand, and 8.5% for commercial use. This will correspond to water demand of 237 liters/capita-day, which is significantly lower than the typical municipal water use in California. Specifically, for Sonoma County the average water use in 2005 was 605 L/cap-day (160 gpd). Of the 237 liters/cap-day, 22% is reclaimed water from treated effluent and stormwater; hence, the average demand on the municipal grid is 185 liters/capita-day.

Stormwater   Throughout the village, stormwater management practices will be used to reduce pollutants and runoff coming from the development. Rain gardens and biofiltration swales are to be used as the initial primary catchment for the runoff from the main street network and from roof downspouts on large buildings. These systems will drain filtered water to the underlying aquifers, reducing runoff volumes while increasing groundwater recharge (Sonoma Mountain, 2010). Alleyways will be constructed with pervious pavements and combined with under-drained substrate to reduce the amount of impervious surfaces in the development. Street trees will provide additional areas for the transient storage and percolation of stormwater in the soil structure (Sonoma Mountain, 2010)

Underground infiltration galleries will also be used to store and percolate runoff where space restrictions or other land use considerations limit the use of biofiltration or rain gardens. A channel corridor will also be constructed running the length of the village along an existing railroad track. Along this corridor trails and attractive landscaping will be used with a channel system that will have overbank storage for flood flows to transport stormwater out of the village. In order to control peak runoff flows, stormwater detentions will also be used (Sonoma Mountain, 2010). Throughout the development, stormwater will mainly flow on the surface and through the soil, rather than in pipes (Sonoma Mountain Village, 2008).

In order to reduce the amount of pollutants from contaminated stormwater, all homeowners will receive a manual welcoming them to the neighborhood and describing how to maintain their home. This manual will contain a section detailing all prohibited materials and explaining why they cannot be used. These materials will prohibit use of synthetic fertilizers, but compost and naturally derived fertilizers will be allowed and used extensively (Sonoma Mountain, 2010).

Waste Management   Waste management throughout the community will begin with the construction phase and continue through the life of the village. Existing buildings will remain, simply incorporated into the design. All asphalt and concrete removed from the area was stockpiled and reused during construction (Sonoma Mountain Village, 2008). The home manufacturing is being done on-site in a near-zero-waste panel-home production facility. All of the new construction will utilize recycled steel framing from an on-site factory run by Codding Steel Frame Solutions. This new technology will allow for building a 200-m^2 home with recycled steel from used cars rather than trees (Sonoma Mountain Village, 2009). This facility will be run on solar power and create zero waste, with the final steel frame products being 100% recyclable (Peters, 2009). Twenty percent of materials for the entire construction process will be manufactured on-site, and 60% will come from within 500 miles (Sonoma Mountain Village, 2008). Overall, the amount of CO_2 equivalent greenhouse gas emission during the construction period of the development will be reduced from a California average of 113,400 tons for a similar community to 39,690 tons.

In terms of waste management, after the completion of development, an intensive recycling program will be put into place, resulting in only 2% of the waste entering landfills by 2020. This plan includes addressing retail and grocery packaging, food waste composting, school education, and creative contests to promote waste-free living (Sonoma Mountain Village, 2008). Town-wide composting is used to create soils for the community gardens, small parks, and fruit trees throughout the village (Kraemer, 2008).

Summary   Sonoma Mountain Village will incorporate the 10 One Planet Living principles into the design of a small 5000-person village north of San Francisco. The community has applied for inclusion in the LEED Neighborhood Development pilot program to obtain Platinum LEED certification for the entire village, as well as LEED certification for each individual building (Carlsen, 2007). The development is scheduled to be completed after 2020 and upon completion could become the first development in North America to be certified as a One Planet Living community.

XI.2.8 Dockside Green

Dockside Green is a mixed-use, brownfield redevelopment project, in the city of Victoria, British Columbia, whose design and planning process is based upon achieving both LEED Platinum-level certification for all buildings and the landscape-scale LEED Neighborhood Development designation, as defined by the Canada Green Building Council. The Dockside Green site is a former heavy industrial, contaminated site, designated by the city for development. The resulting development concept was based upon an IRM construct and features district energy, on-site wastewater collection and treatment, and a biomass gasification facility (O'Riordan et al., 2008). Figure 11.19 shows the aerial view of the development and Figure 11.20 is

Figure 11.19 Aerial view of Dockside Green in 2009, with the first two development phases complete (Courtesy W.P. Lucey, Aqua-Tex Scientific Consulting, Ltd, Victoria, British Columbia).

Figure 11.20   Artist's rendering of Dockside Green with the central waterway as a major feature (Dockside Green, 2005).

an artist's rendering of the central waterway. See also Figures 10.3 and 10.6. In 2010 Dockside Green development received platinum LEED certification.

Water   The core design concept of the on-site treatment train process is a closed-loop cycle minimizing operating costs, while providing a fit-for-purpose, reclaimed water supply used for toilet flushing, landscape irrigation, green roof watering, and an on-site natural stream/pond complex. The used water treatment plant is situated beneath some of the most desirable of Dockside's residential units and is immediately adjacent to a well-known Victoria bakery; it has a low visual impact and produces little to no odor or noise (see Figure 10.3). The treatment technology is a Zenon system (GE Water & Process Technologies), incorporating suspended-growth, activated sludge, coupled with hollow-fiber ultrafiltration membranes, UV disinfection, and effluent tank storage, prior to reclaimed water use. The reclaimed water meets the province's Municipal Sewage Regulation criteria for unrestricted public access. Dockside Green uses the reclaimed water for toilet flushing, green roof and balcony planter irrigation, and to augment the on-site watercourse. The discussion in this section is based on O'Riordan et al. (2008).

The domestic wastewater treatment system consists of two similar parallel facilities rated for 190 m^3/day (one system installed with each of the two development phases). This parallel treatment train redundancy design enables periodic maintenance of each system (one system being operational), while all system processes and mechanical elements have full redundancy. There is an emergency bypass that connects this facility with the city's large sewerage system, which lies adjacent to the development site. There is a natural stream channel and pond complex that has been constructed through the central longitudinal axis of the site, providing direct residential access to a natural landscape feature, maintained in proper functioning condition. This feature has significantly enhanced the valuation appraisal of those residential units fronting the stream/pond system, while significantly increasing the biological diversity of the site. The increased economic return from enhanced real estate values exceeds the cost of the construction of an ecologically viable stream/pond complex.

Rainwater capture further augments the closed-loop design process, as rainwater routes are aligned to the stream/pond complex, providing for the design-with-nature concept of "capture, store, beneficial use." Stormwater is captured and biofiltered on-site by green roofs, cisterns, and bioswales and channeled into the waterway for storage. Water within the stream/pond complex can be recycled to provide a range of flow rates to optimize seasonal water quality, as the stream facility provides a natural polishing process. Finally, the implementation of a closed-loop water management and treatment train process permits the green roofs to be managed such that soil moisture is optimal for plant growth and high rates of evapotranspiration, resulting in significant cooling of buildings and the potential for an urban agriculture-based production of high-value crops, as a further revenue generation mechanism that can offset long-term building management and maintenance costs.

The latter design process shifts building shell (roof) design from a "green building/green roof" stormwater management concept to an integrated Engineered Ecology design, whereby ecological function is preserved. The shift in design thinking requires the natural, terrestrial landscape (contiguous horizontal and vertical connection via the soils) to be replaced with an "Island Archipelago" ecology (buildings with ecologically functional green roofs), with the roads and other ground-level structures forming the media within which the "islands" are situated: thus, the essential sustainable design concept that an ecosystem's proper functioning condition must not be lost but may be changed from one functional condition to another (O'Riordan et al., 2008).

Energy   Dockside Green utilizes an on-site biomass gasification technology to enable the community to generate clean, low-cost heat using locally sourced wood waste (Figure 8.26). Biomass gasification is a process whereby organic wastes—for example construction wood waste—are converted into synthetic natural gas, or "syngas." The resulting syngas is GHG neutral and can be used for any natural gas application: the generation of heat, electricity, or both (cogeneration). Dockside Green will use the syngas as the principal energy source for the district heating system (Ministry of Community and Rural Development).

Gasification is a thermal-chemical process that uses heat to convert any organic (carbon containing) fuel into syngas. The biomass gasification plant uses locally sourced wood fuel, including construction wood debris and municipal tree trimmings. Dockside's gasifier is a fixed-bed, bottom-fed, updraft design. Before the fuel arrives, it is broken down to pieces roughly 3 inches in diameter to prevent equipment jamming. The fuel is then fed into the bottom of a tall cylindrical gasifier where the wood is heated in an oxygen-starved environment to induce release of volatile gases. During the process, pyrolysis and gasification convert the fuel to syngas, leaving behind noncombustible ash. The temperature must be precisely controlled (1500–1800°F) for efficient gasification and to ensure that the ash does not melt, and remains granular and free flowing (Sparica, 2008).

Following gasification the syngas travels through an oxidizer (combustion), where the gases are combusted and directed through a boiler. The boiler will produce hot water for the majority of Dockside heating requirements, and it can be sold off-site to neighboring hotels and businesses. Remaining particulates are removed from the flue gas prior to release, using an electrostatic precipitator. The whole facility is centrally located in the development and housed in an acoustically isolated building to avoid any noise disturbance. The energy system is backed using a natural gas boiler (Sparica, 2008).

Dockside's other sustainable features/initiatives:

- Energy/water meters in suites, which detail personal daily use
- Dual-flush toilets and low-flow fixtures
- Efficient washing machines that use less water and require less drying
- Energy supply supplemented with photovoltaic panels and wind turbines
- Buildings that provide 100% fresh air by utilizing central or individual heat recovery ventilators
- Building materials with low or no volatile organic compounds (VOC), i.e., carpets, paints, adhesives etc.
- Encouragement of exclusive use of green cleaning products
- Recycled content, sustainably harvested materials, and rapidly renewable resources to be used whenever possible, for example:
  - Carpets sourced from a GHG-neutral business
  - Use of rapidly renewable bamboo flooring
  - Selection of salvaged wood products
- Establishment of a mini-transit system, linking Dockside to popular downtown destinations
- Creation of pedestrian- and bicycle-friendly infrastructure
- Implementation of a car-share program
- Efficient lighting, occupancy sensors, solar lighting, and a design that enhances daylight
- Organics collection initiative
- Commitment to education using a variety of communication channels and strategic relationships

XI.3 BRIEF SUMMARY

The pressing global concerns of a changing climate, a post-peak oil world, declining freshwater resources, degraded and de-evolving marine ecosystems, together with the challenge of a steadily increasing population becoming increasingly urbanized, demand the accelerated adoption of regenerative adaptive design-based development. Twenty-first century design will require innovation, imagination, and courage to overcome the barriers of regulation, financial risk, and loss of natural capital. There are projects on the immediate horizon that are seeking this scale of change, including Montreal, Victoria, and Vancouver, Canada; Malmö, Sweden; 2012 Olympic sites in London, UK; Oslo, Norway; Portugal, Australia, South Africa, and Turkey, in addition to those described in this chapter and throughout this book. Perhaps in addition to whole-city change, it will be essential to focus on restructuring educational institutions to accelerate the change required for integrated sustainable water and landscape management.

Currently, there are dozens of urban developments throughout the world claiming to be or become "ecocities." A few, with various degrees of success, are striving to become certified as One Planet Living communities. In our analysis we have compared only a fraction of the most publicized developments. The analysis was hampered by a lack of peer-reviewed technical articles; the majority of information sources were web-based articles or gray documents by the developers and/or media, admiring or criticizing such developments. Nevertheless, the picture that is appearing is far reaching, especially in the context of sustainability and reducing global warming. As a matter of fact, accepting the One Planet Living criteria will lead to dramatic reduction of GHG emissions from urban areas, including transportation, which represent the major cause of global warming. Several of the assessed ecocities are near compliance with the OPL criteria.

Because of their frugality with respect to energy and water requirements, the "new cities" can be built in "hostile" environments such as arid or desert areas or decontaminated brownfields. This may relieve the pressure on valuable agricultural lands, wetlands, and forests, even in countries still undergoing excessive population growth. With the exception of Dongtan, all the ecocities analyzed herein have been planned for, or are being built on, previously contaminated brownfields or poor-quality, inhospitable lands.

In developed countries, the direction is more towards retrofitting the existing cities and reducing, or even reversing, urban sprawl by bringing people back from distant energy-gobbling low-density suburbs to the retrofitted and water- and energy-efficient cites or build satellite medium density communities interconnected by efficient and good pubic transporation instead of low density urban sprawl surrounding the major cities relying on long commutes by automobiles on congested highways. Unfortunately, in old municipalities, bringing new sustainable concepts into rebuilding and retrofitting may run into resistance and obstacles caused by existing regulations and traditions.

These concepts are also applicable to developing countries. Cities or parts thereof may not have sewers; stormwater drainage is on the surface—in most cases in lined channels also carrying gray water. These systems can be improved without destroying them and replacing them with hard underground sewers and infrastructures. It is perfectly acceptable to replace latrines with flushless composting toilets or to collect fecal matter for community biogas production, and implement urine collection for fertilizer recovery and simple treatment (e.g., by wetlands) of gray water for reuse. Use of photovoltaics for TVs and light, and concentrated heat solar panels for hot water is becoming common even in some shantytowns (perhaps sponsored by international organizations) that do not have electricity from the grid. We have also documented that the use of fecal matter (both human and animal) and vegetation for biogas production is also taking root in some developing countries. The degree of recycling is also very high. As a result, annual GHG emissions are less than 1 ton/person and should stay that way.

The future of the Cities of the Future starts now.

REFERENCES

GENERAL

Environmental Objectives Council (2006) *Sweden's environmental objectives – Buying into a better future, de Facto 2006.* Swedish Environmental Objectives Council, Stockholm, pdf: www.miljomalen.nu

Ma, Q. (2009) "Eco-city and Eco-planning in China: Taking an example for Caofeidian Eco-city," Proceedings of the 4th International Conference of the International Forum on Urbanism, Amsterdam 2009, http://newurbanquestion.ifou.org/proceedings/index.html (accessed December 2009)

Malmqvist, P. A. and G. Heinicke (2007) "Strategic planning of the sustainable future wastewater and biowaste system in Göteborg, Sweden," Chapter 18 in *Cities of the Future: Towards Integrated Sustainable Water and Landscape Management* (V. Novotny and P. Brown, eds.), IWA Publishing, London, UK, pp. 284–299

Novotny, V. and E. V. Novotny (2009) "Water centric ecocities – Towards macroscale assessment of sustainability," *Water Practice and Technology* 44)**4** (4):, ISSN Online: 1751-231X

Register, R. (1987) *Ecocity Berkeley: Building Cities for a Healthy Future*, North Atlantic Books, Berkeley, CA

CASE STUDIES

Hammarby Sjöstad

GlashusEtt (2007) *Hammarby Sjöstad – a unique environmental project in Stockholm*, GlashusEtt, Stockholm, http://www.hammarbysjostad.se/inenglish/pdf/HS_miljo_bok_eng_ny.pdf; http://www.hammarbysjostad.se/frameset.asp?target=inenglish/inenglish_project.asp

Vestbro, D. U. (2005) "Conflicting perspectives in the development of Hammarby Sjöstad, Stockholm," paper online, http://www.infra.kth.se/bba/HamSjostad.pdf http://en.wikipedia.org/wiki/Hammarby_Sj%C3%B6stad

Dongtan

Anon. (2009a) "A Chinese eco-city – City of dreams," *The Economist*, March 21

Arup (2008) "Dongtan Key Facts," http://www.worldarchitecturenews.com/news_images/Dongtan.pdf

Head, P. R. and J. G. Lawrence (2008) "Urban development to combat climate change: Dongtan eco-city and risk management," CTBUH 8th World Congress, Dubai, March 2008

Langellier, L. P. and B. Pedroletti (2006) "China to build first Eco-city," *Guardian Weekly*, May 7

Moore, M. (2008) "China pioneering ecocity of Dongtan stalls," *Sunday Telegraph*, October 19

SIIC (2003) "About Dongtan," Shanghai Chongming Dongtan Investment & Development Co., http://www.dongtan.biz/english/zhdt

Urban Agent (2008) "Sustainable City Race, Part 3: Dongtan," *Agents of Urbanism*, April 16, 2008, http://agentsofurbanism.com/2008/04/16/sustainable-city-race-part-3-dongtan

Qingdao

Fraker, H., Jr. (2006) "Unforbidden Cities: Can a new type of 'gated community' reverse China's ecological debacle?" *California Magazine* 118(5), http://www.alumni.berkeley.edu/calmag/200609/fraker.asp

Fraker, H., Jr. (2008) "The Ecoblock-China Sustainable Neighborhood Project," PowerPoint presentation, Connected Urban Development Conference, September 24, 2008, Amsterdam, http://bie.berkeley.edu/ecoblocks

Green Dragon Media Project (2008), www.greendragonfilm.com/qingdao_ecoblock_project.html

Schlaikjer, E. (2007) "EcoBlocks in China," ResponsibleChina.com, October 30, 2007, responsiblechina.com/2007/10/30/ecoblocks-in-china

Tianjin
Source Web Sites

Clean energy (2008) http://www.cleanergreenchina.com/2008/11/23/US-china-green-energy-conference-beijing-conference-notes-1/

http://www.eco-city.gov.cn/eco/shouye/zongtiguihua_en/Part_5/index.html

http://www.kepcorp.com/press/press.asp?RID=1994&L=&Y=2009&Q=1

http://www.Luxuryasiahome.wordpress.com/2008/5/07/97b-price-tag-for-landmark-tianjin-eco-city/

http://www.tianjinecocity.gov.sg

http://www.wikipedia.org/wiki/sino-singapore_tianjin_eco-city

Masdar

Al-Jaber, S. (2008) Select Committee Hearing: "Planning Communities for a Changing Climate—Smart Growth, Public Demand and Private Opportunity," Washington, D.C. June 18, 2008, http://globalwarming.house.gov/pubs/?id=0044#main_content

Alnaser, W. E. (2008) "Major Solar and Wind Energy Projects in GCC Countries," World Renewable Energy Congress, http://www.see.ed.ac.uk/comp/pcsoft/apps/shire/.n/The-School/RIs/ESI/Conf_proceedings/WREC_X_2008/DATA/INVP81.PDF.

BioRegional (2008) One Planet Living: United Arab Emirates endorsed community – Masdar City, http://www.bioregional.com/oneplanetliving/uae/masdar

Corporate Counsel Center (2009) "Suntech Solar Panels to Power Largest PV Solar Project in the Middle East," http://news.corporate.findlaw.com/prnewswire/20090119/19jan20090030.html

CH2M HILL (2008) *Water Portfolio Management – Understanding Challenges Around the World*, http://www.ch2m.com/corporate/wfes/assets/water/BROCH_WPM_WFES.pdf.

Crampsie, S. (2008) "City of Dreams," *Engineering and Technology* 3(15):50–55

Foster + Partners (2007) "Foster + Partners create the world's first zero carbon, zero waste city in Abu Dhabi," press release, London

Hahn, M. (2009) "Masdar City Abu Dhabi," AGS Annual Meeting, ETH Zurich (Swiss Federal Institute of Technology, Zurich), January 26–29, 2009, Maxmakers Ltd., Zurich

Hartman, M., M.B. Knell, and J.W. Witherspoon (2010) Masdar City's integrated approach to sustainability, Proc. Cities of the Future 2010 Conference, March 7-10, Cambridge, MA. PP. 104-117

Palca, J. (2009) "Abu Dhabi aims to build first carbon-neutral city," http://www.npr.org/templates/story/story.php?storyId=90042092 (accessed February 2, 2009)

Todorova, V. (2008) "Abu Dhabi green city Masdar sets water efficiency targets," *The National*, May 03, http://www.questjournalists.com/index.php/site/abu_dhabi_green_city/

WSP (2009) "Masdar City Case Study," http://www.wspgroup.com/upload/documents/PDF/news%20attachments/Masdar%20Case%20Study.pdf

World Wildlife Fund (2008) "WWf, Abu Dhabi unveil plans for sustainable city," http://www.panda.org/index.cfm?uNewsID=121361

Treasure Island

Biello, D. (2008) "Eco-Cities: Urban Planning for the Future," *Scientific American: Earth 3.0*, September 2008, http://www.sciam.com/article.cfm?id=eco-cities-urban-planning&sc=SE-Earth3_20080924

Sylvan, J. (2008) "New City-Scale Ecocity Initiative: Treasure Island redevelopment project," Ecocity World Summit, April 25, 2008, San Francisco, California, http://www.ecocityworldsummit.org/presentations.fri.2.htm

TIDA, TICD (2006) *Development Plan and Term Sheet for the redevelopment of Naval Station Treasure Island*, http://www.sfgov.org/site/uploadedfiles/treasureisland/Treasure_Island_Development_Plan/FinalDevPlanDec06.pdf (accessed December 2006)

TICD (2006). "A sustainable future for Treasure Island," http://www.sfgov.org/site/uploadedfiles/treasureisland/Treasure_Island_Development_Plan/Part1SustPlan10-24-06.pdf (accessed October 2006)

Ward, L. (January 2008) "The City of the Future," *Popular Mechanics* 185, pp. 76–81

Sonoma Mountain Village

Carlsen, Robert (Spring 2007) "Sonoma Mountain Village Project Gets Solar Energy Boost," *Green Building Quarterly*, http://california.construction.com/news/2007_spring_greenbuilding.pdf

Codding Enterprises (2007) "Sonoma Mountain Village Water Plan," Codding Enterprises, Rohnert, CA Kraemer, S. (2008) "From Industrial Park to Zero Carbon Town," http://featured.matternetwork.com/2008/10/from-industrial-park-zero-carbon.cfm (accessed March 12, 2008)

Peters, A. (2009) "Sonoma Mountain village: Is Green Suburbia Possible?" Worldchanging, Seattle, WA, http://www.worldchanging.com/archives/009448.html

Sonoma Mountain Village (2008) http://www.ci.rohnert-park.ca.us/Modules/ShowDocument.aspx?documentid=113

Sonoma Mountain Village (2010) http://www.sonomamountainvillage.com/community/index.php (accessedMay 30, 2010)

Dockside Green

Corps, C., S. Salter, W. P. Lucey, and J. O'Riordan (2008) *Resources from Waste: Integrated Resource Management Phase I Study Report*, British Columbia Ministry of Community Services, Victoria, BC, Canada

Dockside Green Website, www.docksidegreen.com

O'Riordan, J., W. P. Lucey, C. L. Barraclough, and C. G. Corps (September, 2008) "Resources from waste: An integrated approach to managing municipal water and waste systems," *Industrial Biotechnology* 4(3):238–245

Sparica, D. (2008) "Biomass Gasification Anchors Dockside Green," *Municipal World: Canada's Municipal Magazine*, January 2008, pp. 13–14

APPENDIX

Conversion Factors: US Customary to SI (Metric)

US Customary Units	Multiply by:	To obtain SI Units	
		Symbol	Name
acres	0.405	ha	hectares[a]
acre-ft	1233.5	m^3	cubic meters[b]
acre-in.	102.79	m^3	
feet (ft)	0.3048	m	meters[c]
ft/s (fps)	0.3048	m/s	meters per second
ft^2 (sq ft)	0.0920	m^2	square meters[d]
ft^3 (cu ft)	0.0283	m^3	cubic meters
$ft^{3/}s$ (cfs)	0.0283	m^3/s	cubic meters per second
°F	0.555 (°F - 32)	°C	degrees Celsius
gallons, U.S. (gal)	3.785	L	liters
gal/acre	9.353	L/ha	liters per hectare
gal/ft^2	40.743	L/m^2	liters per square meter
gal/ft^2	0.0407	$m^3/\,m^2 = m$	meters
gal/ft^2-day	4.72×10^{-7}	m/s	meters per second
gal/ft-day	1.438×10^{-7}	m^2/s	square meters per second
hp	0.746	kW	kilowatts
inches (in.)	2.54	cm	centimeters
in^2	6.452	cm^2	square centimeters
in^3	16.39	cm^3	cubic centimeters
in^3	0.0164	L	liters
pounds (lb)	0.454	kg	kilograms[e]
lb	454	g	grams
lb/acre	1.121	kg/ha	kilograms per hectare

(*Continued*)

Conversion Factors: US Customary to SI (Metric) (*Continued*)

US Customary Units	Multiply by:	To obtain ST Units	
		Symbol	Name
lb/ft^3	16,042	g/m^3	grams per cubic meter[f]
lb/in^2 (psi)	0.0703	kg/cm^2	kilograms per square centimeter [g]
lb/Mgal	0.120	mg/L	miligrams per liter
lb/mi	0.282	g/m	grams per meter
miles (mi)	1.609	km	kilometers
miles/hour (mph)	0.447	m/s	meters per second
million gallons (Mgal)	3785	m^3	cubic meters
million gallons per day (Mgd)	0.0438	m^3/s	cubic meters per second
mi^2 (sq mi)	2.59	km^2	square kilometers
parts per billion (ppb) in water	1.0	$\mu g/L$	micrograms per liter
parts per million (ppm) in water	1.0	mg/L	miligrams per liter
tons (short)	0.907	t	tonnes
tons/acre	2240	kg/ha	kilograms per hectare
tons/sq mi	3.503	kg/ha	kilograms per hectare
yards (yd)	0.914	m	meters
yd^3	0.765	m^3	cubic meters

[a] $1\ ha = 10,000\ m^2 - 0.01\ km^2$.
[b] $1\ m^3 = 1000$.
[c] $1\ m = 100\ cm = 0.001\ km = 1000\ mm$.
[d] $1\ m^2 - 10,000\ cm^2 = 10^{'6}\ km^2$.
[e] $1\ kg - 1000\ g = 0.001$ tonne.
$1\ g - 1000\ mg - IO^6//g = 109$ ng.
[f] $1\ g/m^3 = 1\ mg/L$.
[g] $1\ kg/cm^2 = 0.968$ atm $- 0.981$ bar.

INDEX

Accra, Ghana, 314
Acid precipitation, 179, 253
Activated sludge process, 316, 318–319
Adaptive capacity, 154–155
Adaptive planning, 155, 157–158
Aerobic treatments, 29, 315–316, 346, 347
African countries, 45, 50
Aging infrastructure, 59–60
Agricultural water use, 234, 240, 289–293. *See also* Irrigation
Albedo, 52
Algae to biogas processes, 327–329
Algal blooms, 78, 287, 337
Anaerobic digestion, 332–334, 346, 392–393, 409
Anaerobic treatment and energy recovery reactors (ATERRs), 351
Anaerobic treatments, 29, 316, 398–399
Ancient cities, water management in, 3, 8–12, 185
Anoxic-oxic treatment processes, 331
Anoxic treatments, 29, 315–316
Antidegradation rules, 82
Arcadia Creek (Michigan), 436
Arizona, 231, 254, 450–453
Asian countries, 41, 42, 45, 50, 237, 287, 384
Atlanta, Georgia, 229–230
Augustenborg (Malmö, Sweden), 164–167
Australia, 50, 231, 254–255, 360
Austria, 37, 55
Automobile use, 32–38, 40, 418–419

Bardenpho systems, 29
Basic water supply paradigm, 6, 8–9
Battery Park (New York City), 256, 509
Beijing, China, 448, 456
Beijing Olympic Park, 289
Berlin, Germany, 149
Best Available Treatment Economically Achievable (BATEA), 120

Best management practices (BMPs):
    for Cities of the Future, 117–118
    for diffuse pollution abatement, 29–31
    for green infrastructure, 158–170
    nature-mimicking, 115–116
    for release of stored water into receiving water, 266–267
    for runoff control, 30, 31, 110–113, 189–212
Biodiversity, 149–152
Bioelectrochemically assisted microbial reactors (BEAMRs), 405, 409, 413, 415, 416
Biofilters, 200–201
Biogas, 334, 393
    algae to biogas processes, 327–329
    electric energy production from, 399–407
    from water reclamation process, 276
Biological integrity of water, 76
Biological used water treatments, 324–329
Bioretention, 195
Black water, 106, 278, 313
    in closed cycle water reclamation, 351
    resource recovery from, 409–411
    separation of, 314–315
    treatment of, 294
Blue water, 107, 278
Boston, Massachusetts, 17–18, 51
Bottled water, 241
Brackish water, desalination of, 260–266
Brazil, 37
Bremen Street Park (East Boston, Massachusetts), 160
Brentwood, British Columbia, 153, 154
British Columbia, 411
Brownfield remediation/development, 189–190
Brown water, 243, 278, 313
Buffer zones, 199–200
Business as usual (BAU) scenario, 60–61, 67, 156

Cairo, Egypt, 48
California, 231, 245–247, 258, 384
Canada, 50, 336, 360
Canals, 13, 15, 231, 238–239

Canonical Correspondence Analysis (CCA), 444–447
Cao Fei Dian (China), 123, 541
Cap and trade system, 94, 524–525
Carbon footprint, 94, 365–371, 484–485
Carbon sequestering, 323–324, 348
Cedar Creek (Wisconsin), 467–469
Central American countries, 384
Central Arizona Project (CAP) canal, 238–239
Central government, role of, 526–527
Centralized systems, 61–62, 124, 125, 281–282
Changi Water Reclamation Plant (Singapore), 393, 407
Channel drops, 463
Charles River (Boston), 28
Chemical integrity of water, 76
Cheonggyecheon River (Seoul, Korea), 429–433, 435–437
Chicago, Illinois, 18–21, 27, 28, 149, 251–252, 283, 342, 407, 429, 450, 451, 509, 527
Chicago City Hall, 191
Chicago Sanitary and Ship Canal (CSSC), 18–20, 28
China, 37, 38, 41–42, 44–47, 55, 78, 122–123, 231, 287, 335, 360, 371, 384, 540–541
Cities. *See also* Urban ... *entries; specific cities*
    implications of global warming for, 56–59
    retrofitting, 457–476
    and sustainable development, 4
    water for, xii–xiii
Cities of the Future (COTF, COF), 114–124
    best management practices, 117–118
    costs of utilities for, 523–524
    drainage and water management, 114–116
    integrated resource management clusters, 120–123
    ISO environmental performance criteria, 117
    LEED criteria, 116–117
    low impact development, 117
    macroscale goals, 118–119
    microscale measures, 116–120
    One Planet Living, 119–120
    planning and management, 483, 485, 486
    spatial integration, 123–124
    surface water bodies in, 427–428
    SWITCH program approach to, 533, 534
    traditional cities vs., 99, 100
    utilities management in, 528
Clean Water Act (CWA), 23, 25, 26, 28, 30, 32, 65, 66, 82–83, 94, 101–102, 177, 443, 448–449, 527, 528
Closed cycle water reclamation, 348–354
Cluster-based distributed systems, 455–457, 497–499, 503–505
Cluster (ecoblock) management, 501. *See also* Ecoblocks
    gray water reclamation, 258–259
    rainwater harvesting, 253, 255–256
    water reclamation and reuse, 283–286
Cognitive values of sustainability, 83, 86
Combined sewers, 182–183
Combined sewer overflows (CSOs), 27, 28, 182–183, 267, 509
Commercial water use, 240, 249–250
Concentrated heat plants, 383
Concentrators (stormwater), 216, 217
Connectivity, 114, 123, 152–154
Conservation:
    energy, 371–380, 419–421
    as LID design practice, 187
Constructed wetlands, 123, 208, 259, 316, 319–323
Contingent valuation (DV), 522–523
Conversion factors, 595–596
Cost of sustainability, 521–525
Culverts, 463
Cuyahoga River (Ohio), 20, 21, 28
Czech Republic, 78

Dams, 15, 463–467
Daylighting, 108, 109, 436–438, 441, 469–476, 510
Decentralized systems:
    for ecocities, 124–126, 500–503
    privatization scenario for, 62
    for water reclamation, 281–282, 348–354
    for water use, 497–499
De-icing chemicals, 180, 212–213, 216, 268
Desalination, 260–266
Designer wetlands, 209
Des Plaines River (Illinois), 18–20, 28, 342, 451
Direct electric energy production, 327, 399–407
Direct hydrogen production, 327
Direct reuse, 278, 281, 297–299
Discharge permits, 528
Diseconomics, 77
Disinfection, 340–346
Distributed systems, 497–514
    cluster creation and size, 503–505
    conversion for sustainability, 511–514
    integrated resources tasks for, 499–503
    need for decentralized water use, 497–499
    regionalized vs. cluster-based, 455–457
    types of water/energy reclamations, 505–514
Diversity, 149–152
Dockside Green (Victoria, British Columbia), 282, 498, 504, 506, 585–588
Domestic water use, *see* Residential water use
Dongtan (China), 290, 489–492, 541, 549–556
Drainage:
    for Cities of the Future, 114–116
    combined sewers, 182–183
    for ecocities/ecovillages, 126–129
    in emerging paradigm, 110–113

fast-conveyance infrastructure, xiii
  natural, 183, 184
  separated sewer systems, 183, 184
Drinking water quality, 25, 234–235, 249–250,
  287. *See also* Potable water
Drivers of sustainability movement, 42–65
  aging infrastructure, 59–60
  greenhouse emissions and global warming,
    51–59
  impossibility of status quo and business as
    usual, 60–65
  population increases/pressures, 44–49
  water scarcity, 49–51
Droughts, 52, 229–231
Dry ponds, 201, 203, 205

Ecoblocks. *See also* Cluster (ecoblock)
  management
  closed cycle water reclamation, 348–354
  gray water reclamation, 258–259
  IRMCs as, 120–123
  private operation of, 62
  Qingdao, 122–123, 556–560
  rainwater harvesting, 253, 255–256
  water reclamation and reuse, 283–286
Ecocities, xv, 124–129, 539–590. *See also* Cities
  of the Future; *specific cities*
  in China, 42, 122–123, 540–541
  Dockside Green, 585–588
  Dongtan, 549–556
  Hammarby Sjöstad, 542–549
  low impact development, 489
  Masdar, 566–573
  population density effects, 489–493
  Qingdao, 556–560
  sanitary sewage conveyance, 124, 126
  smart growth, 488–489
  Sonoma Mountain Village, 579–585
  surface drainage, 126–129
  and Sweden's environmental quality
    objectives, 539–540
  Tianjin, 560–566
  Treasure Island, 573–578
Ecological corridors, 486
Ecological footprint, 485–487
Ecological habitats, fragmentation of, 178–180
Ecological (natural) uses of water, 232, 233
Ecology:
  as component of sustainability, 87–90
  and economics/society, 73, 74
  in triple bottom line concept, 104, 105, 493,
    495
Eco-mimicry, 189
Economic development (in TBL concept), 104,
  105, 493, 495
Economic footprint, 486
Economic sustainability, 82–85

Economic value of water, 94–97
Economy of scale, 97
Ecosystem services, 138–143
Ecotones, 88, 113–114, 123, 437
Effluents:
  BATEA standards for, 120
  reuse of, 106–108
  from traditional biological treatments, 311
  treated effluent discharges for ecocities,
    126–129
  water-sewage-water cycle, 280–281
Effluent dominated/dependent water bodies, 17,
  28, 280–281, 447–453
Electricity demand, 419
Electricity production, 418
  direct, 327, 399–407
  from water reclamation process, 276
  water use in, 240
  from wind, 391
Emerald Necklace (Boston), 88, 89, 459
Emerging (fifth) paradigm, 65–67, 73–75,
  97–114
  restoring urban streams, 108–110
  stormwater pollution and flood abatement,
    110–113
  triple bottom line–life cycle assessment,
    104–106
  urban landscape, 113–114
  water reclamation and reuse, 106–108
  water systems, 99–104
Emscher River (Germany), 436
End-of-pipe treatment, 7, 25–32
Energy. *See also* Water-energy nexus; *specific
  types of energy*
  from renewable sources, 380–392
  in water reclamation plants, 346–348
Energy conservation, 371–380, 419–421
Energy outlook, 36–38, 416–422
Energy recovery/reclamation, 106–108
  from biogas and used water, 399–407
  costs of, 334
  for ecocities, 125, 126
  from sludge, 334–336
  for sustainable communities, 505–514
  from used water and waste organic solids,
    392–399
  from water reclamation process, 276
Energy recovery units (ERUs), 498–499
Energy storage, 421–422
Engineered water supply and runoff conveyance
  paradigm, 6, 9–15
Environment:
  as component of sustainability, 87–90
  and economics/society, 73, 74
  in triple bottom line concept, 104, 105, 493,
    495
Environmental awakening, 23–24

Environmental corridors, 199–200
Environmental ethics, 87
Environmental policy, 94
Epidemics, 18, 19, 87
Equilibrium paradigm, 141, 142
European countries, 25, 37, 45, 50, 51, 78, 79, 96, 183, 237, 335, 361, 371, 383, 384, 390–391, 463
Eutrophication, 24, 293, 294, 336, 485
Extended aeration, 330, 331
Extended dry ponds, 201, 203, 205, 206
Externalities, 77–79, 95–96

Fast-conveyance drainage, 186
  with end of pipe treatment, 7, 25–32
  infrastructure for, xiii, 3–4, 16, 17
  with no minimum treatment, 7, 15–25
Feces (brown) water, 243, 278, 313
Federal government, role of, 526–527
Filter strips, grass, 196–198
Filtration systems, 188
  biofilters, 200–201
  grass filter strips, 196–198
  membrane filtration, 339–340
Flood control, 15, 110–113
Flooding, 22, 49, 206, 458, 521
Florida, 261
Footprints, 484–487
France, 37
Free water surface (FWS) wetlands systems, 208, 209

Ganges River (India), 28
Germany, 280, 531
Ghent, Belgium, 433–435, 509, 510
Gila River (Arizona), 450–452
Global climatic change, xiv, 51, 56, 59, 92–93
Global scale ecological footprints, 485
Global warming, 4, 24
  as driver of change toward sustainability, 51–59
  and extreme hydrologic events, 2
  and flooding, 49
  and long-distance transfers and pumping, 237–239
  and population numbers, 45
  reduction of/adaptation to, 59–60, 85
  and wetlands, 209
Göteborg, Sweden, 314–315
Gray water, 106, 278
  concentration strength of, 314
  for lawn irrigation, 247, 259
  reclamation/reuse of, 149, 256–259
  resource recovery from, 409, 410
  from showers, 243–244
  treatment of, 294

Green Alleys Program (Chicago), 152
Green development, 4, 116–117, 520
Green heart concept, 156–157
Greenhouse gases (GHGs), xiv
Greenhouse gas emissions:
  "bubble" approach for, 528
  as driver of change toward sustainability, 51–59
  in emerging countries, 42
  as externality, 79–80
  from gas-powered lawn mowers, 247
  global warming potential of, 276
  from supplying water, 231
  from urban areas, 360–361
  and use/disposal of water, 348–360
  from vehicular traffic, 35
  in water reclamation plants, 346–348
  from wetlands, 209, 259, 324, 348
Green infrastructure, 158–170
  Malmö, Sweden, 164–170
  SEA Street Seattle, 159–162
  Staten Island Bluebelt, 162–165
  Westergasfabriek Park, 162, 163
Green roofs, 190–193
Green scenario (infrastructure improvement), 63–65
Green space, 189
Green Streets (Portland, Oregon), 147–148
Green water, 278
Groundwater, 21–22, 228, 260
Groundwater recharge, 21–22, 267, 300–304
Gulf of Mexico, 78–79

Habitat degradation, 109, 439
Hammarby Sjöstad (Sweden), 51, 128, 489, 490, 503, 506–507, 530–531, 542–549
Hard infrastructure, 216–218
Heating:
  of buildings, 419
  of water, 244, 372–379
Heat pumps, 375–379
Heat recovery, 276, 366, 379–380
High-efficiency treatment plants, 124, 126
High-performance green buildings, 365, 366
Hong Kong, 491
Human (anthropogenic) uses of water, 232–233
Hydraulic energy, 406–407
Hydraulic loading, 210, 211
Hydrogen fuel cells, 276, 399–403
Hydrogen production, 327, 335
Hydrologic connectivity, 153, 154
Hydrology:
  of cluster watersheds, 503–504
  restoring, 440–443, 486
  and stormwater runoff, 177–178
  and urbanization, 440–443
Hydroperiod, 209, 211

Imperviousness, 91–92
Impervious pavements, 21, 22
Index of Biotic Integrity (IBI), 91–92, 443–445
India, 41, 42, 47
Indirect reuse, 278, 297–299, 301–302, 312
Infiltration, 186, 187, 194
Infrastructure, xiv
  aging of, 5, 59–60
  for fast-conveyance drainage, xiii, 3–4, 7, 15–32
  green, best practices for, 158–170
Institutions and governance, 525–535
  achieving multibenefit objectives, 533–535
  enhanced private sector, 532
  for integrated resource management, 526–532
Integrated analysis, 514
Integrated planning and management, 482–487
Integrated resource management (IRM), 75, 493–497, 499–503, 526–532
Integrated resource management clusters (IRMCs), 120–123, 497, 498, 529
Integrated resource recovery facilities (IRRFs), 411–416, 497–499, 529
Integrated water reclamation/reuse (Singapore), 304–308
Integrated water resources management (IWRM), 75, 98–99
Integrity of water, 76, 443
International water quality guidelines, 235
Irrigation, 50, 51
  domestic water use for, 245–247
  gray water for, 257, 258
  reclaimed water for, 282, 287
  used clean water for, 107
  using potable water for, 233–234
  water quality goals/limits for, 289–293
ISO environmental performance criteria, 117

Jackson Creek (Wisconsin), 440
Japan, 19, 41

Kallang River (Singapore), 459, 462–463
Kay Bailey Hutchison Desalination Plant (Texas), 264–266
Kitazawa Creek (Tokyo), 472–474
Kyoto Protocol, 56, 335

Lake Ontario, 219, 220
Landscapes:
  non-equilibrium theory of, 142
  as systems, 153
Landscape ecology, 140–141
Landscape irrigation, 289–293
Landscape management, 125, 127

Landscaping:
  low-impact, 188
  xeriscapes, 247–248, 289
Land use limits, 93
Latin American countries, 41, 42, 45, 50, 384, 512
Lawn irrigation, 51, 246–247, 259, 289–290
Leaks in water systems, 245, 251–252
Learning alliances, 533–535
LEED standards/certification, 116–118, 483
Libya, 231
Life cycle assessment (LCA), 367, 493
Limits, living within, 90–94
Lincoln Creek (Wisconsin), 436, 440, 459–462
Loading capacity, 94
Local ecological footprints, 485
Los Angeles, California, 48
Los Angeles River, 22, 23
Low-energy secondary treatment, 315–324
Low Impact Development (LID), 63
  for Cities of the Future, 117
  diversity of, 151–152
  for ecocities, 489–493
  natural drainage in, 183, 184
  portfolio approach toward, 221
  runoff control for reuse, 190–201
  stormwater, 186–189

Macroscale goals, 118–119
Malmö, Sweden, 164–170, 509, 510
Maryland, 444, 445
Masdar (United Arab Emirates), 51, 286, 290, 489, 490, 503, 508, 566–573
Massachusetts, 234–236, 456, 527–528
Maximum Contaminant Levels (MCL), 235
Maximum Contaminant Level Goals (MCLG), 235
Measurement conversion factors, 595–596
Medieval cities, water management in, 12–15
Megacities/megalopoli, xiii–xiv, 45, 47–49
Membrane bioreactors (MBRs), 317–319
Membrane filtration, 319, 339–340
Mexico, 335
Mexico City, Mexico, 49, 106
Microbial electrolysis cells (MECs), 405–406, 409
Microbial fuel cells (MFCs), 276, 327, 400, 403–406, 416
Microfiltration (MF), 339–340
Microscale measures, 116–120
Middle East countries, 260, 261, 287, 360, 361
Milwaukee, Wisconsin, 27, 28, 32, 108, 109, 178, 407, 429, 527
Modified dry ponds, 201, 203, 205
Modularization, 148–149
Monaco, 282
Muhlheim, Germany, 255

Multifunctionality, 146–148
Multiscale networks, 153–154
Municipal-scale utilities, 527–532
Municipal water use, 231, 235–240, 282

Namibia, Africa, 298
Nanofiltration (NF), 339, 340
National Environmental Policy (U.S.), 63
National Environmental Protection Policy Act of 1969, 80
National Pollution Discharge Elimination System (NPDES), 28–29, 120
National Primary Drinking Water Standards, 235
National Recommended Water Quality Criteria, 235
National Urban Runoff Project (NURP), 101–102, 111
Natural Biological Mineralization Process, 412–413
Natural drainage, 183, 184
Nature, protection of, 93
Netherlands, 156–157
Networks, 153–154
Net zero carbon footprint, 59, 365–371, 484–485
Net zero energy building, 365
NEWater, 231, 297, 305–307, 359
New Orleans (Louisiana), 3, 280
New Zealand, 50
Nitrogen, 313, 321–323, 337
Non-equilibrium paradigm, 141–143
Nonpoint (diffuse) pollution, 24, 25, 29–32
North American countries, 50
North Avenue Dam (Milwaukee), 464–467
Nuclear energy, 38, 54, 55
Nutrient recovery, 336–339
    for ecocities, 125, 126
    from gray water, 259
    from urine (yellow water), 243, 313–314
    during water reclamation, 275, 287

Ohio, 444–445
Oil supply, 36–37
Olmsted, Frederick Law, 87–88
Once-through flow water/wastewater management systems, 26
One Planet Living (OPL), 119–120, 290, 360, 365, 484
Ontario, Canada, 246
Orange County, California, 303–304
Ozone, disinfection with, 342–344

Paradigms of water management, 1–67. *See also* Emerging (fifth) paradigm
    from ancient cities to 20th century, 5–42
    basic paradigm, 6, 8–9
    competition for resources, 40–42

and drivers of change toward sustainability, 42–65
    engineered supply and runoff conveyance, 6, 9–15
    fast conveyance with end of pipe treatment, 7, 25–32
    fast conveyance with no minimum treatment, 7, 15–25
    impact of automobile use, 32–38
    urban sprawl, 38–40
Paradigm shifts, xvi
Peat-sand filters, 218
Pervious pavements, 193–195
Phosphorus, 313, 318–319, 337, 485
Photovoltaic plants, 383–385
Physical integrity of water, 76
Planning and design, 135–171
    achieving sustainability, 135–137
    ecosystem services, 138–143
    green infrastructure best practices, 158–170
    planning process, 143–158
    resilience strategies, 144–155
    urban sustainability through, 137–138
Planning process, 143–158
    adaptive planning, 157–158
    ecosystem service goals/assessments, 143–144
    resilience strategies, 144–155
    scenario planning, 155–157
    as transdisciplinary process, 157
Points of reuse, 498
Point source pollution control, 28–29, 101, 109
Pollution, 76–80
    causes of, xiii
    defining, 76–77
    legacy, 206–207, 467–469
    public protest against, 22–23
    solving problem of, 77–80
    from stormwater, 110–113, 177, 252–253
    from urbanization, 438–439
    from urban runoff, 179–182
Pollution export, 79
Pollution Prevention Act of 1990, 102
Ponds (runoff control), 201–207
Population density (ecocities), 489–493
Population increase/pressure, 44–49, 60, 72, 231–232
Porous pavements, 193–195
Potable water:
    direct and indirect uses, 240
    irrigation uses, 233–234
    non-drinking uses, 250
    quality standards/criteria/guidelines, 234–235
    and reclaimed water, 108
    residential use, 249
    reuse of, 278, 297–302, 312
Private sector development/governance, 532

Privatization scenario (infrastructure improvement), 61–62
Providence River (Rhode Island), 436
Public health, xv–xvi, 85–87, 93
Public water use and conservation, 240, 249–250, 289

Qanads, 9
Qingdao (China) ecoblock/ecocity, 122–123, 282–286, 489, 490, 503, 508, 556–560

Rain gardens, 126, 195
Rainwater, 179, 185–186, 279, 352
Rainwater harvesting (RWH), 10, 195, 252–256, 279–280
  for ecocities, 125
  for lawn irrigation, 247
Real time control (RTC), 123–124
Reclaimed water. *See also* Water reclamation and reuse
  defined, 278
  disinfection of, 340–346
  excess, discharge of, 311
  sources of, 275
Redundancy, 148–149
Regenerative adaptive design (RAD), 493–494
Regenerative fuel cells, 401
Regional ecological footprints, 485
Regionalized systems, 97, 120, 455–457
Regulators (institutions), 527
Regulators (stormwater), 216, 217
Renewable energy sources, 37–38
  solar energy, 380–387
  wind power, 387–392
Republic of Korea, 41
Residential water use, 239, 241–249
  gray water reuse, 258
  rainwater harvesting, 253–254
Resilience, 136, 145
  adaptive planning for, 155, 158
  as dimension of sustainability, 84–85
  in non-equilibrium paradigm, 142–143
Resilience strategies, 144–155
  adaptive capacity, 154–155
  diversity/biodiversity, 149–152
  multifunctionality, 146–148
  multiscale networks and connectivity, 152–154
  redundancy and modularization, 148–149
Resource optimization, xv
Resource preservation, 82–85
Resource recovery, 329–336. *See also* Water reclamation and reuse
  from residual solids, 334–336
  traditional technologies for, 311–329
Retrofitting cities, 75, 97, 98, 457–476
  for damages from changed hydrology, 458–463
  daylighting, 469–476

legacy pollution remediation, 467–469
  to remove fragmentation of channel, 463–467
Reverse osmosis (RO), 261–264, 339, 340, 354
Roofs, green, 190–193
Runoff:
  pollution from, 179–182
  storage and conveyance of, 187
Runoff control for reuse, 189–212
  biofilters, 200–201
  environmental corridors and buffer zones, 199–200
  filter strips, 196–198
  grass swales, 196, 197
  green roofs, 190–193
  hard infrastructure, 216–218
  LID soft surface approaches, 190–201
  LID urban drainage, 218–221
  ponds, 201–207
  porous pavements, 193–195
  rain gardens, 195
  rainwater harvesting, 195
  wetlands, 207–212
  winter limitations, 212–216

Safe Drinking Water Act of 1974 (amended 1986, 1996), 235
St. Louis, Missouri, 47
San Antonio River (Texas), 433–436
Sand filters, 217–218
San Diego, California, 301–302
Sanitary sewage conveyance, 124, 126
Sanitary sewers, 182, 183
Sanitary sewer overflows (SSOs), 27, 28, 182, 183, 267, 509
São Paulo, Brazil, 48
Satellite withdrawal and treatment, 282–283
Scenario planning, 155–157
Schwab Rehabilitation Hospital (Chicago), 191
Sea level rise, 56–58
SEA Street Seattle, 159–162, 186–187, 189, 196, 508–509
Secondary Drinking Water Regulations, 235
Self-organizing mapping (SOM), 444–446
Seoul, Korea, 256, 433
Separated sewer systems, 183, 184
Separators (stormwater), 216, 217
Sequencing batch reactors (SBRs), 319
Sewage conveyance, 124, 126
Sewers, 182–184
Shortages of water, *see* Water scarcity/shortages
Singapore, 123, 230–231, 304–308, 371, 429, 491, 509, 510, 527, 541–542
Sludge, as resource, 334–336
Sludge handling, 329–334
Smart growth, 63, 65, 488–489
Social equity (in TBL concept), 104, 105, 493, 495

Social policy imperatives, 85–87
Solaire Apartments (New York City), 191, 295–296
Solar energy, 380–387, 403
Solar water heaters, 373–375
Solid waste:
  energy from, 392–399
  management, reuse, disposal of, 530–531
Sonoma Mountain Village (California, U.S.), 363, 489, 490, 579–585
Source separation, 125, 312–315
South America, 335, 371, 384, 512
Spatial integration, 123–124
Staten Island Bluebelt (New York), 162–165
Status quo, impossibility of, 60–65
Stockholm, Sweden, 51, 144
Stony Brook (Boston), 18, 19, 109–110, 429
Storage-oriented, slow-release systems, 186
Storm sewers, 182, 183, 216
Stormwater, 177–221
  aesthetic problems from, 178
  as asset and resource, 184–186
  current urban drainage, 182–184
  in emerging (fifth) paradigm, 110–113
  and fragmentation of habitats, 178–180
  hydrologic problems from, 177–178
  in Low Impact Development, 186–189
  pollution from, 110–111, 179–182
  and quality of water bodies, 177
  rainwater vs., 279
  reclamation and reuse of, 279–280
  runoff control for reuse, 189–212
Strawberry Creek (California), 436, 441
Stream restoration, 88, 427–478
  in emerging (fifth) paradigm, 108–110
  goals of, 437–438, 453–455
  rediscovering urban streams, 427–437
  regionalized vs. cluster-based distributed systems, 455–457
  to remedy impacts of urbanization, 438–453
  retrofitting older cities, 457–458
  and water centric (eco)cities, 453–476
Street Edge Alternative project, see SEA Street Seattle
Struvite, 313, 337–339
Subsurface flow wetlands, 319–323
Surface drainage, 126–129
Sustainability, 4, 493–497. See also Urban sustainability
  achieving, 135–137
  and aging infrastructure, 59–60
  conversion of distributed systems for, 511–514
  definitions of, 4, 80–82
  drivers of change toward, 42–65
  and greenhouse emissions/global warming, 51–59
  and population increases/pressures, 44–49

and status quo/business as usual, 60–65
  water scarcity and flooding of large cities, 49–51
Sustainability science, 83, 99, 101
Sustainable communities, 482–535. See also Ecocities; Urban sustainability
  distributed systems for, 497–514
  institutions and governance for, 525–535
  integrated planning and management, 482–487
  integrated resources management, 493–497
  system analysis and modeling, 514–525
  urban planning, 487–493
Sustainable Urban Drainage Systems (SUDS), 29–30, 63, 118, 186. See also Low Impact Development (LID)
Sustainable water centric ecocities, xv
Swales, 112, 123, 126, 196, 197
Sweden, 539–540
SWITCH program, 533, 534
Sydney, Australia, 429
Syngas, 393
System analysis and modeling, 514–525

Take, make, waste approach, 495
Tampa Bay Seawater Desalination Plant (Florida), 265–266
Tankless water heaters, 373
Technocratic scenario (infrastructure improvement), 62–63
Thames River (London), 23, 28
Thermal gasification, 335
Tianjin (China), 123, 282, 285–286, 489, 490, 503, 541, 560–566
Tokyo, Japan, 18, 22, 30
Topaz solar power plant (California), 384
Toronto, Ontario, 429
Total Maximum Daily Load (TMDL) process, 94, 449
Transit-oriented development (TOD), 488
Transportation, 32–39, 418–419, 487–489
Treasure Island (California, U.S.), 489, 490, 573–578
Treatment and resource recovery unit processes, 311–354
  algae to biogas, 327–329
  anaerobic sludge blanket reactors, 324–327
  biological treatments, 324–329
  decentralized water reclamation technologies, 348–354
  direct electricity and hydrogen production, 327
  disinfection, 340–346
  energy and GHG emission issues, 346–348
  membrane filtration and reverse osmosis, 339–340
  nutrient recovery, 336–339
  sludge handling and resource recovery, 329–336

source separation, 312–315
traditional water/resource reclamation
    technologies, 311–324
Trickling filters, 318, 415
Trinity River (Texas), 280–281, 456
Triple bottom line (TBL), 104, 136, 493–495,
    518–525
Triple bottom line–life cycle assessment
    (TBL-LCA), 64, 104–106, 493, 518
Tucson, Arizona, 26, 106

Ultrafiltration (UF), 339–340
Ultraviolet radiation, disinfection by, 343–346
United Kingdom, 369
United Nations Millennium Ecosystem
    Assessment, 138, 143
United States, 50, 51, 56, 94, 183, 236, 237,
    246–247, 258, 260–261, 287, 360, 361, 363,
    371, 387, 390, 391, 417–418, 509, 524–527,
    531, 532
Upflow anaerobic sludge blanket (UASB)
    reactors, 324–327, 398–399, 409, 413–414
Urban ecosystems, 88, 150, 151
Urbanization, 42, 46–47
    adverse impacts of, 438–453
    new paradigm of, xv–xvi
Urban landscape, 113–114
Urban metabolism (pollution), 272–274
Urban planning, 137, 487–493. *See also* Planning
    and design
Urban sprawl, 38–40
Urban sustainability, 72–129. *See also*
    Sustainable communities
    Cities of the Future–water centric ecocities,
        114–124
    components of, 73, 74
    definitions of sustainability, 80–82
    ecocity/ecovillage concepts, 124–129
    economic vs. resources preservation, 82–85
    and economy, 94–97
    emerging water systems, 99–104
    and environmental policy, 94
    environment and ecology components of,
        87–90
    and living within limits, 90–94
    new paradigm for, 73–75, 97–114
    and pollution, 76–80
    resilience dimension of, 84–85
    society as component of, 85–87
    stormwater pollution and flood abatement,
        110–113
    stream restoration, 108–110
    through planning and design, 137–138
    triple bottom line–life cycle assessment,
        104–106
    urban landscape, 113–114
    vision of, 72–73
    water reclamation and reuse, 106–108

Urban water supply grid, 252
Urine (yellow) water, 106, 242–243, 313, 314,
    409
Use Attainability Analysis (UAA), 448–449, 451
Used water, 75. *See also* Water reclamation and
    reuse
    energy from, 126, 392–407
    heat recovery from, 379–380
    reclamation processes for, 408–411

Venice, Italy, 57–58
Virtual water, 240–241
Vision of sustainability, 72–73
Vltava River (Czech Republic), 28

Wastewater, 121. *See also* Used water
    contaminants in, 299–300
    historic management of, 8, 10–11, 15, 16
    long-distance transfers of, 26–28
    untreated, 18
Wastewater treatment, 16
    Bardenpho system for, 29
    conventional vs. reclamation and reuse,
        276–278. *See also* Water reclamation and
        reuse
Water, 228–232
    and ecological footprint, 485, 486
    reclamation and reuse of, *see* Water
        reclamation and reuse
    virtual, 240–241
Water centric (eco)cities, 453–476. *See also*
    Cities of the Future
    distributed systems for, 455–457, 497–500
    goals of, 453–455
    retrofitting older cities, 457–476
Water centric urbanism, 5. *See also* Urban
    sustainability
Water conservation, 241–252. *See also* Water use
    commercial and public, 249–250
    leaks and other losses, 251–252
    residential water use, 241–249
Water-energy nexus, 358–371
    direct electric energy production from biogas
        and used water, 399–407
    energy conservation, 371–380
    energy from renewable sources, 380–392
    energy from used water and waste organic
        solids, 392–399
    GHG emissions from urban areas, 360–361
    GHGs and energy for use/disposal of water,
        348–360
    integrated resource recovery facilities,
        411–416
    net zero carbon footprint goal, 365–371
    overall energy outlook, 416–422
    on regional and cluster scale, 362–364
    and used water reclamation processes,
        408–411

Water footprint, 240, 486–487
Water Framework Directive (WFD), 25, 28, 66, 443
Water heaters, 372–379
Water management:
  for Cities of the Future, 114–116
  paradigms of, *see* Paradigms of water management
Water pollution control regulations, xiii
Water quality:
  goals and limits for, 286–308
  for groundwater recharge, 300–304
  for integrated reclamation/reuse, 304–308
  for landscape and agricultural irrigation, 289–293
  for potable reuse, 297–300
  standards/criteria/guidelines for, 234–235
  for uses other than irrigation and potable water, 293–297
  of water bodies, 177
Water Quality Standards Regulation, 449
Water reclamation and reuse, xv, 272–308
  centralized vs. decentralized reclamation, 281–282
  cluster water reclamation units, 282–286
  concept of, 274–279
  decentralized reclamation technologies, 348–354
  in emerging paradigm, 106–108
  energy and GHG emission issues, 346–348
  gray water, 256–259
  groundwater recharge, 300–304
  integrated reclamation/reuse in Singapore, 304–308
  landscape and agricultural irrigation, 289–293
  potable reuse, 234, 297–300
  rainwater and stormwater, 279–280. *See also* Rainwater harvesting (RWH)
  for sustainable communities, 505–514
  traditional technologies for, 311–312
  types of solids from, 331–334
  uses other than irrigation and potable water, 293–297
  water quality goals/limits, 286–308
  water-sewage-water cycle, 280–281
Water reclamation plants (WRPs), 107, 282–286, 498–499
Water reuse, 278. *See also* Water reclamation and reuse
Water scarcity/shortages, 49–51, 60, 228–234
Water Sensitive Urban Design (WSUD), 63, 186. *See also* Low Impact Development (LID)
Water-sewage-water (WSW) cycle, 20, 280–281

Watershed management agency, 528, 531–532
Watershed-scale utilities, 527–532
Watershed-wide goals, 118–119
Water sources, 252–268
  desalination of seawater and brackish water, 260–266
  rainwater harvesting, 252–256
  stormwater and other freshwater flows, 266–268
  urban water supply grid, 252
Water supply grid, 252
Water systems:
  in emerging (fifth) paradigm, 99–104
  leaks and other losses in, 251–252
  in the United States, 532
Water use, 228–250
  commercial and public, 249–250
  decentralized, 497–499
  drinking water quality standards, 234–235
  fundamentals of, 232–235
  minimum criteria for, 234
  municipal, 235–240
  virtual water, 240–241
  water on Earth, 228–232
Weirs, 463
Westergasfabriek Park (Amsterdam), 162, 163
Western Harbor (Malmö, Sweden), 167–170
Wet detention ponds, 201, 202, 204–207
Wetlands, 207–212, 259, 319–323, 348. *See also* Constructed wetlands
White water, 106, 278
Wildlife corridors, 154
Willingness to pay (WTP), 521–525
Willow Brook (British Columbia), 219
Wind power, 55, 387–392
Wisconsin, 444
Wood gas, 392
Woodlands (Texas), 521–522
World Commission on Environment and Development, 81
World Wildlife Fund (WWF), 119

Xeriscapes, 247–248, 289

Yangtze River (China), 28
Yellow water, 278. *See also* Urine (yellow) water
Yokohama, Japan, 48

Zero energy building (ZEB), 365
Zhiangijawo New Town (China), 220, 504–505, 507
Zhuan River (China), 470–471
Zurich, Switzerland, 436, 472, 474–476